SCHAUM'S OUTLINE OF

THEORY AND PROBLEMS

OF

STATISTICS
IN SI UNITS

FIRST EDITION

BY

MURRAY R. SPIEGEL, PhD

Professor of Mathematics,
Rensselaer Polytechnic Institute

SI EDITION adapted by

R. W. BOXER, BSc, AFIMA

Farnborough Technical College

McGraw-Hill International Book Company, New York
co-published with
McGraw-Hill Book Company (UK) Limited, London
McGraw-Hill Book Company GmbH, Düsseldorf
McGraw-Hill Book Company Australia Pty, Limited
McGraw-Hill Book Company (SA) (Pty) Limited, Johannesburg

07 084399 1

Preface

Statistics, or statistical methods as it is sometimes called, is playing an increasingly important role in nearly all phases of human endeavour. Formerly dealing only with affairs of the state, thus accounting for its name, the influence of statistics has now spread to agriculture, biology, business, chemistry, communications, economics, education, electronics, medicine, physics, political science, psychology, sociology and numerous other fields of science and engineering.

The purpose of this book is to present an introduction to the general statistical principles which will be found useful to all individuals regardless of their fields of specialization. It has been designed for use either as a supplement to all current standard texts or as a textbook for a formal course in statistics. It should also be of considerable value as a book of reference for those presently engaged in applications of statistics to their own special problems of research.

Each chapter begins with clear statements of pertinent definitions, theorems and principles together with illustrative and other descriptive material. This is followed by graded sets of solved and supplementary problems which in many instances use data drawn from actual statistical situations. The solved problems serve to illustrate and amplify the theory, bring into sharp focus those fine points without which the student continually feels himself on unsafe ground, and provide the repetition of basic principles so vital to effective teaching. Numerous derivations of formulae are included among the solved problems. The large number of supplementary problems with answers serve as a complete review of the material of each chapter.

The only mathematical background needed for an understanding of the entire book is arithmetic and the elements of algebra. A review of important mathematical concepts used in the book is presented in the first chapter which may either be read at the beginning of the course or referred to later as the need arises.

The early part of the book deals with the analysis of frequency distributions and associated measures of central tendency, dispersion, skewness and kurtosis. This leads quite naturally to a discussion of elementary probability theory and applications, which paves the way for a study of sampling theory. Techniques of large sampling theory, which involve the normal distribution, and applications to statistical estimation and tests of hypotheses and significance are treated first. Small sampling theory, involving Student's t and the chi-square distributions together with applications, follows in separate chapters. A chapter on curve fitting and the method of least squares, which is of interest in itself, leads logically to the topics of correlation and regression involving two variables. Multiple and partial correlation involving more than two variables are treated in a separate chapter. Two final chapters deal with analysis of time series and index numbers respectively.

Considerably more material has been included here than can be covered in most first courses. This has been done to make the book more flexible, to provide a more useful book of reference and to stimulate further interest in the topics. In using the book it is possible to change the order of many later chapters or even to omit certain chapters without difficulty. For example, Chapters 13-17 can for the most part be introduced immediately after Chapter 5 if it is desired to treat correlation, regression, time series and index numbers before sampling theory. Similarly, most of Chapter 6 may be omitted if one does not wish to devote too much time to probability. In a first course all of Chapter 15 may be omitted. The present order has been used because there is an increasing tendency in modern courses to introduce sampling theory and statistical inference as early as possible.

I wish to thank the various agencies, both governmental and private, for their co-operation in supplying data for tables. Appropriate references to such sources are given throughout the book. In particular, I am indebted to Professor Sir Ronald A. Fisher, F.R.S., Cambridge and to Dr. Frank Yates, F.R.S., Rothamsted, also to Messrs. Oliver and Boyd Ltd., Edinburgh, for permission to use data from Table III of their book "Statistical Tables for Biological, Agricultural and Medical Research".

I also wish to express my gratitude to the staff of the Schaum Publishing Company for their fine spirit of co-operation in meeting the seemingly endless attempts at perfection by the author.

M. R. SPIEGEL

Rensselaer Polytechnic Institute

October, 1961

Contents

CHAPTER 1

Variables and Graphs

STATISTICS

Statistics is concerned with scientific methods for collecting, organizing, summarizing, presenting and analysing data, as well as drawing valid conclusions and making reasonable decisions on the basis of such analysis.

In a narrower sense the term is used to denote the data themselves or numbers derived from the data as, for example, averages. Thus we speak of employment statistics, accident statistics, etc.

POPULATION AND SAMPLE. DESCRIPTIVE AND INDUCTIVE STATISTICS

In collecting data concerning characteristics of a group of individuals or objects, such as heights and weights of students in a university or numbers of defective and non-defective bolts produced in a factory on a given day, it is often impossible or impractical to observe the entire group, especially if it is large. Instead of examining the entire group, called the *population* or *universe*, one examines a small part of the group called a *sample*.

A population can be *finite* or *infinite*. For example, the population consisting of all bolts produced in a factory on a given day is finite, whereas the population consisting of all possible outcomes (heads, tails) in successive tosses of a coin is infinite.

If a sample is representative of a population, important conclusions about the population can often be inferred from analysis of the sample. The phase of statistics dealing with conditions under which such inference is valid is called *inductive statistics* or *statistical inference*. Because such inference cannot be absolutely certain, the language of *probability* is often used in stating conclusions.

The phase of statistics which seeks only to describe and analyse a given group without drawing any conclusions or inferences about a larger group is called *descriptive* or *deductive statistics*.

Before proceeding with the study of statistics we review some important mathematical concepts.

DISCRETE AND CONTINUOUS VARIABLES

A *variable* is a symbol, such as X, Y, H, x, B, which can assume any of a prescribed set of values, called the *domain* of the variable. If the variable can assume only one value it is called a *constant*.

A variable which can theoretically assume any value between two given values is called a *continuous variable*, otherwise it is called a *discrete variable*.

Example 1. The number N of children in a family, which can assume any of the values 0, 1, 2, 3, . . . but cannot be 2·5 or 3·842, is a discrete variable.

Example 2. The age A of an individual, which can be 62 years, 63·8 years or 65·8341 years, depending on accuracy of measurement, is a continuous variable.

Data which can be described by a discrete or continuous variable are called *discrete data* or *continuous data* respectively. The number of children in each of 1000 families is an example of discrete data, while the heights of 100 university students is an example of continuous data. In general, *measurements* give rise to continuous data while *enumerations* or *countings* give rise to discrete data.

It is sometimes convenient to extend the concept of variable to non-numerical entities. For example, colour C in a rainbow is a variable which can take on the "values" red, orange, yellow, green, blue, indigo and violet. It is generally possible to replace such variables by numerical quantities. For example, denote red by 1, orange by 2, etc.

ROUNDING OF DATA

The result of rounding a number such as 72·8 to the nearest unit is 73, since 72·8 is closer to 73 than to 72. Similarly, 72·8146 rounded to the nearest hundredth or to two decimal places is 72·81, since 72·8146 is closer to 72·81 than to 72·82.

In rounding 72·465 to the nearest hundredth, however, we are faced with a dilemma since 72·465 is *just as far* from 72·46 as from 72·47. It has become the practice in such cases to round to the *even integer* preceding the 5. Thus 72·465 is rounded to 72·46, 183·575 is rounded to 183·58, 116 500 000 rounded to the nearest million is 116 000 000. This practice is especially useful in minimizing *cumulative rounding errors* when a large number of operations is involved (see Problem 1.4).

SCIENTIFIC NOTATION

When writing numbers, especially those involving many zeros before or after the decimal point, it is convenient to employ the scientific notation using powers of 10.

Example 1. $10^1 = 10, 10^2 = 10 \times 10 = 100, 10^5 = 10 \times 10 \times 10 \times 10 \times 10 = 100\,000, 10^8 = 100\,000\,000$

Example 2. $10^0 = 1, 10^{-1} = 0{\cdot}1, 10^{-2} = 0{\cdot}01, 10^{-5} = 0{\cdot}000\,01$

Example 3. $864\,000\,000 = 8{\cdot}64 \times 10^8, 0{\cdot}000\,034\,16 = 3{\cdot}416 \times 10^{-5}$

Note that multiplying a number by 10^8, for example, has the effect of moving the decimal point of the number 8 places *to the right*. Multiplying a number by 10^{-6} has the effect of moving the decimal point of the number 6 places *to the left*.

Often we use parentheses or dots to show multiplications of two or more numbers. Thus $(5)(3) = 5 . 3 = 5 \times 3 = 15, (10)(10)(10) = 10 . 10 . 10 = 10 \times 10 \times 10 = 1000$. When letters are used to represent numbers the parentheses or dots are often omitted. For example $ab = (a)(b) = a . b = a \times b$.

The scientific notation is often useful in computation, especially in locating decimal points. Use is then made of the rules

$$(10^p)(10^q) = 10^{p+q}, \qquad \frac{10^p}{10^q} = 10^{p-q}$$

where p and q are any numbers.

In 10^p, p is called the *exponent* and 10 is called the *base*.

Example 1. $(10^3)(10^2) = 1000 \times 100 = 100\,000 = 10^5$ (i.e. 10^{3+2}),

$\dfrac{10^6}{10^4} = \dfrac{1\,000\,000}{10\,000} = 100 = 10^2$ (i.e. 10^{6-4})

Example 2. $(4\,000\,000)(0{\cdot}000\,000\,000\,2) = (4 \times 10^6)(2 \times 10^{-10}) = (4)(2)(10^6)(10^{-10}) = 8 \times 10^{6-10}$
$= 8 \times 10^{-4} = 0{\cdot}000\,8$

Example 3. $\dfrac{(0{\cdot}006)(80\,000)}{0{\cdot}04} = \dfrac{(6 \times 10^{-3})(8 \times 10^4)}{4 \times 10^{-2}} = \dfrac{48 \times 10^1}{4 \times 10^{-2}} = \left(\dfrac{48}{4}\right) \times 10^{1-(-2)}$
$= 12 \times 10^3 = 12\,000$

SIGNIFICANT FIGURES

If a mass is accurately recorded as 65·4 kg it means that the true mass lies between 65·35 and 65·45 kg. The accurate digits, apart from zeros needed to locate the decimal point, are called the *significant digits* or *significant figures* of the number.

Example 1.   65·4 has 3 significant figures.

Example 2.   4·5300 has 5 significant figures.

Example 3.   0·0018 = 1·8 × 10^{-3} has 2 significant figures.

Example 4.   0·001 800 = 1·800 × 10^{-3} has 4 significant figures.

Numbers associated with enumerations or countings, as opposed to measurements, are of course exact and so have an unlimited number of significant figures. In some of these cases, however, it may be difficult to decide which figures are significant without further information. For example, the number 186 000 000 may have 3, 4, . . ., 9 significant figures. If it is known to have 5 significant figures, it would be better to record the number as 186·00 million or 1·8600 × 10^8.

COMPUTATIONS

In performing calculations involving multiplication, division and extraction of roots of numbers, the final result can have no more significant figures than the numbers with the fewest significant figures (see Problem 1.9).

Examples: 1. 73·24 × 4·52 = (73·24)(4·52) = 331   **3.** $\sqrt{38·7} = 6·22$

2. 1·648/0·023 = 72   **4.** (8·416)(50) = 420·8, if 50 is exact.

In performing additions and subtractions of numbers, the final result has no more significant figures after the decimal point than the numbers with fewest significant figures after the decimal point (see Problem 1.10).

Examples: 1. 3·16 + 2·7 = 5·9   **2.** 83·42 − 72 = 11   **3.** 47·816 − 25 = 22·816, if 25 is exact.

The above rule for addition and subtraction can be extended (see Problem 1.11).

FUNCTIONS

If to each value which a variable X can assume there corresponds one or more values of a variable Y, we say that Y is a *function* of X and write $Y = F(X)$ (read "Y equals F of X") to indicate this functional dependence. Other letters such as G, φ, etc., can be used instead of F.

The variable X is called the *independent variable* and Y is called the *dependent variable*.

If only one value of Y corresponds to each value of X, we call Y a *single-valued function* of X; otherwise it is called a *multiple-valued function* of X.

Example 1.   The total population P of the British Isles is a function of the time t, and we write $P = F(t)$.

Example 2.   The stretch S of a vertical spring is a function of the weight W placed on the end of the spring. In symbols, $S = G(W)$.

The functional dependence or correspondence between variables is often depicted in a *table*. However, it can also be indicated by an equation connecting the variables, such as $Y = 2X - 3$, from which Y can be determined corresponding to various values of X.

If $Y = F(X)$, it is customary to let $F(3)$ denote "the value of Y when $X = 3$", $F(10)$ denote "the value of Y when $X = 10$", etc. Thus if $Y = F(X) = X^2$, then $F(3) = 3^2 = 9$ is the value of Y when $X = 3$.

The concept of function can be extended to two or more variables (see Problem 1.17).

RECTANGULAR CO-ORDINATES

Consider two mutually perpendicular lines $X'OX$ and $Y'OY$, called the *x and y axes* respectively (see Fig. 1-1), on which appropriate scales are indicated. These lines divide the plane determined by them, called the *xy plane*, into four regions denoted by I, II, III and IV, and called the first, second, third and fourth *quadrants*, respectively.

Point O is called the *origin* or *zero point*. Given any point P, drop perpendiculars to the x and y axes from P. The values of x and y at the points where the perpendiculars meet these axes are called the *rectangular co-ordinates* or simply the *co-ordinates* of P, and are denoted by (x, y). The co-ordinate x is sometimes called the *abscissa*, and y is the *ordinate* of the point. In Fig. 1-1 the abscissa of point P is 2, the ordinate is 3, and the co-ordinates of P are (2, 3).

Conversely, given the co-ordinates of a point, we can locate or *plot* the point. Thus points with co-ordinates $(-4, -3)$, $(-2\cdot3, 4\cdot5)$ and $(3\cdot5, -4)$ are represented by Q, R and S respectively in the figure on the right.

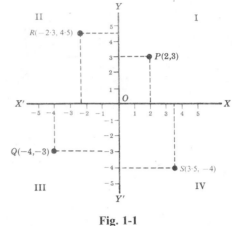

Fig. 1-1

By constructing a *z axis* through O and perpendicular to the *xy* plane, we can easily extend the above ideas. In such case the co-ordinates of a point P would be denoted by (x, y, z).

GRAPHS

A *graph* is a pictorial presentation of the relationship between variables. Many types of graphs are employed in statistics, depending on the nature of the data involved and the purpose for which the graph is intended. Among these are *bar graphs*, *pie graphs*, *pictographs*, etc. These graphs are sometimes referred to as *charts* or *diagrams*. Thus we speak of bar charts, pie diagrams, etc. (see Problems 1.23, 1.24, 1.26 and 1.27).

EQUATIONS

Equations are statements of the form $A = B$, where A is called the *left-hand member* or *side* of the equation and B the *right-hand member* or *side*. So long as we apply the *same* operations to both members of an equation we obtain *equivalent equations*. Thus we can add, subtract, multiply or divide both members of an equation by the same value and obtain an equivalent equation, the only exception being that *division by zero is not allowed*.

Example: Given the equation $2X + 3 = 9$.

Subtract 3 from both members: $2X + 3 - 3 = 9 - 3$ or $2X = 6$.
Divide both members by 2: $2X/2 = 6/2$ or $X = 3$.

This value of X is a *solution* of the given equation, as seen by replacing X by 3, obtaining $2(3) + 3 = 9$ or $9 = 9$ which is an *identity*. The process of obtaining solutions of an equation is called *solving the equation*.

The above ideas can be extended to finding solutions of two equations in two unknowns, three equations in three unknowns, etc. Such equations are called *simultaneous equations* (see Problem 1.30).

INEQUALITIES

The symbols < and > mean "less than" and "greater than", respectively. The symbols ≤ and ≥ mean "less than or equal to" and "greater than or equal to", respectively. They are known as *inequality symbols*.

Example 1. 3 < 5   is read "3 is less than 5".

Example 2. 5 > 3   is read "5 is greater than 3".

Example 3. $X < 8$ is read "X is less than 8".

Example 4. $X \geq 10$ is read "X is greater than or equal to 10".

Example 5. $4 < Y \leq 6$ is read "4 is less than Y which is less than or equal to 6", or "Y is between 4 and 6 excluding 4 but including 6", or "Y is greater than 4 and less than or equal to 6".

Relations involving inequality symbols are called *inequalities*. Just as we speak of members of an equation, so we can speak of *members of an inequality*. Thus in the inequality $4 < Y \leq 6$, 4, Y and 6 are the members.

A valid inequality remains valid when:

(*a*) the same number is added to or subtracted from each member.

Examples: Since $15 > 12$, $15 + 3 > 12 + 3$ (i.e. $18 > 15$) and $15 - 3 > 12 - 3$ (i.e. $12 > 9$).

(*b*) each member is multiplied or divided by the same *positive* number.

Examples: Since $15 > 12$, $(15)(3) > (12)(3)$ (i.e. $45 > 36$) and $\dfrac{15}{3} > \dfrac{12}{3}$ (i.e. $5 > 4$).

(*c*) each member is multiplied or divided by the same *negative* number provided the inequality symbols are reversed.

Examples: Since $15 > 12$, $(15)(-3) < (12)(-3)$ (i.e. $-45 < -36$) and $\dfrac{15}{-3} < \dfrac{12}{-3}$ (i.e. $-5 < -4$).

LOGARITHMS

Every positive number N can be expressed as a power of 10, i.e. we can always find p such that $N = 10^p$. We call p the *logarithm of N to the base 10* or the *common logarithm* of N, and write briefly $p = \log N$ or $p = \log_{10} N$. For example, since $1000 = 10^3$, $\log 1000 = 3$. Similarly since $0.01 = 10^{-2}$, $\log 0.01 = -2$.

When N is a number between 1 and 10, i.e. 10^0 and 10^1, $p = \log N$ is a number between 0 and 1 and can be found from tables of logarithms in the Appendix, Pages 346 and 347.

Example 1.   To find log 2.36, we glance down the *left* column headed N until we come to the first two digits, 23. Then we proceed *right* to the column headed 6. We find the entry 3729. Thus log $2.36 = 0.3729$, i.e. $2.36 = 10^{0.3729}$.

Logarithms of *all* positive numbers can be found from logarithms of numbers between 1 and 10.

Example 2.   From Example 1, $2.36 = 10^{0.3729}$. Multiplying successively by 10,
$$23.6 = 10^{1.3729}, \quad 236 = 10^{2.3729}, \quad 2360 = 10^{3.3729}, \dots$$
Thus log $2.36 = 0.3729$, log $23.6 = 1.3729$, log $236 = 2.3729$, log $2360 = 3.3729$.

Example 3.   Since $2.36 = 10^{0.3729}$, we find on successive divisions by 10,
$$0.236 = 10^{0.3729-1} = 10^{-0.6271}, \quad 0.0236 = 10^{0.3729-2} = 10^{-1.6271}, \dots$$
Often we write $0.3729 - 1$ as $9.3729 - 10$ or $\bar{1}.3729$, and $0.3729 - 2$ as $8.3729 - 10$ or $\bar{2}.3729$, etc. With this notation we have,
$$\log 0.236 = 9.3729 - 10 = \bar{1}.3729 = -0.6271$$
$$\log 0.0236 = 8.3729 - 10 = \bar{2}.3729 = -1.6271, \text{ etc.}$$

The decimal part 0.3729 in all these logarithms is called the *mantissa*. The remaining part, before the decimal point of the mantissa, i.e. 1, 2, 3, and $\bar{1}$, $\bar{2}$ or $9 - 10$, $8 - 10$, is called the *characteristic*.

The following rules are easily demonstrated:

(*1*) For a number greater than 1, the characteristic is positive and is one *less* than the number of digits before the decimal point.

Thus the characteristics of the logarithms of 2360, 236, 23·6, 2·36 are 3, 2, 1, 0; and the required logarithms are 3·3729, 2·3729, 1·3729, 0·3729.

(*2*) For a number less than 1, the characteristic is negative and is one *more* than the number of zeros immediately following the decimal point.

Thus the characteristics of the logarithms of 0·236, 0·0236, 0·002 36 are -1, -2, -3, and the required logarithms are $\bar{1}$·3729, $\bar{2}$·3729, $\bar{3}$·3729 or 9·3729$-$10, 8·3729$-$10, 7·3729$-$10 respectively.

If logarithms of four-digit numbers such as 2·364 and 758·2 are required, the method of *interpolation* can be used (see Problem 1.36).

ANTILOGARITHMS

In the exponential form $2\cdot36 = 10^{0\cdot3729}$, the number 2·36 is called the *antilogarithm* of 0·3729, or antilog 0·3729. It is the number whose logarithm is 0·3729. It follows at once that

antilog 1·3729 = 23·6, antilog 2·3729 = 236, antilog 3·3729 = 2360, . . .

antilog 9·3729$-$10 = antilog $\bar{1}$·3729 = 0·236,

antilog 8·3729$-$10 = antilog $\bar{2}$·3729 = 0·0236, . . .

The antilog of any number can be found by reference to tables in the Appendix.

Example: To find antilog 8·6284$-$10, look up the mantissa 0·6284 in the body of the table. Since it appears in the row marked 42 and the column headed 5, the required digits of the number are 425. Since the characteristic is 8$-$10, the required number is 0·0425.

Similarly, antilog 3·6284 = 4250, antilog 5·6284 = 425 000.

If the mantissas are not found in the table, interpolation can be used (see Problem 1.37).

COMPUTATIONS USING LOGARITHMS employ the following properties:

$$\log MN = \log M + \log N$$

$$\log \frac{M}{N} = \log M - \log N$$

$$\log M^p = p \log M$$

By combining these results we find, for example,

$$\log \frac{A^p B^q C^r}{D^s E^t} = p \log A + q \log B + r \log C - s \log D - t \log E$$

See Problems 1.38–1.45.

Solved Problems

VARIABLES

1.1. State which of the following represent discrete data and which represent continuous data.

(a)	Numbers of shares sold each day in the stock market.	*Ans.* discrete
(b)	Temperatures recorded every half hour at a weather bureau.	*Ans.* continuous
(c)	Lifetimes of television tubes produced by a company.	*Ans.* continuous
(d)	Yearly incomes of college professors.	*Ans.* discrete
(e)	Lengths of 1000 bolts produced in a factory.	*Ans.* continuous

1.2. Give the domain of each of the following variables and state whether the variables are continuous or discrete.

(a) Number V of litres of water in a washing machine.
Domain: Any value from zero litres to the capacity of the machine.
Variable is continuous.

(b) Number B of books on a library shelf.
Domain: 0, 1, 2, 3, . . . up to the largest number of books which can fit on a shelf.
Variable is discrete.

(c) Sum S of points obtained in tossing a pair of dice.
Domain: Points obtained on a single die can be 1, 2, 3, 4, 5, or 6. Then the sum of points on a pair of dice can be 2, 3, 4, 5, 6, 7, 8, 9, 10, 11, 12, which is the domain of S.
Variable is discrete.

(d) Diameter d of a sphere.
Domain: If we consider a point as a sphere of zero diameter, the domain of d is all values from zero upward.
Variable is continuous.

(e) Country C in Europe.
Domain: England, France, Germany, etc., which can be represented numerically by 1, 2, 3, etc.
Variable is discrete.

ROUNDING OF DATA

1.3. Round each of the following numbers to the indicated accuracy.

(a)	48·6	nearest unit	49		(f)	143·95	nearest tenth	144·0
(b)	136·5	nearest unit	136		(g)	368	nearest hundred	400
(c)	2·484	nearest hundredth	2·48		(h)	24 448	nearest thousand	24 000
(d)	0·0435	nearest thousandth	0·044		(i)	5·565 00	nearest hundredth	5·56
(e)	4·500 01	nearest unit	5		(j)	5·565 01	nearest hundredth	5·57

1.4. Add the numbers 4·35, 8·65, 2·95, 12·45, 6·65, 7·55, 9·75 (a) directly, (b) by rounding to the nearest tenth according to the "even integer" convention, (c) by rounding so as to increase the digit before the 5.

Solution:

(a)		(b)		(c)	
	4·35		4·4		4·4
	8·65		8·6		8·7
	2·95		3·0		3·0
	12·45		12·4		12·5
	6·65		6·6		6·7
	7·55		7·6		7·6
	9·75		9·8		9·8
Total	52·35	Total	52·4	Total	52·7

Note that procedure (b) is superior to procedure (c), since *cumulative rounding errors* are minimized in procedure (b).

SCIENTIFIC NOTATION AND SIGNIFICANT FIGURES

1.5.   Express each of the following numbers without using powers of 10.

     (a) $4 \cdot 823 \times 10^7$. Move decimal point 7 places to the right and obtain 48 230 000.
     (b) $8 \cdot 4 \times 10^{-6}$. Move decimal point 6 places to the left and obtain 0·000 008 4.
     (c) $3 \cdot 80 \times 10^{-4} = 0 \cdot 000\,380$ (e) $300 \times 10^8 = 30\,000\,000\,000$
     (d) $1 \cdot 86 \times 10^5 = 186\,000$ (f) $70\,000 \times 10^{-10} = 0 \cdot 000\,007\,000\,0$

1.6.   How many significant figures are in each of the following, assuming the numbers are accurately recorded?

(a)	149·8 mm	**four**	(d)	0·002 80 m	**three**	(g)	9 houses	**unlimited**
(b)	149·80 mm	**five**	(e)	1·002 80 m	**six**	(h)	$4 \cdot 0 \times 10^3$ g	**two**
(c)	0·0028 m	**two**	(f)	9 g	**one**	(i)	$7 \cdot 584\,00 \times 10^{-5}$ N	**six**

1.7.   What is the maximum error in each of the following measurements assumed to be accurately recorded? Give the number of significant figures in each case.

     (a)   **73·854 mm.** The measurement could have ranged anywhere from 73·8535 mm to 73·8545 mm, and so the maximum error is 0·0005 mm. Five significant figures are present.
     (b)   **0·098 00 m³.** The number of m³ can be anywhere from 0·097 995 to 0·098 005, so that the maximum error is 0·000 005 m³. Four significant figures are present.
     (c)   **$3 \cdot 867 \times 10^8$ km.** The actual number of kilometres is greater than $3 \cdot 8665 \times 10^8$ but less than $3 \cdot 8675 \times 10^8$; hence the maximum error is $0 \cdot 0005 \times 10^8$ or 50 000 km. Four significant figures are present.

1.8.   Write each number using the scientific notation. Assume all figures significant unless otherwise indicated.

     (a)   24 380 000 (four sig. fig.) $= 2 \cdot 438 \times 10^7$ (c)   7 300 000 000 (five sig. fig.) $= 7 \cdot 3000 \times 10^9$
     (b)   0·000 009 851 $= 9 \cdot 851 \times 10^{-6}$ (d)   0·000 184 00 $= 1 \cdot 8400 \times 10^{-4}$

COMPUTATIONS

1.9.   Show that the product of the numbers 5·74 and 3·8, assumed to have three and two significant figures respectively, cannot be accurate to more than two significant figures.

First method:
     $5 \cdot 74 \times 3 \cdot 8 = 21 \cdot 812$, but not all figures in this product are significant. To determine how many figures are significant, observe that 5·74 stands for any number between 5·735 and 5·745 while 3·8 stands for any number between 3·75 and 3·85. Then the smallest possible value of the product is $5 \cdot 735 \times 3 \cdot 75 = 21 \cdot 506\,25$, and the largest possible value is $5 \cdot 745 \times 3 \cdot 85 = 22 \cdot 118\,25$.
     Since the possible range of values is 21·506 25 to 22·118 25, it is clear that no more than the first two figures in the product can be significant, the result being written as 22. Note that the number 22 stands for any number between 21·5 and 22·5.

Second method:
     Considering an italic figure as one which may be doubtful, the product can be computed as shown here.

$$\begin{array}{r} 5 \cdot 74 \\ 3 \cdot 8 \\ \hline 4592 \\ 1722 \\ \hline 21 \cdot 812 \end{array}$$

We should keep no more than one doubtful figure in the answer, which is therefore 22 to two significant figures.

     Note that it is unnecessary to carry more significant figures than are present in the least accurate factor. Thus if 5·74 is rounded to 5·7, the product is $5 \cdot 7 \times 3 \cdot 8 = 21 \cdot 66 = 22$ to two significant figures, agreeing with the above results.
     In calculating without use of machinery, labour can be saved by not keeping more than one or two figures beyond that of the least accurate factor and rounding to the proper number of significant figures in the final answer.

1.10.   Add the numbers 4·193 55, 15·28, 5·9561, 12·3, 8·472 assuming all figures to be significant.

Solution:

In (a), the doubtful figures in the addition are in italic type. The final answer with no more than one doubtful figure is recorded as 46·2.

(a)		(b)	
	4·193 *55*		4·19
	15·*28*		15·28
	5·956 *1*		5·96
	12·*3*		12·3
	8·*472*		8·47
	46·*201 65*		46·*20*

Some labour can be saved by proceeding as in (b), where we have kept one more significant decimal place than that in the least accurate number. The final answer, rounded to 46·2, agrees with (a).

1.11.   Calculate 475 000 000 + 12 684 000 − 1 372 410 if these numbers have 3, 5 and 7 significant figures respectively.

Solution:

In calculation (a), all figures are kept and the final answer is rounded. In (b), a method similar to that of Problem 1.10(b) is used. In both cases doubtful figures are in italic type.

(a)
$$475\,000\,000 + 12\,684\,000 = 487\,684\,000$$
$$487\,684\,000 - 1\,372\,410 = 486\,311\,590$$

(b)
$$475\,000\,000 + 12\,700\,000 = 487\,700\,000$$
$$487\,700\,000 - 1\,400\,000 = 486\,300\,000$$

The final result is rounded to 486 000 000 or better yet, to show that there are 3 significant figures, it is written as 486 million or $4·86 \times 10^8$.

1.12.   Perform each of the indicated operations.

(a) $48·0 \times 943 = (48·0)(943) = 45\,300$ (b) $8·35/98 = 0·085$

(c) $(28)(4193)(182) = (2·8 \times 10^1)(4·193 \times 10^3)(1·82 \times 10^2)$
$$= (2·8)(4·193)(1·82) \times 10^{1+3+2} = 21 \times 10^6 = 2·1 \times 10^7$$

This can also be written as 21 million to show the two significant figures.

(d) $\dfrac{(526·7)(0·001\,280)}{0·000\,034\,921} = \dfrac{(5·267 \times 10^2)(1·280 \times 10^{-3})}{3·4921 \times 10^{-5}} = \dfrac{(5·267)(1·280)}{3·4921} \times \dfrac{(10^2)(10^{-3})}{10^{-5}}$

$$= 1·931 \times \dfrac{10^{2-3}}{10^{-5}} \quad = \quad 1·931 \times \dfrac{10^{-1}}{10^{-5}}$$

$$= 1·931 \times 10^{-1+5} \quad = \quad 1·931 \times 10^4$$

This can also be written as 19·31 thousand to show the four significant figures.

(e) $\dfrac{(1·475\,62 - 1·473\,22)(4895·36)}{0·000\,159\,180} = \dfrac{(0·002\,40)(4895·36)}{0·000\,159\,180} = \dfrac{(2·40 \times 10^{-3})(4·895\,36 \times 10^3)}{1·591\,80 \times 10^{-4}}$

$$= \dfrac{(2·40)(4·895\,36)}{1·591\,80} \times \dfrac{(10^{-3})(10^3)}{10^{-4}} = 7·38 \times \dfrac{10^0}{10^{-4}} = 7·38 \times 10^4$$

This can also be written as 73·8 thousand to show the three significant figures.

Note that although six significant figures were originally present in all numbers, some of these were lost in subtracting 1·473 22 from 1·475 62.

(f)   If denominators 5 and 6 are exact, $\dfrac{(4·38)^2}{5} + \dfrac{(5·482)^2}{6} = 3·84 + 5·009 = 8·85$

(g) $3·1416 \sqrt{71·35} = (3·1416)(8·447) = 26·54$ (h) $\sqrt{128·5 - 89·24} = \sqrt{39·3} = 6·27$

1.13. Evaluate each of the following, given that $X = 3$, $Y = -5$, $A = 4$, $B = -7$ where all numbers are supposed exact.

(a) $2X - 3Y = 2(3) - 3(-5) = 6 + 15 = 21$

(b) $4Y - 8X + 28 = 4(-5) - 8(3) + 28 = -20 - 24 + 28 = -16$

(c) $\dfrac{AX + BY}{BX - AY} = \dfrac{(4)(3) + (-7)(-5)}{(-7)(3) - (4)(-5)} = \dfrac{12 + 35}{-21 + 20} = \dfrac{47}{-1} = -47$

(d) $X^2 - 3XY - 2Y^2 = (3)^2 - 3(3)(-5) - 2(-5)^2 = 9 + 45 - 50 = 4$

(e) $2(X + 3Y) - 4(3X - 2Y) = 2[(3) + 3(-5)] - 4[3(3) - 2(-5)]$
$$= 2[3 - 15] - 4[9 + 10] = 2(-12) - 4(19)$$
$$= -24 - 76 = -100$$

Another method:

$2(X + 3Y) - 4(3X - 2Y) = 2X + 6Y - 12X + 8Y = -10X + 14Y = -10(3) + 14(-5)$
$$= -30 - 70 = -100$$

(f) $\dfrac{X^2 - Y^2}{A^2 - B^2 + 1} = \dfrac{(3)^2 - (-5)^2}{(4)^2 - (-7)^2 + 1} = \dfrac{9 - 25}{16 - 49 + 1} = \dfrac{-16}{-32} = \dfrac{1}{2} = 0.5$

(g) $\sqrt{2X^2 - Y^2 - 3A^2 + 4B^2 + 3} = \sqrt{2(3)^2 - (-5)^2 - 3(4)^2 + 4(-7)^2 + 3}$
$$= \sqrt{18 - 25 - 48 + 196 + 3} = \sqrt{144} = 12$$

(h) $\sqrt{\dfrac{6A^2}{X} + \dfrac{2B^2}{Y}} = \sqrt{\dfrac{6(4)^2}{3} + \dfrac{2(-7)^2}{-5}} = \sqrt{\dfrac{96}{3} + \dfrac{98}{-5}} = \sqrt{12.4} = 3.52$, approx.

FUNCTIONS

1.14. Table 1.1 shows the number of tonnes of swedes and beet produced on the PQR Farm during the years 1950 to 1960.

With reference to this table determine the year or years during which:

(a) the least number of tonnes of swedes was produced.
(b) the greatest number of tonnes of beet was produced.
(c) the greatest decline in swede production occurred.
(d) the beet production decreased while the swede production increased over that of the preceding year.
(e) the same numbers of tonnes of swedes were produced.
(f) the combined production of swedes and beet was a maximum.

Year	Number of tonnes of Swedes (to nearest 5 tonnes)	Number of tonnes of Beat (to nearest 5 tonnes)
1950	200	75
1951	185	90
1952	225	100
1953	250	85
1954	240	80
1955	195	100
1956	210	110
1957	225	105
1958	250	95
1959	230	110
1960	235	100

Table 1.1

Ans. (a) 1951 (b) 1956 and 1959 (c) 1955 (d) 1953, 1957, 1958, 1960 (e) 1952, 1957; 1953, 1958 (f) 1958.

1.15. Let W and C denote respectively the number of tonnes of swedes and beet produced during the year t on the PQR Farm of Problem 1.14. It is clear that W and C are both functions of t; this we can indicate by $W = F(t)$ and $C = G(t)$.

(a) Find W when $t = 1956$. *Ans.* 210
(b) Find C when $t = 1953$ and 1959. *Ans.* 85 and 110 respectively
(c) Find t when $W = 225$. *Ans.* 1952 and 1957
(d) Find $F(1954)$. *Ans.* 240
(e) Find $G(1958)$. *Ans.* 95
(f) Find C when $W = 210$. *Ans.* 110
(g) What is the domain of the variable t? *Ans.* The years 1950, 1951, . . ., 1960.

(*h*) Is W a single-valued function of t?

Yes, since to each value which t can assume (i.e. in the domain of t) there corresponds one and only one value of W.

(*i*) Is t a function of W? If so is it single-valued?

Yes, t is a function of W since to each value which W can assume there corresponds one or more values of t which we can find from the table.

Since there may be more than one value of t corresponding to a value of W (e.g. when $W = 225$, $t = 1952$ or 1957), the function is multiple-valued. This functional dependence of t on W can be written $t = H(W)$.

(*j*) Is C a function of W?

Yes, since to each value which W can assume there corresponds one or more values of C as determined from Table 1.1. Similarly, W is a function of C.

(*k*) Which variable is independent, t or W?

Physically, it is customary to think of W as determined from t rather than t as determined from W. Thus physically, t is the independent variable and W the dependent variable.

Mathematically, however, either variable can be considered the independent variable and the other the dependent variable. The one which is assigned various values is the independent variable; the one which is then determined as a result is the dependent variable.

1.16. A variable Y is determined from a variable X according to the equation $Y = 2X - 3$ (where the 2 and 3 are exact).

(*a*) Find Y when $X = 3$, -2 and $1 \cdot 5$.

When $X = 3$, $Y = 2X - 3 = 2(3) - 3 = 6 - 3 = 3$.
When $X = -2$, $Y = 2X - 3 = 2(-2) - 3 = -4 - 3 = -7$.
When $X = 1 \cdot 5$, $Y = 2X - 3 = 2(1 \cdot 5) - 3 = 3 - 3 = 0$.

(*b*) Construct a table showing the values of Y corresponding to $X = -2, -1, 0, 1, 2, 3, 4$.

The values of Y, computed as in part (*a*), are shown in the table.

Note that by using other values of X we can construct many tables. The relation $Y = 2X - 3$ is equivalent to the collection of *all* such possible tables.

X	-2	-1	0	1	2	3	4
Y	-7	-5	-3	-1	1	3	5

(*c*) If the dependence of Y on X is denoted by $Y = F(X)$, determine $F(2 \cdot 4)$ and $F(0 \cdot 8)$.

$F(2 \cdot 4) = 2(2 \cdot 4) - 3 = 4 \cdot 8 - 3 = 1 \cdot 8$, $F(0 \cdot 8) = 2(0 \cdot 8) - 3 = 1 \cdot 6 - 3 = -1 \cdot 4$.

(*d*) What value of X corresponds to $Y = 15$?

Substitute $Y = 15$ in $Y = 2X - 3$. Then $15 = 2X - 3$, $2X = 18$, $X = 9$.

(*e*) Can X be expressed as a function of Y?

Yes. Since $Y = 2X - 3$, $Y + 3 = 2X$ or $X = \frac{1}{2}(Y + 3)$. This expresses X *explicitly* as a function of Y.

(*f*) Is Y a single-valued function of X?

Yes, since for each value which X can assume (and there are an infinite number of these values) there corresponds one and only one value of Y.

(*g*) Is X a single-valued function of Y?

Yes, since from part (*e*), $X = \frac{1}{2}(Y + 3)$, so that corresponding to each value assumed by Y there is one and only one value of X.

1.17. If $Z = 16 + 4X - 3Y$, find the value of Z corresponding to:
(*a*) $X = 2$, $Y = 5$; (*b*) $X = -3$, $Y = -7$; (*c*) $X = -4$, $Y = 2$.

Solution:
(*a*) $Z = 16 + 4(2) - 3(5) = 16 + 8 - 15 = 9$
(*b*) $Z = 16 + 4(-3) - 3(-7) = 16 - 12 + 21 = 25$
(*c*) $Z = 16 + 4(-4) - 3(2) = 16 - 16 - 6 = -6$

Given values of X and Y, there corresponds a value of Z. We can denote this dependence of Z on X and Y by writing $Z = F(X, Y)$, read "Z is a function of X and Y". $F(2, 5)$ denotes the value of Z when $X = 2$ and $Y = 5$ and is 9, from (*a*). Similarly, $F(-3, -7) = 25$ and $F(-4, 2) = -6$, from (*b*) and (*c*) respectively.

The variables X and Y are called *independent variables* and Z is called a *dependent variable*.

GRAPHS

1.18. Locate on the x axis of a co-ordinate system the points corresponding to (*a*) $x = 4$, (*b*) $x = -3$, (*c*) $x = 2 \cdot 5$, (*d*) $x = -4 \cdot 3$ and (*e*) $x = 0 \cdot 4$, assuming these values to be exact.

Solution:

Each exact value of x corresponds to one and only one point on the axis. Conversely it is proved in advanced mathematics that to each point on the axis there corresponds one and only one value of x.

Thus theoretically there is a point corresponding to $x = 22/7 = 3 \cdot 142\,857\,142\,857\ldots$ or $x = \pi = 3 \cdot 141\,592\,653\,58\ldots$. Practically, of course, we can never hope to locate a point exactly since the pencil mark which we make has a thickness and covers an infinite number of points. The x axis itself has a thickness. Thus the diagram above is a physical representation of the actual mathematical situation.

1.19. Let x denote the diameter of a ball bearing in millimetres. If $x = 4 \cdot 58$ to three significant figures, how should this be represented on the x axis?

Solution:

The indicated measurement, $4 \cdot 58$ mm, shows that the true measurement lies between $4 \cdot 575$ mm and $4 \cdot 585$ mm. Thus the measurement should be represented by the heavy line segment shown.

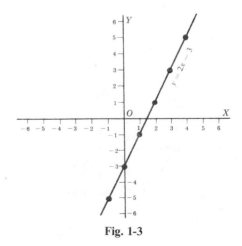

1.20. Locate on a rectangular co-ordinate system the points having co-ordinates (*a*) (5, 2), (*b*) (2, 5), (*c*) (−3, 1), (*d*) (1, −3), (*e*) (3, −4), (*f*) (−2·5, −4·8), (*g*) (0, −2·5), (*h*) (4, 0). Assume all given numbers are exact. See adjoining Fig. 1-2 for solution.

Solution:

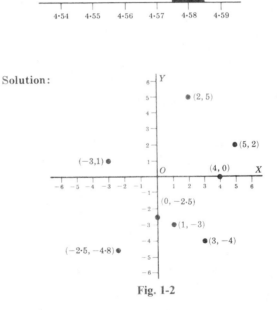

Fig. 1-2

1.21. Graph the equation $y = 2x - 3$.

Solution:

Placing $x = -2, -1, 0, 1, 2, 3, 4$, we find $y = -7, -5, -3, -1, 1, 3, 5$ respectively (see Prob. 1.16(*b*)). Then points on the graph are given by $(-2, -7), (-1, -5), (0, -3), (1, -1), (2, 1),$ $(3, 3)$ and $(4, 5)$ which are plotted on a rectangular co-ordinate system as shown in Fig. 1-3. All of these points, as well as points obtained by using other values of x, lie on a straight line which is the required graph.

Because the graph of $y = 2x - 3$ is a straight line, we sometimes call $F(x) = 2x - 3$ a *linear function*. In general $F(x) = ax + b$, where a and b are any constants, is a linear function whose graph is a straight line.

Note that only two points are actually needed to graph a linear function, since two points determine a line.

Fig. 1-3

1.22. Graph the equation $y = x^2 - 2x - 8$.

Solution:

The values of y corresponding to various values of x are shown in the table. For example, when $x = 2$,
$y = (-2)^2 - 2(-2) - 8 = 4 + 4 - 8 = 0$.

x	-3	-2	-1	0	1	2	3	4	5
y	7	0	-5	-8	-9	-8	-5	0	7

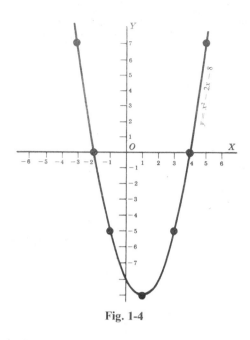

From the table, points on the graph are given by $(-3, 7)$, $(-2, 0)$, $(-1, -5)$, $(0, -8)$, $(1, -9)$, $(2, -8)$, $(3, -5)$, $(4, 0)$, $(5, 7)$. These points, as well as others obtained by using different values of x, are seen to lie on the curve shown in Fig. 1-4.

The curve is called a *parabola*. The function $F(x) = x^2 - 2x - 8$ is called a *quadratic function*.

In general the graph of an equation $y = a + bx + cx^2$, where a, b and c are constants and $c \neq 0$, is a parabola. If $c = 0$, the graph is a straight line as in Prob. 1.21.

Fig. 1-4

1.23. Table 1.2 gives the population of the United States (in millions) for the years 1840, 1850, . . ., 1960. Graph these data.

Table 1.2   Population (in millions) of the United States, 1840-1960.

Year	1840	1850	1860	1870	1880	1890	1900	1910	1920	1930	1940	1950	1960
Population (in millions)	17·1	23·2	31·4	39·8	50·2	62·9	76·0	92·0	105·7	122·8	131·7	151·1	179·3

Source: Bureau of the Census.

First method:

Refer to Fig. 1-5. In this graph, Population, denoted by P, is the dependent variable and time, denoted by t, is the independent variable. Points are located as usual by co-ordinates read from the table, for example (1880, 50·2). Successive points are then connected by straight lines, since no information is given as to the population during intermediate years. For this reason this graph is called a *line graph*.

Note that units on the axes are not equal as they were in graphing $y = 2x - 3$. This is perfectly justified since the two variables represent essentially different quantities.

Note also that the zero has been indicated on the vertical axis but (for obvious reasons) not on the horizontal axis. In general the zero should be indicated whenever possible especially on the vertical axis. If it is impossible for some reason to indicate the zero and if such omission might lead to any erroneous conclusions drawn by the reader, then it is wise to call attention to the omission by some means such as is indicated in Problem 1.26.

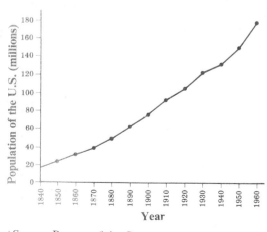

(*Source*: Bureau of the Census)

Fig. 1-5

A table or graph showing the distribution of a variable as a function of time is called a *time series*.

Second method:

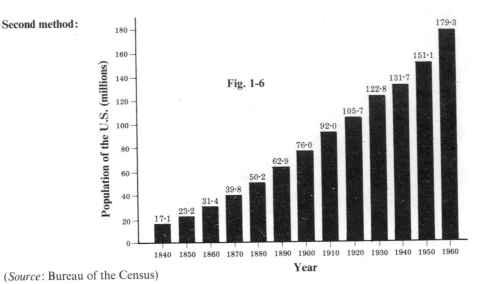

Fig. 1-6

(*Source*: Bureau of the Census)

Fig. 1-6 is called a *bar graph*, *bar chart* or *bar diagram*. The widths of the bars, which are all equal, have no significance in this case and can be taken any convenient size so long as the bars do not overlap.

The numbers at the tops of the bars may or may not be omitted. If they are kept, the vertical scale at the left is unnecessary and may be omitted.

Third method:

Year		Population
1840	🚹🚹	17·1 million
1850	🚹🚹🚹	23·2 million
1860	🚹🚹🚹	31·4 million
1870	🚹🚹🚹🚹	39·8 million
1880	🚹🚹🚹🚹🚹	50·2 million
1890	🚹🚹🚹🚹🚹🚹	62·9 million
1900	🚹🚹🚹🚹🚹🚹🚹🚹	76·0 million
1910	🚹🚹🚹🚹🚹🚹🚹🚹🚹	92·0 million
1920	🚹🚹🚹🚹🚹🚹🚹🚹🚹🚹	105·7 million
1930	🚹🚹🚹🚹🚹🚹🚹🚹🚹🚹🚹🚹	122·8 million
1940	🚹🚹🚹🚹🚹🚹🚹🚹🚹🚹🚹🚹	131·7 million
1950	🚹🚹🚹🚹🚹🚹🚹🚹🚹🚹🚹🚹🚹🚹🚹	151·1 million
1960	🚹🚹🚹🚹🚹🚹🚹🚹🚹🚹🚹🚹🚹🚹🚹🚹🚹🚹	179·3 million

**POPULATION
OF THE
UNITED STATES**
DURING THE YEARS
1840 to 1960

Each Figure Represents 10 000 000 people

Fig. 1-7

(*Source*: Bureau of the Census)

Charts or diagrams, such as shown in Fig. 1-7, are called *pictographs* or *pictograms* (short for picture diagrams) and are often used to present statistical data in a manner appealing to the general public. Many of these pictographs display a great deal of originality and ingenuity in the art of presentation of data.

The numbers to the right of the figures in the previous pictograph may or may not be included. Even if they are not shown the reader can still estimate the population to the nearest 5 million people.

1.24. Graph the data of Problem 1.14 using (*a*) line graphs, (*b*) bar graphs.

Solution:

(*a*)

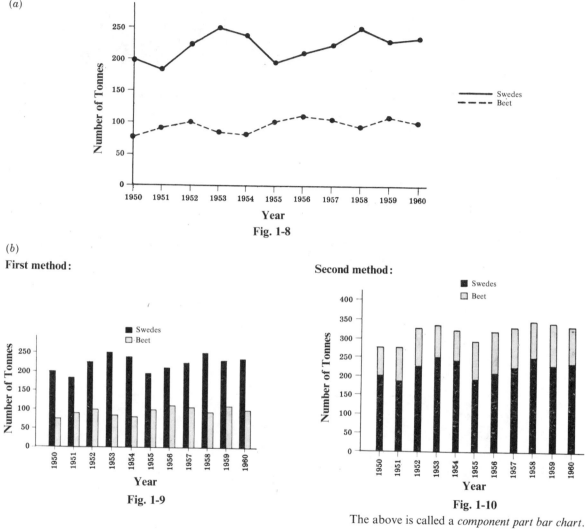

Fig. 1-8

(*b*)

First method:

Fig. 1-9

Second method:

Fig. 1-10

The above is called a *component part bar chart*.

1.25. (*a*) Express the yearly number of tonnes of swedes and beet of Problem 1.14 as percentages of total annual production. (*b*) Graph the percentages in (*a*).

Solution:

(*a*) For 1950, percentage of swedes $= \dfrac{200}{200 + 75} = 72 \cdot 7\%$, percentage of beet $= 100\% - 72 \cdot 7\% = 27 \cdot 3\%$.

Table 1.3

Year	1950	1951	1952	1953	1954	1955	1956	1957	1958	1959	1960
Percentage Swedes	72·7	67·3	69·2	74·6	75·0	66·1	65·6	68·2	72·5	67·6	70·1
Percentage Beet	27·3	32·7	30·8	25·4	25·0	33·9	34·4	31·8	27·5	32·4	29·9

(b) The graph of the percentages in (a), shown in Fig. 1-11, is called a *percentage component part graph*. A graph similar to that of the first method of Problem 1.24(b) can also be used.

1.26. Using a line graph, represent graphically only the wheat production in Table 1.1 of Problem 14.

Solution:

The required line graph is obtained from Problem 1.24(a) by removing the lower line graph. As a result much wasted space appears between the upper line graph and the horizontal axis. To avoid this we could start our vertical scale at 150 instead of 0 bushels. This may, however, lead to erroneous conclusions on the part of a reader who may not have noticed omission of the zero. To draw attention to this omission, we can construct the graph as in Fig. 1-12 below.

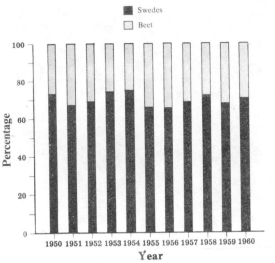

Fig. 1-11

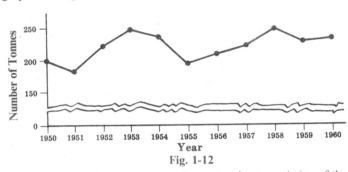

Fig. 1-12

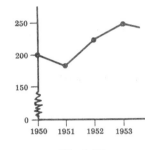

Fig. 1-13

Another device often employed to draw attention to omission of the zero is the use of a zigzag line on one of the axes as shown in Fig. 1-13 above.

1.27. The areas of the various continents of the world in millions of square kilometres are presented in Table 1.4. Graph the data.

First method:

Fig. 1-14

AREAS OF CONTINENTS OF THE WORLD
(from data supplied by the United Nations)

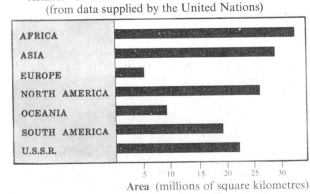

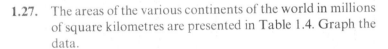
Area (millions of square kilometres)

The above is a bar graph where the bars are horizontal instead of vertical. Note that the continents have been listed in alphabetical order. If desired they could be listed in increasing or decreasing order of area.

Table 1.4

AREAS OF CONTINENTS OF THE WORLD

Continent	Area (millions of square kilometres)
Africa	30·3
Asia	26·9
Europe	4·9
North America	24·3
Oceania	8·5
South America	17·9
U.S.S.R.	20·5

Total 133·3

Source: United Nations.

Note 1.   Europe excludes Russia and countries under Russian domination which are included in the U.S.S.R.

Note 2.   Europe excludes Turkey which is included in Asia.

Second method:

Fig. 1-15 is called a *pie graph, circular graph* or *pie chart*. To construct it we use the fact that the total area, 133·3 million square kilometres, corresponds to the total number of degrees in the circular arc, namely 360°. Thus 1 million square kilometres corresponds to 360°/133·3. It follows that Africa, with 30·3 million square kilometres corresponds to an arc of 30·3(360°/133·3) = 82°, while Asia, Europe, North America, Oceania, South America and U.S.S.R. correspond to arcs of 73°, 13°, 66°, 23°, 48° and 55° respectively. Using a protractor the required dividing lines can be drawn.

1.28. The time t (in seconds) required for one complete vibration of a simple pendulum of length l (in millimetres) is given by the following observations obtained in a physics laboratory. (*a*) Exhibit graphically t as a function of l. (*b*) From the graph estimate t for a pendulum whose length is 400 millimetres.

AREAS OF CONTINENTS OF THE WORLD
(millions of square kilometres)

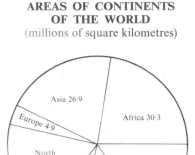

Fig. 1-15

l	101	162	222	338	420	534	667	745	866	1000
t	0·64	0·81	0·95	1·17	1·30	1·47	1·65	1·74	1·87	2·01

Solution:

(*a*) The graph shown in Fig. 1-16 has been obtained by connecting the observation points by a smooth curve.

Fig. 1-16

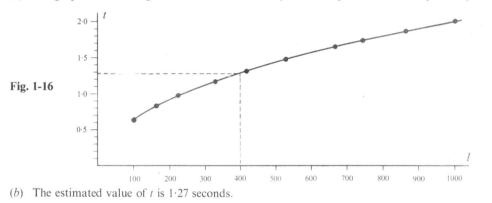

(*b*) The estimated value of t is 1·27 seconds.

EQUATIONS

1.29. Solve each of the following equations.

(*a*) $4a - 20 = 8$.

Add 20 to both members: $4a - 20 + 20 = 8 + 20$ or $4a = 28$.
Divide both sides by 4: $4a/4 = 28/4$ and $a = 7$.
Check: $4(7) - 20 = 8$, $28 - 20 = 8$, $8 = 8$.

(*b*) $3X + 4 = 24 - 2X$.

Subtract 4 from both members: $3X + 4 = 24 - 2X - 4$ or $3X = 20 - 2X$.
Add $2X$ to both sides: $3X + 2X = 20 - 2X + 2X$ or $5X = 20$.
Divide both sides by 5: $5X/5 = 20/5$ and $X = 4$.
Check: $3(4) + 4 = 24 - 2(4)$, $12 + 4 = 24 - 8$, $16 = 16$.

The result can be obtained much more quickly by realising that any term can be moved or *transposed* from one member of an equation to the other simply by changing its sign. Thus we can write
$$3X + 4 = 24 - 2X, \quad 3X + 2X = 24 - 4, \quad 5X = 20, \quad X = 4$$

(*c*) $18 - 5b = 3(b + 8) + 10$.

$$18 - 5b = 3b + 24 + 10, \ 18 - 5b = 3b + 34$$
Transposing, $-5b - 3b = 34 - 18$ or $-8b = 16$.

Dividing by -8, $\dfrac{-8b}{-8} = \dfrac{16}{-8}$ and $b = -2$.
Check: $18 - 5(-2) = 3(-2 + 8) + 10$, $18 + 10 = 3(6) + 10$, $28 = 28$.

(d) $\dfrac{Y+2}{3} + 1 = \dfrac{Y}{2}$.

    First multiply both members by 6, the lowest common denominator.

$$6\left(\dfrac{Y+2}{3}+1\right)=6\left(\dfrac{Y}{2}\right), \quad 6\left(\dfrac{Y+2}{3}\right)+6(1)=\dfrac{6Y}{2}, \quad 2(Y+2)+6=3Y$$

$$2Y+4+6=3Y, \quad 2Y+10=3Y, \quad 10=3Y-2Y, \quad Y=10$$

Check: $\dfrac{10+2}{3}+1=\dfrac{10}{2}, \dfrac{12}{3}+1=\dfrac{10}{2}, 4+1=5, 5=5.$

1.30. Solve each set of the following simultaneous equations.

(a) $\begin{cases} 3a - 2b = 11 \\ 5a + 7b = 39 \end{cases}$

Multiply the first equation by 7:	$21a - 14b = 77$ **(1)**
Multiply the second equation by 2:	$10a + 14b = 78$ **(2)**
Add:	$31a = 155$
Divide by 31:	$a = 5$

    Note that by multiplying each of the given equations by suitable numbers, we are able to write two *equivalent equations*, (**1**) and (**2**), where the coefficients of the unknown b are numerically equal. Then by addition we are able to *eliminate* the unknown b and thus find a.

    Substitute $a = 5$ in the first equation: $3(5) - 2b = 11, -2b = -4, b = 2$.
Thus $a = 5$ and $b = 2$.

    Check: $3(5) - 2(2) = 11, 15 - 4 = 11, 11 = 11$.
              $5(5) + 7(2) = 39, 25 + 14 = 39, 39 = 39$.

(b) $\begin{cases} 5X + 14Y = 78 \\ 7X + 3Y = -7 \end{cases}$

Multiply the first equation by 3:	$15X + 42Y = 234$ **(1)**
Multiply the second equation by -14:	$-98X - 42Y = 98$ **(2)**
Add:	$-83X = 332$
Divide by -83:	$X = -4$

    Substitute $X = -4$ in the first equation: $5(-4) + 14Y = 78, 14Y = 98, Y = 7$.
Thus $X = -4, Y = 7$.

    Check: $5(-4) + 14(7) = 78, -20 + 98 = 78, 78 = 78$.
              $7(-4) + 3(7) = -7, -28 + 21 = -7, -7 = -7$.

(c) $\begin{cases} 3a + 2b + 5c = 15 \\ 7a - 3b + 2c = 52 \\ 5a + b - 4c = 2 \end{cases}$

Multiply the first equation by 2:	$6a + 4b + 10c = 30$
Multiply the second equation by -5:	$-35a + 15b - 10c = -260$
Add:	$-29a + 19b = -230$ **(1)**
Multiply the second equation by 2:	$14a - 6b + 4c = 104$
Repeat the third equation:	$5a + b - 4c = 2$
Add:	$19a - 5b = 106$ **(2)**

    We have thus eliminated c and are left with two equations, (**1**) and (**2**), to be solved simultaneously for a and b.

Multiply equation (**1**) by 5:	$-145a + 95b = -1150$
Multiply equation (**2**) by 19:	$361a - 95b = 2014$
Add:	$216a = 864$
Divide by 216:	$a = 4$

    Substituting $a = 4$ in (**1**) or (**2**), we find $b = -6$.
    Substituting $a = 4$ and $b = -6$ in any of the given equations, we obtain $c = 3$.

    Thus $a = 4, b = -6, c = 3$.

    Check: $3(4) + 2(-6) + 5(3) = 15, 15 = 15. \ 7(4) - 3(-6) + 2(3) = 52, 52 = 52$.
              $5(4) + (-6) - 4(3) = 2, 2 = 2$.

INEQUALITIES

1.31. Express in words the meaning of each of the following.

    (a) $N > 30$. N is greater than 30.

    (b) $X \leq 12$. X is less than or equal to 12.

    (c) $0 < p \leq 1$. p is greater than 0 but less than or equal to 1.

    (d) $\mu - 2t < X < \mu + 2t$. X is greater than $\mu - 2t$ but less than $\mu + 2t$.

1.32. Translate the following into symbols.

    (a)   The variable X has values between 2 and 5 inclusive: $2 \leq X \leq 5$.

    (b)   The arithmetic mean $\bar{X}$ is greater than 28·42 but less than 31·56: $28·42 < \bar{X} < 31·56$.

    (c) m is a positive number less than or equal to 10: $0 < m \leq 10$.

    (d) P is a non-negative number: $P \geq 0$.

1.33. Using inequality symbols, arrange the numbers $3·42$, $-0·6$, $-2·1$, $1·45$, -3 in (a) increasing and (b) decreasing order of magnitude.

    Solution:

    (a) $-3 < -2·1 < -0·6 < 1·45 < 3·42$.   (b) $3·42 > 1·45 > -0·6 > -2·1 > -3$.

           Note that when the numbers are plotted as points on a line (see Problem 1.18) they increase from left to right.

1.34. In each of the following find a corresponding inequality for X, i.e. solve each inequality for X.

    (a) $2X < 6$.   Divide both sides by 2 obtain $X < 3$.

    (b) $3X - 8 \geq 4$.   Adding 8 to both sides, $3X \geq 12$; dividing both sides by 3, $X \geq 4$.

    (c) $6 - 4X < -2$.   Adding -6 to both sides, $-4X < -8$; dividing both sides by -4, $X > 2$.

           Note that, as in equations, we can transpose a term from one side of an inequality to the other simply by changing the sign of the term, e.g. $3X \geq 8 + 4$.

    (d) $-3 < \dfrac{X - 5}{2} < 3$.   Multiplying by 2, $-6 < X - 5 < 6$; adding 5, $-1 < X < 11$.

    (e) $-1 \leq \dfrac{3 - 2X}{5} \leq 7$.   Multiplying by 5, $-5 \leq 3 - 2X \leq 35$; adding -3, $-8 \leq -2X \leq 32$; dividing by -2, $4 \geq X \geq -16$ or $-16 \leq X \leq 4$.

LOGARITHMS AND ANTILOGARITHMS

1.35. Determine the characteristic of the common logarithm (base 10) of each of the following numbers.

(a) 57	(c) 5·63	(e) 982·5	(g) 186 000	(i) 0·7314	(k) 0·0071
(b) 57·4	(d) 35·63	(f) 7824	(h) 0·71	(j) 0·0325	(l) 0·0003
(a) 1	(c) 0	(e) 2	(g) 5	(i) 9 − 10	(k) 7 − 10
(b) 1	(d) 1	(f) 3	(h) 9 − 10	(j) 8 − 10	(l) 6 − 10

1.36. Verify each of the following logarithms.

    (a) $\log 87·2$. Mantissa $= 0·9405$, characteristic $= 1$; then $\log 87·2 = 1·9405$.

    (b) $\log 37\,300 = 4·5717$ (d) $\log 9·21 \;\;= 0·9643$

    (c) $\log 753 \;\;\;\;= 2·8768$ (e) $\log 54·50 = 1·7364$

    (f) $\log 0·382$. Mantissa $= 0·5821$, characteristic $= 9 - 10$; then $\log 0·382 = 9·5821 - 10$.

    (g) $\log 0·001\,59 = 7·2014 - 10$ (i) $\log 0·000\,827 = 6·9175 - 10$

    (h) $\log 0·0753 \;\;= 8·8768 - 10$ (j) $\log 0·0503 \;\;\;\;= 8·7016 - 10$

(*k*) log 4·638. Mantissa of log 4638 is 0·8 of the way between the mantissas of log 4630 and log 4640.

Mantissa of log 4640 = 0·6665	Mantissa of log 4·638 = 0·6656 + (0·8)(0·0009)
Mantissa of log 4630 = 0·6656	= 0·6663 to four digits.
Tabular difference = 0·0009	Then log 4·638 = 0·6663.
	This process is called *linear interpolation*.

If desired, the proportional parts table on Pages 346 and 347 can be used to give the mantissa directly (6656 + 7).

(*l*)	log 6·753 = 0·8295 (8293 + 2)	(*p*)	log 0·2548 = 9·4062 − 10 (4048 + 14)
(*m*)	log 183·2 = 2·2630 (2625 + 5)	(*q*)	log 0·043 72 = 8·6407 − 10 (6405 + 2)
(*n*)	log 43·15 = 1·6350 (6345 + 5)	(*r*)	log 0·009 848 = 7·9933 − 10 (9930 + 3)
(*o*)	log 876 400 = 5·9427 (9425 + 2)	(*s*)	log 0·000 178 8 = 6·2524 − 10 (2504 + 20)

1.37. Verify each of the following antilogarithms.

(*a*) antilog 1·9058.
   In the table the mantissa 0·9058 corresponds to the number 805. Since the characteristic is 1, the number must have two digits before the decimal point; then the required number is 80·5, i.e. antilog 1·9058 = 80·5.

(*b*) antilog 3·8531 = 7130, antilog 2·1875 = 154, antilog 0·4997 = 3·16, antilog 4·9360 = 86 300

(*c*) antilog 7·8657 − 10.
   In the table the mantissa 0·8657 corresponds to the number 734. Since the characteristic is 7 − 10, the number must have two zeros immediately following the decimal point. Hence the required number is 0·007 34, i.e. antilog 7·8657 − 10 = 0·007 34
   The proportional parts table on Pages 346 and 347 can also be used.

(*d*) antilog 9·8267 − 10 = 0·671, antilog 2·3927 = 0·0247, antilog 7·7443 − 10 = 0·005 55

(*e*) antilog 9·3842 − 10.
   Since the mantissa is not found in the tables, interpolation must be used.

Mantissa of log 2430 = 0·3856		Given mantissa = 0·3842
Mantissa of log 2420 = 0·3838		Next smaller mantissa = 0·3838
Tabular difference = 0·0018		Difference = 0·0004

   Then 2420 + (4/18)(2430 − 2420) = 2422 to four digits, and the required number is 0·2422.

(*f*) antilog 2·6715 = 469·3   (3/9 × 10 = 3 approx.)
       antilog 4·1853 = 15 320   (6/28 × 10 = 2 approx.)
       antilog 0·9245 = 8·404   (2/5 × 10 = 4)

(*g*) antilog 1·6089 = 0·4064   (4/11 × 10 = 4 approx.)
       antilog 8·8907 − 10 = 0·077 75   (3/6 × 10 = 5)
       antilog 1·2000 = 15·85   (13/27 × 10 = 5 approx.)

COMPUTATIONS USING LOGARITHMS

Calculate each of the following using logarithms.

1.38. $P = (3·81)(43·4)$. log P = log 3·81 + log 43·4

log 3·81 = 0·5809	
(+) log 43·4 = 1·6375	
log P = 2·2184.	Then P = antilog 2·2184 = 165·3, or 165 to 3 significant figures.

Note the exponential significance of the computation. Thus,
$$(3·81)(43·4) = (10^{0·5809})(10^{1·6375}) = 10^{0·5809 + 1·6375} = 10^{2·2184} = 165·3$$

1.39. $P = (73·42)(0·004\ 620)(0·5143)$.   log P = log 73·42 + log 0·004 620 + log 0·5143

log 73·42 = 1·8658	
(+) log 0·004 620 = 7·6646 − 10	
(+) log 0·5143 = 9·7112 − 10	
log P = 19·2416 − 20 = 9·2416 − 10. Then P = 0·1744.	

1.40. $P = \dfrac{(784 \cdot 6)(0 \cdot 0431)}{28 \cdot 23}$. $\qquad \log P = \log 784 \cdot 6 + \log 0 \cdot 0431 - \log 28 \cdot 23$

$$
\begin{array}{lll}
\log 784 \cdot 6 & = & 2 \cdot 8947 \\
(+) \ \log 0 \cdot 0431 & = & 8 \cdot 6345 - 10 \\
& & \overline{11 \cdot 5292 - 10} \\
(-) \ \log 28 \cdot 23 & = & 1 \cdot 4507 \\
\log P & = & \overline{10 \cdot 0785 - 10} = 0 \cdot 0785.
\end{array}
$$
Then $P = 1 \cdot 198$ or $1 \cdot 20$ to 3 significant figures.

Note the exponential significance of the computation. Thus,

$$\frac{(784 \cdot 6)(0 \cdot 0431)}{28 \cdot 23} = \frac{(10^{2 \cdot 8947})(10^{8 \cdot 6345 - 10})}{10^{1 \cdot 4507}} = 10^{2 \cdot 8947 + 8 \cdot 6345 - 10 - 1 \cdot 4507} = 10^{0 \cdot 0785} = 1 \cdot 198$$

1.41. $P = (5 \cdot 395)^8$. $\qquad \log P = 8 \log 5 \cdot 395 = 8(0 \cdot 7320) = 5 \cdot 8560$, and $P = 717\,800$ or $7 \cdot 178 \times 10^5$.

1.42. $P = \sqrt{387 \cdot 2} = (387 \cdot 2)^{1/2}$. $\qquad \log P = \frac{1}{2} \log 387 \cdot 2 = \frac{1}{2}(2 \cdot 5879) = 1 \cdot 2940$, and $P = 19 \cdot 68$.

1.43. $P = (0 \cdot 083\,17)^{1/5}$. $\qquad \log P = \ \log 0 \cdot 083\,17 = \frac{1}{5}(8 \cdot 9200 - 10) = \frac{1}{5}(48 \cdot 9200 - 50) = 9 \cdot 7840 - 10$, and $P = 0 \cdot 6081$.

1.44. $P = \dfrac{\sqrt{0 \cdot 003\,654}\,(18 \cdot 37)^3}{(8 \cdot 724)^4 \sqrt[4]{743 \cdot 8}}$. $\qquad \log P = \frac{1}{2} \log 0 \cdot 003\,654 + 3 \log 18 \cdot 37 - (4 \log 8 \cdot 724 + \frac{1}{4} \log 743 \cdot 8)$

Numerator N $\qquad\qquad\qquad\qquad\qquad\qquad\qquad$ *Denominator D*

$$
\begin{array}{ll}
\frac{1}{2} \log 0 \cdot 003\,654 = \frac{1}{2}(7 \cdot 5628 - 10) \\
\qquad\qquad\quad = \frac{1}{2}(17 \cdot 5628 - 20) \quad = \quad 8 \cdot 7814 - 10 \\
3 \log 18 \cdot 37 \quad = 3(1 \cdot 2641) \qquad\quad = \quad 3 \cdot 7923 \\
\text{Add:} \qquad\qquad\qquad\qquad\quad \log N = \overline{12 \cdot 5737 - 10} \\
\qquad\qquad\qquad (-) \ \log D = \quad 4 \cdot 4806 \\
\qquad\qquad\qquad\qquad \log P = \overline{\ 8 \cdot 0931 - 10} \\
\qquad\qquad\qquad\qquad\quad\ P = \quad 0 \cdot 012\,39
\end{array}
$$

$$
\begin{array}{l}
4 \log 8 \cdot 724 = 4(0 \cdot 9407) = 3 \cdot 7628 \\
\frac{1}{4} \log 743 \cdot 6 = \frac{1}{4}(2 \cdot 8714) = 0 \cdot 7178 \\
\text{Add:} \qquad\qquad \log D = \overline{4 \cdot 4806}
\end{array}
$$

1.45. $P = \sqrt{\dfrac{(874 \cdot 3)(0 \cdot 038\,16)(28 \cdot 53)^3}{(1 \cdot 754)^4 \,(0 \cdot 007\,352)}}$.

$$
\begin{array}{llll}
\log 874 \cdot 3 & = 2 \cdot 9417 & = 2 \cdot 9417 \\
\log 0 \cdot 038\,16 & = 8 \cdot 5816 - 10 & = 8 \cdot 5816 - 10 \\
3 \log 28 \cdot 53 & = 3(1 \cdot 4553) & = 4 \cdot 3659 \\
\text{Add:} & & \overline{15 \cdot 8892 - 10} \\
& (-) & \underline{\ 8 \cdot 8424 - 10} \\
& & 7 \cdot 0468
\end{array}
$$

$$
\begin{array}{ll}
4 \log 1 \cdot 754 = 4(0 \cdot 2440) = 0 \cdot 9760 \\
\log 0 \cdot 007\,352 \qquad\quad = 7 \cdot 8664 - 10 \\
\text{Add:} \qquad\qquad\qquad\ \ \overline{8 \cdot 8424 - 10}
\end{array}
$$

Then $\log P = \frac{1}{2}(7 \cdot 0468) = 3 \cdot 5234$, and $P = 3338$.

Supplementary Problems

VARIABLES

1.46. State which of the following represent discrete data and which represent continuous data.

    (*a*) Number of millimetres of rainfall in a city during various months of the year.

    (*b*) Speed of an automobile in kilometres per hour.

    (*c*) Number of £5 notes circulating in the United Kingdom at any time.

    (*d*) Total value of shares sold each day in the Stock Market.

    (*e*) Student enrolment in a university over a number of years.

    Ans. (*a*) continuous, (*b*) continuous, (*c*) discrete, (*d*) discrete, (*e*) discrete

1.47. Give the domain of each of the following variables and state whether the variables are continuous or discrete.

    (*a*) Number W of kilogrammes of wheat produced per hectare on a farm over a number of years.

    (*b*) Number N of individuals in a family.   (*d*) Time t of flight of a missile.

    (*c*) Marital status of an individual.   (*e*) Number P of petals on a flower.

    Ans. (*a*) Zero upward; continuous. (*b*) 2, 3, . . .; discrete. (*c*) Single, married, divorced, separated, widowed; discrete.

        (*d*) Zero upward; continuous. (*e*) 0, 1, 2, . . .; discrete.

ROUNDING OF DATA, SCIENTIFIC NOTATION AND SIGNIFICANT FIGURES

1.48. Round each of the following numbers to the indicated accuracy.

(*a*)	3256	nearest hundred	(*f*)	3 502 378	nearest million
(*b*)	5·781	nearest tenth	(*g*)	148·475	nearest unit
(*c*)	0·0045	nearest thousandth	(*h*)	0·000 098 501	nearest millionth
(*d*)	46·7385	nearest hundredth	(*i*)	2184·73	nearest ten
(*e*)	125·9995	two decimal places	(*j*)	43·875 00	nearest hundredth

    Ans. (*a*) 3300, (*b*) 5·8, (*c*) 0·004, (*d*) 46·74, (*e*) 126·00, (*f*) 4 000 000, (*g*) 148, (*h*) 0·000 099, (*i*) 2180, (*j*) 43·88

1.49. Express each number without using powers of 10.

    (*a*) $132·5 \times 10^4$, (*b*) $418·72 \times 10^{-5}$ (*c*) 280×10^{-7}, (*d*) 7300×10^6, (*e*) $3·487 \times 10^{-4}$, (*f*) $0·000 185 0 \times 10^5$.

    Ans. (*a*) 1 325 000, (*b*) 0·004 187 2, (*c*) 0·000 028 0, (*d*) 7 300 000 000, (*e*) 0·000 348 7, (*f*) 18·50

1.50. How many significant figures are in each of the following, assuming the numbers are accurately recorded?

(*a*)	2·54 mm	(*d*)	3·51 million litres	(*f*)	378 people	(*h*)	$4·50 \times 10^{-3}$ km
(*b*)	0·004 500 m	(*e*)	10·000 100 m	(*g*)	378 g	(*i*)	$500·8 \times 10^5$ kg
(*c*)	3 510 000 litres					(*j*)	100·00 km

    Ans. (*a*) 3, (*b*) 4, (*c*) 7, (*d*) 3, (*e*) 8, (*f*) unlimited, (*g*) 3, (*h*) 3, (*i*) 4, (*j*) 5

1.51. What is the maximum error in each of the following measurements, assumed to be accurately recorded? Give the number of significant figures in each case.

(*a*)	7·20 million litres	(*c*)	5280 metres	(*e*)	186 000 metres per second
(*b*)	0·000 048 35 millimetres	(*d*)	$3·0 \times 10^8$ metres	(*f*)	186 thousand metres per second

    Ans.

(*a*)	0·005 million or 5000 litres; 3	(*c*)	0·5 m; 4	(*e*)	0·5 m/s; 6
(*b*)	0·000 000 005 or 5×10^{-9} mm; 4	(*d*)	$0·05 \times 10^8$ or 5×10^6 m; 2	(*f*)	0·5 thousand or 500 m/s; 3

1.52. Write each of the following numbers using the scientific notation. Assume all figures significant unless otherwise indicated.

    (*a*) 0·000 317, (*b*) 428 000 000 (four sig. fig.), (*c*) 21 600·00, (*d*) 0·000 009 810, (*e*) 732 thousand, (*f*) 18·0 ten thousandths.

    Ans. (*a*) $3·17 \times 10^{-4}$, (*b*) $4·280 \times 10^8$, (*c*) $2·160 000 \times 10^4$, (*d*) $9·810 \times 10^{-6}$, (*e*) $7·32 \times 10^5$, (*f*) $1·80 \times 10^{-3}$

COMPUTATIONS

1.53. Show that (*a*) the product, (*b*) the quotient of the numbers 72·48 and 5·16, assumed to have four and three significant figures respectively, cannot be accurate to more than three significant figures. Write the accurately recorded product and quotient.

Ans. (*a*) 374, (*b*) 14·0

1.54. Perform each indicated operation. Unless otherwise specified, assume the numbers are accurately recorded.

(*a*) 0.36×781.4

(*c*) $5.78 \times 2700 \times 16.00$

(*e*) $\sqrt{120 \times 0.5386 \times 0.4614}$ (120 exact)

(*b*) $\dfrac{873.00}{4.881}$

(*d*) $\dfrac{0.004\,80 \times 2300}{.2084}$

(*f*) $\dfrac{(416\,000)(0.000\,187)}{\sqrt{73.84}}$

(*g*) $14.8641 + 4.48 - 8.168 + 0.361\,25$

(*h*) $4\,173\,000 - 170\,264 + 1\,820\,470 - 78\,320$ (Numbers are respectively accurate to 4, 6, 6 and 5 sig. fig.)

(*i*) $\sqrt{\dfrac{7(4.386)^2 - 3(6.47)^2}{6}}$ (3, 6 and 7 are exact.)

(*j*) $4.120\sqrt{\dfrac{3.1416[(9.483)^2 - (5.075)^2]}{0.000\,198\,0}}$

Ans. (*a*) 280 (two sig. fig.), or 2·8 hundred, or 2.8×10^2. (*b*) 178·9. (*c*) 250 000 (three sig. fig.), or 250 thousand, or 2.50×10^5. (*d*) 53·0. (*e*) 5·461. (*f*) 9·05. (*g*) 11·54. (*h*) 5 745 000 (four sig. fig.), or 5·745 thousand, or 5·745 million, 5.745×10^6. (*i*) 1·2. (*j*) 4157

1.55. Evaluate each of the following, given that $U = -2$, $V = \frac{1}{2}$, $W = 3$, $X = -4$, $Y = 9$, $Z = \frac{1}{6}$, where all numbers are supposed exact.

(*a*) $4U + 6V - 2W$

(*e*) $\sqrt{U^2 - 2UV + W}$

(*h*) $\dfrac{X - 3}{\sqrt{(Y - 4)^2 + (U + 5)^2}}$

(*b*) $\dfrac{XYZ}{UVW}$

(*f*) $3X(4Y + 3Z) - 2Y(6X - 5Z) - 25$

(*i*) $X^3 + 5X^2 - 6X - 8$

(*c*) $\dfrac{2X - 3Y}{UW + XV}$

(*g*) $\sqrt{\dfrac{(W - 2)^2}{V} + \dfrac{(Y - 5)^2}{Z}}$

(*j*) $\dfrac{U - V}{\sqrt{U^2 + V^2}}[U^2 V(W + X)]$

(*d*) $3(U - X)^2 + Y$

Ans. (*a*) -11

(*e*) 3

(*h*) $-7/\sqrt{34}$, or $-1.200\,49$ approx.

(*b*) 2

(*f*) -16

(*i*) 32

(*c*) 35/8 or 4·375

(*g*) $\sqrt{98}$, or 9·899 61 approx.

(*j*) $10/\sqrt{17}$, or 2·425 36 approx.

(*d*) 21

FUNCTIONS, TABLES AND GRAPHS

1.56. A variable Y is determined from a variable X according to the equation $Y = 10 - 4X$.
(*a*) Find Y when $X = -3, -2, -1, 0, 1, 2, 3, 4, 5$, showing the results in a table.
(*b*) Find Y when $X = -2.4, -1.6, -0.8, 1.8, 2.7, 3.5, 4.6$.
(*c*) If the dependence of Y on X is denoted by $Y = F(X)$, find $F(2.8)$, $F(-5)$, $F(\sqrt{2})$, $F(-\pi)$.
(*d*) What value of X corresponds to $Y = -2, 6, -10, 1.6, 16, 0, 10$?
(*e*) Express X explicitly as a function of Y.

Ans. (*a*) 22, 18, 14, 10, 6, 2, -2, -6, -10.   (*c*) -1.2, 30, $10 - 4\sqrt{2} = 4.34$ approx., $10 + 4\pi = 22.57$ approx.
(*b*) 19·6, 16·4, 13·2, 2·8, -0.8, -4, -8.4.   (*d*) 3, 1, 5, 2·1, -1.5, 2·5, 0.   (*e*) $X = \frac{1}{4}(10 - Y)$

1.57. If $Z = X^2 - Y^2$, find Z when: (*a*) $X = -2$, $Y = 3$; (*b*) $X = 1$, $Y = 5$. (*c*) If using the functional notation $Z = F(X, Y)$, find $F(-3, -1)$.   *Ans.* (*a*) -5, (*b*) -24, (*c*) 8

1.58. If $W = 3XZ - 4Y^2 + 2XY$, find W when: (*a*) $X = 1$, $Y = -2$, $Z = 4$; (*b*) $X = -5$, $Y = -2$, $Z = 0$. (*c*) If using the functional notation $W = F(X, Y, Z,)$, find $F(3, 1, -2)$.   *Ans.* (*a*) -8, (*b*) 4, (*c*) -16

1.59. Locate on a rectangular co-ordinate system the points having co-ordinates (*a*) (3, 2), (*b*) (2, 3), (*c*) (-4, 4), (*d*) (4, -4), (*e*) (-3, -2), (*f*) (-2, -3), (*g*) (-4.5, 3), (*h*) (-1.2, -2.4), (*i*) (0, -3), (*j*) (1·8, 0).

1.60. Graph the equations   (*a*) $y = 10 - 4x$ (see Problem 1.56.),   (*b*) $y = 2x + 5$,   (*c*) $y = \frac{1}{3}(x - 6)$,   (*d*) $2x + 3y = 12$, (*e*) $3x - 2y = 6$.

1.61. Graph the equations (*a*) $y = 2x^2 + x - 10$, (*b*) $y = 6 - 3x - x^2$.

1.62.   Graph $y = x^3 - 4x^2 + 12x - 6$.

1.63.   The following table shows the numbers of agricultural and non-agricultural workers in the United States for the years 1840–1950. Graph the data using (*a*) line graphs, (*b*) bar charts, (*c*) component bar charts.

Year	1840	1850	1860	1870	1880	1890	1900	1910	1920	1930	1940	1950
Agricultural workers (millions)	3·7	4·9	6·2	6·9	8·6	9·9	10·9	11·6	11·4	10·5	8·8	6·8
Non-agricultural workers (millions)	1·7	2·8	4·3	6·1	8·8	13·4	18·2	25·8	31·0	38·4	42·9	52·2

Source: Department of Commerce, Bureau of the Census.

1.64.   Design an appropriate pictograph to show the variation in numbers of (*a*) agricultural, (*b*) non-agricultural workers for the data of the preceding problem. Can you design a pictograph which will show the variation in both (*a*) and (*b*)?

1.65.   Using the data of Problem 1.63 construct a graph showing the percentage of all workers who are (*a*) agricultural, (*b*) non-agricultural. Can you design a graph which will show parts (*a*) and (*b*) simultaneously?

1.66.   The following table shows the birth and death rates per 1000 people in the United States for the years 1915–1955. Graph the data using an appropriate type of graph.

Year	1915	1920	1925	1930	1935	1940	1945	1950	1955
Birth rate (per 1000 people)	25·0	23·7	21·3	18·9	16·9	17·9	19·5	23·6	24·6
Death rate (per 1000 people)	13·2	13·0	11·7	11·3	10·9	10·8	10·6	9·6	9·3

Source: Department of Health, Education and Welfare.

1.67.   The following table shows the heights of the seven tallest buildings and structures in the world. Graph the data using an appropriate type of graph.

Building or Structure	Location	Height (metres)
Empire State Building	New York	381
Chrysler Building	New York	319
Eiffel Tower	Paris	300
Wall Street Building	New York	290
Bank of Manhattan	New York	283
R.C.A. Building, Rockefeller Centre	New York	259
Woolworth Building	New York	241

1.68.   The following table shows the orbital velocities of the planets in our solar system. Graph the data.

Planet	Mercury	Venus	Earth	Mars	Jupiter	Saturn	Uranus	Neptune	Pluto
Velocity (km/s)	47·8	35·1	29·8	24·1	13·0	9·7	6·8	5·5	4·8

1.69. The following table shows the marital status of males and females (14 years and older) in the United States as of the year 1958. Graph the data using two pie charts having the same diameter.

Marital Status	Male (per cent of total)	Female (per cent of total)
Single	24·5	18·8
Married	69·8	66·0
Widowed	3·9	12·8
Divorced	1·8	2·3

Source: Bureau of the Census.

1.70. The following table shows the areas in millions of square kilometres of the oceans of the world. Graph the data using (*a*) a bar chart, (*b*) a pie chart.

Ocean	Pacific	Atlantic	Indian	Antarctic	Arctic
Area (million km²)	183·4	106·7	73·8	19·7	12·4

EQUATIONS

1.71. Solve each of the following equations

(*a*) $16 - 5c = 36$ (*c*) $4(X - 3) - 11 = 15 - 2(X + 4)$ (*e*) $3[2(X + 1) - 4] = 10 - 5(4 - 2X]$

(*b*) $2Y - 6 = 4 - 3Y$ (*d*) $3(2U + 1) = 5(3 - U) + 3(U - 2)$ (*f*) $\frac{2}{3}(12 + Y) = 6 - \frac{1}{4}(9 - Y)$

Ans. (*a*) -4 (*b*) 2   (*c*) 5   (*d*) $\frac{3}{4}$ (*e*) 1   (*f*) -7

1.72. Solve each of the following simultaneous equations

(*a*) $\begin{cases} 2a + b = 10 \\ 7a - 3b = 9 \end{cases}$ (*b*) $\begin{cases} 3a + 5b = 24 \\ 2a + 3b = 14 \end{cases}$ (*c*) $\begin{cases} 8X - 3Y = 2 \\ 3X + 7Y = -9 \end{cases}$ (*d*) $\begin{cases} 5A - 9B = -10 \\ 3A - 4B = 16 \end{cases}$

(*e*) $\begin{cases} 2a + b - c = 2 \\ 3a - 4b + 2c = 4 \\ 4a + 3b - 5c = -8 \end{cases}$ (*f*) $\begin{cases} 5X + 2Y + 3Z = -5 \\ 2X - 3Y - 6Z = 1 \\ X + 5Y - 4Z = 22 \end{cases}$ (*g*) $\begin{cases} 3U - 5V + 6W = 7 \\ 5U + 3V - 2W = -1 \\ 4U - 8V + 10W = 11 \end{cases}$

Ans.

(*a*) $a = 3, b = 4$ (*b*) $a - -2, b = 6$ (*c*) $X = -0·2, Y = -1·2$

(*d*) $A = 184/7 = 26·285\ 71$ approx., $B = 110/7 = 15·714\ 29$ approx.

(*e*) $a = 2, b = 3, c = 5$ (*f*) $X = -1, Y = 3, Z = -2$ (*g*) $U = 0·4, V = -0·8, W = 0·3$

1.73. (*a*) Graph the equations $5x + 2y = 4$ and $7x - 3y = 23$ using the same set of co-ordinates axes.
(*b*) From the graphs determine the simultaneous solution of the two equations.   (*c*) Use this method to obtain the simultaneous solutions of equations (*a*)–(*d*) of Problem 1.72.

Ans. (*b*) $(2, -3)$, i.e. $x = 2, \ y = -3$

1.74. (*a*) Use the graph of Problem 1.61(*a*) to solve the equation $2x^2 + x - 10 = 0$. (Hint: Find the values of x where the parabola intersects the x axis, i.e. where $y = 0$.)   (*b*) Use the method in (*a*) to solve $3x^2 - 4x - 5 = 0$.

Ans. (*a*) 2, $-2·5$; (*b*) 2·1 and $-0·8$ approx.

1.75. The solutions of the quadratic equation $\ \ aX^2 + bX + c = 0 \ \ $ are given by the *quadratic formula:*

$X = \dfrac{-b \pm \sqrt{b^2 - 4ac}}{2a}$. Use this formula to find the solutions of (*a*) $3X^2 - 4X - 5 = 0$, (*b*) $2X^2 + X - 10 = 0$,

(*c*) $5X^2 + 10X = 7$,   (*d*) $X^2 + 8X + 25 = 0$.

Ans.   (*a*) $\dfrac{4 \pm \sqrt{76}}{6}$ or 2·12 and $-0·79$ approx.   (*b*) 2, $-2·5$.   (*c*) 0·549, $-2·549$ approx.

(*d*) $\dfrac{-8 \pm \sqrt{-36}}{2} = \dfrac{-8 \pm \sqrt{36}\sqrt{-1}}{2} = \dfrac{-8 \pm 6\sqrt{-1}}{2} = -4 \pm 3\sqrt{-1} = -4 \pm 3i$

where $i = \sqrt{-1}$. These roots are *complex numbers* and will not show up when a graphical procedure is employed.

INEQUALITIES

1.76. Using inequality symbols, arrange the numbers $-4.3, -6.15, 2.37, 1.52, -1.5$ in (a) increasing, (b) decreasing order of magnitude.

    Ans. (a) $-6.15 < -4.3 < -1.5 < 1.52 < 2.37$, (b) $2.37 > 1.52 > -1.5 > -4.3 > -6.15$

1.77. Use inequality symbols to express each of the following statements.   (a) The number N of children is between 30 and 50 inclusive.   (b) The Sum S of points on the pair of dice is not less than 7.   (c) X is greater than or equal to -4 but less than 3.   (d) P is at most 5.   (e) X exceeds Y by more than 2.

    Ans. (a) $30 \leqq N \leqq 50$, (b) $S \geqq 7$, (c) $-4 \leqq X < 3$, (d) $P \leqq 5$, (e) $X - Y > 2$

1.78. Solve each of the following inequalities.

    (a) $3X \geqq 12$ (d) $3 + 5(Y - 2) \leqq 7 - 3(4 - Y)$ (g) $-2 \leqq 3 + \frac{1}{2}(a - 12) < 8.$
    (b) $4X < 5X - 3$ (e) $-3 \leqq \frac{1}{5}(2X + 1) \leqq 3$
    (c) $2N + 15 > 10 + 3N$ (f) $0 < \frac{1}{2}(15 - 5N) \leqq 12$

    Ans. (a) $X \geqq 4$, (b) $X > 3$, (c) $N < 5$, (d) $Y \leqq 1$, (e) $-8 \leqq X \leqq 7$, (f) $-1.8 \leqq N < 3$, (g) $2 \leqq a < 22$

LOGARITHMS

1.79. Find the common logarithm of each of the following numbers.

    (a) 387   (c) 0.0792   (e) 0.6042   (g) 476.3   (i) 7.146   (k) 0.000 98
    (b) 0.387   (d) 14 630   (f) 0.002 795   (h) 1.007   (j) 71.46   (l) 84.620 000

    Ans.
    (a) 2.5877   (c) $8.8987-10$ (e) $9.7812-10$ (g) 2.6779   (i) 0.8541   (k) $6.9912-10$
    (b) $9.5877-10$ (d) 4.1653   (f) $7.4464-10$ (h) 0.0030   (j) 1.8541   (l) 7.9275

1.80. Find the antilogarithm of each of the following.

    (a) 3.5611   (c) 1.7045   (e) 2.4700   (g) 2.8003   (i) 0.0800
    (b) $9.8293-10$ (d) $8.9266-10$ (f) $6.4700-10$ (h) 3.7072   (j) 6.3841

    Ans.
    (a) 3640   (c) 50.64   (e) 295.1   (g) 0.063 14   (i) 1.202
    (b) 0.675   (d) 0.084 45   (f) 0.000 295 1 (h) 5096   (j) 2 422 000
                                                                             or 2.422×10^6

1.81. Evaluate each of the following by use of logarithms.

    (a) $(783.6)(1654)$ (e) $\dfrac{(0.3854)^4 (12.48)^2}{(0.043\,82)^3}$ (h) $\sqrt[5]{(21.63)(33.81)(47.53)(65.28)(87.47)}$

    (b) $\dfrac{21.7}{378.2}$ (f) $0.041\,82\sqrt{0.6758}$ (i) $\sqrt{\dfrac{(48.79)(0.005\,74)^3}{(2.143)^5}}$

    (c) $\dfrac{(0.045\,56)(624.1)}{(14.32)(0.003\,572)}$ (g) $\sqrt[3]{3728}$ (j) $\dfrac{3.781}{0.018\,73}\sqrt{\dfrac{(43.25)(0.087\,43)}{(0.002\,356)(6.824)}}$

    (d) $(1.562)^{15}$

    Ans. (a) 1 296 000 or 1.296×10^6. (b) 0.057 39, or 0.0574 to 3 sig. fig. (c) 556.0. (d) 804.4. (e) 40 820. (f) 0.034 38.
    (g) 15.51. (h) 45.67. (i) 0.000 451 9 $= 4.519 \times 10^{-4}$ or 4.52×10^{-4} to 3 sig. fig. (j) 3096.

1.82. Graph (a) $y = \log x$, (b) $y = 10^x$ and discuss the similarities between the two graphs.

1.83. Write the equation (a) $2 \log X - 3 \log Y = 2$, (b) $\log Y + 2X = \log 3$ in a form free of logarithms.

    Ans. (a) $X^2 = 100 Y^3$, (b) $Y = 3(10^{-2X})$

1.84. If $a^p = N$, where a and p are positive numbers and $a \neq 1$, we call p the *logarithm of N to the base a* and write $p = \log_a N$. Evaluate (a) $\log_2 8$, (b) $\log_{25} 125$, (c) $\log_4 1/16$, (d) $\log_{1/2} 32$, (e) $\log_5 1$.

    Ans. (a) 3, (b) 3/2, (c) -2, (d) -5, (e) 0

1.85. Show that $\log_e N = 2.303 \log_{10} N$ approximately, where $e = 2.718\,28\ldots$ is called the *natural base* of logarithms and $N > 0$.

1.86. Show that $(\log_b a)(\log_a b) = 1$ where $a > 0, b > 0, a \neq 1, b \neq 1$.

CHAPTER 2

Frequency Distributions

RAW DATA

Raw data are collected data which have not been organized numerically. An example is the set of masses of 100 male students obtained from an alphabetical listing of university records.

ARRAYS

An array is an arrangement of raw numerical data in ascending or descending order of magnitude. The difference between the largest and smallest numbers is called the *range* of the data. For example, if the largest mass of 100 male students is 74 kg and the smallest mass is 60 kg, the range is $74 - 60 = 14$ kg.

FREQUENCY DISTRIBUTIONS

When summarizing large masses of raw data it is often useful to distribute the data into *classes* or *categories* and to determine the number of individuals belonging to each class, called the *class frequency*. A tabular arrangement of data by classes together with the corresponding class frequencies is called a *frequency distribution* or *frequency table*. Table 2.1 is a frequency distribution of masses (recorded to the nearest kg) of 100 male students at XYZ University.

Table 2.1

Masses of 100 Male Students at XYZ University

Mass (kilogrammes)	Number of Students
60–62	5
63–65	18
66–68	42
69–71	27
72–74	8
	Total 100

The first class or category, for example, consists of masses from 60 to 62 kg and is indicated by the symbol 60–62. Since 5 students have masses belonging to this class, the corresponding class frequency is 5.

Data organized and summarized as in the above frequency distribution are often called *grouped data*. Although the grouping process generally destroys much of the original detail of the data, an important advantage is gained in the clear "overall" picture which is obtained and in vital relationships which are thereby made evident.

CLASS INTERVALS AND CLASS LIMITS

A symbol defining a class such as 60–62 in the above table is called a *class interval*. The end numbers, 60 and 62, are called *class limits*; the smaller number 60 is the *lower class limit* and the larger number 62 is the *upper class limit*. The terms class and class interval are often used interchangeably, although the class interval is actually a symbol for the class.

A class interval which, at least theoretically, has either no upper class limit or no lower class limit indicated is called an *open class interval*. For example, referring to age groups of individuals, the class interval "65 years and over" is an open class interval.

CLASS BOUNDARIES

If masses are recorded to the nearest kg, the class interval 60–62 theoretically includes all measurements from 59·5000. . . to 62·5000. . . kg. These numbers, indicated briefly by the exact numbers 59·5 and 62·5, are called *class boundaries* or *true class limits*; the smaller number 59·5 is the *lower class boundary* and the larger number 62·5 is the *upper class boundary*.

In practice, the class boundaries are obtained by adding the upper limit of one class interval to the lower limit of the next higher class interval and dividing by 2.

Sometimes, class boundaries are used to symbolize classes. For example, the various classes in the first column of Table 2.1 could be indicated by 59·5–62·5, 62·5–65·5, etc. To avoid ambiguity in using such notation, class boundaries should not coincide with actual observations. Thus if an observation were 62·5 it would not be possible to decide whether it belonged to the class interval 59·5–62·5 or 62·5–65.5.

THE SIZE OR WIDTH OF A CLASS INTERVAL

The size or width of a class interval is the difference between the lower and upper class boundaries and is also referred to as the *class width*, *class size*, or *class strength*. If all class intervals of a frequency distribution have equal widths, this common width is denoted by c. In such case c is equal to the difference between two successive lower limits or two successive upper class limits. For the data of Table 2.1, for example, the class interval is $c = 62·5 - 59·5 = 65·5 - 62·5 = 3$.

THE CLASS MARK

The class mark is the midpoint of the class interval and is obtained by adding the lower and upper class limits and dividing by two. Thus the class mark of the interval 60–62 is $(60 + 62)/2 = 61$. The class mark is also called the *class midpoint*.

For purposes of further mathematical analysis, all observations belonging to a given class interval are assumed to coincide with the class mark. Thus all masses in the class interval 60–62 kg are considered as 61 kg.

GENERAL RULES FOR FORMING FREQUENCY DISTRIBUTIONS

1. Determine the largest and smallest numbers in the raw data and thus find the range (difference between largest and smallest numbers).

2. Divide the range into a convenient number of class intervals having the same size. If this is not feasible, use class intervals of different sizes or open class intervals (see Problem 2.12). The number of class intervals is usually taken between 5 and 20, depending on the data. Class intervals are also chosen so that the class marks or midpoints coincide with actually observed data. This tends to lessen the so-called *grouping error* involved in further mathematical analysis. However, class boundaries should not coincide with actually observed data.

3. Determine the number of observations falling into each class interval, i.e. find the class frequencies. This is best done by using a *tally* or *score sheet* (see Problem 2.8).

HISTOGRAMS AND FREQUENCY POLYGONS are two graphical representations of frequency distributions.

1. A *histogram* or *frequency histogram* consists of a set of rectangles having

    (*a*) bases on a horizontal axis (the *x* axis) with centres at the class marks and lengths equal to the class interval sizes,

    (*b*) areas proportional to class frequencies.

    If the class intervals all have equal size, the heights of the rectangles are proportional to the class frequencies and it is then customary to take the heights numerically equal to the class frequencies. If class intervals do not have equal size, these heights must be adjusted (see Problem 2.13).

2. A *frequency polygon* is a line graph of class frequency plotted against class mark. It can be obtained by connecting midpoints of the tops of the rectangles in the histogram.

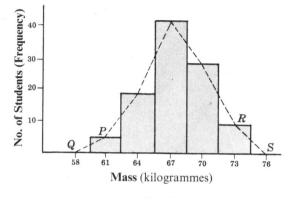

Fig. 2-1

The histogram and frequency polygon corresponding to the above frequency distribution of masses are shown on the same set of axes in Fig. 2-1. It is customary to add the extensions *PQ* and *RS* to the next lower and higher class marks which have corresponding class frequency of zero. In such case the sum of the areas of the rectangles in the histogram equals the total area bounded by the frequency polygon and the *x* axis (see Problem 2.11).

RELATIVE FREQUENCY DISTRIBUTIONS

The *relative frequency* of a class is the frequency of the class divided by the total frequency of all classes and is generally expressed as a percentage. For example, the relative frequency of the class 66–68 in Table 2.1 is $42/100 = 42\%$. The sum of the relative frequencies of all classes is clearly 1 or 100%.

If frequencies in the above frequency table are replaced by corresponding relative frequencies, the resulting table is called a *relative frequency distribution*, *percentage distribution*, or *relative frequency table*.

Graphical representations of relative frequency distributions can be obtained from the histogram or frequency polygon by simply changing the vertical scale from frequency to relative frequency, keeping exactly the same diagram. The resulting graphs are called *relative frequency histograms* or *percentage histograms* and *relative frequency polygons* or *percentage polygons,* respectively.

CUMULATIVE FREQUENCY DISTRIBUTIONS.   OGIVES

The total frequency of all values less than the upper class boundary of a given class interval is called the *cumulative frequency* up to and including the class interval. For example, the cumulative frequency up to and including the class interval 66–68 in Table 2.1 is $5 + 18 + 42 = 65$, signifying that 65 students have masses less than 68·5 kg.

A table presenting such cumulative frequencies is called a *cumulative frequency distribution, cumulative frequency table*, or briefly a *cumulative distribution*, and is shown in Table 2.2 for the student height distribution.

Table 2.2

Mass (kilogrammes)	Number of Students
less than 59·5	0
less than 62·5	5
less than 65·5	23
less than 68·5	65
less than 71·5	92
less than 74·5	100

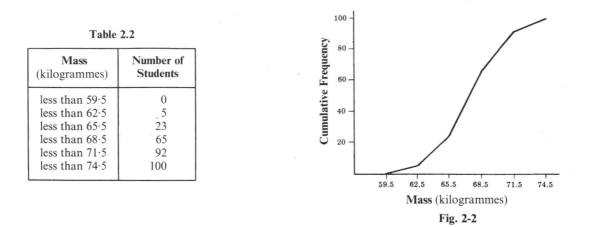

Fig. 2-2

A graph showing the cumulative frequency less than any upper class boundary plotted against the upper class boundary is called a *cumulative frequency polygon* or *ogive* and is shown in Fig. 2-2 for the student height distribution.

For some purposes it is desirable to consider a cumulative frequency distribution of all values greater than or equal to the lower class boundary of each class interval. Because we consider in this case masses of 59·5 kg or more, 62·5 kg or more, etc., this is sometimes called an "*or more*" *cumulative distribution* while the one considered above is a "*less than*" *cumulative distribution*. One is easily obtained from the other (see Problem 2.15). The corresponding ogives are then called "or more" and "less than" ogives. Whenever we refer to cumulative distributions or ogives without qualification, the "less than" type is implied.

RELATIVE CUMULATIVE FREQUENCY DISTRIBUTIONS.   PERCENTAGE OGIVES

The *relative cumulative frequency* or *percentage cumulative frequency* is the cumulative frequency divided by the total frequency. For example, the relative cumulative frequency of masses less than 68·5 kg is 65/100 = 65%, signifying that 65% of the students have masses less than 68·5 kg.

If relative cumulative frequencies are used in Table 2.2 and Fig. 2-2 in place of cumulative frequencies, the results are called *relative cumulative frequency distributions* or *percentage cumulative distributions* and *relative cumulative frequency polygons* or *percentage ogives,* respectively.

FREQUENCY CURVES.   SMOOTHED OGIVES

Collected data can usually be considered as belonging to a sample drawn from a large population. Since so many observations are available in the population, it is theoretically possible (for continuous data) to choose class intervals very small and still have sizeable numbers of observations falling within each class. Thus one would expect the frequency polygon or relative frequency polygon for a large population to have so many small broken line segments that they closely approximate curves, which we call *frequency curves* or *relative frequency curves* respectively.

It is reasonable to expect that such theoretical curves can be approximated by smoothing the frequency polygons or relative frequency polygons of the sample, the approximation improving as the sample size is increased. For this reason a frequency curve is sometimes called a *smoothed frequency polygon*.

In a similar manner *smoothed ogives* are obtained by smoothing the cumulative frequency polygons or ogives. It is usually easier to smooth an ogive than a frequency polygon (see Problem 2.18).

TYPES OF FREQUENCY CURVES

Frequency curves arising in practice take on certain characteristic shapes as indicated in Fig. 2-3.

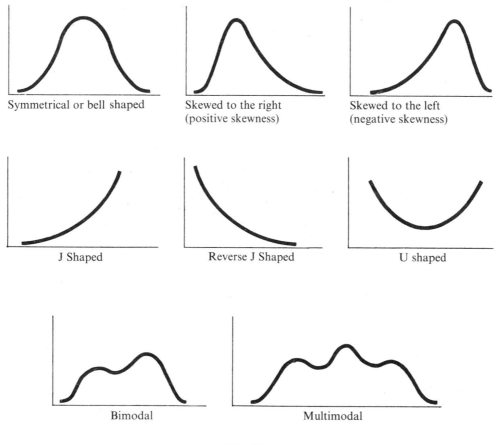

Fig. 2-3

(*a*) The *symmetrical* or *bell shaped* frequency curves are characterised by the fact that observations equidistant from the central maximum have the same frequency. An important example is the *normal curve*.

(*b*) In the *moderately asymmetrical* or *skewed* frequency curves the tail of the curve to one side of the central maximum is longer than that to the other. If the longer tail occurs to the right the curve is said to *be skewed to the right* or to have *positive skewness*, while if the reverse is true the curve is said to be *skewed to the left* or to have *negative skewness*.

(*c*) In a *J shaped* or *reverse J shaped* curve a maximum occurs at one end.

(*d*) A *U shaped* frequency curve has maxima at both ends.

(*e*) A *bimodal* frequency curve has two maxima.

(*f*) A *multimodal* frequency curve has more than two maxima.

Solved Problems

ARRAYS

2.1. (*a*) Arrange the numbers 17, 45, 38, 27, 6, 48, 11, 57, 34, 22 in an array and

(*b*) determine the range.

Solution:

(*a*) In ascending order of magnitude the array is: 6, 11, 17, 22, 27, 34, 38, 45, 48, 57.
In descending order of magnitude the array is: 57, 48, 45, 38, 34, 27, 22, 17, 11, 6.

(*b*) Since the smallest number is 6 and the largest number is 57, the range is $57 - 6 = 51$.

2.2. The final marks in mathematics of 80 students at State University are recorded in the accompanying table

68	84	75	82	68	90	62	88	76	93
73	79	88	73	60	93	71	59	85	75
61	65	75	87	74	62	95	78	63	72
66	78	82	75	94	77	69	74	68	60
96	78	89	61	75	95	60	79	83	71
79	62	67	97	78	85	76	65	71	75
65	80	73	57	88	78	62	76	53	74
86	67	73	81	72	63	76	75	85	77

With reference to this table find (*a*) the highest mark, (*b*) the lowest mark, (*c*) the range,

(*d*) the marks of the five highest ranking students,

(*e*) the marks of the five lowest ranking students,

(*f*) the marks of the student ranking tenth highest,

(*g*) how many students received marks of 75 or higher,

(*h*) how many students received marks below 85,

(*i*) what percentage of students received marks higher than 65 but not higher than 85,

(*j*) which marks did not appear at all.

Solution:

Some of these questions are so detailed that they are best answered by first constructing an array. This can be done by subdividing the data into convenient classes and placing each number taken from the table into the appropriate class as in Table 2.3 below, called an *entry table*. By then arranging the numbers of each class into an array as in Table 2.4, the required array is obtained.

Table 2.3

50–54	53
55–59	59, 57
60–64	62, 60, 61, 62, 63, 60, 61, 60, 62, 62, 63
65–69	68, 68, 65, 66, 69, 68, 67, 65, 65, 67
70–74	73, 73, 71, 74, 72, 74, 71, 71, 73, 74, 73, 72
75–79	75, 76, 79, 75, 75, 78, 78, 75, 77, 78, 75, 79, 79, 78, 76, 75, 78, 76, 76, 75, 77
80–84	84, 82, 82, 83, 80, 81
85–89	88, 88, 85, 87, 89, 85, 88, 86, 75
90–94	90, 93, 93, 94
95–99	95, 96, 95, 97

Table 2.4

50–54	53
55–59	57, 59
60–64	60, 60, 60, 61, 61, 62, 62, 62, 62, 63, 63
65–69	65, 65, 65, 66, 67, 67, 68, 68, 69
70–74	71, 71, 71, 72, 72, 73, 73, 73, 73, 74, 74, 74
75–79	75, 75, 75, 75, 75, 75, 75,76, 76, 76, 76, 77, 77, 78, 78, 78, 78, 78, 79, 79, 79
80–84	80, 81, 82, 82, 83, 84
85–89	85, 85, 85, 86, 87, 88, 88, 88, 89
90–94	90, 93, 93, 94
95–99	95, 95, 96, 97

From Table 2.4 it is relatively easy to answer the above questions. Thus

(*a*) the highest mark is 97

(*b*) the lowest mark is 53

(*c*) the range is 97 − 53 = 44

(*d*) the five highest ranking students have marks 97, 96, 95, 95, 94

(*e*) the five lowest ranking students have 53, 57, 59, 60, 60

(*f*) the mark of the student ranking tenth highest is 88

(*g*) the number of students receiving marks of 75 or higher is 44

(*h*) the number of students receiving marks below 85 is 63

(*i*) the percentage of students recieving marks higher than 65 but not higher than 85 is 49/80 = 61·2%

(*j*) the marks which did not appear are 0 through 52, 54, 55, 56, 58, 64, 70, 91, 92, 98, 99, 100.

FREQUENCY DISTRIBUTIONS, HISTOGRAMS AND FREQUENCY POLYGONS

2.3. Table 2.5 shows a frequency distribution of the monthly wages in pounds sterling of 65 employees at the P and R Company.

With reference to this table determine:

Table 2.5

Wages (£)	Number of Employees
£50·00–£59·99	8
60·00– 69·99	10
70·00– 79·99	16
80·00– 89·99	14
90·00– 99·99	10
100·00–109·99	5
110·00–119·99	2
	Total 65

(*a*) The lower limit of the sixth class. *Ans.* £100·00

(*b*) The upper limit of the fourth class. *Ans.* £89·99

(*c*) The class mark (or class midpoint) of the third class.
Class mark of 3rd class = $\frac{1}{2}$(£70·00 + £79·99) = £74·995.
For most practical purposes this is rounded to £75·00.

(*d*) The class boundaries of the fifth class.
Lower class boundary of 5th class = $\frac{1}{2}$(£90·00 + £89·99) = £89·995.
Upper class boundary of 5th class = $\frac{1}{2}$(£99·99 + £100·00) = £99·995.

(*e*) The size of the fifth class interval.
Size of 5th class interval = upper boundary of 5th class − lower boundary of 5th class
= £99·995 − £88·995 = £10·00.

In this case all class intervals have the same size, £10·00.

(f)   The frequency of the third class.   *Ans.* 16

(g)   The relative frequency of the third class.   *Ans.* $16/65 = 0.246 = 24.6\%$

(h)   The class interval having the largest frequency.   *Ans.* £70.00–£79.99
        This is sometimes called the *modal class interval*. Its frequency is then called the *modal class frequency*.

(i)   The percentage of employees earning less than £80.00 per month.
        Total number of employees earning less than £80.00 per month $= 16 + 10 + 8 = 34$.
        Percentage of employees earning less than £80.00 per month $= 34/65 = 52.3\%$.

(j)   Percentage of employees earning less than £100.00 but at least £60.00 per month.
        Number of employees earning less than £100.00 but at least £60.00 per month $= 10 + 14 + 16 + 10 = 50$.
        Percentage of employees earning less than £100.00 but at least £60.00 per month $= 50/65 = 76.9\%$.

2.4.   If the class marks in a frequency distribution of lengths of laurel leaves are 128, 137, 146, 155, 164, 173 and
        182 mm, find (a) the class interval size, (b) the class boundaries and (c) the class limits, assuming the
        lengths were measured to the nearest millimetre.

Solution:
(a)   Class interval size = common difference between successive class marks $= 137 - 128 = 146 - 137 =$ etc. $= 9$ mm.

(b)   Since class intervals all have equal size, the class boundaries are midway between the class marks and so have the
        values
$$\tfrac{1}{2}(128 + 137), \tfrac{1}{2}(137 + 146), \ldots, \tfrac{1}{2}(173 + 182) \text{ or } 132.5, 141.5, 150.5, \ldots, 177.5 \text{ mm.}$$
        The first class boundary is $132.5 - 9 = 123.5$ and the last class boundary is $177.5 + 9 = 186.5$, since the common
        class interval size is 9 mm. Then all the class boundaries are given by
$$123.5, 132.5, 141.5, 150.5, 159.5, 168.5, 177.5, 186.5 \text{ mm.}$$

(c)   Since the class limits are integers, we choose them as the integers nearest to the class boundaries namely 123, 124,
        132, 133, 141, 142, . . .
        Then the first class has limits 124–132, the next 133–141, etc.

2.5.   Represent graphically the results of the preceding problem.

Solution:

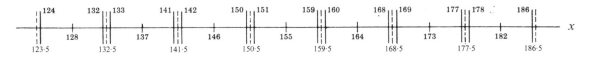

        The class marks 128, 137, 146, . . ., 182 are located on the x axis.
        The class boundaries are indicated by the long vertical dashed lines, and the class limits are indicated by the long
vertical heavy lines.

2.6.   The smallest of 150 measurements is 5.18 mm and the largest is 7.44 mm. Determine a suitable set of
        (a) class intervals, (b) class boundaries, and (c) class marks which might be used in forming a frequency
        distribution of these measurements.

Solution:

        The range is $7.44 - 5.18 = 2.26$ mm. For a minimum of 5 class intervals, the class interval size is $2.26/5 = 0.45$
approximately; and for a maximum of 20 class intervals, the class interval size is $2.26/20 = 0.11$ approximately.
Convenient choices of class interval sizes lying between 0.11 and 0.45 would be 0.20, 0.30 or 0.40.

(a) In columns I, II and III are indicated appropriate class intervals having sizes 0·20, 0·30 and 0·40 respectively.

I	II	III
5·10–5·29	5·10–5·39	5·10–5·49
5·30–5·49	5·40–5·69	5·50–5·89
5·50–5·69	5·70–5·99	5·90–6·29
5·70–5·89	6·00–6·29	6·30–6·69
5·90–6·09	6·30–6·59	6·70–7·09
6·10–6·29	6·60–6·89	7·10–7·49
6·30–6·49	6·90–7·19	
6·50–6·69	7·20–7·49	
6·70–6·89		
6·90–7·09		
7·10–7·29		
7·30–7·49		

Note that the lower limit in each first class could have been different from 5·10. For example, in column I if we had started with 5·15 as lower limit the first class interval could have been written 5·15–5·34.

(b) The class boundaries corresponding to columns I, II and III of (a) are given respectively by
> I 5·095–5·295, 5·295–5·495, 5·495–5·695, . . ., 7·295–7·495
> II 5·095–5·395, 5·395–5·695, 5·695–5·995, . . ., 7·195–7·495
> III 5·095–5·495, 5·495–5·895, 5·895–6·295, . . ., 7·095–7·495

Note that these class boundaries are suitable since they cannot coincide with observed measurements.

(c) The class marks corresponding to columns I, II and III of (a) are given respectively by
> I 5·195, 5·395, . . ., 7·395 II 5·245, 5·545, . . ., 7·345 III 5·295, 5·695, . . ., 7·295

These class marks have the disadvantage of not coinciding with observed measurements.

2.7. In answering the preceding problem a student chose the class intervals 5·10–5·40, 5·40–5·70, . . ., 6·90–7·20, 7·20–7·50. Was there anything wrong with this choice?

Solution:

These class intervals overlap at 5·40, 5·70, . . ., 7·20. Thus a measurement recorded as 5·40, for example, could be placed in either of the first two class intervals. Some statisticians justify this choice by agreeing to place half of such ambiguous cases in one class and half in the other.

The ambiguity is removed by writing the class intervals as 5·10 – under 5·40, 5·40 – under 5·70, etc. In this case limits coincide with class boundaries and class marks can coincide with observed data.

In general it is desirable to avoid overlapping class intervals whenever possible and to choose them so that class boundaries are values not coinciding with actual observed data. For example, the class intervals for the previous problem could have been chosen as 5·095 – 5·395, 5·395 – 5·695, etc., without ambiguity. A disadvantage of this particular choice is that class marks do not coincide with observed data.

2.8. In the following table the lengths of 40 laurel leaves are recorded to the nearest millimetre. Construct a frequency distribution.

138	164	150	132	144	125	149	157
146	158	140	147	136	148	152	144
168	126	138	176	163	119	154	165
146	173	142	147	135	153	140	135
161	145	135	142	150	156	145	128

Solution:

The largest weight is 176 mm and the smallest weight is 119 mm, so that the range is 176 – 119 = 57 mm.

If 5 class intervals are used, the class interval size is 57/5 = 11 approximately.

If 20 class intervals are used, the class interval size is 57/20 = 3 approximately.

One convenient choice for the class interval size is 5 mm. Also, it is convenient to choose the class marks as 120, 125, 130, 135, . . . millimetres. Thus the class intervals can be taken as 118–122, 123–127, 128–132, . . . With this choice the class boundaries are 117·5, 122·5, 127·5, . . . which do not coincide with observed data.

The required frequency distribution is shown in Table 2.6. The centre column, called a *tally* or *score sheet*, is used to tabulate the class frequencies from the raw data and is usually omitted in the final presentation of the frequency distribution. It is unnecessary to make an array, although if it is available it can be used in tabulation of frequencies.

Table 2.6

Length (mm)	Tally	Frequency
118–122	/	1
123–127	//	2
128–132	//	2
133–137	////	4
138–142	TTTT /	6
143–147	TTTT ///	8
148–152	TTTT	5
153–157	////	4
158–162	//	2
163–167	///	3
168–172	/	1
173–177	//	2
		Total 40

Another method:

Of course, other possible frequency distributions exist. Table 2.7, for example, shows a frequency distribution with only 7 classes in which the class interval is 9 mm.

Table 2.7

Length (mm)	Tally	Frequency
118–126	///	3
127–135	TTTT	5
136–144	TTTT ////	9
145–153	TTTT TTTT //	12
154–162	TTTT	5
163–171	////	4
172–180	//	2
		Total 40

2.9.   Construct (*a*) a histogram, (*b*) a frequency polygon for the length distribution in Problem 2.8.

Solution:

The histogram and frequency polygon for each of the cases considered in Problem 2.8 are given in Figs. 2-4(*a*) and 2-4(*b*).

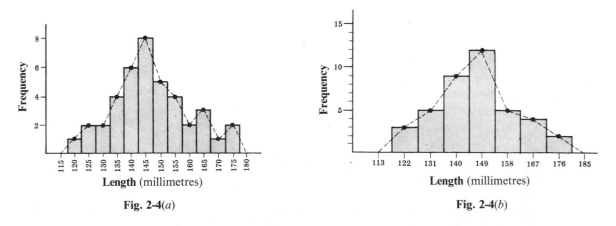

Fig. 2-4(*a*)   **Fig. 2-4(*b*)**

Note that the centres of the bases of the rectangles are located at the class marks.

2.10. For the data of Problem 2.3 construct (*a*) a relative or percentage frequency distribution, (*b*) a histogram, (*c*) a relative frequency histogram, (*d*) a frequency polygon, (*e*) a relative frequency polygon.

Solution:

(*a*) The relative frequency distribution shown in Table 2.8 is obtained from the frequency distribution of Problem 2.3 by dividing each class frequency by the total frequency (65) and expressing the result as a per cent.

(*b*) and (*c*). The histogram and relative frequency histogram are shown in Fig. 2-5. Note that to convert from a histogram to a relative frequency histogram it is only necessary to add to the histogram a vertical scale showing the relative frequencies as indicated on the right.

(*d*) and (*e*). The frequency polygon and relative frequency polygon are indicated by the dashed line graph in Fig. 2-5. Thus to convert from a frequency polygon to a relative frequency polygon, one need only add a vertical scale showing the relative frequencies.

Note that if only a relative frequency polygon, for example, is desired, the adjoining figure would not contain the histogram, and the relative frequency axis would be shown at the left in place of the frequency axis.

Table 2.8

Wages	Relative Frequency (as a per cent)
£50·00–£59·99	12·3
60·00– 69·99	15·4
70·00– 79·99	24·6
80·00– 89·99	21·5
90·00– 99·99	15·4
100·00–109·99	7·7
110·00–119·99	3·1
	Total 100·0%

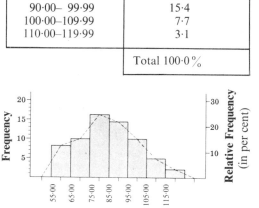

Wages (in £)

Fig. 2-5

2.11. Prove that the total area of the rectangles in a histogram is equal to the total area bounded by the corresponding frequency polygon and the *x* axis.

Solution:

The proof will be given for the case of a histogram consisting of three rectangles, as shown, and the corresponding frequency polygon, shown dashed.

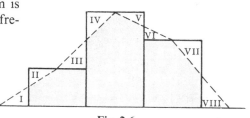

Fig. 2-6

Total area of rectangles = shaded area + area II + area IV + area V + area VII
= shaded area + area I + area III + area II + area VIII
= total area bounded by frequency polygon and *x* axis

since area I = area II, area III = area IV, area V = area VI, and area VII = area VIII.

2.12. At the P and R Company (Problem 2.3), five new employees were hired at monthly wages of £85·34, £116·83, £135·78, £156·21 and £174·50. Construct a frequency distribution of wages for the 70 employees.

Solution:

Possible frequency distributions are shown in Tables (*a*), (*b*), (*c*) and (*d*) below.

In (*a*), the same class interval size, £10·00, has been maintained throughout the table. As a result there are too many empty classes and the detail is much too fine at the upper end of the wage scale.

In (*b*), empty classes and fine detail have been avoided by use of the open class interval "£120·00 and over". A disadvantage of this is that the table becomes useless in performing certain mathematical calculations. For example, it is impossible to determine the total amount of wages paid per week, since "over £120·00" might conceivably imply that individuals could earn as high as £1200·00 per month.

In (c), a class interval size of £20·00 has been used. A disadvantage is that much information is lost at the lower end of the wage scale and the detail is still fine at the upper end of the scale.

In (d), unequal class interval sizes have been used. A disadvantage is that certain mathematical calculations to be made later lose a simplicity which is available when class intervals have the same size. Also, the larger the class interval size the greater will be the grouping error.

(a)

Wages	Frequency
£50·00 − £59·99	8
60·00 − 69·99	10
70·00 − 79·99	16
80·00 − 89·99	15
90·00 − 99·99	10
100·00 − 109·99	5
110·00 − 119·99	3
120·00 − 129·99	0
130·00 − 139·99	1
140·00 − 149·99	0
150·00 − 159·99	1
160·00 − 169·99	0
170·00 − 179·99	1
	Total 70

(b)

Wages	Frequency
£50·00 − £59·99	8
60·00 − 69·99	10
70·00 − 79·99	16
80·00 − 89·99	15
90·00 − 99·99	10
100·00 − 109·99	5
110·00 − 119·99	3
120·00 and over	3
	Total 70

(c)

Wages	Frequency
£50·00 − £69·99	18
70·00 − 89·99	31
90·00 − 109·99	15
110·00 − 129·99	3
130·00 − 149·99	1
150·00 − 169·99	1
170·00 − 189·99	1
	Total 70

(d)

Wages	Frequency
£50·00 − £59·99	8
60·00 − 69·99	10
70·00 − 79·99	16
80·00 − 89·99	15
90·00 − 99·99	10
100·00 − 119·99	8
120·00 − 179·99	3
	Total 70

2.13. Construct a histogram for the frequency distribution in Table (d) of Problem 2.12.

Solution:

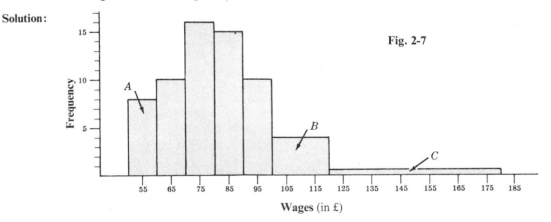

Fig. 2-7

The required histogram is shown in Fig. 2-7. To construct it we use the fact that area is proportional to frequency. Suppose rectangle A corresponds to the first class (see Table (d) of Problem 2.12) with class frequency 8. Since the sixth class of Table (d) also has class frequency 8, rectangle B, which represents this class, should have the same area as A. Then since B is twice as wide as A, it must be half as high, as indicated.

Similarly, rectangle C, representing the last class in Table (d), has height of a half unit on the vertical scale.

CUMULATIVE FREQUENCY DISTRIBUTIONS AND OGIVES

2.14. Construct (a) a cumulative frequency distribution, (b) a percentage cumulative distribution, (c) an ogive, and (d) a percentage ogive from the frequency distribution of Problem 2.3.

Solution:

(a), (b). The cumulative frequency distribution and percentage cumulative distribution (or cumulative relative frequency distribution) are shown combined in Table 2.9.

Note that each entry in column 2 is obtained by adding successive entries in column 2 of the table of Problem 2.3. Thus $18 = 8 + 10$, $34 = 8 + 10 + 16$, etc.

Each entry in column 3 is obtained from the previous column by dividing by 65, the total frequency, and expressing the result as a percentage. Thus $34/65 = 52\cdot3\%$. Entries in this column can also be obtained by adding successive entries in column 2 of the table of Problem 2.10(a). Thus $27\cdot7 = 12\cdot3 + 15\cdot4$, $52\cdot3 = 12\cdot3 + 15\cdot4 + 24\cdot6$, etc.

(c), (d). The ogive (or cumulative frequency polygon) and percentage ogive (or cumulative relative frequency polygon) are both indicated in Fig. 2-8. The vertical scale at the left enables us to read the cumulative frequency, while the vertical scale at the right indicates the percentage cumulative frequency.

The above are sometimes referred to as "less than" cumulative frequency distributions and ogives because of the manner in which the frequencies have been cumulated.

Table 2.9

Wages	Cumulative Frequency	Percentage Cumulative Frequency
Less than £50·00	0	0·0
Less than 60·00	8	12·3
Less than 70·00	18	27·7
Less than 80·00	34	52·3
Less than 90·00	48	73·8
Less than 100·00	58	89·2
Less than 110·00	63	96·9
Less than 120·00	65	100·0

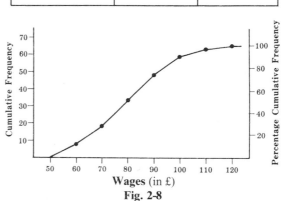

Fig. 2-8

2.15. Construct (a) an "or more" cumulative frequency distribution, (b) an "or more" ogive from the frequency distribution of Problem 2.3.

Solution:

(a) Note that each entry in column 2 of Table 2.10 is obtained by adding successive entries in column 2 of Table 2.5 of Problem 2.3, *starting at the bottom* of that table. Thus $7 = 2 + 5$, $17 = 2 + 5 + 10$, etc.

These entries can also be obtained by subtracting each entry in column 2 of the table of Problem 2.14 from the total frequency, 65. Thus $57 = 65 - 8$, $47 = 65 - 18$, etc.

Table 2.10

Wages	"Or More" Cumulative Frequency
£50·00 or more	65
60·00 or more	57
70·00 or more	47
80·00 or more	31
90·00 or more	17
100·00 or more	7
110·00 or more	2
120·00 or more	0

(b)

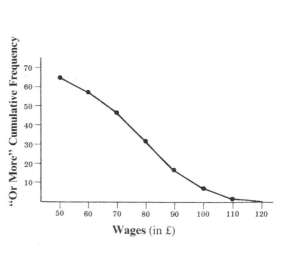

Fig. 2-9

2.16. From the ogives in Problems 2.14 or 2.15, estimate the number of employees earning (*a*) less than £88·00 per month, (*b*) £96·00 or more per month, (*c*) at least £63·00 but less than £75·00 per month.

Solution:

(*a*) Referring to the "less than" ogive of Problem 2.14, construct a vertical line intersecting the "wages" axis at £88·00. This line meets the ogive at the point with co-ordinates (88, 45); hence 45 employees earn less than £88·00 per month.

(*b*) In the "or more" ogive of Problem 2.15, construct a vertical line at £96·00. This line meets the ogive at the point (96, 11); then 11 employees earn £96·00 or more.

This can also be obtained from the "less than" ogive. By constructing a vertical line at £96·00 we find that 54 employees earn less than £96·00; hence $65 - 54 = 11$ employees earn £96·00 or more.

(*c*) Using the "less than" ogive of Problem 2.14, we have:

Required number of employees = number earning less than £75·00 − number earning less than £63·00
$= 26 - 11 = 15$

Note that the above results could just as well have been obtained by the process of *interpolation* in the cumulative frequency tables. For example, in part (*a*) since £88·00 is 8/10 or 4/5 of the way between £80 and £90, the required number of employees should be 4/5 of the way between the corresponding values 34 and 48 (see table of Problem 2.14). But 4/5 of the way between 34 and 48 is $\frac{4}{5}(48 - 34) = 11$. Then the required number of employees is $34 + 11 = 45$.

2.17. Five pennies were tossed 1000 times, and at each toss the number of heads was observed. The number of tosses during which 0, 1, 2, 3, 4, and 5 heads were obtained is shown in the Table 2.11. (*a*) Graph the data. (*b*) Construct a table showing the percentage of tosses resulting in a number of heads less than 0, 1, 2, 3, 4, 5 or 6. (*c*) Graph the data of the table in (*b*).

Table 2.11

Number of Heads	Number of Tosses (Frequency)
0	38
1	144
2	342
3	287
4	164
5	25
	Total 1000

Solution:

(*a*) The data can be shown graphically either as in Figs. 2-10 or 2-11.

Fig. 2-10 seemes to be a more natural graph to use, since for example the number of heads cannot be 1·5 or 3·2. This graph is a form of bar graph where the bars have zero width, and is sometimes called a *rod graph*. It is especially used when the data are discrete.

Fig. 2-11 shows a histogram of the data. Note that the total area of the histogram is the total frequency 1000, as it should be. In using the histogram representation or the corresponding frequency polygon we are essentially

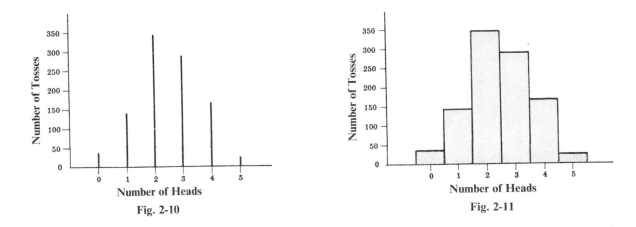

Fig. 2-10   Fig. 2-11

treating the data *as if* they were continuous. This will later be found useful. Note that we have already used the histogram and frequency polygon for discrete data in Problem 2.10.

(b) Referring to the required Table 2.12, note that it shows simply a cumulative frequency distribution and percentage cumulative distribution of the number of heads. It should be observed that the entries "less than 1", "less than 2", etc., could just as well have been "less than or equal to 0", "less than or equal to 1", etc.

Table 2.12

Number of Heads	Number of Tosses (Cumulative Frequency)	Percentage Number of Tosses (Percentage Cumulative Frequency)
Less than 0	0	0·0
Less than 1	38	3·8
Less than 2	182	18·2
Less than 3	524	52·4
Less than 4	811	81·1
Less than 5	975	97·5
Less than 6	1000	100·0

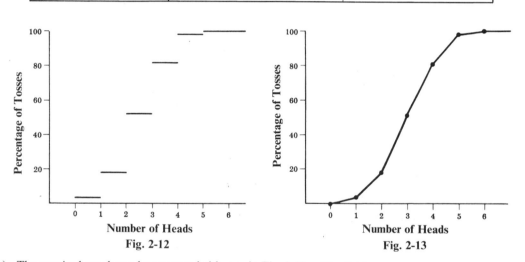

Fig. 2-12   Fig. 2-13

(c) The required graph can be presented either as in Fig. 2-12 or Fig. 2-13.

Fig. 2-12 is most natural for presenting discrete data, since for example the percentage of tosses in which there will be less than 2 heads is equal to the percentage in which there will be less than 1·75, 1·56 or 1·23 heads, so that the same percentage, 18·2%, should be shown for these values (indicated by the horizontal line).

Fig. 2-13 shows the cumulative frequency polygon or ogive for the data and essentially treats the data as if they were continuous.

Note that Figs. 2-12 and 2-13 correspond respectively to Figs. 2-10 and 2-11 of part (a).

FREQUENCY CURVES AND SMOOTHED OGIVES

2.18. The 100 male students at XYZ University (see Page 27) actually constituted a sample of 1546 male students at the university. From the data provided in the sample,

(a) construct a smoothed percentage frequency polygon (frequency curve) and
(b) construct a smoothed "less than" percentage ogive.
(c) From the results of (a) or (b), estimate the number of students at the university having masses between 65 and 70 kg. What assumptions must you make?
(d) Can the results be used to estimate the proportion of males in the United States having masses between 65 and 70 kg?

Solution:

(*a*), (*b*). In Figs. 2-14 and 2-15, the dashed graphs represent the frequency polygons and ogives and are obtained from those given respectively on Pages 29–30. The required smoothed graphs (shown heavy) are obtained by approximating these by smooth curves.

In practice it is easier to smooth an ogive, so that very often the smoothed ogive is first obtained and then the smoothed frequency polygon is obtained by reading off values from the smooth ogive.

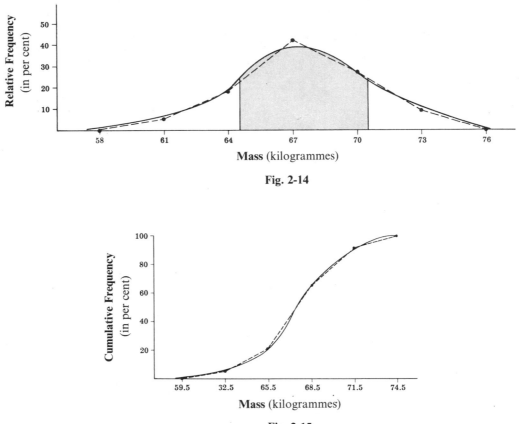

Fig. 2-14

Fig. 2-15

(*c*) If the sample of 100 students is representative of the population of 1546 students, the smoothed curves of parts (*a*) and (*b*) can be assumed to be the percentage frequency curve and percentage ogive for this population. This assumption is correct only if the sample is *random*, i.e. if each student has as much chance of being selected in the sample as any other student.

Since masses between 65 and 70 kg recorded to the nearest kilogramme actually represent masses between 64·5 and 70·5 kg, the percentage of students in the population having these masses can be found by dividing the shaded area in Fig. 2-14 by the total area bounded by the smoothed curve and the *x* axis.

It is simpler, however, to use Fig. 2-15 from which we see that

percentage of students with masses less than 70·5 kg = 82%
percentage of students with masses less than 64·5 kg = 18%

so that the percentage of students with masses between 64·5 and 70·5 kg = 82% − 18% = 64%. Then the number of students in the university having masses between 65 and 70 kg to the nearest kilogramme = 64% of 1546 = 989.

Another way of saying this is that the *probability* or *chance* that a person selected at random from the 1546 students has a mass between 65 and 70 kg is 64%, 0·64, or 64 out of 100. Because of the relationship to probability (considered in Chapter 6), relative frequency curves are often called *probability curves* or *probability distributions*.

(*d*) We could consider the required proportion to be 64% (with a much greater uncertainty than before) only if we were convinced that the sample of 100 students from the total male population of the United States was truly a random sample. However, this is somewhat unlikely for several reasons, such as (1) some college students may not have reached their maximum masses, (2) the younger generation may tend to be more massive than their parents.

Supplementary Problems

2.19. (a) Arrange the numbers 12, 56, 42, 21, 5, 18, 10, 3, 61, 34, 65, 24 in an array and (b) determine the range.
Ans. (b) 62

2.20. Table 2.13 shows a frequency distribution of the lifetimes of 400 radio tubes tested at the L & M Tube Company. With reference to this table determine the
(a) upper limit of the fifth class,
(b) lower limit of the eighth class,
(c) class mark of the seventh class,
(d) class boundaries of the last class,
(e) class interval size,
(f) frequency of the fourth class,
(g) relative frequency of the sixth class,
(h) percentage of tubes whose lifetimes do not exceed 600 hours,
(i) percentage of tubes with lifetimes greater than or equal to 900 hours,
(j) percentage of tubes whose lifetimes are at least 500 but less than 1000 hours.

Table 2.13

Lifetime (hours)	Number of Tubes
300 – 399	14
400 – 499	46
500 – 599	58
600 – 699	76
700 – 799	68
800 – 899	62
900 – 999	48
1000 – 1099	22
1100 – 1199	6
Total 400	

Ans. (a) 799 (b) 1000 (c) 949·5 (d) 1099·5, 1199·5 (e) 100 hours (f) 76 (g) 62/400 = 0·155 or 15·5% (h) 29·5% (i) 19·0% (j) 78·0%

2.21. Construct (a) a histogram and (b) a frequency polygon corresponding to the frequency distribution of the preceding problem.

2.22. For the data of Problem 2.20 construct (a) a relative or percentage frequency distribution, (b) a relative frequency histogram, (c) a relative frequency polygon.

2.23. For the data of Problem 2.20 construct (a) a cumulative frequency distribution, (b) a percentage or relative cumulative distribution, (c) an ogive, (d) a percentage ogive. (Note that unless otherwise specified a cumulative distribution refers to one made on a "less than" basis.)

2.24. Work the preceding problem when frequencies are cumulated on an "or more" basis.

2.25. Estimate the percentage of tubes of Problem 2.20 with lifetimes of (a) less than 560 hours, (b) 970 or more hours, (c) between 620 and 890 hours. *Ans.* (a) 24% (b) 11% (c) 46%

2.26. The inner diameters of washers produced by a company can be measured to the nearest hundredth of a millimetre. If the class marks of a frequency distribution of these diameters are given in millimetres by 3·21, 3·24, 3·27, 3·30, 3·33 and 3·36, find (a) the class interval size (b) the class boundaries, (c) the class limits.
Ans. (a) 0·03 mm (b) 3·195, 3·225, 3·255, . . ., 3·375 mm (c) 3·20–3·22, 3·23–3·25, 3·26–3·28, . . ., 3·35–3·37

2.27. The following table shows the diameters in millimetres of a sample of 60 ball bearings manufactured by a company. Construct a frequency distribution of the diameters using appropriate class intervals.

7·38	7·29	7·43	7·40	7·36	7·41	7·35	7·31	7·26	7·37
7·28	7·37	7·36	7·35	7·24	7·33	7·42	7·36	7·39	7·35
7·45	7·36	7·42	7·40	7·28	7·38	7·25	7·33	7·34	7·32
7·33	7·30	7·32	7·30	7·39	7·34	7·38	7·39	7·27	7·35
7·35	7·32	7·35	7·27	7·34	7·32	7·36	7·41	7·36	7·44
7·32	7·37	7·31	7·46	7·35	7·35	7·29	7·34	7·30	7·40

2.28. For the data of the preceding problem construct (a) a histogram, (b) a frequency polygon, (c) a relative frequency distribution, (d) a relative frequency histogram, (e) a relative frequency polygon, (f) a cumulative frequency distribution, (g) a percentage cumulative distribution, (h) an ogive, (i) a percentage ogive.

2.29. From the results in Problem 2.28 determine the percentage of ball bearings having diameters (*a*) exceeding 0·732 mm, (*b*) not more than 0·736 mm, (*c*) between 0·730 and 0·738 mm. Compare your results with those obtained directly from the raw data of Problem 2.27.

2.30. Work Problem 2.28 for the data of Problem 2.20.

2.31. Table 2.14 shows the percentage distribution of total income of males 14 years old or over in the United States in 1956. Using this table, answer the following questions.

Table 2.14

Income (dollars)	Per cent of Males
Under $1000	17·2
1000 – 1999	11·7
2000 – 2999	12·1
3000 – 3999	14·8
4000 – 4999	15·9
5000 – 5999	11·9
6000 – 9999	12·7
10000 and over	3·6

Source: Bureau of the Census.

    (*a*) What is the width or size of the second class interval? the seventh class interval?

    (*b*) How may different class interval sizes are there?

    (*c*) How many open class intervals are there?

    (*d*) How should the first class interval be written so that its class width will equal that of the second class interval?

    (*e*) What is the class mark of the second class interval? the seventh class interval?

    (*f*) What are the class boundaries of the fourth class interval?

    (*g*) What percentage of the males earned $4000 and over? under $3000?

    (*h*) What percentage of the males earned at least $3000 but not more than $5000?

    (*i*) What percentage of the males earned between $3000 and $6300? What assumptions are made in this calculation?

    (*j*) Why don't the percentages total 100 %?

    Ans. (*a*) $1000, $4000. (*b*) Four (although strictly speaking the last class has no specified size). (*c*) One (although the first class appears to be an open class interval, it actually is a substitute for 0–$999·99). (*d*) 0–$999. (*e*) $1499·50; for most practical purposes these can be written $1500 and $8000 respectively. (*f*) $2999·50 and $3999·50. (*g*) 44·1 %, 41·0 %. (*h*) 30·7 %. (*i*) 42·0 %. (*j*) Because of rounding errors in computing percentages.

2.32. (*a*) Why is it impossible to construct a percentage histogram or frequency polygon for the distribution in the preceding problem? (*b*) How would you modify the distribution so that a percentage histogram and frequency polygon could be constructed? (*c*) Perform the construction using the modification in (*b*).

2.33. (*a*) Construct a smoothed percentage frequency polygon and smoothed percentage ogive corresponding to the data of Problem 2.20.

    (*b*) From the results in (*a*) estimate the probability that a tube will·burn out before 600 hours.

    (*c*) Discuss the risk or chance which the manufacturer takes in guaranteeing that a tube will last 425 hours? 875 hours?

    (*d*) If the manufacturer offers a 90-day money-back guarantee on a tube, what is the probability that he will have to refund the money assuming that a tube is in use 4 hours per day? 8 hours per day?

    Ans. (*b*) 0·30   (*d*) 0·008, 0·52

2.34. (*a*) Toss 4 coins fifty times and tabulate the number of heads at each toss. (*b*) Construct a frequency distribution showing the number of tosses in which 0, 1, 2, 3, 4 heads appeared. (*c*) Construct a percentage distribution corresponding to (*b*). (*d*) Compare the percentage obtained in (*c*) with the theoretical ones 6·25 %, 25 %, 37·5 %, 25 %, 6·25 % (proportional to 1, 4, 6, 4, 1) arrived at by rules of probability. (*e*) Construct graphical presentation of the distributions in (*b*) and (*c*). (*f*) Construct a percentage ogive for the data.

2.35. Work the previous problem with fifty more tosses of the 4 coins and see if the experiment is more in agreement with theoretical expectation. If not, give possible reasons for the differences.

CHAPTER 3

The Mean, Median, Mode and other Measures of Central Tendency

INDEX OR SUBSCRIPT NOTATION

Let the symbol X_j (read "X sub j") denote any of the N values $X_1, X_2, X_3, \ldots, X_N$ assumed by a variable X. The letter j in X_j, which can stand for any of the numbers $1, 2, 3, \ldots, N$ is called a *subscript* or *index*. Clearly any letter other than j, such as i, k, p, q and s could have been used as well.

SUMMATION NOTATION

The symbol $\sum\limits_{j=1}^{N} X_j$ is used to denote the sum of all the X_j's from $j = 1$ to $j = N$, i.e. by definition

$$\sum_{j=1}^{N} X_j = X_1 + X_2 + X_3 + \ldots + X_N$$

When no confusion can result we shall often denote this sum simply by ΣX, ΣX_j or $\sum\limits_j X_j$. The symbol Σ is the Greek capital letter *sigma*, denoting sum.

Example 1. $\sum\limits_{j=1}^{N} X_j Y_j = X_1 Y_1 + X_2 Y_2 + X_3 Y_3 + \ldots + X_N Y_N$

Example 2. $\sum\limits_{j=1}^{N} aX_j = aX_1 + aX_2 + \ldots + aX_N$
$$= a(X_1 + X_2 + \ldots + X_N) = a\sum_{j=1}^{N} X_j$$

where a is a constant. More simply, $\Sigma aX = a\Sigma X$.

Example 3. If a, b, c are any constants,
$\Sigma(aX + bY - cZ) = a\Sigma X + b\Sigma Y - c\Sigma Z$. See Problem 3.3.

AVERAGES AND MEASURES OF CENTRAL TENDENCY

An *average* is a value which is typical or representative of a set of data. Since such typical values tend to lie centrally within a set of data arranged according to magnitude, averages are also called *measures of central tendency*.

Several types of averages can be defined, the most common being the *arithmetic mean* or briefly the *mean*, the *median*, the *mode*, the *geometric mean*, and the *harmonic mean*. Each has advantages and disadvantages depending on the data and the intended purpose.

THE ARITHMETIC MEAN

The arithmetic mean or the *mean* of a set of N numbers $X_1, X_2, X_3, \ldots, X_N$ is denoted by $\bar{X}$ (read "X bar") and is defined as

$$\bar{X} = \frac{X_1 + X_2 + X_3 + \ldots + X_N}{N} = \frac{\sum\limits_{j=1}^{N} X_j}{N} = \frac{\Sigma X}{N} \qquad (1)$$

Example: The arithmetic mean of the numbers 8, 3, 5, 12, 10 is

$$\bar{X} = \frac{8 + 3 + 5 + 12 + 10}{5} = \frac{38}{5} = 7 \cdot 6$$

If the numbers $X_1, X_2, \ldots, X_K$ occur $f_1, f_2, \ldots, f_K$ times respectively (i.e. occur with frequencies $f_1, f_2, \ldots, f_K$), the arithmetic mean is

$$\bar{X} = \frac{f_1 X_1 + f_2 X_2 + \ldots + f_K X_K}{f_1 + f_2 + \ldots + f_K} = \frac{\sum_{j=1}^{K} f_j X_j}{\sum_{j=1}^{K} f_j} = \frac{\Sigma fX}{\Sigma f} = \frac{\Sigma fX}{N} \qquad (2)$$

where $N = \Sigma f$ is the *total frequency*, i.e. the total number of cases.

Example: If 5, 8, 6 and 2 occur with frequencies 3, 2, 4 and 1 respectively, the arithmetic mean is

$$\bar{X} = \frac{(3)(5) + (2)(8) + (4)(6) + (1)(2)}{3 + 2 + 4 + 1} = \frac{15 + 16 + 24 + 2}{10} = 5{\cdot}7$$

WEIGHTED ARITHMETIC MEAN

Sometimes we associate with the numbers $X_1, X_2, \ldots, X_K$ certain *weighting factors* or *weights* $w_1, w_2, \ldots, w_K$ depending on the significance or importance attached to the numbers. In this case

$$\bar{X} = \frac{w_1 X_1 + w_2 X_2 + \ldots + w_K X_K}{w_1 + w_2 + \ldots + w_K} = \frac{\Sigma wX}{\Sigma w} \qquad (3)$$

is called the *weighted arithmetic mean*. Note the similarity to (2), which can be considered a weighted arithmetic mean with weights $f_1, f_2, \ldots, f_K$.

Example: If a final examination in a course is weighted three times as much as a quiz and a student has a final examination grade of 85 and quiz grades 70 and 90, the mean grade is

$$\bar{X} = \frac{(1)(70) + (1)(90) + (3)(85)}{1 + 1 + 3} = \frac{415}{5} = 83$$

PROPERTIES OF THE ARITHMETIC MEAN

(a)   The algebraic sum of the deviations of a set of numbers from their arithmetic mean is zero.

Example: The deviations of the numbers 8, 3, 5, 12, 10 from their arithmetic mean 7·6 are $8 - 7{\cdot}6$, $3 - 7{\cdot}6$, $5 - 7{\cdot}6$, $12 - 7{\cdot}6$, $10 - 7{\cdot}6$ or 0·4, −4·6, −2·6, 4·4, 2·4 with algebraic sum $0{\cdot}4 - 4{\cdot}6 - 2{\cdot}6 + 4{\cdot}4 + 2{\cdot}4 = 0$.

(b)   The sum of the squares of the deviations of a set of numbers X_j from any number a is a minimum if and only if $a = \bar{X}$. See Problem 4.27, Chapter 4.

(c)   If f_1 numbers have mean m_1, f_2 numbers have mean m_2, \ldots, f_K numbers have mean m_K, then the mean of all the numbers is

$$\bar{X} = \frac{f_1 m_1 + f_2 m_2 + \ldots + f_K m_K}{f_1 + f_2 + \ldots + f_K} \qquad (4)$$

i.e. a weighted arithmetic mean of all the means. See Problem 3.12.

(d)   If A is any *guessed* or *assumed arithmetic mean* (which may be any number) and if $d_j = X_j - A$ are the deviations of X_j from A, then equations (1) and (2) become respectively

$$\bar{X} = A + \frac{\sum_{j=1}^{N} d_j}{N} = A + \frac{\Sigma d}{N} \qquad (5)$$

$$\bar{X} = A + \frac{\sum_{j=1}^{K} f_j d_j}{\sum_{j=1}^{K} f_j} = A + \frac{\Sigma fd}{N} \qquad (6)$$

where $N = \sum\limits_{j=1}^{K} f_j = \Sigma f$. Note that (5) and (6) are summarized in the equation $\bar{X} = A + \bar{d}$. (See Problem 3.18.)

ARITHMETIC MEAN COMPUTED FROM GROUPED DATA

When data are presented in a frequency distribution, all values falling in a given class interval are considered as coincident with the class mark or midpoint of the interval. Formulae (2) and (6) are valid for such grouped data if we interpret X_j as the class mark, f_j its corresponding class frequency, A any guessed or assumed class mark, and $d_j = X_j - A$ the deviations of X_j from A.

Computations using formulae (2) and (6) are sometimes called the *long* and *short methods* respectively (see Problems 3.15 and 3.20).

If class intervals all have equal size c, the deviation $d_j = X_j - A$ can all be expressed as cu_j where u_j can be positive or negative integers or zero, i.e. $0, \pm 1, \pm 2, \pm 3, \ldots$, and formula (6) becomes

$$\bar{X} = A + \left(\frac{\sum\limits_{j=1}^{K} f_j u_j}{N}\right) c = A + \left(\frac{\Sigma fu}{N}\right) c \tag{7}$$

which is equivalent to the equation $\bar{X} = A + c\bar{u}$ (see Problem 3.21). This is called the *coding method* for computing the mean. It is a very short method and should always be used for grouped data where class interval sizes are equal (see Problems 3.22 and 3.23). Note that in the coding method the values of the variable X are *transformed* into the values of the variable u according to $X = A + cu$.

THE MEDIAN

The median of a set of numbers arranged in order of magnitude (i.e. in an array) is the middle value or the arithmetic mean of the two middle values.

Example 1. The set of numbers 3, 4, 4, 5, 6, 8, 8, 8, 10 has median 6.
Example 2. The set of numbers 5, 5, 7, 9, 11, 12, 15, 18 has median $\frac{1}{2}(9 + 11) = 10$.

For grouped data the median, obtained by interpolation, is given by

$$\text{Median} = L_1 + \left(\frac{\frac{N}{2} - (\Sigma f)_1}{f_{\text{median}}}\right) c \tag{8}$$

where
L_1 = lower class boundary of the median class (i.e. the class containing the median)

N = number of items in the data (i.e. total frequency)

$(\Sigma f)_1$ = sum of frequencies of all classes lower than the median class

f_{median} = frequency of median class

c = size of median class interval.

Geometrically the median is the value of X (abscissa) corresponding to that vertical line which divides a histogram into two parts having equal areas. This value of X is sometimes denoted by $\tilde{X}$.

THE MODE

The mode of a set of numbers is that value which occurs with the geatest frequency, i.e. it is the most common value. The mode may not exist, and even if it does exist it may not be unique.

Example 1. The set 2, 2, 5, 7, 9, 9, 9, 10, 10, 11, 12, 18 has mode 9.

Example 2. The set 3, 5, 8, 10, 12, 15, 16 has no mode.

Example 3. The set 2, 3, 4, 4, 4, 5, 5, 7, 7, 7, 9 has two modes, 4 and 7, and is called *bimodal*.

A distribution having only one mode is called *unimodal*.

In the case of grouped data where a frequency curve has been constructed to fit the data, the mode will be the value (or values) of X corresponding to the maximum point (or points) on the curve. This value of X is sometimes denoted by $\hat{X}$.

From a frequency distribution or histogram the mode can be obtained from the formula

$$\text{Mode} = L_1 + \left(\frac{\Delta_1}{\Delta_1 + \Delta_2} \right) c \tag{9}$$

where
L_1 = lower class boundary of modal class (i.e. class containing the mode)

Δ_1 = excess of modal frequency over frequency of next lower class

Δ_2 = excess of modal frequency over frequency of next higher class

c = size of modal class interval.

EMPIRICAL RELATION BETWEEN MEAN, MEDIAN AND MODE

For unimodal frequency curves which are moderately skewed (asymmetrical), we have the empirical relation

$$\text{Mean} - \text{Mode} = 3(\text{Mean} - \text{Median}) \tag{10}$$

In Figs. 3-1 and 3-2 below are shown the relative positions of the mean, median and mode for frequency curves which are skewed to the right and left respectively. For symmetrical curves the mean, mode and median all coincide.

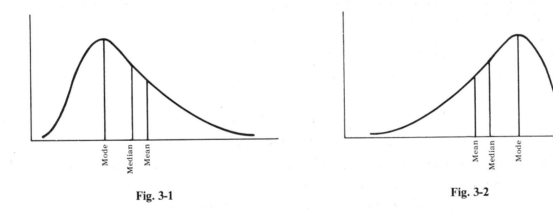

Fig. 3-1   Fig. 3-2

THE GEOMETRIC MEAN G

The geometric mean G of a set of N numbers $X_1, X_2, X_3, \ldots, X_N$ is the Nth root of the product of the numbers:

$$G = \sqrt[N]{X_1 X_2 X_3 \ldots X_N} \tag{11}$$

Example: The geometric mean of the numbers 2, 4, 8 is $G = \sqrt[3]{(2)(4)(8)} = \sqrt[3]{64} = 4$.

In practice, G is computed by logarithms (see Problem 3.35). For the geometric mean from grouped data, see Problems 3.36 and 3.91.

THE HARMONIC MEAN H

The harmonic mean H of a set of N numbers $X_1, X_2, X_3, \ldots, X_N$ is the reciprocal of the arithmetic mean of the reciprocals of the numbers:

$$H = \frac{1}{\dfrac{1}{N} \sum\limits_{j=1}^{N} \dfrac{1}{X_j}} = \frac{N}{\Sigma \dfrac{1}{X}} \tag{12}$$

In practice it may be easier to remember that

$$\frac{1}{H} = \frac{\Sigma \dfrac{1}{X}}{N} = \frac{1}{N} \Sigma \frac{1}{X} \tag{13}$$

Example: The harmonic mean of the numbers 2, 4, 8 is $H = \dfrac{3}{\frac{1}{2} + \frac{1}{4} + \frac{1}{8}} = \dfrac{3}{\frac{7}{8}} = 3 \cdot 43$.

For the harmonic mean from grouped data, see Problems 3.99 and 3.100.

RELATION BETWEEN ARITHMETIC, GEOMETRIC AND HARMONIC MEANS

The geometric mean of a set of positive numbers $X_1, X_2, \ldots, X_N$ is less than or equal to their arithmetic mean but is greater than or equal to their harmonic mean. In symbols,

$$H \leqq G \leqq \bar{X} \tag{14}$$

The equality signs hold only if all the numbers $X_1, X_2, \ldots, X_N$ are identical.

Example: The set 2, 4, 8 has arithmetic mean 4·67, geometric mean 4, and harmonic mean 3·43.

THE ROOT MEAN SQUARE (R.M.S.)

The root mean square (R.M.S.) or *quadratic mean* of a set of numbers $X_1, X_2, \ldots, X_N$ is sometimes denoted by $\sqrt{\overline{X^2}}$ and is defined by

$$\text{R.M.S.} = \sqrt{\overline{X^2}} = \sqrt{\frac{\sum\limits_{j=1}^{N} X_j^2}{N}} = \sqrt{\frac{\Sigma X^2}{N}} \tag{15}$$

This type of average is frequently used in physical applications.

Example: The R.M.S. of the set of numbers 1, 3, 4, 5, 7 is

$$\sqrt{\frac{1^2 + 3^2 + 4^2 + 5^2 + 7^2}{5}} = \sqrt{20} = 4 \cdot 47.$$

QUARTILES, DECILES AND PERCENTILES

If a set of data is arranged in order of magnitude, the middle value (or arithmetic mean of the two middle values) which divides the set into two equal parts is the median. By extending this idea we can think of those values which divide the set into four equal parts. These values, denoted by Q_1, Q_2 and Q_3, are called the first, second and third *quartiles* respectively, the value Q_2 being equal to the median.

Similarly the values which divide the data into ten equal parts are called *deciles* and are denoted by $D_1, D_2, \ldots, D_9$, while the values dividing the data into one hundred equal parts are called *percentiles* and are denoted by $P_1, P_2, \ldots, P_{99}$. The 5th decile and the 50th percentile correspond to the median. The 25th and 75th percentiles correspond to the first and third quartiles respectively.

Collectively, quartiles, deciles, percentiles and other values obtained by equal subdivisions of the data are called *quantiles*. For computations of these from grouped data, see Problems 3.44 to 3.46.

Solved Problems

SUMMATION NOTATION

3.1. Write the terms in each of the following indicated sums.

(a) $\sum_{j=1}^{6} X_j$ $X_1 + X_2 + X_3 + X_4 + X_5 + X_6$

(b) $\sum_{j=1}^{4} (Y_j - 3)^2$ $(Y_1 - 3)^2 + (Y_2 - 3)^2 + (Y_3 - 3)^2 + (Y_4 - 3)^2$

(c) $\sum_{j=1}^{N} a$ $a + a + a + \cdots + a = Na$

(d) $\sum_{k=1}^{5} f_k X_k$ $f_1 X_1 + f_2 X_2 + f_3 X_3 + f_4 X_4 + f_5 X_5$

(e) $\sum_{j=1}^{3} (X_j - a)$ $(X_1 - a) + (X_2 - a) + (X_3 - a) = X_1 + X_2 + X_3 - 3a$

3.2. Express each of the following using the summation notation.

(a) $X_1^2 + X_2^2 + X_6^2 + \ldots + X_{10}^2$ $\sum_{j=1}^{10} X_j^2$

(b) $(X_1 + Y_1) + (X_2 + Y_2) + \ldots + (X_8 + Y_8)$ $\sum_{j=1}^{8} (X_j + Y_j)$

(c) $f_1 X_1^3 + f_2 X_2^3 + \ldots + f_{20} X_{20}^3$ $\sum_{j=1}^{20} f_j X_j^3$

(d) $a_1 b_1 + a_2 b_2 + a_3 b_3 + \ldots + a_N b_N$ $\sum_{j=1}^{N} a_j b_j$

(e) $f_1 X_1 Y_1 + f_2 X_2 Y_2 + f_3 X_3 Y_3 + f_4 X_4 Y_4$ $\sum_{j=1}^{4} f_j X_j Y_j$

3.3. Prove that $\sum_{j=1}^{N} (aX_j + bY_j - cZ_j) = a\sum_{j=1}^{N} X_j + b\sum_{j=1}^{N} Y_j - c\sum_{j=1}^{N} Z_j$, where a, b and c are any constants.

Solution:

$$\sum_{j=1}^{N} (aX_j + bY_j - cZ_j) = (aX_1 + bY_1 - cZ_1) + (aX_2 + bY_2 - cZ_2) + \ldots + (aX_N + bY_N - cZ_N)$$

$$= (aX_1 + aX_2 + \ldots + aX_N) + (bY_1 + bY_2 + \ldots + bY_N) - (cZ_1 + cZ_2 + \ldots + cZ_N)$$

$$= a(X_1 + X_2 + \ldots + X_N) + b(Y_1 + Y_2 + \ldots + Y_N) - c(Z_1 + Z_2 + \ldots + Z_N)$$

$$= a\sum_{j=1}^{N} X_j + b\sum_{j=1}^{N} Y_j - c\sum_{j=1}^{N} Z_j$$

or briefly $\Sigma(aX + bY - cZ) = a\Sigma X + b\Sigma Y - c\Sigma Z$.

3.4. Two variable X and Y assume the values $X_1 = 2, X_2 = -5, X_3 = 4, X_4 = -8$ and $Y_1 = -3, Y_2 = -8, Y_3 = 10, Y_4 = 6$ respectively. Calculate (a) ΣX, (b) ΣY, (c) ΣXY, (d) ΣX^2, (e) ΣY^2, (f) $(\Sigma X)(\Sigma Y)$, (g) ΣXY^2, (h) $\Sigma(X + Y)(X - Y)$.

Solution:

Note that in each case the subscript j on X and Y has been omitted and Σ is understood as $\sum_{j=1}^{4}$.

Thus ΣX, for example, is short for $\sum_{j=1}^{4} X_j$.

(a) $\Sigma X = (2) + (-5) + (4) + (-8) = 2 - 5 + 4 - 8 = -7$
(b) $\Sigma Y = (-3) + (-8) + (10) + (6) = -3 - 8 + 10 + 6 = 5$
(c) $\Sigma XY = (2)(-3) + (-5)(-8) + (4)(10) + (-8)(6) = -6 + 40 + 40 - 48 = 26$
(d) $\Sigma X^2 = (2)^2 + (-5)^2 + (4)^2 + (-8)^2 = 4 + 25 + 16 + 64 = 109$
(e) $\Sigma Y^2 = (-3)^2 + (-8)^2 + 10^2 + (6)^2 = 9 + 64 + 100 + 36 = 209$
(f) $(\Sigma X)(\Sigma Y) = (-7)(5) = -35$, using (a) and (b). Note that $(\Sigma X)(\Sigma Y) \neq \Sigma XY$.
(g) $\Sigma XY^2 = (2)(-3)^2 + (-5)(-8)^2 + (4)(10)^2 + (-8)(6)^2 = -190$
(h) $\Sigma(X + Y)(X - Y) = \Sigma(X^2 - Y^2) = \Sigma X^2 - \Sigma Y^2 = 109 - 209 = -100$, using (d) and (e).

3.5. If $\sum_{j=1}^{6} X_j = -4$ and $\sum_{j=1}^{6} X_j^2 = 10$, calculate (a) $\sum_{j=1}^{6} (2X_j + 3)$, (b) $\sum_{j=1}^{6} X_j(X_j - 1)$, (c) $\sum_{j=1}^{6} (X_j - 5)^2$.

Solution:

(a) $\sum_{j=1}^{6} (2X_j + 3) = \sum_{j=1}^{6} 2X_j + \sum_{j=1}^{6} 3 = 2 \sum_{j=1}^{6} X_j + (6)(3) = 2(-4) + 18 = 10$

(b) $\sum_{j=1}^{6} X_j(X_j - 1) = \sum_{j=1}^{6} (X_j^2 - X_j) = \sum_{j=1}^{6} X_j^2 - \sum_{j=1}^{6} X_j = 10 - (-4) = 14$

(c) $\sum_{j=1}^{6} (X_j - 5)^2 = \sum_{j=1}^{6} (X_j^2 - 10X_j + 25) = \sum_{j=1}^{6} X_j^2 - 10 \sum_{j=1}^{6} X_j + 25(6) = 10 - 10(-4) + 25(6) = 200$

If desired, we can omit the subscript j and use Σ in place of $\sum_{j=1}^{6}$ so long as these abbreviations are understood.

THE ARITHMETIC MEAN

3.6. The grades of a student on six examinations were 84, 91, 72, 68, 87 and 78. Find the arithmetic mean of the grades.

Solution:

$$\bar{X} = \frac{\Sigma X}{N} = \frac{84 + 91 + 72 + 68 + 87 + 78}{6} = \frac{480}{6} = 80$$

Frequently one uses the term *average* as synonymous with arithmetic mean. Strictly speaking, however, this is incorrect since there are averages other than the arithmetic mean.

3.7. Ten measurements of the diameter of a cylinder were recorded by a scientist as 38·8, 40·9, 39·2, 39·7, 40·2, 39·5, 40·3, 39·2, 39·8 and 40·6 millimetres. Find the arithmetic mean of the measurements.

Solution:

$$\bar{X} = \frac{\Sigma X}{N} = \frac{38 \cdot 8 + 40 \cdot 9 + 39 \cdot 2 + 39 \cdot 7 + 40 \cdot 2 + 39 \cdot 5 + 40 \cdot 3 + 39 \cdot 2 + 39 \cdot 8 + 40 \cdot 6}{10} = \frac{398 \cdot 2}{10} = 39 \cdot 8 \text{ mm}$$

3.8. The annual salaries of four men were \$5000, \$6000, \$6500 and \$30 000. (a) Find the arithmetic mean of their salaries. (b) Would you say that this average is typical of the salaries?

Solution:

(a) $\bar{X} = \dfrac{\$5000 + \$6000 + \$6500 + \$30\,000}{4} = \dfrac{\$45\,500}{4} = \$11\,875$

(assuming all figures significant in the reported salaries).

(b) The mean \$11 875 is certainly not typical of the salaries, and presenting this figure as an *average* salary without further comment would be grossly misleading.

A great disadvantage of the mean is that it is strongly affected by extreme values.

3.9. Find the arithmetic mean of the numbers 5, 3, 6, 5, 4, 5, 2, 8, 6, 5, 4, 8, 3, 4, 5, 4, 8, 2, 5, 4.

Method 1:

$$\bar{X} = \frac{\Sigma X}{N} = \frac{5 + 3 + 6 + 5 + 4 + 5 + 2 + 8 + 6 + 5 + 4 + 8 + 3 + 4 + 5 + 4 + 8 + 2 + 5 + 4}{20} = \frac{96}{20} = 4 \cdot 8$$

Method 2:

There are six 5's, two 3's, two 6's, five 4's, two 2's and three 8's. Then

$$\bar{X} = \frac{\Sigma fX}{\Sigma f} = \frac{\Sigma fX}{N} = \frac{(6)(5) + (2)(3) + (2)(6) + (5)(4) + (2)(2) + (3)(8)}{6 + 2 + 2 + 5 + 2 + 3} = \frac{96}{20} = 4 \cdot 8$$

3.10. Out of 100 numbers, 20 were 4's, 40 were 5's, 30 were 6's and the remainder were 7's. Find the arithmetic mean of the numbers.

$$\bar{X} = \frac{\Sigma fX}{\Sigma f} = \frac{\Sigma fX}{N} = \frac{(20)(4) + (40)(5) + (30)(6) + (10)(7)}{100} = \frac{530}{100} = 5 \cdot 30$$

3.11. A student's final marks in Mathematics, Physics, English and Hygiene are respectively 82, 86, 90 and 70. If the respective credits received for these courses are 3, 5, 3 and 1, determine an approximate average mark.

Solution:

We use a weighted arithmetic mean, the weights associated with each mark being taken as the number of credits received. Then

$$\bar{X} = \frac{\Sigma wX}{\Sigma w} = \frac{(3)(82) + (5)(86) + (3)(90) + (1)(70)}{3 + 5 + 3 + 1} = 85$$

3.12. In a company having 80 employees, 60 earn \$3·00 per hour and 20 earn \$2·00 per hour. (*a*) Determine the mean earnings per hour. (*b*) Would the answer to (*a*) be the same if the 60 employees earn a *mean* hourly wage of \$3·00 per hour and 20 employees earn a mean hourly wage of \$2·00 per hour? Prove your answer. (*c*) Do you believe the mean hourly wage to be typical?

Solution:

(*a*) $\bar{X} = \dfrac{\Sigma fX}{N} = \dfrac{(60)(\$3 \cdot 00) + (20)(\$2 \cdot 00)}{60 + 20} = \dfrac{\$220 \cdot 00}{80} = \$2 \cdot 75.$

(*b*)   Yes, the result is the same. To prove this, suppose that f_1 numbers have mean m_1 and f_2 numbers have mean m_2. We must show that the mean of all the numbers is

$$\bar{X} = \frac{f_1 m_1 + f_2 m_2}{f_1 + f_2}$$

Let the f_1 numbers add up to M_1 and the f_2 numbers add up to M_2. Then by definition of arithmetic mean,

$$m_1 = \frac{M_1}{f_1} \qquad m_2 = \frac{M_2}{f_2}$$

or $M_1 = f_1 m_1$, $M_2 = f_2 m_2$. Since all $(f_1 + f_2)$ numbers add up to $(M_1 + M_2)$, the arithmetic mean of all numbers is

$$\bar{X} = \frac{M_1 + M_2}{f_1 + f_2} = \frac{f_1 m_1 + f_2 m_2}{f_1 + f_2}$$

as required. The result is easily extended.

(*c*)   We can say that \$2·75 is a "typical" hourly wage in the sense that most of the employees earn \$3·00 per hour, which is not far from \$2·75 per hour. It must be remembered that whenever we summarize numerical data into a single number (as is true in an average), we are bound to make some error. Certainly, however, the result is not as misleading as that of Problem 3.8.

Actually, to be on safe ground some estimate of the "spread" or "variation" of the data about the mean (or other average) should be given. This is called the *dispersion* of the data. Various measures of this are given in Chapter 4.

3.13. Four groups of students, consisting of 15, 20, 10 and 18 individuals, reported mean heights of 1·62, 1·48, 1·53 and 1·40 metres respectively. Find the mean height of all the students.

Solution:

$$\bar{X} = \frac{\Sigma fX}{\Sigma f} = \frac{(15)(1·62) + (20)(1·48) + (10)(1·53) + (18)(1·40)}{15 + 20 + 10 + 18} = 1·50 \text{ m}$$

3.14. If the mean annual incomes of agricultural and non-agricultural workers in the United States are respectively $3500 and $4500, would the mean annual income of both groups together be $4000?

Solution:

It would be $4000 only if the numbers of agricultural and non-agricultural workers were the same. To determine the true mean annual income, we would have to know the numbers of workers in each group. If, for example, there is 1 agricultural worker for every 11 non-agricultural workers the mean would be

$$\bar{X} = \frac{(1)(\$3500) + (11)(\$4500)}{1 + 11} = \$4400$$

to the nearest $100. This is a weighted arithmetic mean.

3.15. Use the frequency distribution of masses in the table on Page 27 to find the mean mass of the 100 male students at XYZ University.

Solution:

The work is outlined in Table 3.1. Note that all students having masses 60–62 kg, 63–65 kg, etc., are considered as having masses 61, 64, etc., kg. The problem then reduces to finding the mean mass of 100 students, if 5 students have mass 61 kg, 18 have mass 64 kg, etc.

Table 3.1

Mass (kg)	Class Mark (X)	Frequency (f)	fX
60 – 62	61	5	305
63 – 65	64	18	1152
66 – 68	67	42	2814
69 – 71	70	27	1890
72 – 74	73	8	584
		$N = \Sigma f = 100$	$\Sigma fX = 6745$

$$\bar{X} = \frac{\Sigma fX}{\Sigma f} = \frac{\Sigma fX}{N} = \frac{6745}{100} = 67·45 \text{ kg}$$

The computations involved can become tedious especially for cases in which the numbers are large and many classes are present. Short techniques are available for lessening the labour in such cases. See Problems 3.20 and 3.22, for example.

PROPERTIES OF THE ARITHMETIC MEAN

3.16. Prove that the sum of the deviations of $X_1, X_2, \ldots, X_N$ from their mean $\bar{X}$ is equal to zero.

Solution:

Let $d_1 = X_1 - \bar{X}, d_2 = X_2 - \bar{X}, \ldots, d_N = X_N - \bar{X}$ be the deviations of $X_1, X_2, \ldots, X_N$ from their mean $\bar{X}$. Then

$$\text{sum of deviations} = \Sigma d_j = \Sigma(X_j - \bar{X}) = \Sigma X_j - N\bar{X}$$

$$= \Sigma X_j - N\left(\frac{\Sigma X_j}{N}\right) = \Sigma X_j - \Sigma X_j = 0$$

where we have used Σ in place of $\sum_{j=1}^{N}$. We could, if desired, have omitted the subscript j in X_j provided it is *understood*.

3.17. If $Z_1 = X_1 + Y_1, Z_2 = X_2 + Y_2, \ldots, Z_N = X_N + Y_N$, prove that $\bar{Z} = \bar{X} + \bar{Y}$.

Solution:

By definition, $\bar{X} = \dfrac{\Sigma X}{N}, \bar{Y} = \dfrac{\Sigma Y}{N}, \bar{Z} = \dfrac{\Sigma Z}{N}$. Then

$$\bar{Z} = \frac{\Sigma Z}{N} = \frac{\Sigma(X + Y)}{N} = \frac{\Sigma X + \Sigma Y}{N} = \frac{\Sigma X}{N} + \frac{\Sigma Y}{N} = \bar{X} + \bar{Y}$$

where we have omitted subscripts j on X, Y, Z and Σ means $\displaystyle\sum_{j=1}^{N}$.

3.18. (a) If N numbers $X_1, X_2, \ldots, X_N$ have deviations from any number A given respectively by $d_1 = X_1 - A$, $d_2 = X_2 - A, \ldots, d_N = X_N - A$, prove that

$$\bar{X} = A + \frac{\displaystyle\sum_{j=1}^{N} d_j}{N} = A + \frac{\Sigma d}{N}$$

(b) In case $X_1, X_2, \ldots, X_K$ have respective frequencies $f_1, f_2, \ldots, f_K$ and $d_1 = X_1 - A, \ldots, d_K - A$, show that the result in (a) is replaced by

$$\bar{X} = A + \frac{\displaystyle\sum_{j=1}^{K} f_j d_j}{\displaystyle\sum_{j=1}^{K} f_j} = A + \frac{\Sigma fd}{N} \qquad \text{where} \qquad \sum_{j=1}^{K} f_j = \Sigma f = N.$$

(a)
Method 1:

Since $d_j = X_j - A$ and $X_j = A + d_j$, then
$$\bar{X} = \frac{\Sigma X_j}{N} = \frac{\Sigma(A + d_j)}{N} = \frac{\Sigma A + \Sigma d_j}{N} = \frac{NA + \Sigma d_j}{N} = A + \frac{\Sigma d_j}{N}$$

where we have used Σ in place of $\displaystyle\sum_{j=1}^{N}$ for brevity.

Method 2:

We have $d = X - A$ or $X = A + d$, omitting subscript on d and X. Then by Problem 3.17,

$$\bar{X} = \bar{A} + \bar{d} = A + \frac{\Sigma d}{N}$$

since the mean of a number of constants all equal to A is A.

(b)
$$\bar{X} = \frac{\displaystyle\sum_{j=1}^{K} f_j X_j}{\displaystyle\sum_{j=1}^{K} f_j} = \frac{\Sigma f_j X_j}{N} = \frac{\Sigma f_j(A + d_j)}{N} = \frac{\Sigma A f_j + \Sigma f_j d_j}{N} = \frac{A \Sigma f_j + \Sigma f_j d_j}{N}$$

$$= \frac{AN + \Sigma f_j d_j}{N} = A + \frac{\Sigma f_j d_j}{N} = A + \frac{\Sigma fd}{N}$$

Note that *formally* the result is obtained from (a) by replacing d_j by $f_j d_j$ and summing from $j = 1$ to K instead of $j = 1$ to N. The result is equivalent to $\bar{X} = A + \bar{d}$, where $\bar{d} = (\Sigma fd)/N$.

COMPUTATIONS OF ARITHMETIC MEAN FROM GROUPED DATA

3.19. Use the method of Problem 3.18(a) to find the arithmetic mean of the numbers 5, 8, 11, 9, 12, 6, 14 and 10, choosing as "guessed mean" A the value (a) 9, (b) 20.

Solution:

(a) The deviations of the given numbers from 9 are $-4, -1, 2, 0, 3, -3, 5, 1$; and the sum of the deviations is $\Sigma d = -4 - 1 + 2 + 0 + 3 + 3 + 5 + 1 = 3$. Then
$$\bar{X} = A + \frac{\Sigma d}{N} = 9 + \frac{3}{8} = 9 \cdot 375$$

(b) The deviations of the given numbers from 20 are $-15, -12, -9, -11, -8, -14, -6, -10$; and $\Sigma d = -85$. Then
$$\bar{X} = A + \frac{\Sigma d}{N} = 20 + \frac{(-85)}{8} = 9 \cdot 375$$

3.20. Use the method of Problem 3.18(b) to find the arithmetic mean of the masses of the 100 male students at XYZ University (see Problem 3.15).

Solution:

The work may be arranged as in Table 3.2. We take the guessed mean A as the class mark 67 (which has the largest frequency), although any class mark can be used for A. Note that the computations are simpler than those in Problem 3.15. To shorten the labour even more, we can proceed as in Problem 3.22 where use is made of the fact that the deviations (column 2 in the table) are all integer multiples of the class interval size.

Table 3.2

Class Mark X	Deviation $d = X - A$	Frequency (f)	fd
61	-6	5	-30
64	-3	18	-54
$A \rightarrow$ 67	0	42	0
70	3	27	81
73	6	8	48
		$N = \Sigma f = 100$	$\Sigma fd = 45$

$$\bar{X} = A + \frac{\Sigma fd}{N} = 67 + \frac{45}{100} = 67\cdot45 \text{ kg}$$

3.21. Let $d_j = X_j - A$ denote the deviations of any class mark X_j in a frequency distribution from a given class mark A. Show that if all class intervals have equal size c, then: (a) the deviations are all multiples of c, i.e. $d_j = cu_j$ where $u_j = 0, \pm1, \pm2, \ldots$; (b) the arithmetic mean can be computed from the formula

$$\bar{X} = A + \left(\frac{\Sigma fu}{N}\right)c.$$

Solution:

(a) The result is illustrated in the table of Problem 3.20 where it is observed that the deviations in column 2 are all multiples of the class interval size $c = 3$ kg.

To see that the result is true in general, note that if $X_1, X_2, X_3, \ldots$ are successive class marks their common difference will for this case be equal to c, so that $X_2 = X_1 + c$, $X_3 = X_1 + 2c$, and in general $X_j = X_1 = (j - 1)c$. Then any two class marks X_p and X_q, for example, will differ by

$$X_p - X_q = [X_1 + (p - 1)c] - [X_1 + (q - 1)c] = (p - q)c$$

which is a multiple of c.

(b) By (a) the deviations of all the class marks from any given one are multiples of c, i.e. $d_j = cu_j$. Then using Problem 3.18(b), we have

$$\bar{X} = A + \frac{\Sigma f_j d_j}{N} = A + \frac{\Sigma f_j(cu_j)}{N} = A + c\frac{\Sigma f_j u_j}{N} = A + \left(\frac{\Sigma fu}{N}\right)c$$

Note that this is equivalent to the result $\bar{X} = A + c\bar{u}$ which can be obtained from $\bar{X} = A + \bar{d}$ by placing $d = cu$ and observing that $\bar{d} = c\bar{u}$ (see Problem 3.18).

3.22. Use the result of Problem 3.21(b) to find the mean mass of the 100 male students at XYZ University (see Problem 3.20).

Solution:

The work may be arranged as in Table 3.3. The method is called the *coding method* and should be employed whenever possible.

Table 3.3

X	u	f	fu
61	-2	5	-10
64	-1	18	-18
$A \rightarrow$ 67	0	42	0
70	1	27	27
73	2	8	16
		$N = 100$	$\Sigma fu = 15$

$$\bar{X} = A + \left(\frac{\Sigma fu}{N}\right)c = 67 + \left(\frac{15}{100}\right)(3) = 67\cdot45 \text{ kg}$$

3.23. Compute the mean monthly wage of the 65 employees at the P and R Company from the frequency distribution on Page 33 using (a) the long method, (b) the coding method.

Solution:

(a)   **Table 3.4**

X	f	fX
£55·00	8	£440·00
65·00	10	650·00
75·00	16	1200·00
85·00	14	1190·00
95·00	10	950·00
105·00	5	525·00
115·00	2	230·00
	N = 65	ΣfX = £5185·00

$$\bar{X} = \frac{\Sigma fX}{N} = \frac{£5185·00}{65} = £79·77$$

(b)   **Table 3.5**

X	u	f	fu
£55·00	−2	8	−16
65·00	−1	10	−10
A→ 75·00	0	16	0
85·00	1	14	14
95·00	2	10	20
105·00	3	5	15
115·00	4	2	8
		N = 65	Σfu = 31

$$\bar{X} = A + \left(\frac{\Sigma fu}{N}\right)c = £75·00 - \left(\frac{31}{65}\right)(£10·00)$$
$$= £79·77$$

It might be supposed that error would be introduced in the above tables since the class marks are actually £54·995, £64·995, etc., instead of £55·00, £65·00, etc. If in Table 3.4 these true class marks are used instead, $\bar{X}$ turns out to be £79·76 instead of £79·77 and the difference is negligible.

3.24. Find the mean wage of the 70 employees at the P and R Company using Table (d) on Page 38.

Solution:
In this case the class intervals do not have equal size and we must use the long method as shown in Table 3.6.

$$\bar{X} = \frac{\Sigma fX}{N} = \frac{£5845·00}{70} = £83·50$$

Table 3.6

X	f	fX
£55·00	8	£440·00
65·00	10	650·00
75·00	16	1200·00
85·00	15	1275·00
95·00	10	950·00
110·00	8	880·00
150·00	3	450·00
	N = 70	ΣfX = £5845·00

THE MEDIAN

3.25. The grades of a student on six examinations were 84, 91, 72, 68, 87 and 78. Find the median of the grades.

Solution:
Arranged in an array, the grades are 68, 72, 78, 84, 87, 91.

Since there is an even number of items there are two middle values, 78 and 84 whose arithmetic mean ½(78 + 84) = 81 is the required median grade. Compare with Problem 3.6 where the arithmetic mean = 80.

3.26. The hourly wages of five employees in an office are $2·52, $3·96, $3·28, $9·20, $3·75. Find (a) the median hourly wage, (b) the mean hourly wage.

Solution:
(a) Arranged in an array, the wages are $2·52, $3·28, $3·75, $3·96, $9·20. Since there is an odd number of items there is only one middle value, $3·75, which is the required median.

(b) The arithmetic mean is $\dfrac{\$2·52 + \$3·96 + \$3·28 + \$9·20 + \$3·75}{5} = \$4·54$.

Note that the median is not affected by the extreme value $9·20 while the mean is affected by it. In this case the median gives a better indication of the average hourly wage than the mean.

3.27. If (a) 85, (b) 150 numbers are arranged in an array, how would you find the median of the numbers?

Solution:

(a) Since there are 85 items, an odd number, there is only one middle value with 42 numbers below and 42 numbers above it. Then the median is the 43rd number in the array.

(b) Since there are 150 items, an even number, there are two middle values with 74 numbers below them and 74 numbers above them. The two middle values are the 75th and 76th numbers in the array and their arithmetic mean is the required median.

3.28. Find the median lengths of the 40 laurel leaves (see Problem 3.8, Page 35) by using (a) the second frequency of Problem 3.8, reproduced here, (b) the original data.

Solution:

(a) **Method 1, using interpolation:**

The lengths in the frequency distribution shown at the right are assumed to be continuously distributed. In such case the median is that length for which half the total frequency (40/2 = 20) lies above it and half lies below it.

Now the sum of the first three class frequencies is $3 + 5 + 9 = 17$. Thus to give the desired 20, we require 3 more of the 12 cases in the fourth class. Since the fourth class interval, 145–153, actually corresponds to lengths 144·5 to 153·5, the median must lie 3/12 of the way between 144·5 and 153·5, i.e. the median is

Table 3.7

Length (mm)	Frequency
118–126	3
127–135	5
136–144	9
145–153	12
154–162	5
163–171	4
172–180	2
	Total 40

$$144·5 + \frac{3}{12}(153·5 - 144·5) = 144·5 + \frac{3}{12}(9) = 146·8 \text{ mm}$$

Method 2, using formula:

Since the sum of the first three and first four class frequencies are respectively $3 + 5 + 9 = 17$ and $3 + 5 + 9 + 12 = 29$, it is clear that the median lies in the fourth class which is therefore the median class. Then

$$L_1 = \text{lower class boundary of median class} = 144·5,$$
$$N = \text{number of items in the data} = 40,$$
$$(\Sigma f)_1 = \text{sum of all classes lower than the median class} = 3 + 5 + 9 = 17,$$
$$f_{median} = \text{frequency of median class} = 12,$$
$$c = \text{size of median class interval} = 9, \text{ and so}$$

$$\text{Median} = L_1 + \left(\frac{N/2 - (\Sigma f)_1}{f_{median}}\right)c = 144·5 + \left(\frac{40/2 - 17}{12}\right)(9) = 146·8 \text{ mm}$$

(b) Arranged in an array, the original lengths are

119, 125, 126, 128, 132, 135, 135, 135, 136, 138, 138, 140, 140, 142, 142, 144, 144, 145, 145, 146, 146, 147, 147, 148, 149, 150, 150, 152, 153, 154, 156, 157, 158, 161, 163, 164, 165, 168, 173, 176

The median is the arithmetic mean of the 20th and 21st length in this array and is equal to 146 mm.

3.29. Show how the median length in the previous problem can be obtained from (a) a histogram, (b) a percentage ogive.

Solution:

(a) In Fig. 3-3(a) is shown the histogram corresponding to the lengths in the preceding problem. The median is the abscissa corresponding to the line LM which divides the histogram into two equal areas. Since area corresponds to frequency in a histogram, LM is such that the total area to the right and left of it is half the total frequency, or 20. Thus areas AMLD and MBEL correspond to frequencies of 3 and 9. Then $AM = \frac{3}{12}AB = \frac{3}{12}(9) = 2·25$, and the median has the value $144·5 + 2·25 = 146·75$, or 146·8 mm to the nearest tenth of a millimetre. The value can also be read approximately directly from the graph.

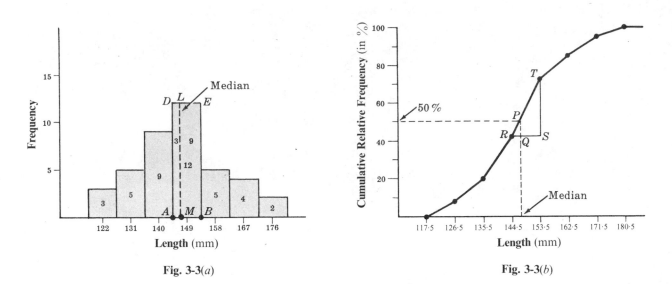

Fig. 3-3(a)   **Fig. 3-3(b)**

(b) In Fig. 3-3(b) is shown the cumulative relative frequency polygon or percentage ogive corresponding to the weights in the previous problem. The median is the abscissa of point P on this ogive whose ordinate is 50%. To compute its value we see from the similar triangles PQR and RST that

$$\frac{RQ}{RS} = \frac{PQ}{ST} \quad \text{or} \quad \frac{RQ}{9} = \frac{50\% - 42\cdot5\%}{72\cdot5\% - 42\cdot5\%} = \frac{1}{4} \quad \text{so that} \quad RQ = \frac{9}{4} = 2\cdot25$$

Then

$$\text{Median} = 144\cdot5 + RQ = 144\cdot5 + 2\cdot25 = 146\cdot75 \text{ mm}$$

or 146·8 mm to the nearest tenth of a millimetre. This value can also be read approximately directly from the graph.

3.30. Find the median wage of the 65 employees at the P and R Company (see Chap. 2, Prob. 3.3, Page 33).

Solution:

Here $N = 65$, $N/2 = 32\cdot5$. Since the sum of the first two and first three class frequencies are respectively $8 + 10 = 18$ and $8 + 10 + 16 = 34$, the median class is the third class. Using the formula,

$$\text{Median} = L_1 + \left(\frac{N/2 - (\Sigma f)_1}{f_{\text{median}}}\right) c = £69\cdot995 + \left(\frac{32\cdot5 - 18}{16}\right)(£10\cdot00) = £79\cdot06$$

THE MODE

3.31. Find the mean, median and mode for the set of numbers: (a) 3, 5, 2, 6, 5, 9, 5, 2, 8, 6; (b) 51·6, 48·7, 50·3, 49·5, 48·9.

Solution:

(a) Arranged in an array, the numbers are 2, 2, 3, 5, 5, 5, 6, 6, 8, 9.

Mean $= \frac{1}{10}(2 + 2 + 3 + 5 + 5 + 5 + 6 + 6 + 8 + 9) = 5\cdot1$,

Median $=$ arithmetic mean of two middle numbers $= \frac{1}{2}(5 + 5) = 5$.

Mode $=$ number occurring most frequently $= 5$.

(b) Arranged in an array, the numbers are 48·7, 48·9, 49·5, 50·3, 51·6.

Mean $= \frac{1}{5}(48\cdot7 + 48\cdot9 + 49\cdot5 + 50\cdot3 + 51\cdot6) = 49\cdot8$.

Median $=$ middle number $= 49\cdot5$.

Mode $=$ number occurring most frequently; it does not exist here.

3.32. Develop a formula for determining the mode from data presented in a frequency distribution.

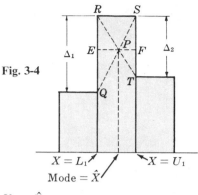

Fig. 3-4

Solution:

Assume Fig. 3-4 represents three rectangles of the histogram of the frequency distribution, the central rectangle corresponding to the modal class. Assume also that the class intervals have equal size.

We define the mode as the abscissa $\hat{X}$ of the point of intersection P of the constructed lines QS and RT.

Let $X = L_1$ and $X = U_1$ represent the lower and upper class boundaries of the modal class, and Δ_1 and Δ_2 represent respectively the excess of the modal class frequency over the class frequencies to the left and right of the modal class.

From similar triangles PQR and PST, we have $\dfrac{EP}{RQ} = \dfrac{PF}{ST}$ or $\dfrac{\hat{X} - L_1}{\Delta_1} = \dfrac{U_1 - \hat{X}}{\Delta_2}$. Then

$$\Delta_2(\hat{X} - L_1) = \Delta_1(U_1 - \hat{X}), \quad \Delta_2\hat{X} - \Delta_2 L_1 = \Delta_1 U_1 + \Delta_1\hat{X}, \quad (\Delta_1 + \Delta_2)\hat{X} = \Delta_1 U_1 + \Delta_2 L_1$$

or

$$\hat{X} = \frac{\Delta_1 U_1 + \Delta_2 L_1}{\Delta_1 + \Delta_2}$$

Since $U_1 = L_1 + c$, where c is the class interval size, this becomes

$$\hat{X} = \frac{\Delta_1(L_1 + c) + \Delta_2 L_1}{\Delta_1 + \Delta_2} = \frac{(\Delta_1 + \Delta_2)L_1 + \Delta_1 c}{\Delta_1 + \Delta_2} = L_1 + \left(\frac{\Delta_1}{\Delta_1 + \Delta_2}\right)c$$

The result has the following interesting interpretation. If a parabola is constructed so as to pass through the three midpoints of the tops of the rectangles in the figure, the abscissa of the maximum of this parabola will be the mode as obtained above.

3.33. Find the modal wage of the 65 employees at the P and R Company (see Problem 3.23) by using the formula developed in Problem 3.32.

Solution:

Here $L_1 = £69{\cdot}995$, $\Delta_1 = 16 - 10 = 6$, $\Delta_2 = 16 - 14 = 2$, $c = £10{\cdot}00$. Then

$$\text{Mode} = L_1 + \left(\frac{\Delta_1}{\Delta_1 + \Delta_2}\right)c = £69{\cdot}995 + \left(\frac{6}{2 + 6}\right)(£10{\cdot}00) = £77{\cdot}50$$

EMPIRICAL RELATION BETWEEN MEAN, MEDIAN AND MODE

3.34. (a) Use the empirical formula Mean − Mode = 3(Mean − Median) to find the modal wage of the 65 employees at the P and R Company.

(b) Compare your result with the mode obtained in Problem 3.33.

Solution:

(a) From Problems 3.23 and 3.30, we have Mean = £79·77, Median = £79·06. Then

$$\text{Mode} = \text{Mean} - 3(\text{Mean} - \text{Median}) = £79{\cdot}77 - 3(£79{\cdot}77 - £79{\cdot}06) = £77{\cdot}64$$

(b) From Problem 3.33 the modal wage is £77·50, so there is good agreement with the empirical result in this case.

THE GEOMETRIC MEAN

3.35. Find (a) the geometric mean and (b) the arithmetic mean of the numbers 3, 5, 6, 6, 7, 10, 12. Assume the numbers are exact.

Solution:

(a) Geometric mean $= G = \sqrt[7]{(3)(5)(6)(6)(7)(10)(12)} = \sqrt[7]{453\,600}$. Using common logarithms,

$$\log G = \tfrac{1}{7}\log 453\,600 = \tfrac{1}{7}(5{\cdot}6567) = 0{\cdot}8081 \quad \text{and} \quad G = 6{\cdot}43 \text{ (to nearest hundredth)}$$

Another Method:
$$\log G = \tfrac{1}{7}(\log 3 + \log 5 + \log 6 + \log 6 + \log 7 + \log 10 + \log 12)$$
$$= \tfrac{1}{7}(0\cdot4771 + 0\cdot6990 + 0\cdot7782 + 0\cdot7782 + 0\cdot8451 + 1\cdot0000 + 1\cdot0792)$$
$$= 0\cdot8081, \quad G = 6\cdot43$$

(b) Arithmetic mean $= \bar{X} = \tfrac{1}{7}(3 + 5 + 6 + 6 + 7 + 10 + 12) = 7$

This illustrates the fact that the geometric mean of a set of unequal positive numbers is less than the arithmetic mean.

3.36. The numbers $X_1, X_2, \ldots, X_K$ occur with frequencies $f_1, f_2, \ldots, f_K$, where $f_1 + f_2 + \ldots + f_K = N$ is the total frequency. (a) Find the geometric mean G of the numbers. (b) Derive an expression for $\log G$. (c) How can the result be used in finding the geometric mean for data grouped into a frequency distribution?

Solution:

(a) $G = \sqrt[N]{\underbrace{X_1 X_1 \ldots X_1}_{f_1 \ times} \ \underbrace{X_2 X_2 \ldots X_2}_{f_2 \ times} \ \ldots \ \underbrace{X_K X_K \ldots X_K}_{f_K \ times}} = \sqrt[N]{X_1^{f_1} X_2^{f_2} \ldots X_K^{f_K}}$

where $N = \Sigma f$. This is sometimes called the *weighted geometric mean*.

(b) $\log G = \dfrac{1}{N} \log(X_1^{f_1} X_2^{f_2} \ldots X_K^{f_K}) = \dfrac{1}{N}(f_1 \log X_1 + f_2 \log X_2 + \ldots + f_K \log X_K)$

$$= \dfrac{1}{N}\sum_{j=1}^{K} f_j \log X_j = \dfrac{\Sigma f \log X}{N}$$

where we assume all the numbers are positive, otherwise the logarithms are not defined.

Note that the logarithm of the geometric mean of a set of positive numbers is the arithmetic mean of the logarithms of the numbers.

(c) The result can be used in finding the geometric mean for grouped data by taking $X_1, X_2, \ldots, X_K$ as class marks and $f_1, f_2, \ldots, f_K$ as the corresponding class frequencies.

3.37. During one year the ratio of milk prices per litre to bread prices per loaf was $3\cdot00$, whereas during the next year the ratio was $2\cdot00$. (a) Find the arithmetic mean of these ratios for the two-year period. (b) Find the arithmetic mean of the ratios of bread prices to milk prices for the two-year period. (c) Discuss the advisability of using the arithmetic mean for averaging ratios. (d) Discuss the suitability of the geometric mean for averaging ratios.

Solution:

(a)   Mean ratio of milk to bread prices $= \tfrac{1}{2}(3\cdot00 + 2\cdot00) = 2\cdot50$

(b) Since the ratio of milk to bread prices for the first year is $3\cdot00$, the ratio of bread to milk prices is $1/3\cdot00 = 0\cdot333$. Similarly the ratio of bread to milk prices for the second year is $1/2\cdot00 = 0\cdot500$. Then

Mean ratio of bread to milk prices $= \tfrac{1}{2}(0\cdot333 + 0\cdot500) = 0\cdot417$

(c) We would expect the mean ratio of milk to bread prices to be the reciprocal of the mean ratio of bread to milk prices if the mean is an appropriate average. However $1/0\cdot417 = 2\cdot40 \neq 2\cdot50$.

This shows that the arithmetic mean is a poor average to use for ratios.

(d) Geometric mean of ratios of milk to bread prices $= \sqrt{(3\cdot00)(2\cdot00)} = \sqrt{6\cdot00}$.

Geometric mean of ratios of bread to milk prices $= \sqrt{(0\cdot333)(0\cdot500)} = \sqrt{0\cdot0167} = 1/\sqrt{6\cdot00}$.

Since these averages are reciprocals, our conclusion is that the geometric mean is more suitable than the arithmetic mean for averaging ratios for this type of problem.

3.38. The bacterial count in a certain culture increased from 1000 to 4000 in three days. What was the average percentage increase per day?

Solution:

Since an increase from 1000 to 4000 is a 300% increase, one might be led to conclude that the average percentage increase per day would be 300%/3 = 100%. This, however, would imply that during the first day the count went from 1000 to 2000, during the second day from 2000 to 4000 and during the third day from 4000 to 8000, which is contrary to the facts.

To determine this average percentage increase, let us denote it by r. Then

Total bacterial count after 1 day $= 1000 + 1000r = 1000(1 + r)$
Total bacterial count after 2 days $= 1000(1 + r) + 1000(1 + r)r = 1000(1 + r)^2$
Total bacterial count after 3 days $= 1000(1 + r)^2 + 1000(1 + r)^2 r = 1000(1 + r)^3$

This last expression must equal 4000, so that

$$1000(1 + r)^3 = 4000, \quad (1 + r)^3 = 4, \quad 1 + r = \sqrt[3]{4} \quad \text{and} \quad r = \sqrt[3]{4} - 1$$

Using logarithms, we find that $\sqrt[3]{4} = 1.587$, so that $r = 0.587 = 58.7\%$.

In general, if we start with a quantity P and increase it at a constant rate r per unit of time, we will have after n units of time the amount

$$A = P(1 + r)^n$$

This is called the *compound interest formula*. See Problems 3.94 and 3.95.

THE HARMONIC MEAN

3.39. Find the harmonic mean H of the numbers 3, 5, 6, 6, 7, 10, 12.

Solution:

$$\frac{1}{H} = \frac{1}{N} \Sigma \frac{1}{X} = \frac{1}{7} \left(\frac{1}{3} + \frac{1}{5} + \frac{1}{6} + \frac{1}{6} + \frac{1}{7} + \frac{1}{10} + \frac{1}{12} \right) = \frac{1}{7} \left(\frac{140 + 84 + 70 + 70 + 60 + 42 + 35}{420} \right)$$

$$= \frac{501}{2940} \quad \text{and} \quad H = \frac{2940}{501} = 5.87$$

It is often convenient to express the fractions in decimal form first. Thus

$$\frac{1}{H} = \tfrac{1}{7} (0.3333 + 0.2000 + 0.1667 + 0.1667 + 0.1429 + 0.1000 + 0.0833)$$

$$= \tfrac{1}{7} (1.1929) \quad \text{and} \quad H = 7/1.1929 = 5.87$$

Comparison with Problem 3.35 illustrates the fact that the harmonic mean of several positive numbers not all equal is less than their geometric mean, which is in turn less than their arithmetic mean.

3.40. During four successive years a home owner purchased oil for his furnace at respective costs of 1·6, 1·8 2·1 and 2·5 per litre. What was the average cost of oil over the four-year period?

Solution:
Case 1:

Suppose the home owner purchases the same quantity of oil each year, say 1000 litres. Then

$$\text{Average cost} = \frac{\text{total cost}}{\text{total quantity purchased}} = \frac{£16 + £18 + £21 + £25}{4000 \text{ litres}} = 2.00\text{p/l}$$

This is the same as the arithmetic mean of the costs per litre, i.e., $\tfrac{1}{4}(16 + 18 + 21 + 25) = 2.0\text{p/l}$
This result would be the same even if x litres were used each year.

Case 2:

Suppose the home owner spends the same amount of money each year, say £200. Then

$$\text{Average cost} = \frac{\text{total cost}}{\text{total quantity purchased}} = \frac{£800}{12\,500 + 11\,111 + 9524 + 8000 \text{ litres}} = 1.94\text{p/l}$$

This is the same as the harmonic mean of the costs per litre, i.e., $\dfrac{4}{\frac{1}{16} + \frac{1}{18} + \frac{1}{21} + \frac{1}{25}} = 1.94\text{p/l}$

This result would be the same even if £y were spent each year.

Both averaging processes are correct, each average being computed under different prevailing conditions.

It should be noted that in case the number of litres used changes from one year to another instead of remaining the same, the ordinary arithmetic mean of Case 1 is replaced by a weighted arithmetic mean. Similarly if the total amount spent changes from one year to another, the ordinary harmonic mean of Case 2 is replaced by a weighted harmonic mean.

3.41. A man travels from A to B at an average speed of 30 km/h and returns from B to A along the same route at an average speed of 60 km/h. Find the average speed for the entire trip.

Solution:

Assume the distance from A to B is 60 km (although any distance can be assumed). Then

$$\text{Time to move from } A \text{ to } B = \frac{60 \text{ km}}{30 \text{ km/h}} = 2 \text{ h}, \quad \text{time from } B \text{ to } A = \frac{60 \text{ km}}{60 \text{ km/h}} = 1 \text{ h}$$

and
$$\text{Average speed for round trip} = \frac{\text{total distance}}{\text{total time}} = \frac{120 \text{ km}}{3 \text{ h}} = 40 \text{ km/h}.$$

The above average is the harmonic mean of 30 and 60, i.e., $\dfrac{2}{1/30 + 1/60} = 40$ km/h. If distances travelled are not all equal, a weighted harmonic mean of the speeds would be used where the weights are the respective distances. (See Problem 3.102.)

Note that one might very well be tempted to take the arithmetic mean of 30 and 60 km/h to obtain 45 km/h, but this is wrong.

THE QUADRATIC MEAN, OR ROOT MEAN SQUARE

3.42. Find the quadratic mean of the numbers 3, 5, 6, 6, 7, 10, 12.

Solution:

$$\text{Quadratic mean} = \text{R.M.S.} = \sqrt{\frac{3^2 + 5^2 + 6^2 + 6^2 + 7^2 + 10^2 + 12^2}{7}} = \sqrt{57} = 7.55$$

3.43. Prove that the quadratic mean of two positive unequal numbers, a and b, is greater than their geometric mean.

Solution:

We are required to show that $\sqrt{\frac{1}{2}(a^2 + b^2)} > \sqrt{ab}$. If this is true, then by squaring both sides, $\frac{1}{2}(a^2 + b^2) > ab$ so that $a^2 + b^2 > 2ab$, $a^2 - 2ab + b^2 > 0$, or $(a - b)^2 > 0$. But this last inequality is true since the square of any real number not equal to zero must be positive.

The proof consists in establishing the reversal of the above steps. Thus starting with $(a - b)^2 > 0$, which we know to be true, we can show that $a^2 + b^2 > 2ab$, $\frac{1}{2}(a^2 + b^2) > ab$, and finally $\sqrt{\frac{1}{2}(a^2 + b^2)} > \sqrt{ab}$ as required.

Note that $\sqrt{\frac{1}{2}(a^2 + b^2)} = \sqrt{ab}$ if and only if $a = b$.

QUARTILES, DECILES AND PERCENTILES

3.44. Find (a) the quartiles Q_1, Q_2, Q_3 and (b) the deciles $D_1, D_2, \ldots, D_9$ for the wages of the 65 employees at the P and R Company (Problem 2.3, Chap. 2).

Solution:

(a) The first quartile Q_1 is that wage obtained by counting $N/4 = 65/4 = 16.25$ of the cases beginning with the first (lowest) class. Since the first class contains 8 cases, we must take 8.25 (16.25 − 8) of the 10 cases from the second class. Using the method of linear interpolation, we have

$$Q_1 = £59.995 + \frac{8.25}{10} (£10.00) = £68.25$$

The second quartile Q_2 is obtained by counting the first $2N/4 = N/2 = 65/2 = 32.5$ of the cases. Since the first two classes comprise 18 cases, we must take $32.5 - 18 = 14.5$ of the 16 cases from the third class; then

$$Q_2 = £69.995 + \frac{14.5}{16}(£10.00) = £79.06$$

Note that Q_2 is actually the median.

The third quartile Q_3 is obtained by counting the first $3N/4 = \frac{3}{4}(65) = 48.75$ of the cases. Since the first four classes comprise 48 cases, we must take $48.75 - 48 = 0.75$ of the 10 cases from the fifth class; then

$$Q_3 = £89.995 + \frac{0.75}{10}(£10.00) = £90.75$$

Hence 25% of the employees earn £68.25 or less, 50% earn £79.06 or less, 75% earn £90.75 or less.

(b) The first, second, ..., ninth deciles are obtained by counting $N/10$, $2N/10$, ..., $9N/10$ of the cases beginning with the first (lowest) class. This

$$D_1 = £49.995 + \frac{6.5}{8}(£10.00) = £58.12 \qquad D_6 = £79.995 + \frac{5}{14}(£10.00) = £83.57$$

$$D_2 = £59.995 + \frac{5}{10}(£10.00) = £65.00 \qquad D_7 = £79.995 + \frac{11.5}{14}(£10.00) = £88.21$$

$$D_3 = £69.995 + \frac{1.5}{16}(£10.00) = £70.94 \qquad D_8 = £89.995 + \frac{4}{10}(£10.00) = £94.00$$

$$D_4 = £69.995 + \frac{8}{16}(£10.00) = £75.00 \qquad D_9 = £99.995 + \frac{0.5}{5}(£10.00) = £101.00$$

$$D_5 = £69.995 + \frac{14.5}{16}(£10.00) = £79.06$$

Thus 10% of the employees earn £58.12 or less, 20% earn £65.00 or less, ..., 90% earn £101.00 or less.

Note that the fifth decile is the median. The second, fourth, sixth and eighth deciles, which divide the distribution into five equal parts, are called *quintiles* which are sometimes used in practice.

3.45. Determine the (a) 35th and (b) 60th percentile for the distribution in the preceding problem.

Solution:

(a) The 35th percentile, denoted by P_{35}, is obtained by counting the first $35N/100 = 35(65)/100 = 22.75$ of the cases beginning with the first (lowest) class. Then, as in Problem 3.44, $P_{35} = £69.995 + \frac{4.75}{16}(£10.00) = £72.97$. This means that 35% of the employees earn £72.97 or less.

(b) The 60th percentile is $P_{60} = £79.995 + \frac{5}{14}(£10.00) = £83.57$. Note that this is the same as the 6th decile or 3rd quintile.

3.46. Show how the results of Problems 3.44 and 3.45 can be obtained from a percentage ogive.

Solution:

The percentage ogive corresponding to the data of Problems 3.44 and 3.45 is shown below.

The first quartile is the abscissa of that point on the ogive whose ordinate is 25%. Similarly the second and third quartiles are the abscissas of those points on the ogive with ordinates 50% and 75% respectively.

The deciles and percentiles can be similarly obtained. For example the 7th decile and 35th percentile are the abscissas of those points on the ogive corresponding to ordinates of 70% and 35% respectively.

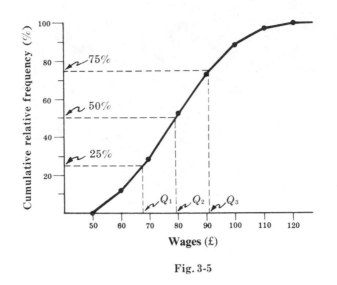

Fig. 3-5

Supplementary Problems

SUMMATION NOTATION

3.47. Write the terms in each of the following indicated sums.

(a) $\sum_{j=1}^{4}(X_j+2)$, (b) $\sum_{j=1}^{5} f_j X_j^2$, (c) $\sum_{j=1}^{3} U_j(U_j+6)$, (d) $\sum_{k=1}^{N}(Y_k^2-4)$, (e) $\sum_{j=1}^{4} 4X_j Y_j$.

Ans. (a) $X_1 + X_2 + X_3 + X_4 + 8$

    (b) $f_1 X_1^2 + f_2 X_2^2 + f_3 X_3^2 + f_4 X_4^2 + f_5 X_5^2$ (d) $Y_1^2 + Y_2^2 + \ldots + Y_N^2 - 4N$

    (c) $U_1(U_1 + 6) + U_2(U_2 + 6) + U_3(U_3 + 6)$ (e) $4X_1 Y_1 + 4X_2 Y_2 + 4X_3 Y_3 + 4X_4 Y_4$

3.48. Express each of the following using the summation notation.

(a) $(X_1 + 3)^3 + (X_2 + 3)^3 + (X_3 + 3)^3$ (d) $(X_1/Y_1 - 1)^2 + (X_2/Y_2 - 1)^2 + \ldots + (X_8/Y_8 - 1)^2$

(b) $f_1(Y_1 - a)^2 + f_2(Y_2 - a)^2 + \ldots + f_{15}(Y_{15} - a)^2$

(c) $(2X_1 - 3Y_1) + (2X_2 - 3Y_2) + \ldots + (2X_N - 3Y_N)$ (e) $\dfrac{f_1 a_1^2 + f_2 a_2^2 + \ldots + f_{12} a_{12}^2}{f_1 + f_2 + \ldots + f_{12}}$

Ans. (a) $\sum_{j=1}^{3}(X_j+3)^3$, (b) $\sum_{j=1}^{15} f_j(Y_j-a)^2$, (c) $\sum_{j=1}^{N}(2X_j-3Y_j)$, (d) $\sum_{j=1}^{8}\left(\dfrac{X_j}{Y_j}-1\right)^2$, (e) $\dfrac{\sum_{j=1}^{12} f_j a_j^2}{\sum_{j=1}^{12} f_j}$

3.49. Prove that $\sum_{j=1}^{N}(X_j-1)^2 = \sum_{j=1}^{N} X_j^2 - 2\sum_{j=1}^{N} X_j + N$.

3.50. Prove that $\Sigma(X + a)(Y + b) = \Sigma XY + a\Sigma Y + b\Sigma X + Nab$, where a and b are constants. What subscript notation is implied?

3.51. Two variables U and V assume the values $U_1 = 3$, $U_2 = -2$, $U_3 = 5$ and $V_1 = -4$, $V_2 = -1$, $V_3 = 6$ respectively. Calculate (a) ΣUV, (b) $\Sigma(U + 3)(V - 4)$, (c) ΣV^2, (d) $(\Sigma U)(\Sigma V)^2$, (e) ΣUV^2, (f) $\Sigma(U_2 - 2v^2 + 2)$, (g) $\Sigma(U/V)$.
Ans. (a) 20, (b) -37, (c) 53, (d) 6, (e) 226, (f) -62, (g) 25/12

3.52. Given $\sum_{j=1}^{4} X_j = 7$, $\sum_{j=1}^{4} = -3$ and $\sum_{j=1}^{4} X_j Y_j = 5$, find (a) $\sum_{j=1}^{4}(2X_j + 5Y_j)$ and (b) $\sum_{j=1}^{4}(X_j - 3)(2Y_j + 1)$.
Ans. (a) -1, (b) 23

THE ARITHMETIC MEAN

3.53. A student received marks of 85, 76, 93, 82 and 96 in five subjects. Determine the arithmetic mean of the marks.
Ans. 86

3.54. The reaction times of an individual to certain stimuli were measured by a psychologist to be 0·53, 0·46, 0·50, 0·49, 0·52, 0·53, 0·44 and 0·55 seconds respectively. Determine the mean reaction time of the individual to the stimuli.
Ans. 0·50 s

3.55. A set of numbers consists of six 6's, seven 7's, eight 8's, nine 9's and ten 10's. What is the arithmetic mean of the numbers?
Ans. 8·25

3.56. A student's marks in the laboratory, lecture and recitation parts of a physics course were 71, 78 and 89 respectively. (*a*) If the weights accorded these marks are 2, 4 and 5 respectively, what is an appropriate average mark? (*b*) What is the average mark if equal weights are used?   *Ans.* (*a*) 82, (*b*) 79

3.57. Three teachers of economics reported mean examination marks of 79, 74 and 82 in their classes which consisted of 32, 25 and 17 students respectively. Determine the mean mark for all the classes.
Ans. 78

3.58. The mean annual salary paid to all employees in a company was £1500. The mean annual salaries paid to male and female employees of the company were £1560 and £1260 respectively. Determine the percentages of males and females employed by the company.   *Ans.* 80%, 20%.

3.59. Table 3.8 shows the distribution of the maximum loads in kilonewtons supported by certain cables produced by a company. Determine the mean maximum loading using (*a*) the "long method", (*b*) the coding method.
Ans. 110·9 kN

Table 3.8

Maximum Load (kN)	Number of Cables
93 – 97	2
98 – 102	5
103 – 107	12
108 – 112	17
113 – 117	14
118 – 122	6
123 – 127	3
128 – 132	1
	Total 60

3.60. Find $\bar{X}$ for the data in Table 3.9 using (*a*) the "long method", (*b*) the coding method.   *Ans.* 501·0

Table 3.9

X	462	480	498	516	534	552	570	588	606	624
f	98	75	56	42	30	21	15	11	6	2

3.61. Table 3.10 below shows the distribution of the diameters of the heads of rivets manufactured by a company. Compute the mean diameter.   *Ans.* 7·2642 mm

Table 3.10

Diameter (mm)	Frequency
7·247 – 7·249	2
7·250 – 7·252	6
7·253 – 7·255	8
7·256 – 7·258	15
7·259 – 7·261	42
7·262 – 7·264	68
7·265 – 7·267	49
7·268 – 7·270	25
7·271 – 7·273	18
7·274 – 7·276	12
7·277 – 7·279	4
7·280 – 7·282	1
	Total 250

Table 3.11

Class	Frequency
10 – under 15	3
15 – under 20	7
20 – under 25	16
25 – under 30	12
30 – under 35	9
35 – under 40	5
40 – under 45	2
	Total 54

3.62. Compute the mean for the data in Table 3.11 above.   *Ans.* 26·2

3.63. Compute the mean lifetimes of tubes manufactured by the L and M Tube Company of Prob. 2.20, Chap. 2. *Ans.* 715 hours

3.64. (*a*) Use the frequency distribution obtained in Prob. 2.27, Chap. 2, to compute the mean diameter of the ball bearings. (*b*) Compute the mean directly from the raw data and compare with (*a*), explaining any discrepancy which may occur. *Ans.* (*b*) 7·349 mm

THE MEDIAN

3.65. Find the mean and median of the sets of numbers:
(*a*) 5, 4, 8, 3, 7, 2, 9; (*b*) 18·3, 20·6, 19·3, 22·4, 20·2, 18·8, 19·7, 20·0.
Ans. (*a*) Mean = 5·4, median = 5. (*b*) Mean = 19·91, median = 19·85

3.66. Find the median mark of Problem 3.53.   *Ans.* 85

3.67. Find the median reaction time of Problem 3.54.   *Ans.* 0·51 seconds

3.68. Find the median of the set of numbers in Problem 3.55.   *Ans.* 8

3.69. Find the median of the maximum loads of the cables of Problem 3.59.   *Ans.* 110·7 kN

3.70. Find the median $\tilde{X}$ for the distribution in Problem 3.60.   *Ans.* 490·6

3.71. Find the median diameter of the rivet heads in Problem 3.61.   *Ans.* 7·2638 mm

3.72. Find the median of the distribution in Problem 3.62.   *Ans.* 25·4

3.73. Table 3.12 shows the age distribution of heads of families in the United States during the year 1957. (*a*) Find the median age. (*b*) Why is the median a more suitable measure of central tendency than the mean in this case?   *Ans.* 45·1

3.74. Find the median income for the data of Problem 2.31, Chap. 2. *Ans.* $3608

3.75. Find the median lifetime of the tubes in Problem 2.20, Chap. 2. *Ans.* 708·3 hours

Table 3.12

Age of Head of Family (years)	Number (in millions)
Under 25	2·22
25–29	4·05
30–34	5·08
35–44	10·45
45–54	9·47
55–64	6·63
65–74	4·16
75 and over	1·66
	Total 43·72

Source: Bureau of the Census.

THE MODE

3.76. Find the mean, median and mode for the set of numbers:
(*a*) 7, 4, 10, 9, 15, 12, 7, 9, 7;   (*b*) 8, 11, 4; 3, 2, 5, 10, 6, 4, 1, 10, 8, 12, 6, 5, 7.
Ans. (*a*) Mean = 8·9, median = 9, mode = 7. (*b*) Mean = 6·4, median = 6. Since each of the numbers 4, 5, 6, 8 and 10 occurs twice, we can consider that these are the five modes. However, it is more reasonable to conclude in this case that no mode exists.

3.77. Find the modal mark in Problem 3.53.   *Ans.* It does not exist

3.78. Find the modal reaction time in Problem 3.54.   *Ans.* 0·53 seconds

3.79. Find the mode of the set of numbers in Problem 3.55.   *Ans.* 10

3.80. Find the mode of the maximum loads of the cables of Problem 3.59.   *Ans.* 110·6 kN

3.81. Find the mode $\hat{X}$ for the distribution in Problem 3.60.   *Ans.* 462

3.82. Find the modal diameter of the rivet heads in Problem 3.61.   *Ans.* 7·2632 mm

3.83. Find the mode of the distribution in Problem 3.62.   *Ans.* 23·5

3.84. Find the modal lifetime of the tubes in Problem 2.20, Chap. 2.   *Ans.* 668·7 hours

3.85. Is it possible to determine the mode for the distributions of
(a) Prob. 3.73 of this chapter? (b) Prob. 2.31 of Chap. 2? Give reasons for your answer.

3.86. Use the empirical formula, Mean − Mode = 3(Mean − Median), to compute the mode for the distributions of (a) Prob. 3.59, (b) Prob. 3.60, (c) Prob. 3.61, (d) Prob. 3.62, (e) Prob. 2.20 of Chap. 2. Compare results with those obtained from Formula (9), Page 48, explaining any agreement or disagreement.

3.87. Prove the statement made at the end of Problem 3.32.

THE GEOMETRIC MEAN

3.88. Find the geometric mean of the numbers (a) 4·2 and 16·8, (b) 3·00 and 6·00.   *Ans.* (a) 8·4, (b) 4·23

3.89. Find (a) the geometric mean G and (b) the arithmetic mean $\bar{X}$ of the numbers 2, 4, 8, 16, 32.
Ans. (a) $G = 8$, (b) $\bar{X} = 12·4$

3.90. Find the geometric mean of the numbers (a) 3, 5, 8, 3, 7, 2 and (b) 28·5, 73·6, 47·2, 31·5, 64·8.
Ans. (a) 4·14, (b) 45·8

3.91. Find the geometric mean for the distributions in (a) Prob. 59, (b) Prob. 60. Verify that the geometric mean is less than or equal to the arithmetic mean for these cases.   *Ans.* (a)110·7 kN, (b) 499·5

3.92. If the price of a commodity doubles in a period of 4 years, what is the average percentage increase per year?
Ans. 18·9%

3.93. In 1950 and 1960 the population of the United States (including Alaska and Hawaii) was 151·3 and 179·3 million respectively. (a) What was the average percentage increase per year? (b) Estimate the population in 1954. (c) If the average percentage increase of population per year from 1960 to 1970 is the same as in (a) what would be the population in 1970?   *Ans.* (a) 1·71%, (b) 161·9 million, (c) 212·5 million

3.94. A principal of £1000 is invested at 4% annual rate of interest. What will be the total amount after 6 years if the original principal is not withdrawn?   *Ans.* £1265·30

3.95. If in the previous problem interest is compounded quarterly (i.e. there is a 1% increase in the money every 3 months), what will be the total amount after 6 years?   *Ans.* £1269·70

3.96. Find two numbers whose arithmetic mean is 9·0 and whose geometric mean is 7·2.   *Ans.* 3·6, 14·4

THE HARMONIC MEAN

3.97. Find the harmonic mean of the numbers: (a) 2, 3, 6; (b) 3·2, 5·2, 4·8, 6·1, 4·2.   *Ans.* (a) 3·0, (b) 4·48

3.98. Find the (a) arithmetic mean, (b) geometric mean and (c) harmonic mean of the numbers 0, 2, 4 and 6.
Ans. (a) 3, (b) 0, (c) 0

3.99. If $X_1, X_2, X_3, \ldots$ represent the class marks in a frequency distribution with corresponding class frequencies $f_1, f_2, f_3, \ldots$ respectively, prove that the harmonic mean H of the distribution is given by

$$\frac{1}{H} = \frac{1}{N} \left(\frac{f_1}{X_1} + \frac{f_2}{X_2} + \frac{f_3}{X_3} + \ldots \right) = \frac{1}{N} \sum \frac{f}{X} \qquad \text{where } N = f_1 + f_2 + \ldots = \Sigma f.$$

3.100. Use the preceding problem to find the harmonic mean of the distributions in (a) Prob. 3.59, (b) Prob. 3.60. Compare with Prob. 3.91.   *Ans.* (a) 110·4, (b) 498·2

3.101. Cities A, B and C are equidistant from each other. A motorist travels from A to B at 30 km/h, from B to C at 40 km/h, and from C to A at 50 km/h. Determine his average speed for the entire trip.   *Ans.* 38·3 km/h

3.102. (a) An airplane travels distances of d_1, d_2 and d_3 km/h at speeds v_1, v_2 and v_3 km/h respectively. Show that the average speed is given by V, where $\dfrac{d_1 + d_2 + d_3}{V} = \dfrac{d_1}{v_1} + \dfrac{d_2}{v_2} + \dfrac{d_3}{v_3}$. This is a weighted harmonic mean.
(b) Find V if $d_1 = 2500$, $d_2 = 1200$, $d_3 = 500$, $v_1 = 500$, $v_2 = 400$, $v_3 = 250$.   *Ans.* (b) 420 km/h

3.103. Prove that the geometric mean of two positive numbers a and b is (a) less than or equal to the arithmetic mean, (b) greater than or equal to the harmonic mean of the numbers. Can you extend the proof to more than two numbers?

ROOT MEAN SQUARE OR QUADRATIC MEAN

3.104. Find the root mean square or quadratic mean of the numbers: (a) 11, 23, 35; (b) 2·7, 3·8, 3·2, 4·3.
   Ans. (a) 25, (b) 3·55

3.105. Show that the root mean square of two positive numbers a and b is (a) greater than or equal to the arithmetic mean, (b) greater than or equal to the harmonic mean. Can you extend the proof to more than two numbers?

3.106. Derive a formula which can be used to find the root mean square for grouped data and apply it to one of the frequency distributions already considered.

Table 3.13

Mark	Number of Students
90–100	9
80–89	32
70–79	43
60–69	21
50–59	11
40–49	3
30–39	1
Total 120	

QUARTILES, DECILES, PERCENTILES

3.107. Table 3.13 shows a frequency distribution of marks on a final examination in college algebra.

(a) Find the quartiles of the distribution and (b) interpret clearly the significance of each.
   Ans. (a) Lower quartile $= Q_1 = 67$,
   middle quartile $= Q_2 =$ median $= 75$,
   upper quartile $= Q_3 = 83$.
   (b) 25% scored 67 or lower (or 75% scored 67 or higher), 50% scored 75 or lower (or 50% scored 75 or higher), 75% scored 83 or lower (or 25% scored 83 or higher).

3.108. Find the quartiles Q_1, Q_2, Q_3 for the distributions in (a) Prob. 3.59, (b) Prob. 3.60, (c) Prob. 2.31 of Chap. 2. Interpret clearly the significance of each.
   Ans. (a) $Q_1 = 105\cdot5$, $Q_2 = 110\cdot7$, $Q_3 = 115\cdot7$ kN, (b) $Q_1 = 469\cdot3$, $Q_2 = 490\cdot6$, $Q_3 = 523\cdot3$.
   (c) $Q_1 = \$1667$, $Q_2 = \$3608$, $Q_3 = \$5268$.

3.109. Find (a) the 2nd decile, (b) the 4th decile, (c) the 90th percentile and (d) the 68th percentile for the data of Prob. 3.73, interpreting clearly the significance of each.   *Ans.* (a) 32·4, (b) 40·9, (c) 68·5, (d) 53·4

3.110. Find (a) P_{10}, (b) P_{90}, (c) P_{25} and (d) P_{75} for the data of Prob. 3.59, interpreting clearly the significance of each.
   Ans. (a) 10·15, (b) 11·78, (c) 10·55, (d) 11·57 kN

3.111. (a) Can all quartiles and deciles be expressed as percentiles? (b) Can all quantiles be expressed as percentiles? Explain.

3.112. For the data of Prob. 3.107 determine (a) the lowest mark scored by the top 25% of the class, (b) the highest mark scored by the lowest 20% of the class. Interpret your answers in terms of percentiles.   *Ans.* (a) 83, (b) 64

3.113. Interpret the results of Prob. 3.107 graphically by use of (a) a percentage histogram, (b) a percentage frequency polygon, (c) a percentage ogive.

3.114. Answer Prob. 3.113 for the results of Prob. 3.108.

3.115. (a) Develop a formula similar to that of equation (8), Page 47, for computing any percentile from a frequency distribution. (b) Illustrate the use of the formula by applying it to obtain the results of Prob. 3.110.

CHAPTER 4

The Standard Deviation
and other Measures of Dispersion

DISPERSION OR VARIATION

The degree to which numerical data tend to spread about an average value is called the *variation* or *dispersion* of the data. Various measures of dispersion or variation are available, the most common being the range, mean deviation, semi-interquartile range, 10–90 percentile range, and the standard deviation.

THE RANGE

The range of a set of numbers is the difference between the largest and smallest numbers in the set.

Example: The range of the set 2, 3, 3, 5, 5, 5, 8, 10, 12 is $12 - 2 = 10$. Sometimes the range is given by simply quoting the smallest and largest numbers. In the above example, for instance, the range could be indicated as 2 to 12 or 2–12.

THE MEAN DEVIATION, OR AVERAGE DEVIATION, of a set of N numbers $X_1, X_2, \ldots, X_N$ is defined by

$$\text{Mean Deviation} = \text{M.D.} = \frac{\sum_{j=1}^{N} |X_j - \bar{X}|}{N} = \frac{\Sigma |X - \bar{X}|}{N} = \overline{|X - \bar{X}|} \qquad (1)$$

where $\bar{X}$ is the arithmetic mean of the numbers and $|X_j - \bar{X}|$ is the absolute value of the deviation of X_j from $\bar{X}$. (The *absolute value* of a number is the number without the associated sign and is indicated by two vertical lines placed around the number. Thus $|-4| = 4, |+3| = 3, |6| = 6, |-0.84| = 0.84$.)

Example: Find the mean deviation of the set of numbers 2, 3, 6, 8, 11.

$$\text{Arithmetic Mean} = \bar{X} = \frac{2 + 3 + 6 + 8 + 11}{5} = 6$$

$$\text{Mean Deviation} = \text{M.D.} = \frac{|2 - 6| + |3 - 6| + |6 - 6| + |8 - 6| + |11 - 6|}{5}$$

$$= \frac{|-4| + |-3| + |0| + |2| + |5|}{5} = \frac{4 + 3 + 0 + 2 + 5}{5} = 2.8$$

If $X_1, X_2, \ldots, X_K$ occur with frequencies $f_1, f_2, \ldots, f_K$ respectively, the mean deviation can be written as

$$\text{Mean Deviation} = \text{M.D.} = \frac{\sum_{j=1}^{K} f_j |X_j - \bar{X}|}{N} = \frac{\Sigma f |X - \bar{X}|}{N} = \overline{|X - \bar{X}|} \qquad (2)$$

where $N = \sum_{j=1}^{K} f_j = \Sigma f$. This form is useful for grouped data where the X_j's represent class marks and the f_j's are the corresponding frequencies.

69

Occasionally the mean deviation is defined in terms of absolute deviations from the median or other average instead of the mean. An interesting property of the sum $\sum_{j=1}^{N} |X_j - a|$ is that it is a minimum when a is the median, i.e. the mean deviation about the median is a minimum.

Note that it would be more appropriate to use the terminology, *mean absolute deviation* rather than mean deviation.

THE SEMI-INTERQUARTILE RANGE OR QUARTILE DEVIATION of a set of data is defined by

$$\text{Semi-interquartile Range} \; = \; Q \; = \; \frac{Q_3 - Q_1}{2} \tag{3}$$

where Q_1 and Q_3 are the first and third quartiles for the data. See Problems 4.6 and 4.7. The interquartile range $Q_3 - Q_1$ is sometimes used but the semi-interquartile range is more common as a measure of dispersion.

THE 10–90 PERCENTILE RANGE of a set of data is defined by

$$\text{10–90 Percentile Range} = P_{90} - P_{10} \tag{4}$$

where P_{10} and P_{90} are the 10th and 90th percentiles for the data (see Prob. 4.8). The semi-10-90 percentile range, $\frac{1}{2}(P_{90} - P_{10})$, can also be used but is not commonly employed.

THE STANDARD DEVIATION of a set of N numbers $X_1, X_2, \ldots, X_N$ is denoted by s and is defined by

$$s \; = \; \sqrt{\frac{\sum_{j=1}^{N} (X_j - \bar{X})^2}{N}} \; = \; \sqrt{\frac{\Sigma (X - \bar{X})^2}{N}} \; = \; \sqrt{\frac{\Sigma x^2}{N}} \; = \; \sqrt{\overline{(X - \bar{X})^2}} \tag{5}$$

where x represents the deviations of each of the numbers X_j from the mean $\bar{X}$.

Thus s is the root mean square of the deviations from the mean or, as it is sometimes called, the *root mean square deviation* (see Page 49).

If $X_1, X_2, \ldots, X_K$ occur with frequencies $f_1, f_2, \ldots, f_K$ respectively, the standard deviation can be written as

$$s \; = \; \sqrt{\frac{\sum_{j=1}^{K} f_j (X_j - \bar{X})^2}{N}} \; = \; \sqrt{\frac{\Sigma f (X - \bar{X})^2}{N}} \; = \; \sqrt{\frac{\Sigma f x^2}{N}} \; = \; \sqrt{\overline{(X - \bar{X})^2}} \tag{6}$$

where $N = \sum_{j=1}^{K} f_j = \Sigma f$. In this form it is useful for grouped data.

Sometimes the standard deviation for the data of a sample is defined with $(N - 1)$ replacing N in the denominators of the expressions in (5) and (6) because the resulting value represents a better estimate of the standard deviation of a population from which the sample is taken. For large values of N (certainly $N > 30$) there is practically no difference between the two definitions. Also, when the better estimate is needed we can always obtain it by multiplying the standard deviation computed according to the first definition by $\sqrt{N/(N-1)}$. Hence we shall adhere to the definition given above.

THE VARIANCE

The variance of a set of data is defined as the square of the standard deviation and is thus given by s^2 in (5) and (6).

When it is necessary to distinguish the standard deviation of a population from the standard deviation of a sample drawn from this population, we often use the symbol s for the latter and σ for the former. Thus s^2 and σ^2 would represent the *sample variance* and *population variance* respectively.

SHORT METHODS FOR COMPUTING THE STANDARD DEVIATION

The equations (5) and (6) can be written respectively in the equivalent forms

$$s = \sqrt{\frac{\sum\limits_{j=1}^{N} X_j^2}{N} - \left(\frac{\sum\limits_{j=1}^{N} X_j}{N}\right)^2} = \sqrt{\frac{\sum X^2}{N} - \left(\frac{\sum X}{N}\right)^2} = \sqrt{\overline{X^2} - \bar{X}^2} \tag{7}$$

$$s = \sqrt{\frac{\sum\limits_{j=1}^{K} f_j X_j^2}{N} - \left(\frac{\sum\limits_{j=1}^{K} f_j X_j}{N}\right)^2} = \sqrt{\frac{\sum f X^2}{N} - \left(\frac{\sum f X}{N}\right)^2} = \sqrt{\overline{X^2} - \bar{X}^2} \tag{8}$$

where $\overline{X^2}$ denotes the mean of the squares of the various values of X, while $\bar{X}^2$ denotes the square of the mean of the various values of X. See Problems 4.12 to 4.14.

If $d_j = X_j - A$ are the deviations of X_j from some arbitrary constant A, the results (7) and (8) become respectively

$$s = \sqrt{\frac{\sum\limits_{j=1}^{N} d_j^2}{N} - \left(\frac{\sum\limits_{j=1}^{N} d_j}{N}\right)^2} = \sqrt{\frac{\sum d^2}{N} - \left(\frac{\sum d}{N}\right)^2} = \sqrt{\overline{d^2} - \bar{d}^2} \tag{9}$$

$$s = \sqrt{\frac{\sum\limits_{j=1}^{K} f_j d_j^2}{N} - \left(\frac{\sum\limits_{j=1}^{K} f_j d_j}{N}\right)^2} = \sqrt{\frac{\sum f d^2}{N} - \left(\frac{\sum f d}{N}\right)^2} = \sqrt{\overline{d^2} - \bar{d}^2} \tag{10}$$

See Problems 4.15 and 4.17.

When data are grouped into a frequency distribution whose class intervals have equal size c, we have $d_j = c u_j$ or $X_j = A + c u_j$ and (10) becomes

$$s = c\sqrt{\frac{\sum\limits_{j=1}^{K} f_j u_j^2}{N} - \left(\frac{\sum\limits_{j=1}^{K} f_j u_j}{N}\right)^2} = c\sqrt{\frac{\sum f u^2}{N} - \left(\frac{\sum f u}{N}\right)^2} = c\sqrt{\overline{u^2} - \bar{u}^2} \tag{11}$$

This last formula provides a very short method for computing the standard deviation and should always be used for grouped data when class interval sizes are equal. It is called the *coding method* and is exactly analogous to that used in computing the arithmetic mean for grouped data in Chapter 3. See Problems 4.16 to 4.19.

PROPERTIES OF THE STANDARD DEVIATION

1. The standard deviation can be defined as $\quad s = \sqrt{\dfrac{\sum\limits_{j=1}^{N} (X_j - a)^2}{N}}$

    where a is an average besides the arithmetic mean. Of all such standard deviations, the minimum is that for which $a = \bar{X}$, because of Property (b), Chap. 3, Page 46. This property provides an important reason for defining the standard deviation as above. For a proof of this property see Prob. 4.27.

2. For normal distributions (see Chapter 7) it turns out that:

    (a) 68·27% of the cases are included between $\bar{X} - s$ and $\bar{X} + s$
    (i.e. one standard deviation on either side of the mean)

(b) 95·45% of the cases are included between $\bar{X} - 2s$ and $\bar{X} + 2s$

        (i.e. two standard deviations on either side of the mean)

(c) 99·73% of the cases are included between $\bar{X} - 3s$ and $\bar{X} + 3s$

        (i.e. three standard deviations on either side of the mean)

as indicated in Fig. 4-1.

    For moderately skewed distributions the above percentages may hold approximately (see Prob. 4.24).

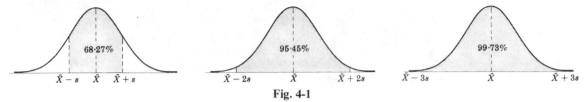

Fig. 4-1

3. Suppose that two sets consisting of N_1 and N_2 numbers (or two frequency distributions with total frequencies N_1 and N_2) have variances given by s_1^2 and s_2^2 respectively and the *same* mean X. Then the *combined* or *pooled variance* of both sets (or both frequency distributions) is given by

$$s^2 = \frac{N_1 s_1^2 + N_2 s_2^2}{N_1 + N_2} \tag{12}$$

Note that this is a weighted arithmetic mean of the variances. This result can be generalized to 3 or more sets.

CHARLIER'S CHECK

    Charlier's check in computations of the mean and standard deviation by the coding method makes use of the identities

$$\Sigma f(u+1) = \Sigma fu + \Sigma f = \Sigma fu + N$$
$$\Sigma f(u+1)^2 = \Sigma f(u^2 + 2u + 1) = \Sigma fu^2 + 2\Sigma fu + \Sigma f = \Sigma fu^2 + 2\Sigma fu + N$$

See Problem 4.20.

SHEPPARD'S CORRECTION FOR VARIANCE

    The computation of the standard deviation is somewhat in error due to grouping of data into classes (grouping error). To adjust for grouping error we use the result

$$\text{Corrected Variance} = \text{Variance from grouped data} - c^2/12 \tag{13}$$

where c is the class interval size. The correction $c^2/12$ which is subtracted is called *Sheppard's correction*. It is used for distributions of continuous variables where the "tails" go gradually to zero in both directions.

    Statisticians differ as to *when* and *whether* Sheppard's corrections should be applied. Certainly they should not be applied without thorough examination of the situation. This is so because often they tend to *overcorrect* and thus replace old errors by new errors. In this book unless otherwise indicated we shall not use these corrections.

EMPIRICAL RELATIONS BETWEEN MEASURES OF DISPERSION

    For moderately skewed distributions we have the empirical formulae

$$\text{Mean Deviation} = \tfrac{4}{5}(\text{Standard Deviation})$$
$$\text{Semi-interquartile Range} = \tfrac{2}{3}(\text{Standard Deviation})$$

    These are consequences of the fact that for the normal distribution we find that the mean deviation and semi-interquartile range are equal respectively to 0·7979 and 0·6745 times the standard deviation.

ABSOLUTE AND RELATIVE DISPERSION. COEFFICIENT OF VARIATION

The actual variation or dispersion as determined from the standard deviation or other measure of dispersion is called the *absolute dispersion*. However, a variation or dispersion of 1 metre in measuring a distance of 1000 metres is quite different in effect from the same variation of 1 metre in a distance of 20 metres. A measure of this effect is supplied by the *relative dispersion* defined by

$$\text{Relative Dispersion} = \frac{\text{Absolute Dispersion}}{\text{Average}} \tag{14}$$

If the absolute dispersion is the standard deviation s and the average is the mean $\bar{X}$, the relative dispersion is called the *coefficient of variation* or *coefficient of dispersion* given by

$$\text{Coefficient of Variation} = V = \frac{s}{\bar{X}} \tag{15}$$

and is generally expressed as a percentage. Other possibilities also occur (see Problem 4.30).

Note that the coefficient of variation is independent of units used. For this reason it is useful in comparing distributions where units may be different. A disadvantage of the coefficient of variation is that it fails to be useful when $\bar{X}$ is close to zero.

STANDARDIZED VARIABLE, STANDARD SCORES

The variable
$$z = \frac{X - \bar{X}}{s} \tag{16}$$

which measures the deviation from the mean in units of the standard deviation is called a *standardized variable* and is a dimensionless quantity (i.e. is independent of units used).

If deviations from the mean are given in units of the standard deviation, they are said to be expressed in *standard units* or *standard scores*. These are of great value in comparison of distributions (see Problem 4.31).

Solved Problems

THE RANGE

4.1. Find the range of each set of numbers: (*a*) 12, 6, 7, 3, 15, 10, 18, 5; (*b*) 9, 3, 8, 8, 9, 8, 9, 18.

Solution:

In both cases, range = largest number − smallest number = 18 − 3 = 15. However, as seen from the arrays of (*a*) and (*b*),

(*a*) 3, 5, 6, 7, 10, 12, 15, 18   (*b*) 3, 8, 8, 8, 9, 9, 9, 18

there is much more variation or dispersion in (*a*) than in (*b*). In fact, (*b*) consists mainly of 8's and 9's.

Since the range indicates no difference between the sets, it is not a very good measure of dispersion in this case. In general, where extreme values are present the range is a poor measure of dispersion.

An improvement is achieved by throwing out the extreme cases 3 and 18. Then for (*a*) the range is (15 − 5) = 10, while for (*b*) the range is (9 − 8) = 1, indicating clearly the fact that (*a*) has greater dispersion than (*b*). However, this is not the way the range is defined. The semi-interquartile range and the 10–90 percentile range were designed to improve on the range by eliminating extreme cases.

4.2.   Find the range of masses of the students at XYZ University as given in Table 2.1, Page 27.

Solution:

There are two ways of defining the range for grouped data.

Method 1:

Range = Class mark of highest class — class mark of lowest class

$= 73 - 61 = 12$ kg

Method 2:

Range = Upper class boundary of highest class — lower class boundary of lowest class

$= 74 \cdot 5 - 59 \cdot 5 = 15$ kg

Method 1 tends to eliminate extreme cases to some extent.

THE MEAN DEVIATION

4.3.   Find the mean deviation of the sets of numbers in Problem 4.1.

Solution:

(a) Arithmetic mean $= \bar{X} = \dfrac{12 + 6 + 7 + 3 + 15 + 10 + 18 + 5}{8} = \dfrac{76}{8} = 9 \cdot 5$

Mean deviation = M.D. $= \dfrac{\Sigma |X - \bar{X}|}{N}$

$= \dfrac{|12 - 9 \cdot 5| + |6 - 9 \cdot 5| + |7 - 9 \cdot 5| + |3 - 9 \cdot 5| + |15 - 9 \cdot 5| + |10 - 9 \cdot 5| + |18 - 9 \cdot 5| + |5 - 9 \cdot 5|}{8}$

$= \dfrac{2 \cdot 5 + 3 \cdot 5 + 2 \cdot 5 + 6 \cdot 5 + 5 \cdot 5 + 0 \cdot 5 + 8 \cdot 5 + 4 \cdot 5}{8} = \dfrac{34}{8} = 4 \cdot 25$

(b) Arithmetic mean $= \bar{X} = \dfrac{9 + 3 + 8 + 8 + 9 + 8 + 9 + 18}{8} = \dfrac{72}{8} = 9$

Mean deviation = M.D. $= \dfrac{\Sigma |X - \bar{X}|}{N}$

$= \dfrac{|9 - 9| + |3 - 9| + |8 - 9| + |8 - 9| + |9 - 9| + |8 - 9| + |9 - 9| + |18 - 9|}{8}$

$= \dfrac{0 + 6 + 1 + 1 + 0 + 1 + 0 + 9}{8} = 2 \cdot 25$

The mean deviation indicates that set (b) shows less dispersion than set (a), as it should.

4.4.   Find the mean deviation of the masses of the 100 male students at XYZ University (see Table 3.2, Page 55).

Solution:

From Problem 3.20 of Chapter 3, arithmetic mean $= \bar{X} = 67 \cdot 45$ kg.

The work can be arranged as in Table 4.1.

Table 4.1

Mass (kg)	Class mark X	$|X - \bar{X}| = |X - 67 \cdot 45|$	Frequency f	$f|X - \bar{X}|$
60–62	61	6·45	5	32·25
63–65	64	3·45	18	62·10
66–68	67	0·45	42	18·90
69–71	70	2·55	27	68·85
72–74	73	5·55	8	44·40
			$N = \Sigma f = 100$	$\Sigma f|X - \bar{X}| = 226 \cdot 50$

Mean deviation = M.D. $= \dfrac{\Sigma f|X - \bar{X}|}{N} = \dfrac{226 \cdot 50}{100} = 2 \cdot 26$ kg

It is possible to devise a coding method for computing the mean deviation (see Problem 4.47).

4.5. Determine the percentage of the students' heights in Problem 4.4 which fall within the ranges (a) $\bar{X} \pm$ M.D., (b) $\bar{X} \pm 2$ MD., (c) $\bar{X} \pm 3$ M.D.

Solution:

(a) $\bar{X} \pm$ M.D. $= 67 \cdot 45 \pm 2 \cdot 26$ is the range from $65 \cdot 19$ kg to $69 \cdot 71$ kg

This range includes all individuals in the third class $+ \frac{1}{3}(65 \cdot 5 - 65 \cdot 19)$ of the students in the second class $+ \frac{1}{3}(69 \cdot 71 - 68 \cdot 5)$ of the students in the fourth class (since class interval size $= 3$ kg, upper class boundary of second class $= 65 \cdot 5$ kg, and lower class boundary of fourth class $= 68 \cdot 5$ kg).

Number of students in range $\bar{X} \pm$ M.D. is

$$42 + \frac{0 \cdot 31}{3}(18) + \frac{1 \cdot 21}{3}(27) = 42 + 1 \cdot 86 + 10 \cdot 89 = 54 \cdot 75, \text{ or } 55$$

which is 55% of the total.

(b) $\bar{X} \pm 2$ M.D. $= 67 \cdot 45 \pm 2(2 \cdot 26) = 67 \cdot 45 \pm 4 \cdot 52$ is the range from $62 \cdot 93$ kg to $71 \cdot 97$ kg

Number of students in range $\bar{X} \pm 2$ M.D. is

$$18 - \left(\frac{62 \cdot 93 - 62 \cdot 5}{3}\right)(18) + 42 + 27 + \left(\frac{71 \cdot 97 - 71 \cdot 5}{3}\right)(8) = 85 \cdot 67, \text{ or } 86$$

which is 86% of the total.

(c) $\bar{X} \pm 3$ M.D. $= 67 \cdot 45 \pm 3(2 \cdot 26) = 67 \cdot 45 \pm 6 \cdot 78$ is the range from $60 \cdot 67$ kg to $74 \cdot 23$ kg

Number of students in range $\bar{X} \pm 3$ M.D. is

$$5 - \left(\frac{60 \cdot 67 - 59 \cdot 5}{3}\right)(5) + 18 + 42 + 27 + \left(\frac{74 \cdot 5 - 74 \cdot 23}{3}\right)(8) = 97 \cdot 33, \text{ or } 97$$

which is 97% of the total.

SEMI-INTERQUARTILE RANGE OR QUARTILE DEVIATION

4.6. Find the semi-interquartile range for the mass distribution of the students at XYZ University (see Table 4.1 of Problem 4.4).

Solution:

Lower and upper quartiles are $Q_1 = 65 \cdot 5 + \frac{2}{42}(3) = 65 \cdot 64$ kg, $Q_3 = 68 \cdot 5 + \frac{10}{27}(3) = 69 \cdot 61$ kg.

Semi-interquartile range or quartile deviation is $Q = \frac{1}{2}(Q_3 - Q_1) = \frac{1}{2}(69 \cdot 61 - 65 \cdot 64) = 1 \cdot 98$ kg.

Note that 50% of the cases lie between Q_1 and Q_3, i.e. 50 students have masses between $65 \cdot 64$ kg and $69 \cdot 61$ kg.

We can consider $\frac{1}{2}(Q_1 + Q_3) = 67 \cdot 63$ kilogrammes as a measure of central tendency, i.e. average mass. It follows that 50% of the masses lie in the range $(67 \cdot 63 \pm 1 \cdot 98)$ kilogrammes.

4.7. Find the semi-interquartile range for the wages of the 65 employees at the P and R Company. See Problem 2.3, Chapter 2, Page 33.

Solution:

From Problem 3.44, Chapter 3, $Q_1 = £68 \cdot 25$ and $Q_3 = £90 \cdot 75$.
Then the semi-interquartile range $Q = \frac{1}{2}(Q_3 - Q_1) = \frac{1}{2}(£90 \cdot 75 - £68 \cdot 25) = £11 \cdot 25$.
Since $\frac{1}{2}(Q_1 + Q_3) = £79 \cdot 50$, we can conclude that 50% of the employees earn wages lying in the range $£79 \cdot 50 \pm £11 \cdot 25$.

10–90 PERCENTILE RANGE

4.8. Find the 10–90 percentile range of masses of the students at XYZ University. See Table 2.1, Page 27.

Solution:

Here $P_{10} = 62 \cdot 5 + \frac{5}{18}(3) = 63 \cdot 33$ kg and $P_{90} = 68 \cdot 5 + \frac{25}{27}(3) = 71 \cdot 27$ kg.
Then 10–90 percentile range $= P_{90} - P_{10} = 71 \cdot 27 - 63 \cdot 33 = 7 \cdot 94$ kg.
Since $\frac{1}{2}(P_{10} + P_{90}) = 67 \cdot 30$ kg and $\frac{1}{2}(P_{90} - P_{10}) = 3 \cdot 97$ kg, we can conclude that 80% of the students have masses in the range $(67 \cdot 30 \pm 3 \cdot 97)$ kilogrammes.

THE STANDARD DEVIATION

4.9. Find the standard deviation of each set of numbers in Problem 4.1.

Solution:

(a) Arithmetic mean $= \bar{X} = \dfrac{\Sigma X}{N} = \dfrac{12 + 6 + 7 + 3 + 15 + 10 + 18 + 5}{8} = \dfrac{76}{8} = 9 \cdot 5$

$$s = \sqrt{\frac{\Sigma (X - \bar{X})^2}{N}}$$

$$= \sqrt{\frac{(12 - 9 \cdot 5)^2 + (6 - 9 \cdot 5)^2 + (7 - 9 \cdot 5)^2 + (3 - 9 \cdot 5)^2 + (15 - 9 \cdot 5)^2 + (10 - 9 \cdot 5)^2 + (18 - 9 \cdot 5)^2 + (5 - 9 \cdot 5)^2}{8}}$$

$$= \sqrt{23 \cdot 75} = 4 \cdot 87.$$

(b) Arithmetic mean $= \bar{X} = \dfrac{9 + 3 + 8 + 8 + 9 + 8 + 9 + 18}{8} = \dfrac{72}{8} = 9$

$$s = \sqrt{\frac{\Sigma (X - \bar{X})^2}{N}}$$

$$= \sqrt{\frac{(9 - 9)^2 + (3 - 9)^2 + (8 - 9)^2 + (8 - 9)^2 + (9 - 9)^2 + (8 - 9)^2 + (9 - 9)^2 + (18 - 9)^2}{8}}$$

$$= \sqrt{15} = 3 \cdot 87.$$

The above results should be compared with those of Problem 4.3. It will be noted that the standard deviation does indicate that the set (b) shows less dispersion than set (a). However, the effect is masked by the fact that extreme values affect the standard deviation much more than the mean deviation. This is of course to be expected, since the deviations are squared in computing the standard deviation.

4.10. Find the variance of the sets of numbers in Problem 4.1.

Solution:

Variance $= s^2$. Then from Problem 4.9 we have: (a) $s^2 = 23 \cdot 75$, (b) $s^2 = 15$.

4.11. Find the standard deviation of the masses of the 100 male students at XYZ University. See Table 2.1, Page 27.

Solution:

From Problems 3.15, 3.20 or 3.22 of Chapter 3, $\bar{X} = 67 \cdot 45$ kg. The work can be arranged as in Table 4.2 below.

Table 4.2

Mass (kg)	Class mark X	$X - \bar{X} = X - 67 \cdot 45$	$(X - \bar{X})^2$	Frequency f	$f(X - \bar{X})^2$
60–62	61	$-6 \cdot 45$	$41 \cdot 6025$	5	$208 \cdot 0125$
63–65	64	$-3 \cdot 45$	$11 \cdot 9025$	18	$214 \cdot 2450$
66–68	67	$-0 \cdot 45$	$0 \cdot 2025$	42	$8 \cdot 5050$
69–71	70	$2 \cdot 55$	$6 \cdot 5025$	27	$175 \cdot 5675$
72–74	73	$5 \cdot 55$	$30 \cdot 8025$	8	$246 \cdot 4200$
				$N = \Sigma f = 100$	$\Sigma f(X - \bar{X})^2$ $= 852 \cdot 7500$

$$s = \sqrt{\frac{\Sigma f(X - \bar{X})^2}{N}} = \sqrt{\frac{852 \cdot 7500}{100}} = \sqrt{8 \cdot 5275} = 2 \cdot 92 \text{ kilogrammes}$$

COMPUTATIONS OF STANDARD DEVIATION FROM GROUPED DATA

4.12. (a) Prove that $s = \sqrt{\dfrac{\Sigma X^2}{N} - \left(\dfrac{\Sigma X}{N}\right)^2} = \sqrt{\overline{X^2} - \bar{X}^2}$.

(b) Use the formula in (a) to find the standard deviation of the set of numbers 12, 6, 7, 3, 15, 10, 18, 5.

Solution:

(a) By definition,

$$s = \sqrt{\frac{\Sigma\,(X - \bar{X})^2}{N}}$$

Then

$$s^2 = \frac{\Sigma\,(X - \bar{X})^2}{N} = \frac{\Sigma\,(X^2 - 2\bar{X}X + \bar{X}^2)}{N} = \frac{\Sigma X^2 - 2\bar{X}\,\Sigma X + N\bar{X}^2}{N}$$

$$= \frac{\Sigma X^2}{N} - 2\bar{X}\frac{\Sigma X}{N} + \bar{X}^2 = \frac{\Sigma X^2}{N} - 2\bar{X}^2 + \bar{X}^2 = \frac{\Sigma X^2}{N} - \bar{X}^2$$

$$= \overline{X^2} - \bar{X}^2 = \frac{\Sigma X^2}{N} - \left(\frac{\Sigma X}{N}\right)^2$$

or

$$s = \sqrt{\frac{\Sigma X^2}{N} - \left(\frac{\Sigma X}{N}\right)^2} = \sqrt{\overline{X^2} - \bar{X}^2}$$

Note that in the above summations we have used the abbreviated form with X replacing X_j and Σ replacing $\sum\limits_{j=1}^{N}$.

Another method: $s^2 = \overline{(X - \bar{X})^2} = \overline{X^2 - 2X\bar{X} + \bar{X}^2} = \overline{X^2} - \overline{2X\bar{X}} + \overline{\bar{X}^2}$

$$= \overline{X^2} - 2\bar{X}\bar{X} + \bar{X}^2 = \overline{X^2} - \bar{X}^2$$

(b) $X^2 = \dfrac{\Sigma X^2}{N} = \dfrac{(12)^2 + (6)^2 + (7)^2 + (3)^2 + (15)^2 + (10)^2 + (18)^2 + (5)^2}{8} = \dfrac{912}{8} = 114$

$\bar{X} = \dfrac{\Sigma X}{N} = \dfrac{12 + 6 + 7 + 3 + 15 + 10 + 18 + 5}{8} = \dfrac{76}{8} = 9{\cdot}5$

Then $\qquad s = \sqrt{\overline{X^2} - \bar{X}^2} = \sqrt{114 - 90{\cdot}25} = \sqrt{23{\cdot}75} = 4{\cdot}87$

This method should be compared with that of Problem 4.9(a).

4.13. Modify the formula of Problem 4.12(a) to allow for frequencies corresponding to the various values of X.

Solution:

The appropriate modification is $\quad s = \sqrt{\dfrac{\Sigma\,fX^2}{N} - \left(\dfrac{\Sigma\,fX}{N}\right)^2} = \sqrt{\overline{X^2} - \bar{X}^2}$.

This can be established as in Problem 4.12(a) by starting with $s = \sqrt{\dfrac{\Sigma f\,(X - \bar{X})^2}{N}}$. Then

$$s^2 = \frac{\Sigma\,f(X - \bar{X})^2}{N} = \frac{\Sigma\,f(X^2 - 2\bar{X}X + \bar{X}^2)}{N} = \frac{\Sigma fX^2 - 2\bar{X}\,\Sigma fX + \bar{X}^2\,\Sigma f}{N}$$

$$= \frac{\Sigma fX^2}{N} - 2\bar{X}\frac{\Sigma fX}{N} + \bar{X}^2 = \frac{\Sigma fX^2}{N} - 2\bar{X}^2 + \bar{X}^2 = \frac{\Sigma fX^2}{N} - \bar{X}^2$$

$$= \frac{\Sigma fX^2}{N} - \left(\frac{\Sigma fX}{N}\right)^2 \qquad \text{or} \qquad s = \sqrt{\frac{\Sigma fX^2}{N} - \left(\frac{\Sigma fX}{N}\right)^2}$$

Note that in the above summations we have used the abbreviated form with X and f replacing X_j and f_j, Σ replacing $\sum\limits_{j=1}^{K}$, and $\sum\limits_{j=1}^{K} f_j = N$.

4.14. Using the formula of Problem 4.13, find the standard deviation for the data of Problem 4.11.

Solution:
The work can be arranged as in Table 4.3.

Table 4.3

Mass (kilogrammes)	Class mark X	X^2	Frequency f	fX^2
60–62	61	3721	5	18 605
63–65	64	4096	18	73 728
66–68	67	4489	42	188 538
69–71	70	4900	27	132 300
72–74	73	5329	8	42 632
			$N = \Sigma f = 100$	$\Sigma fX^2 = 455\ 803$

$$s = \sqrt{\frac{\Sigma fX^2}{N} - \left(\frac{\Sigma fX}{N}\right)^2} = \sqrt{\frac{455\ 803}{100} - (67{\cdot}45)^2} = \sqrt{8{\cdot}5275} = 2{\cdot}92 \text{ kg}$$

where $\bar{X} = \dfrac{\Sigma fX}{N} = 67{\cdot}45$ kg, as obtained in Prob. 3.15 of Chap. 3.

Note that this method, as that of Problem 4.11, entails much tedious computation. In Problem 4.17 it is shown how coding method simplifies the calculations immensely.

4.15. If $d = X - A$ are the deviations of X from an arbitrary constant A, prove that

$$s = \sqrt{\frac{\Sigma fd^2}{N} - \left(\frac{\Sigma fd}{N}\right)^2}$$

Solution:
Since $d = X - A$, $X = A + d$ and $\bar{X} = A + \bar{d}$ as in Prob. 3.18, Chap. 3. Then

$$X - \bar{X} = (A + d) - (A + \bar{d}) = d - \bar{d}$$

so that $\quad s = \sqrt{\dfrac{\Sigma f(X - \bar{X})^2}{N}} = \sqrt{\dfrac{\Sigma f(d - \bar{d})^2}{N}} = \sqrt{\dfrac{\Sigma fd^2}{N} - \left(\dfrac{\Sigma fd}{N}\right)^2}$

using the result of Problem 4.13 with X and $\bar{X}$ replaced by d and $\bar{d}$ respectively.

Another method:

$$s^2 = \overline{(X - \bar{X})^2} = \overline{(d - \bar{d})^2} = \overline{d^2 - 2\bar{d}d + \bar{d}^2}$$
$$= \overline{d^2} - 2\bar{d}^2 + \bar{d}^2 = \overline{d^2} - \bar{d}^2 = \frac{\Sigma fd^2}{N} - \left(\frac{\Sigma fd}{N}\right)^2$$

and the result follows on taking the positive square root.

4.16. Show that if each class mark X in a frequency distribution having class intervals of equal size c is coded into a corresponding value u according to the relation $X = A + cu$, where A is a given class mark, then the standard deviation can be written as

$$s = c\sqrt{\frac{\Sigma fu^2}{N} - \left(\frac{\Sigma fu}{N}\right)^2} = c\sqrt{\overline{u^2} - \bar{u}^2}$$

Solution:
This follows at once from the preceding problem, since $d = X - A = cu$. Thus since c is a constant,

$$s = \sqrt{\frac{\Sigma f(cu)^2}{N} - \left(\frac{\Sigma f(cu)}{N}\right)^2} = \sqrt{c^2\frac{\Sigma fu^2}{N} - c^2\left(\frac{\Sigma fu}{N}\right)^2} = c\sqrt{\frac{\Sigma fu^2}{N} - \left(\frac{\Sigma fu}{N}\right)^2}$$

Another method:

We can also prove the result directly without using Problem 4.15.

Since $X = A + cu$, $\bar{X} = A + c\bar{u}$ and $X - \bar{X} = c(u - \bar{u})$. Then

$$s^2 = \overline{(X - \bar{X})^2} = \overline{c^2(u - \bar{u})^2} = c^2\overline{(u^2 - 2\bar{u}u + \bar{u}^2)} = c^2(\overline{u^2} - 2\bar{u}^2 + \bar{u}^2) = c^2(\overline{u^2} - \bar{u}^2)$$

and

$$s = c\sqrt{\overline{u^2} - \bar{u}^2} = c\sqrt{\frac{\Sigma fu^2}{N} - \left(\frac{\Sigma fu}{N}\right)^2}$$

4.17. Find the standard deviation of the masses of the students at XYZ University by using (*a*) the formula derived in Problem 4.15, (*b*) the coding method of Problem 4.16.

Solution:

In the Tables 4.4 and 4.5 below, we have arbitrarily chosen A as equal to the class mark 67. Note that in Table 4.4 the deviations $d = X - A$ are all multiples of the class interval size $c = 3$. This factor is removed in Table 4.5. As a result, computations in Table 4.5 are greatly simplified. They should be compared with those of Problems 4.11 and 4.14. For this reason the coding method should be used wherever possible.

(*a*)   **Table 4.4**

Class mark X	$d = X - A$	Frequency f	fd	fd^2
61	−6	5	−30	180
64	−3	18	−54	162
$A \longrightarrow$ 67	0	42	0	0
70	3	27	81	243
73	6	8	48	288
		$N = \Sigma f = 100$	$\Sigma fd = 45$	$\Sigma fd^2 = 873$

$$s = \sqrt{\frac{\Sigma fd^2}{N} - \left(\frac{\Sigma fd}{N}\right)^2} = \sqrt{\frac{873}{100} - \left(\frac{45}{100}\right)^2} = \sqrt{8 \cdot 5275} = 2 \cdot 92 \text{ kg}$$

(*b*)   **Table 4.5**

Class mark X	$u = \dfrac{X - A}{c}$	Frequency f	fu	fu^2
61	−2	5	−10	20
64	−1	18	−18	18
$A \longrightarrow$ 67	0	42	0	0
70	1	27	27	27
73	2	8	16	32
		$N = \Sigma f = 100$	$\Sigma fu = 15$	$\Sigma fu^2 = 97$

$$s = c\sqrt{\frac{\Sigma fu^2}{N} - \left(\frac{\Sigma fu}{N}\right)^2} = 3\sqrt{\frac{97}{100} - \left(\frac{15}{100}\right)^2} = \sqrt{0 \cdot 9475} = 2 \cdot 92 \text{ kg}$$

4.18. Find (*a*) the mean and (*b*) the standard deviation for the distribution of wages of the 65 employees at the P and R Company using coding methods. (See Problem 2.3, Chapter 2.)

Solution:

The work can be arranged simply as shown in Table 4.6.

Table 4.6

X	u	f	fu	fu^2
£55·00	−2	8	−16	32
65·00	−1	10	−10	10
75·00	0	16	0	0
85·00	1	14	14	14
95·00	2	10	20	40
105·00	3	5	15	45
115·00	4	2	8	32
		$N = \Sigma f = 65$	$\Sigma fu = 31$	$\Sigma fu^2 = 173$

A points to the row £75·00.

(a) $\bar{X} = A + c\bar{u} = A + c\dfrac{\Sigma fu}{N} = £75\cdot00 + (£10\cdot00)\left(\dfrac{31}{65}\right)$ £79·99

(b) $s = c\sqrt{\overline{u^2} - \bar{u}^2} = c\sqrt{\dfrac{\Sigma fu^2}{N} - \left(\dfrac{\Sigma fu}{N}\right)^2} = (£10\cdot00)\sqrt{\dfrac{173}{65} - \left(\dfrac{31}{65}\right)^2} = (£10\cdot00)\sqrt{2\cdot4341} = £15\cdot60$

4.19. Table 4.7 shows the I.Q.'s of 480 school children at a certain elementary school. Find (a) the mean and (b) the standard deviation using the coding method.

Table 4.7

Class mark X	70	74	78	82	86	90	94	98	102	106	110	114	118	122	126
Frequency f	4	9	16	28	45	66	85	72	54	38	27	18	11	5	2

Solution:

The intelligence quotient (I.Q.) $= \dfrac{\text{mental age}}{\text{chronological age}}$, expressed as a percentage.

For example an 8-year-old child who according to certain educational procedures has a mentality equivalent to that of a 10-year-old child, would have an I.Q. $= 10/8 = 1\cdot25 = 125\%$ or simply 125, the % being understood.

To find the mean and standard deviation of the intelligence quotients, the work can be arranged as in Table 4.8.

Table 4.8

X	u	f	fu	fu^2
70	−6	4	−24	144
74	−5	9	−45	225
78	−4	16	−64	256
82	−3	28	−84	252
86	−2	45	−90	180
90	−1	66	−66	66
94	0	85	0	0
98	1	72	72	72
102	2	54	108	216
106	3	38	114	342
110	4	27	108	432
114	5	18	90	450
118	6	11	66	396
122	7	5	35	245
126	8	2	16	128
		$N = \Sigma f = 480$	$\Sigma fu = 236$	$\Sigma fu^2 = 3404$

A points to the row 94.

(a) $\bar{X} = A + c\bar{u} = A + c\dfrac{\Sigma\,fu}{N} = 94 + 4\left(\dfrac{236}{480}\right) = 95\!\cdot\!97$

(b) $s = c\sqrt{\bar{u^2} - \bar{u}^2} = c\sqrt{\dfrac{\Sigma\,fu^2}{N} - \left(\dfrac{\Sigma\,fu}{N}\right)^2} = 4\sqrt{\dfrac{3404}{480} - \left(\dfrac{236}{480}\right)^2} = 4\sqrt{6\!\cdot\!8499} = 10\!\cdot\!47$

CHARLIER'S CHECK

4.20. Use Charlier's check to help verify computations of (a) the mean and (b) the standard deviation performed in Problem 4.19.

   To supply the required check, we add the columns of Table 4.9 below to those of Table 4.8, with the exception of column 2 which is repeated here for convenience.

Solution:

(a) $\Sigma\,f(u + 1) = 716$, from Table 4.9 below.
   $\Sigma\,fu + N = 236 + 480 = 716$, from Table 4.8 above.
   This provides the required check on the mean.

(b) $\Sigma\,f(u + 1)^2 = 4356$, from Table 4.9 below.
   $\Sigma fu^2 + 2\,\Sigma\,fu + N = 3404 + 2(236) + 480 = 4356$, from Table 4.8 above.
   This provides the required check on the standard deviation.

Table 4.9

$u + 1$	f	$f(u + 1)$	$f(u + 1)^2$
-5	4	-20	100
-4	9	-36	144
-3	16	-48	144
-2	28	-56	112
-1	45	-45	45
0	66	0	0
1	85	85	85
2	72	144	288
3	54	162	486
4	38	152	608
5	27	135	675
6	18	108	648
7	11	77	539
8	5	40	320
9	2	18	162
$N = \Sigma f = 480$		$\Sigma\,f(u + 1) = 716$	$\Sigma\,f(u + 1)^2 = 4356$

SHEPPARD'S CORRECTION FOR VARIANCE

4.21. Apply Sheppard's correction to determine the standard deviation of the data in (a) Prob. 4.17, (b) Prob. 4.18, (c) Prob. 4.19.

Solution:

(a) $s^2 = 8\!\cdot\!5275$, $c = 3$. Corrected variance $= s^2 - c^2/12 = 8\!\cdot\!5275 - 3^2/12 = 7\!\cdot\!7775$.
   Corrected standard deviation $= \sqrt{\text{corrected variance}} = \sqrt{7\!\cdot\!7775} = 2\!\cdot\!79$ kg.

(b) $s^2 = 243\!\cdot\!41$, $c = 10$. Corrected variance $= s^2 - c^2/12 = 243\!\cdot\!41 - 10^2/12 = 235\!\cdot\!08$.
   Corrected standard deviation $= \sqrt{235\!\cdot\!08} = £15\!\cdot\!33$.

(c) $s^2 = 109\!\cdot\!60$, $c = 4$. Corrected variance $= s^2 - c^2/12 = 109\!\cdot\!60 - 4^2/12 = 108\!\cdot\!27$.
   Corrected standard deviation $= \sqrt{108\!\cdot\!27} = 10\!\cdot\!41$.

4.22. For the second frequency distribution of Problem 2.8 of Chap. 2, Page 36, find (*a*) the mean, (*b*) the standard deviation, (*c*) the standard deviation using Sheppard's correction, (*d*) the actual standard deviation from the ungrouped data.

Solution:
The work is arranged in Table 4.10.

Table 4.10

X	u	f	fu	fu^2
122	−3	3	−9	27
131	−2	5	−10	20
140	−1	9	−9	9
A → 149	0	12	0	0 .
158	1	5	5	5
167	2	4	8	16
176	3	2	6	18
		$N = \Sigma f = 40$	$\Sigma\, fu = -9$	$\Sigma\, fu^2 = 95$

(*a*)
$$\bar{X} = A + c\bar{u} = A + c\frac{\Sigma\, fu}{N} = 149 + 9\left(\frac{-9}{40}\right) = 147.0 \text{ mm}$$

(*b*)
$$s = c\sqrt{\bar{u^2} - \bar{u}^2} = c\sqrt{\frac{\Sigma\, fu^2}{N} - \left(\frac{\Sigma\, fu}{N}\right)^2} = 9\sqrt{\frac{95}{40} - \left(\frac{-9}{40}\right)^2} = 9\sqrt{2.324\,375} = 13.7 \text{ mm}$$

(*c*) Corrected variance $= s^2 - c^2/12 = 188.27 - 9^2/12 = 181.52$.
Corrected standard deviation $= 13.5$ mm.

(*d*) To compute the standard deviation from the actual lengths of leaves given in the problem, it is convenient to first subtract a suitable number, say $A = 150$ mm, from each length and then use the method of Problem 4.15. The deviations $d = X - A = X - 150$ are then given in the following table

−12	14	0	−18	−6	−25	−1	7
−4	8	−10	−3	−14	−2	2	−6
18	−24	−12	26	13	−31	4	15
−4	23	−8	−3	−15	3	−10	−15
11	−5	−15	−8	0	6	−5	−22

from which we find $\Sigma d = -128$ and $\Sigma d^2 = 7052$. Then

$$s = \sqrt{\bar{d^2} - \bar{d}^2} = \sqrt{\frac{\Sigma d^2}{N} - \left(\frac{\Sigma d}{N}\right)^2} = \sqrt{\frac{7052}{40} - \left(\frac{-128}{40}\right)^2} = \sqrt{166.06} = 12.9 \text{ mm}$$

Hence Sheppard's correction supplied some improvement in this case.

EMPIRICAL RELATIONS BETWEEN MEASURES OF DISPERSION

4.23. Discuss the validity of the empirical formulae

(*a*) Mean deviation $= \frac{4}{5}$(standard deviation)
(*b*) Semi-interquartile range $= \frac{2}{3}$(standard deviation)

for the distribution of heights of the students at XYZ University.

Solution:
(*a*) From Problems 4.4 and 4.11, $\dfrac{\text{mean deviation}}{\text{standard deviation}} = \dfrac{2.26}{2.92} = 0.77$ which is close to $\frac{4}{5}$.

(*b*) From Problems 4.6 and 4.11, $\dfrac{\text{semi-interquartile range}}{\text{standard deviation}} = \dfrac{1.98}{2.92} = 0.68$ which is close to $\frac{2}{3}$.

Thus the empirical formulae are valid in this case.

Note: We have not used the standard deviation with Sheppard's correction for grouping in the above, since no corresponding correction has been made for the mean deviation or semi-interquartile range.

PROPERTIES OF THE STANDARD DEVIATION

4.24. Determine the percentage of students' I.Q.'s in Problem 4.19 which fall within the ranges (a) $\bar{X} \pm s$, (b) $\bar{X} \pm 2s$, (c) $\bar{X} \pm 3s$.

Solution:

(a) $\bar{X} \pm s = 95.97 \pm 10.47$ is the range of I.Q.'s from 85.5 to 106.4.
   Number of I.Q.'s in range ($\bar{X} \pm s$) is

$$\left(\frac{88 - 85.5}{4}\right)(45) + 66 + 85 + 72 + 54 + \left(\frac{106.4 - 104}{4}\right)(38) = 339$$

   Percentage of I.Q.'s in range $\bar{X} \pm s = 339/480 = 70.6\%$.

(b) $\bar{X} \pm 2s = 95.97 \pm 2(10.47)$ is the range of I.Q.'s from 75.0 to 116.9.
   Number of I.Q.'s in range ($\bar{X} \pm 2s$) is

$$\left(\frac{76 - 75.0}{4}\right)(9) + 16 + 28 + 45 + 66 + 85 + 72 + 54 + 38 + 27 + 18 + \left(\frac{116.9 - 116}{4}\right)(11) = 451$$

   Percentage of I.Q.'s in range ($\bar{X} \pm 2s$) = $451/480 = 94.0\%$.

(c) $\bar{X} \pm 3s = 95.97 \pm 3(10.47)$ is the range of I.Q.'s from 64.6 to 127.4.

   Number of I.Q.'s in range $\bar{X} \pm 3s = 480 - \left(\frac{128 - 127.4}{4}\right)(2) = 479.7$, or 480.

   Percentage of I.Q.'s in range ($\bar{X} \pm 3s$) = $479.7/480 = 99.9\%$, or practically 100%.

The percentages in (a), (b) and (c) agree favourably with those to be expected for a normal distribution, i.e. 68.27%, 95.45% and 99.73% respectively.

Note that we have not used Sheppard's correction for the standard deviation. If this is used the results in this case agree closely with the above. Note also that the above results can also be obtained using the table of Problem 4.32.

4.25. Given the sets of numbers 2, 5, 8, 11, 14 and 2, 8, 14, find (a) the mean of each set, (b) the variance of each set, (c) the mean of the combined or "pooled" sets, (d) the variance of the combined or pooled sets.

Solution:

(a) Mean of 1st set = $\frac{1}{5}(2 + 5 + 8 + 11 + 14) = 8$. Mean of 2nd set = $\frac{1}{3}(2 + 8 + 14) = 8$.

(b) Variance of 1st set = $s_1^2 = \frac{1}{5}[(2 - 8)^2 + (5 - 8)^2 + (8 - 8)^2 + (11 - 8)^2 + (14 - 8)^2] = 18$.
   Variance of 2nd set = $s_2^2 = \frac{1}{3}[(2 - 8)^2 + (8 - 8)^2 - (14 - 8)^2] = 24$.

(c) Mean of combined sets = $\dfrac{2 + 5 + 8 + 11 + 14 + 2 + 8 + 14}{5 + 3} = 8$.

(d) Variance of combined sets is

$$s^2 = \frac{(2 - 8)^2 + (5 - 8)^2 + (8 - 8)^2 + (11 - 8)^2 + (14 - 8)^2 + (2 - 8)^2 + (8 - 8)^2 + (14 - 8)^2}{5 + 3} = 20.25$$

Another method, by formula.

   Variance of combined sets = $s^2 = \dfrac{N_1 s_1^2 + N_2 s_2^2}{N_1 + N_2} = \dfrac{(5)(18) + (3)(24)}{5 + 3} = 20.25$

4.26. Work the preceding problem for the sets of numbers 2, 5, 8, 11, 14 and 10, 16, 22.

Solution:

Here the means of the two sets are 8 and 16 respectively, while the variances are the *same* as the sets of the preceding problem, namely $s_1^2 = 18$ and $s_2^2 = 24$.

   Mean of combined sets = $\dfrac{2 + 5 + 8 + 11 + 14 + 10 + 16 + 22}{5 + 3} = 11$

   Variance of combined sets

$$= \frac{(2 - 11)^2 + (5 - 11)^2 + (8 - 11)^2 + (11 - 11)^2 + (14 - 11)^2 + (10 - 11)^2 + (16 - 11)^2 + (22 - 11)^2}{5 + 3} = 35.25.$$

   Note that the formula $s^2 = \dfrac{N_1 s_1^2 + N_2 s_2^2}{N_1 + N_2}$ which gives the value 20.25 is *not* applicable in this case since the means of the two sets are *not* the same.

4.27. (a) Prove that $w^2 + pw + q$, where q and p are given constants, is a minimum if and only if $w = -\frac{1}{2}p$.

(b) Using (a) prove that $\dfrac{\displaystyle\sum_{j=1}^{N}(X_j - a)^2}{N}$ or briefly $\dfrac{\Sigma(X - a)^2}{N}$ is a minimum if and only if $a = \bar{X}$.

Solution:

(a) We have $w^2 + pw + q = (w + \frac{1}{2}p)^2 + q - \frac{2}{3}p^2$. Since $(q - \frac{1}{4}p^2)$ is a constant, the expression has the least value (i.e. is a minimum) if and only if $w + \frac{1}{2}p = 0$, i.e. $w = -\frac{1}{2}p$.

(b) $\dfrac{\Sigma(X - a)^2}{N} = \dfrac{\Sigma(X^2 - 2aX + a^2)}{N} = \dfrac{\Sigma X^2 - 2a\,\Sigma X + Na^2}{N} = a^2 - 2a\dfrac{\Sigma X}{N} + \dfrac{\Sigma X^2}{N}$

Comparing this last expression with $(w^2 + pw + q)$, we have $w = a$, $p = -2\dfrac{\Sigma X}{N}$, $q = \dfrac{\Sigma X^2}{N}$.

Then the expression is a minimum when $a = -\frac{1}{2}p = (\Sigma X)/N = \bar{X}$, using the result of part (a).

ABSOLUTE AND RELATIVE DISPERSION. COEFFICIENT OF VARIATION

4.28. A manufacturer of television tubes has two types of tubes, A and B. The tubes have respective mean lifetimes $\bar{X}_A = 1495$ hours and $\bar{X}_B = 1875$ hours, and standard deviations $s_A = 280$ hours and $s_B = 310$ hours. Which tube has the greater (a) absolute dispersion, (b) relative dispersion?

Solution:

(a) Absolute dispersion of $A = s_A = 280$ h., of $B = s_B = 310$ h.
     Then tube B has the greater absolute dispersion.

(b) Coefficient of variation of $A = \dfrac{s_A}{\bar{X}_A} = \dfrac{280}{1495} = 18\cdot7\%$, of $B = \dfrac{s_B}{\bar{X}_B} = \dfrac{310}{1875} = 16\cdot5\%$.
     Then tube A has the greater relative variation or dispersion.

4.29. Find the coefficients of variation V for the data of (a) Problem 4.14 and (b) Problem 4.18, using both uncorrected and corrected standard deviations.

Solution:

(a) $V\text{(uncorrected)} = \dfrac{s\text{(uncorrected)}}{\bar{X}} = \dfrac{2\cdot92}{67\cdot45} = 0\cdot0433 = 4\cdot3\%$

$\quad\ \ V\text{(corrected)} = \dfrac{s\text{(corrected)}}{\bar{X}} = \dfrac{2\cdot79}{67\cdot45} = 0\cdot0413 = 4\cdot1\%$, by Problem 4.21(a).

(b) $V\text{(uncorrected)} = \dfrac{s\text{(uncorrected)}}{\bar{X}} = \dfrac{15\cdot60}{79\cdot77} = 0\cdot196 = 19\cdot6\%$

$\quad\ \ V\text{(corrected)} = \dfrac{s\text{(corrected)}}{\bar{X}} = \dfrac{15\cdot33}{79\cdot77} = 0\cdot192 = 19\cdot2\%$, by Problem 4.21(b).

4.30. (a) Define a measure of relative dispersion which could be used for a set of data for which the quartiles were known. (b) Illustrate the calculation of the measure defined in (a) by using the data of Problem 4.6.

Solution:

(a) If Q_1 and Q_3 are given for a set of data, then $\frac{1}{2}(Q_1 + Q_3)$ is a measure of central tendency or average for the data while $Q = \frac{1}{2}(Q_3 - Q_1)$, the semi-interquartile range, is a measure of the dispersion of the data.
     Then we can define a measure of relative dispersion as

$$V_Q = \frac{\frac{1}{2}(Q_3 - Q_1)}{\frac{1}{2}(Q_1 + Q_3)} = \frac{Q_3 - Q_1}{Q_3 + Q_1}$$

which we can call the *quartile coefficient of variation* or *quartile coefficient of relative dispersion*.

(b) $V_Q = \dfrac{Q_3 - Q_1}{Q_3 + Q_1} = \dfrac{69\cdot61 - 65\cdot64}{69\cdot61 + 65\cdot64} = \dfrac{3\cdot97}{135\cdot25} = 0\cdot0293 = 2\cdot9\%$

STANDARDIZED VARIABLES AND STANDARD SCORES

4.31. A student received a mark of 84 on a final examination in mathematics for which the mean mark was 76 and the standard deviation was 10. On the final examination in physics for which the mean mark was 82 and the standard deviation was 16, he received a mark of 90. In which subject was his relative standing higher?

Solution:

The standardised variable $z = (X - \bar{X})/s$ measures the deviation of X from the mean $\bar{X}$ in terms of standard deviation s.

For mathematics, $z = (84 - 76)/10 = 0\cdot8$. For physics, $z = (90 - 82)/16 = 0\cdot5$.

Thus the student had a grade $0\cdot8$ of a standard deviation above the mean in mathematics, but only $0\cdot5$ of a standard deviation above the mean in physics. Thus his relative standing was higher in mathematics.

The variable $z = (X - \bar{X})/s$ is often used in educational testing, where it is known as a *standard score*.

4.32. (*a*) Convert the I.Q.'s of Problem 4.19 into standard scores and (*b*) construct a graph of relative frequency vs. standard score.

Solution:

(*a*) The work of conversion into standard scores can be arranged as in Table 4.11. In this table we have added for use in part (*b*) the I.Q. class marks 66 and 130 which have frequency zero. Also, we have not used Sheppard's correction for the standard deviation. The corrected scores in this case are practically the same as those given here to the indicated accuracy.

Table 4.11 $\bar{X} = 96\cdot0, s = 10\cdot5$

I.Q. (X)	$X - \bar{X}$	$z = \dfrac{X - \bar{X}}{s}$	Frequency f	Relative Frequency f/N (%)
66	$-30\cdot0$	$-2\cdot86$	0	$0\cdot0$
70	$-26\cdot0$	$-2\cdot48$	4	$0\cdot8$
74	$-22\cdot0$	$-2\cdot10$	9	$1\cdot9$
78	$-18\cdot0$	$-1\cdot71$	16	$3\cdot3$
82	$-14\cdot0$	$-1\cdot33$	28	$5\cdot8$
86	$-10\cdot0$	$-0\cdot95$	45	$9\cdot4$
90	$-6\cdot0$	$-0\cdot57$	66	$13\cdot8$
94	$-2\cdot0$	$-0\cdot19$	85	$17\cdot7$
98	$2\cdot0$	$0\cdot19$	72	$15\cdot0$
102	$6\cdot0$	$0\cdot57$	54	$11\cdot2$
106	$10\cdot0$	$0\cdot95$	38	$7\cdot9$
110	$14\cdot0$	$1\cdot33$	27	$5\cdot6$
114	$18\cdot0$	$1\cdot71$	18	$3\cdot8$
118	$22\cdot0$	$2\cdot10$	11	$2\cdot3$
122	$26\cdot0$	$2\cdot48$	5	$1\cdot0$
126	$30\cdot0$	$2\cdot86$	2	$0\cdot4$
130	$34\cdot0$	$3\cdot24$	0	$0\cdot0$
			480	100%

(*b*) The graph of relative frequency vs. z score (relative frequency polygon) is shown in Fig. 4-2. The horizontal axis is measured in terms of the standard deviation s as a unit. Note that the distribution is moderately asymmetrical and slightly skewed to the right.

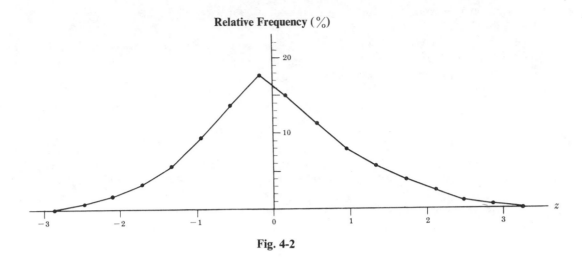

Fig. 4-2

Supplementary Problems

THE RANGE

4.33. Find the range of the sets of numbers: (a) 5, 3, 8, 4, 7, 6, 12, 4, 3; (b) 8·772, 6·453, 10·624, 8·628, 9·434, 6·351.
Ans. (a) 9   (b) 4·273

4.34. Find the range of the maximum loads given in Table 3.8 of Prob. 3.59, Chap. 3.   *Ans.* 40 kN

4.35. Find the range of the rivet diameters in Table 3.10 of Prob. 3.61, Chap. 3.   *Ans.* 0·036 mm

4.36. The largest of 50 measurements is 8·34 kg. If the range is 0·46 kg, find the smallest measurement.   *Ans.* 7·88 kg

4.37. Determine the range of the data in: (a) Prob. 3.62, Chap. 3; (b) Prob. 3.73, Chap. 3; (c) Prob. 2.20, Chap. 2.
Ans. (a) 35   (b) indeterminate   (c) 900 hr

THE MEAN DEVIATION

4.38. Find the absolute value of (a) $-18\cdot2$, (b) $+3\cdot58$, (c) $6\cdot21$, (d) 0, (e) $-\sqrt{2}$, (f) $4\cdot00 - 2\cdot36 - 3\cdot52$.
Ans. (a) 18·2, (b) 3·58, (c) 6·21, (d) 0, (e) $\sqrt{2} = 1\cdot414$ approx., (f) 1·88

4.39. Find the mean deviation of the sets of numbers: (a) 3, 7, 9, 5; (b) 2·4, 1·6, 3·8, 4·1, 3·4.   *Ans.* (a) 2, (b) 0·85

4.40. Find the mean deviation of the sets of numbers in Prob. 4.33.   *Ans.* (a) 2·2, (b) 1·317

4.41. Find the mean deviation of the maximum loads in Table 3.8 of Prob. 3.59, Chap. 3.   *Ans.* 5·76 kN

4.42. (a) Find the mean deviation (M.D.) of the rivet diameters in Table 3.10 of Prob. 3.61, Chap. 3. (b) What percentage of rivet diameters lie between $(\bar{X} \pm \text{M.D.})$, $(\bar{X} \pm 2 \text{ M.D.})$, $(\bar{X} \pm 3 \text{ M.D.})$?
Ans. (a) 0·004 37 mm, (b) 60·0%, 85·2%, 96·4%

4.43. Find the mean deviation (a) from the mean, (b) from the median, of the set of numbers 8, 10, 9, 12, 4, 8, 2. Verify that the mean deviation from the median is not greater than the mean deviation from the mean.   *Ans.* (a) 3·0, (b) 2·8

4.44. Find the mean deviation (a) about the mean, (b) about the median, for the distribution in Prob. 3.60, Chap. 3. Use the result of that problem and also of Prob. 3.70, Chap. 3.   *Ans.* (a) 31·2, (b) 30·6

4.45. Find the mean deviation (a) about the mean, (b) about the median, for the distribution in Prob. 3.62, Chap. 3. Use the result of that problem and also of Prob. 3.72, Chap. 3.   *Ans.* (a) 6·0, (b) 6·0

4.46. Explain why the mean deviation is or is not an appropriate measure of dispersion for the distribution of Prob. 3.73, Chap. 3.

4.47. Derive coding formulae for computing the mean deviation (a) about the mean, (b) about the median, from a frequency distribution. Apply these formulae to verify the results of Problems 4.44 and 4.45.

SEMI-INTERQUARTILE RANGE OR QUARTILE DEVIATION

4.48. Find the semi-interquartile range for the distributions of: (a) Prob. 3.59, Chap. 3; (b) Prob. 3.60, Chap. 3; (c) Prob. 3.107, Chap. 3. Interpret clearly the result in each case.   *Ans.* (a) 5·1 kN, (b) 27·0, (c) 12

4.49. Find the semi-interquartile range for the distributions of (a) Prob. 2.31 of Chap. 2, (b) Prob. 3.73 of Chap. 3, interpreting clearly the results in each case. Explain the advantages of the semi-interquartile range for this type of distribution over other measures of dispersion.   *Ans.* (a) \$1801, (b) 10·8 years

4.50. Prove that for any frequency distribution the total percentage of cases falling in the interval $\frac{1}{2}(Q_1 + Q_3) \pm \frac{1}{2}(Q_3 - Q_1)$ is 50%. Is the same true for the interval $Q_2 \pm \frac{1}{2}(Q_3 - Q_1)$? Explain your answer.

4.51. (a) How would you interpret graphically the semi-interquartile range corresponding to a given frequency distribution? (b) What is the relationship of the semi-interquartile range to the ogive of the distribution? |

10–90 PERCENTILE RANGE

4.52. Find the 10–90 percentile range for the distributions of: (a) Prob. 3.59, Chap. 3; (b) Prob. 3.107, Chap. 3. Interpret clearly the results in each case.   *Ans.* (a) 16·3 kN, (b) 33·6 or 34

4.53. Find the 10–90 percentile range for the distributions of: Prob. 2.31, Chap. 2; (b) Prob. 3.73, Chap. 3. Interpret clearly the results in each case. What advantage does the 10–90 percentile range have over other measures of dispersion? What disadvantage does it have?   *Ans.* (a) \$7402, (b) 40·8

4.54. What advantages or disadvantages would a 20–80 percentile range have in comparison to a 10–90 percentile range?

4.55. Answer Prob. 4.51 with reference to the (a) 10–90 percentile range, (b) 20–80 percentile range, (c) 25–75 percentile range. What is the relationship between (c) and the semi-interquartile range?

THE STANDARD DEVIATION

4.56. Find the standard deviation of the numbers: (a) 3, 6, 2, 1, 7, 5; (b) 3·2, 4·6, 2·8, 5·2, 4·4; (c) 0, 0, 0, 0, 0, 1, 1, 1. *Ans.* (a) 2·16, (b) 0·90, (c) 0·484

4.57. (a) By adding 5 to each of the numbers in the set 3, 6, 2, 1, 7, 5 we obtain the set 8, 11, 7, 6, 12, 10. Show that the two sets have the same standard deviations but different means. How are the means related?
(b) By multiplying each of the numbers 3, 6, 2, 1, 7, 5 by 2 and then adding 5, we obtain the set 11, 17, 9, 7, 19, 15. What is the relationship between the standard deviations and the means for the two sets?
(c) What properties of the mean and standard deviation are illustrated by the particular sets of numbers in (a) and (b)?

4.58. Find the standard deviation of the set of numbers in the arithmetic progression 4, 10, 16, 22, ..., 154.   *Ans.* 45

4.59. Find the standard deviation for the distributions of: (a) Prob. 3.59, Chap. 3; (b) Prob. 3.60, Chap 3; (c) Prob. 3.107, Chap. 3.   *Ans.* (a) 7·33 kN, (b) 38·60, (c) 12·1

4.60. Demonstrate the use of the Charlier check in each part of Prob. 4.59.

4.61. Find (a) the mean and (b) the standard deviation for the distribution of Prob. 2.17 of Chap. 2, explaining the significance of the results obtained.   *Ans.* (a) $\bar{X} = 2\cdot47$, (b) $s = 1\cdot11$

4.62. (a) Explain why the standard deviation is not an appropriate measure of dispersion for the distribution of Prob. 2.31, Chap. 2. (b) What measure of dispersion could be used in its place? Illustrate your answer.

4.63. (a) Find the standard deviation s of the rivet diameters in Table 3.10 of Prob. 3.61, Chap. 3. (b) What percentage of rivet diameters lie between $(\bar{X} \pm s)$, $(\bar{X} \pm 2s)$, $(\bar{X} \pm 3s)$? (c) Compare the percentages in (b) with those which would theoretically be expected if the distribution were normal, and account for any observed differences.
Ans. 0·005 76 mm, (b) 72·1%, 93·3%, 99·76%

4.64. Apply Sheppard's correction to each standard deviation in Prob. 4.59. In each case discuss whether such application is or is not justified.   *Ans.* 7·19, (b) 38·24, (c) 11·8

4.65. What modifications occur in Prob. 4.63 when Sheppard's corrections are applied?
Ans. (*a*) 0·005 69 mm, (*b*) 71·6%, 93·0%, 99·68%

4.66. (*a*) Find the mean and standard deviation for the data of Prob. 2.8, Chap. 2.
(*b*) Construct a frequency distribution for the data and find the standard deviation.
(*c*) Compare the result of (*b*) with that of (*a*). Determine whether an application of Sheppard's corrections produces better results.
Ans. (*a*) 146·8 mm, 12·9 mm

4.67. Work Prob. 4.66 for the data of Prob. 2.27, Chap. 2.   *Ans.* (*a*) 7·349 mm, 0·0495 mm

4.68. (*a*) Of a total of N numbers, the fraction p are ones while the fraction $q = 1 - p$ are zeros. Prove that the standard deviation of the set of numbers is $\sqrt{pq}$. (*b*) Apply the result of (*a*) to Prob. 4.56 (*c*).

4.69. (*a*) Prove that the variance of the set of n numbers $a, a - d, a - 2d, \ldots, a - (n-1)d$ (i.e. an arithmetic progression with first term a and common difference d, is given by $\frac{1}{12}(n^2 - 1)d^2$. (*b*) Use (*a*) in Prob. 4.58 [Hint: Use $1 - 2 - 3 - \cdots - (n-1) = \frac{1}{2}n(n-1)$, $1^2 + 2^2 - 3^2 - \cdots + (n-1)^2 = \frac{1}{6}n(n-1)(2n-1)$.]

4.70. Generalize and prove Property 3 on Page 72.

EMPIRICAL RELATIONS BETWEEN MEASURES OF DISPERSION

4.71. By comparing the standard deviation obtained in Prob. 4.59 with the corresponding mean deviations of Problems 4.41, 4.42 and 4.44, determine whether the following empirical relation holds: Mean deviation = $\frac{4}{5}$(standard deviation). Account for any differences which may occur.

4.72. By comparing the standard deviations obtained in Prob. 4.59 with the corresponding semi-interquartile ranges of Prob. 4.48, determine whether the following empirical relation holds: Semi-interquartile range = $\frac{2}{3}$(standard deviation). Account for any differences which may occur.

4.73. What empirical relation would you expect to exist between the semi-interquartile range and the mean deviation for bell-shaped distributions which are moderately skewed?   *Ans.* semi-interquartile range = $\frac{5}{6}$(mean deviation)

4.74. A frequency distribution which is approximately normal has a semi-interquartile range equal to 10. What values would you expect for (*a*) the standard deviation and (*b*) the mean deviation?   *Ans.* (*a*) 15, (*b*) 12

ABSOLUTE AND RELATIVE DISPERSION. COEFFICIENT OF VARIATION

4.75. On a final examination in statistics the mean mark of a group of 150 students was 78 and the standard deviation was 8·0. In algebra, however, the mean final mark of the group was 73 and the standard deviation was 7·6. In which subject was there the greater (*a*) absolute dispersion, (*b*) relative dispersion?   *Ans.* (*a*) statistics, (*b*) algebra

4.76. Find the coefficient of variation for the data of (*a*) Prob. 3.59 of Chap. 3, (*b*) Prob. 3.107 of Chap. 3.
Ans. (*a*) 6·6%, (*b*) 19·0%

4.77. (*a*) Why is it not possible to calculate the coefficient of variation for the distribution of Prob. 2.31, Chap. 2? (*b*) Calculate the quartile coefficient of relative dispersion for this distribution. (See Prob. 3.108(*c*) of Chap. 3, and also Prob. 4.30)
Ans. 51·9%

4.78. (*a*) Describe a measure of relative dispersion which utilizes the semi-interquartile range. (*b*) Illustrate the calculation of this measure by using the data of Prob. 3.73, Chap. 3.

STANDARDIZED VARIABLES AND STANDARD SCORES

4.79. On the examinations referred to in Prob. 4.75, a student scored 75 in statistics and 71 in algebra. In which examination was his relative standing higher?   *Ans.* algebra

4.80. Convert the set of numbers 6, 2, 8, 7, 5 into standard scores.   *Ans.* 0·19, −1·75, 1·17, 0·68, −0·29

4.81. Prove that the mean and standard deviation of a set of standard scores are equal to zero and one respectively. Illustrate by use of Prob. 4.80.

4.82. (*a*) Convert the marks of Prob. 3.107 of Chap. 3 into standard scores and (*b*) construct a graph of relative frequency vs. standard score.

CHAPTER 5

Moments, Skewness and Kurtosis

MOMENTS

If $X_1, X_2, \ldots, X_N$ are the N values assumed by the variable X, we define the quantity

$$\overline{X^r} = \frac{X_1^r + X_2^r + \ldots + X_N^r}{N} = \frac{\sum_{j=1}^{N} X_j^r}{N} = \frac{\Sigma X^r}{N} \qquad (1)$$

called the rth *moment*. The first moment with $r = 1$ is the arithmetic mean $\bar{X}$.

The rth *moment about the mean* $\bar{X}$ is defined as

$$m_r = \frac{\sum_{j=1}^{N} (X_j - \bar{X})^r}{N} = \frac{\Sigma (X - \bar{X})^r}{N} = \overline{(X - \bar{X})^r} \qquad (2)$$

If $r = 1$, $m_1 = 0$ (see Problem 3.16, Chap. 3). If $r = 2$, $m_2 = s^2$, the variance.

The rth *moment about any origin* A is defined as

$$m_r' = \frac{\sum_{j=1}^{N} (X_j - A)^r}{N} = \frac{\Sigma (X - A)^r}{N} = \frac{\Sigma d^r}{N} = \overline{(X - A)^r} \qquad (3)$$

where $d = X - A$ are the deviations of X from A. If $A = 0$, (3) reduces to (1). For this reason (1) is often called the rth *moment about zero*.

MOMENTS FOR GROUPED DATA

If $X_1, X_2, \ldots, X_K$ occur with frequencies $f_1, f_2, \ldots, f_K$ respectively, the above moments are given by

$$\overline{X^r} = \frac{f_1 X_1^r + f_2 X_2^r + \ldots + f_K X_K^r}{N} = \frac{\sum_{j=1}^{K} f_j X_j^r}{N} = \frac{\Sigma f X^r}{N} \qquad (4)$$

$$m_r = \frac{\sum_{j=1}^{K} f_j (X_j - \bar{X})^r}{N} = \frac{\Sigma f(X - \bar{X})^r}{N} = \overline{(X - \bar{X})^r} \qquad (5)$$

$$m_r' = \frac{\sum_{j=1}^{K} f_j (X_j - A)^r}{N} = \frac{\Sigma f(X - A)^r}{N} = \overline{(X - A)^r} \qquad (6)$$

where $N = \sum_{j=1}^{K} f_j = \Sigma f$. The formulae are suitable for calculating moments from grouped data.

RELATIONS BETWEEN MOMENTS

The following relations exist between moments about the mean m_r and moments about an arbitrary origin m_r'.

$$\left\{ \begin{array}{l} m_2 = m_2' - m_1'^2 \\ m_3 = m_3' - 3m_1' m_2' + 2m_1'^3 \\ m_4 = m_4' - 4m_1' m_3' + 6m_1'^2 m_2' - 3m_1'^4 \end{array} \right. \tag{7}$$

etc. (see Problem 5.5). Note that $m_1' = \bar{X} - A$.

COMPUTATION OF MOMENTS FOR GROUPED DATA

The coding method for computing the mean and standard deviation as given in previous chapters can also be used to provide a short method for calculation of moments. This method uses the fact that $X_j = A + cu_j$ (or briefly $X = A + cu$) so that from equation (6) we have

$$m_r' = c^r \frac{\Sigma f u^r}{N} = c^r \overline{u^r} \tag{8}$$

which can be used to find m_r by applying equations (7).

CHARLIER'S CHECK AND SHEPPARD'S CORRECTIONS

Charlier's check in computing moments by the coding method uses the identities:

$$\left\{ \begin{array}{l} \Sigma f(u+1) = \Sigma fu + N \\ \Sigma f(u+1)^2 = \Sigma fu^2 + 2\Sigma fu + N \\ \Sigma f(u+1)^3 = \Sigma fu^3 + 3\Sigma fu^2 + 3\Sigma fu + N \\ \Sigma f(u+1)^4 = \Sigma fu^4 + 4\Sigma fu^3 + 6\Sigma fu^2 + 4\Sigma fu + N \end{array} \right. \tag{9}$$

Sheppard's corrections for moments (extending the ideas on Page 72) are as follows:

$$\text{Corrected } m_2 = m_2 - \tfrac{1}{12}c^2, \quad \text{Corrected } m_4 = m_4 - \tfrac{1}{2}c^2 m_2 + \tfrac{7}{240}c^4$$

The moments m_1 and m_3 need no correction.

MOMENTS IN DIMENSIONLESS FORM

To avoid particular units we can define the *dimensionless moments* about the mean

$$a_r = \frac{m_r}{s^r} = \frac{m_r}{(\sqrt{m_2})^r} = \frac{m_r}{\sqrt{m_2^r}} \tag{10}$$

where $s = \sqrt{m_2}$ is the standard deviation. Since $m_1 = 0$ and $m_2 = s^2$, we have $a_1 = 0$, $a_2 = 1$.

SKEWNESS

Skewness is the degree of asymmetry, or departure from symmetry, of a distribution. If the frequency curve (smoothed frequency polygon) of a distribution has a longer "tail" to the right of the central maximum than to the left, the distribution is said to be *skewed to the right* or to have *positive skewness*. If the reverse is true it is said to be *skewed to the left* or to have *negative skewness*.

For skewed distributions the mean tends to lie on the same side of the mode as the longer tail (see Figs. 3-1, 3-2, Chap. 3). Thus a measure of the asymmetry is supplied by the difference (Mean − Mode). This can be made dimensionless on division by a measure of dispersion, such as the standard deviation, leading to the definition

$$\text{Skewness} = \frac{\text{mean} - \text{mode}}{\text{standard deviation}} = \frac{\bar{X} - \text{mode}}{s} \tag{11}$$

To avoid use of the mode, we can employ the empirical formula (*10*) on Page 48 and define

$$\text{Skewness} = \frac{3(\text{mean} - \text{median})}{\text{standard deviation}} = \frac{3(\bar{X} - \text{median})}{s} \qquad (12)$$

The above two measures are called, respectively, *Pearson's first and second coefficients of skewness*.

Other measures of skewness defined in terms of quartiles and percentiles are as follows:

$$\text{Quartile coefficient of skewness} = \frac{(Q_3 - Q_2) - (Q_2 - Q_1)}{Q_3 - Q_1} = \frac{Q_3 - 2Q_2 + Q_1}{Q_3 - Q_1} \qquad (13)$$

$$10\text{-}90 \text{ percentile coefficient of skewness} = \frac{(P_{90} - P_{50}) - (P_{50} - P_{10})}{P_{90} - P_{10}} = \frac{P_{90} - 2P_{50} + P_{10}}{P_{90} - P_{10}} \qquad (14)$$

An important measure of skewness uses the third moment about the mean expressed in dimensionless form and is given by

$$\text{Moment coefficient of skewness} = a_3 = \frac{m_3}{s^3} = \frac{m_3}{(\sqrt{m_2})^3} = \frac{m_3}{\sqrt{m_2^3}} \qquad (15)$$

Another measure of skewness is sometimes given by $b_1 = a_3^2$. For perfectly symmetrical curves, such as the normal curve, a_3 and b_1 are zero.

KURTOSIS

Kurtosis is the degree of peakedness of a distribution, usually taken relative to a normal distribution. A distribution having a relatively high peak such as the curve of Fig. 5-1(*a*) is called *leptokurtic*, while the curve of Fig. 5-1(*b*) which is flat-topped is called *platykurtic*. The normal distribution, Fig. 5-1(*c*), which is not very peaked or very flat-topped is called *mesokurtic*.

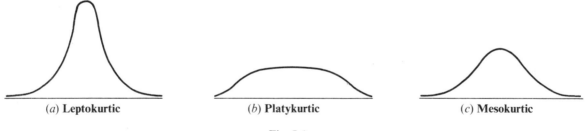

(*a*) **Leptokurtic**   (*b*) **Platykurtic**   (*c*) **Mesokurtic**

Fig. 5-1

One measure of kurtosis uses the fourth moment about the mean expressed in dimensionless form and is given by

$$\text{Moment coefficient of kurtosis} = a_4 = \frac{m_4}{s^4} = \frac{m_4}{m_2^2} \qquad (16)$$

which is often designated by b_2. For the normal distribution, $b_2 = a_4 = 3$. For this reason the kurtosis is sometimes defined by $(b_2 - 3)$ which is positive for a leptokurtic distribution, negative for a platykurtic distribution, and zero for the normal distribution.

Another measure of kurtosis which is also used is based on both quartiles and percentiles and is given by

$$\kappa = \frac{Q}{P_{90} - P_{10}} \qquad (17)$$

where $Q = \frac{1}{2}(Q_3 - Q_1)$ is the semi-interquartile range. We shall refer to this as the *percentile coefficient of kurtosis*. For the normal distribution this has the value 0·263. See Problem 5.14.

POPULATION MOMENTS, SKEWNESS AND KURTOSIS

When it is required to distinguish moments and measures of skewness and kurtosis for a sample from those corresponding to a population of which the sample is a part, it is often the custom to use Latin symbols for the former and Greek symbols for the latter. Thus if sample moments are denoted by m_r and m_r' the corresponding Greek symbols would be μ_r and μ_r' (μ is the Greek letter *mu*). Subscripts are always denoted by Latin symbols.

Similarly if the sample measures of skewness and kurtosis are denoted by a_3 and a_4 respectively, the population skewness and kurtosis would be α_3 and α_4 (α is the Greek letter *alpha*).

We have already mentioned that the standard deviation for sample and population are denoted respectively by s and σ.

Solved Problems

MOMENTS

5.1. Find the (*a*) first, (*b*) second, (*c*) third and (*d*) fourth moments for the set of numbers 2, 3, 7, 8, 10.

Solution:

(*a*) $\bar{X} = \dfrac{\Sigma X}{N} = \dfrac{2 + 3 + 7 + 8 + 10}{5} = \dfrac{30}{5} = 6$ is the first moment or arithmetic mean.

(*b*) $\overline{X^2} = \dfrac{\Sigma X^2}{N} = \dfrac{2^2 + 3^2 + 7^2 + 8^2 + 10^2}{5} = \dfrac{226}{5} = 45 \cdot 2$ is the second moment.

(*c*) $\overline{X^3} = \dfrac{\Sigma X^3}{N} = \dfrac{2^3 + 3^3 + 7^3 + 8^3 + 10^3}{5} = \dfrac{1890}{5} = 378$ is the third moment.

(*d*) $\overline{X^4} = \dfrac{\Sigma X^4}{N} = \dfrac{2^4 + 3^4 + 7^4 + 8^4 + 10^4}{5} = \dfrac{16\,594}{5} = 3318 \cdot 8$ is the fourth moment.

5.2. Find the (*a*) first, (*b*) second, (*c*) third and (*d*) fourth moments about the mean for the set of numbers in Prob. 5.1.

Solution:

(*a*) $m_1 = \overline{(X - \bar{X})} = \dfrac{\Sigma (X - \bar{X})}{N} = \dfrac{(2 - 6) + (3 - 6) + (7 - 6) + (8 - 6) + (10 - 6)}{5} = \dfrac{0}{5} = 0$

m_1 is always equal to zero since $\overline{X - \bar{X}} = \bar{X} - \bar{X} = 0$. (Problem 3.16, Chapter 3.)

(*b*) $m_2 = \overline{(X - \bar{X})^2} = \dfrac{\Sigma (X - \bar{X})^2}{N} = \dfrac{(2 - 6)^2 + (3 - 6)^2 + (7 - 6)^2 + (8 - 6)^2 + (10 - 6)^2}{5} = \dfrac{46}{5} = 9 \cdot 2$

Note that m_2 is the variance s^2.

(*c*) $m_3 = \overline{(X - \bar{X})^3} = \dfrac{\Sigma (X - \bar{X})^3}{N} = \dfrac{(2 - 6)^3 + (3 - 6)^3 + (7 - 6)^3 + (8 - 6)^3 + (10 - 6)^3}{5} = \dfrac{-18}{5} = -3 \cdot 6$

(*d*) $m_4 = \overline{(X - \bar{X})^4} = \dfrac{\Sigma (X - \bar{X})^4}{N} = \dfrac{(2 - 6)^4 + (3 - 6)^4 + (7 - 6)^4 + (8 - 6)^4 + (10 - 6)^4}{5} = \dfrac{610}{5} = 122$

5.3. Find the (*a*) first, (*b*) second, (*c*) third and (*d*) fourth moments about the origin 4 for the set of numbers in Prob. 5.1.

Solution:

(*a*) $m_1' = \overline{(X - 4)} = \dfrac{\Sigma (X - 4)}{N} = \dfrac{(2 - 4) + (3 - 4) + (7 - 4) + (8 - 4) + (10 - 4)}{5} = 2$

(b) $\quad m_2' = \overline{(X-4)^2} = \dfrac{\Sigma (X-4)^2}{N} = \dfrac{(2-4)^2+(3-4)^2+(7-4)^2+(8-4)^2+(10-4)^2}{5} = \dfrac{66}{5} = 13\cdot2$

(c) $\quad m_3' = \overline{(X-4)^3} = \dfrac{\Sigma (X-4)^3}{N} = \dfrac{(2-4)^3+(3-4)^3+(7-4)^3+(8-4)^3+(10-4)^3}{5} = \dfrac{298}{5} = 59\cdot6$

(d) $\quad m_4' = \overline{(X-4)^4} = \dfrac{\Sigma (X-4)^4}{N} = \dfrac{(2-4)^4+(3-4)^4+(7-4)^4+(8-4)^4+(10-4)^4}{5} = \dfrac{1650}{5} = 330$

5.4. Using the results of Problems 5.2 and 5.3, verify the relations between the moments:
(a) $m_2 = m_2' - m_1'^2$, (b) $m_3 = m_3' - 3m_1'm_2' + 2m_1'^3$, (c) $m_4 - 4m_1'm_3' + 6m_1'^2 m_2' - 3m_1'^4$.

Solution:

From Problem 5.3: $m_1' = 2$, $m_2' = 13\cdot2$, $m_3' = 59\cdot6$, $m_4' = 330$. Then:

(a) $\quad m_2 = m_2' - m_1'^2 = 13\cdot2 - (2)^2 = 13\cdot2 - 4 = 9\cdot2$

(b) $\quad m_3 = m_3' - 3m_1' m_2' + 2m_1'^3 = 59\cdot5 - 3(2)(13\cdot2) + 2(2)^3 = 59\cdot6 - 79\cdot2 + 16 = -3\cdot6$

(c) $\quad m_4 = m_4' - 4m_1' m_3' + 6m_1'^2 m_2' - 3m_1'^4 = 330 - 4(2)(59\cdot6) + 6(2)^2(13\cdot2) - 3(2)^4 = 122$

in agreement with Problem 5.2.

5.5. Prove that (a) $m_2 = m_2' - m_1'^2$, (b) $m_3 = m_3' - 3m_1' m_2' + 2m_1'^3$, (c) $m_4 = m_4' - 4m_1' m_3' + 6m_1'^2 m_2' - 3m_1'^4$.

Solution:

(a) If $d = X - A$, then $X = A + d$, $\bar{X} = A + \bar{d}$, and $X - \bar{X} = d - \bar{d}$.

$\begin{aligned} m_2 &= \overline{(X-\bar{X})^2} = \overline{(d-\bar{d})^2} = \overline{d^2 - 2\bar{d}d + \bar{d}^2} \\ &= \overline{d^2} - 2\bar{d}^2 + \bar{d}^2 = \overline{d^2} - \bar{d}^2 = m_2' - m_1'^2 \end{aligned}$

(b) $\begin{aligned} m_3 &= \overline{(X-\bar{X})^3} = \overline{(d-\bar{d})^3} = \overline{(d^3 - 3d^2\bar{d} + 3d\bar{d}^2 - \bar{d}^3)} \\ &= \overline{d^3} - 3\bar{d}\overline{d^2} + 3\bar{d}^3 - \bar{d}^3 = \overline{d^3} - 3\bar{d}\overline{d^2} + 2\bar{d}^3 = m_3' - 3m_1' m_2' + 2m_1'^3 \end{aligned}$

(c) $\begin{aligned} m_4 &= \overline{(X-\bar{X})^4} = \overline{(d-\bar{d})^4} = \overline{(d^4 - 4d^3\bar{d} + 6d^2\bar{d}^2 - 4d\bar{d}^3 + \bar{d}^4)} \\ &= \overline{d^4} - 4\bar{d}\overline{d^3} + 6\bar{d}^2\overline{d^2} - 4\bar{d}^4 + \bar{d}^4 = \overline{d^4} - 4\bar{d}\overline{d^3} + 6\bar{d}^2\overline{d^2} - 3\bar{d}^4 \\ &= m_4' - 4m_1' m_3' + 6m_1'^2 m_2' - 3m_1'^4 \end{aligned}$

By extension of this method we can derive similar results for m_5, m_6, etc.

COMPUTATION OF MOMENTS FROM GROUPED DATA

5.6. Find the first four moments about the mean for the mass distribution of Prob. 3.22, Chap. 3.

Table 5.1

X	u	f	fu	fu^2	fu^3	fu^4
61	-2	5	-10	20	-40	80
64	-1	18	-18	18	-18	18
67	0	42	0	0	0	0
70	1	27	27	27	27	27
73	2	8	16	32	64	128
		$N = \Sigma f = 100$	$\Sigma fu = 15$	$\Sigma fu^2 = 97$	$\Sigma fu^3 = 33$	$\Sigma fu^4 = 253$

Then $m_1' = c\dfrac{\Sigma fu}{N} = (3)\left(\dfrac{15}{100}\right) = 0\cdot45 \qquad\qquad m_3' = c^3\dfrac{\Sigma fu^3}{N} = (3)^3\left(\dfrac{33}{100}\right) = 8\cdot91$

$m_2' = c^2\dfrac{\Sigma fu^2}{N} = (3)^2\left(\dfrac{97}{100}\right) = 8\cdot73 \qquad\qquad m_4' = c^4\dfrac{\Sigma fu^4}{N} = (3)^4\left(\dfrac{253}{100}\right) = 204\cdot93$

so that $m_1 = 0$

$m_2 = m_2' - m_1'^2 = 8\cdot73 - (0\cdot45)^2 = 8\cdot5275$

$m_3 = m_3' - 3m_1' m_2' + 2m_1'^3 = 8\cdot91 - 3(0\cdot45)(8\cdot73) + 2(0\cdot45)^3 = 2\cdot6932$

$m_4 = m_4' - 4m_1' m_3' + 6m_1'^2 m_2' - 3m_1'^4$
$\qquad = 204\cdot93 - 4(0\cdot45)(8\cdot91) + 6(0\cdot45)^2(8\cdot73) - 3(0\cdot45)^4 = 199\cdot3759$

5.7.　Find (a) m_1', (b) m_2', (c) m_3', (d) m_4', (e) m_1, (f) m_2, (g) m_3, (h) m_4, (i) $\bar{X}$, (j) s, (k) $\overline{X^2}$ and (l) $\overline{X^3}$ for the distribution of Problem 4.19, Chapter 4.

Solution:

Table 5.2

X	u	f	fu	fu^2	fu^3	fu^4
70	−6	4	−24	144	−864	5184
74	−5	9	−45	225	−1125	5625
78	−4	16	−64	256	−1024	4096
82	−3	28	−84	252	−756	2268
86	−2	45	−90	180	−360	720
90	−1	66	−66	66	−66	66
A → 94	0	85	0	0	0	0
98	1	72	72	72	72	72
102	2	54	108	216	432	864
106	3	38	114	342	1026	3078
110	4	27	108	432	1728	6912
114	5	18	90	450	2250	11 250
118	6	11	66	396	2376	14 256
122	7	5	35	245	1715	12 005
126	8	2	16	128	1024	8192
		$N = \Sigma f = 480$	$\Sigma fu = 236$	$\Sigma fu^2 = 3404$	$\Sigma fu^3 = 6428$	$\Sigma fu^4 = 74\,588$

(a)　$m_1' = c\dfrac{\Sigma fu}{N} = (4)\left(\dfrac{236}{480}\right) = 1{\cdot}9667$

(c)　$m_3' = c^3\dfrac{\Sigma fu^3}{N} = (4)^3\left(\dfrac{6428}{480}\right) = 857{\cdot}0667$

(b)　$m_2' = c^2\dfrac{\Sigma fu^2}{N} = (4)^2\left(\dfrac{3404}{480}\right) = 113{\cdot}4667$

(d)　$m_4' = c^4\dfrac{\Sigma fu^4}{N} = (4)^4\left(\dfrac{74588}{480}\right) = 39\,780{\cdot}2667$

(e)　$m_1 = 0$

(f)　$m_2 = m_2' - m_1'^2 = 113{\cdot}4667 - (1{\cdot}9667)^2 = 109{\cdot}5988$

(g)　$m_3 = m_3' - 3m_1'\,m_2' + 2m_1'^3 = 857{\cdot}0667 - 3(1{\cdot}9667)(113{\cdot}4667) + 2(1{\cdot}9667)^3 = 202{\cdot}8158$

(h)　$m_4 = m_4' - 4m_1'\,m_3' + 6m_1'^2\,m_2' - 3m_1'^4 = 35\,627{\cdot}2853$

(i)　$\bar{X} = \overline{(A + d)} = A + m_1' = A + c\dfrac{\Sigma fu}{N} = 94 + 1{\cdot}9667 = 95{\cdot}97$

(j)　$s = \sqrt{m_2} = \sqrt{109{\cdot}5988} = 10{\cdot}47$

(k)　$\overline{X^2} = \overline{(A + d)^2} = \overline{(A^2 + 2Ad + d^2)} = A^2 + 2A\bar{d} + \bar{d^2} = A^2 + 2Am_1' + m_2'$

　　　$= (94)^2 + 2(94)(1{\cdot}9667) + 113{\cdot}4667 = 9319{\cdot}2063$, or 9319 to four significant figures

(l)　$\overline{X^3} = \overline{(A + d)^3} = \overline{(A^3 + 3A^2d + 3Ad^2 + d^3)} = A^3 + 3A^2\bar{d} + 3A\bar{d^2} + \bar{d^3}$
　　　$A^3 + 3A^2m_1' + 3Am_2' + m_3' = 915\,571{\cdot}9597$, or 915 600 to four significant figures

CHARLIER'S CHECK

5.8.　Illustrate the use of Charlier's check for the computations in Problem 5.7.

Solution:
　　　To supply the required check we add the following columns to those of Problem 5.7, with the exception of column 2 which is repeated here for convenience.

Table 5.3

$u + 1$	f	$f(u + 1)$	$f(u + 1)^2$	$f(u + 1)^3$	$f(u + 1)^4$
-5	4	-20	100	-500	2500
-4	9	-36	144	-576	2304
-3	16	-48	144	-432	1296
-2	28	-56	112	-224	448
-1	45	-45	45	-45	45
0	66	0	0	0	0
1	85	85	85	85	85
2	72	144	288	576	1152
3	54	162	486	1458	4374
4	38	152	608	2432	9728
5	27	135	675	3375	16 875
6	18	108	648	3888	23 328
7	11	77	539	3773	26 411
8	5	40	320	2560	20 480
9	2	18	162	1458	13 122
$N = \Sigma f = 480$		$\Sigma f(u + 1)$ $= 716$	$\Sigma f(u + 1)^2$ $= 4356$	$\Sigma f(u + 1)^3$ $= 17\,828$	$\Sigma f(u + 1)^4$ $= 122\,148$

In each of the following groupings the first is taken from Table 5.3 and the second is taken from Table 5.2 of Problem 5.7. Equality of results in each grouping provides the required check.

$$\begin{cases} \Sigma f(u + 1) = 716 \\ \Sigma fu + N = 236 + 480 = 716 \end{cases}$$

$$\begin{cases} \Sigma f(u + 1)^2 = 4356 \\ \Sigma fu^2 + 2 \Sigma fu + N = 3404 + 2(236) + 480 = 4356 \end{cases}$$

$$\begin{cases} \Sigma f(u + 1)^3 = 17\,828 \\ \Sigma fu^3 + 3 \Sigma fu^2 + 3 \Sigma fu + N = 6428 + 3(3404) + 3(236) + 480 = 17\,828 \end{cases}$$

$$\begin{cases} \Sigma f(u + 1)^4 = 122\,148 \\ \Sigma fu^4 + 4 \Sigma fu^3 + 6 \Sigma fu^2 + 4 \Sigma fu + N = 74\,588 + 4(6428) + 6(3404) + 4(236) + 480 = 122\,148 \end{cases}$$

SHEPPARD'S CORRECTIONS FOR MOMENTS

5.9. Apply Sheppard's corrections to determine the moments about the mean for the data in (a) Problem 5.6, (b) Problem 5.7.

Solution:

(a) Corrected $m_2 = m_2 - c^2/12 = 8\cdot5275 - 3^2/12 = 7\cdot7775$
Corrected $m_4 = m_4 - \frac{1}{2}c^2 m_2 + \frac{7}{240}c^4$
$\qquad = 199\cdot3759 - \frac{1}{2}(3)^2(8\cdot5275) + \frac{7}{240}(3)^4$
$\qquad = 163\cdot3646$

m_1 and m_3 need no correction.

(b) Corrected $m_2 = m_2 - c^2/12 = 109\cdot5988 - 4^2/12 = 108\cdot2655$
Corrected $m_4 = m_4 - \frac{1}{2}c^2 m_2 + \frac{7}{240}c^4$
$\qquad = 35\,627\cdot2853 - \frac{1}{2}(4)^2(109\cdot5988) + \frac{7}{240}(4)^4$
$\qquad = 34\,757\cdot9616$

SKEWNESS

5.10. Find Pearson's (a) first and (b) second coefficient of skewness for the distribution of wages of the 65 employees at the P and R Company. See Prob. 3.44, Chap. 3 and Prob. 4.18, Chap. 4.

Solution:

Mean = £79·76, median = £79·06, mode = £77·50, standard deviation = s = £15·60.

(a) First coefficient of skewness $= \dfrac{\text{mean} - \text{mode}}{s} = \dfrac{£79·76 - £77·50}{£15·60} = 0·1448$, or $0·14$.

(b) Second coefficient of skewness $= \dfrac{3(\text{mean} - \text{median})}{s} = \dfrac{3(£79·76 - £79·06)}{£15·60} = 0·1346$, or $0·13$.

If the corrected standard deviation is used (see Prob. 4.21(b), Chap. 4) these coefficients became, respectively,

(a) $\dfrac{\text{mean} - \text{mode}}{\text{corrected } s} = \dfrac{£79·76 - £77·50}{£15·33} = 0·1474$ or $0·15$

(b) $\dfrac{3(\text{mean} - \text{median})}{\text{corrected } s} = \dfrac{3(£79·76 - £79·06)}{£15·33} = 0·1370$, or $0·14$

Since the coefficients are positive the distribution is skewed positively, i.e. to the right.

5.11. Find the (a) quartile and (b) percentile coefficients of skewness for the distribution of Problem 5.10 (see Prob. 3.44, Chap. 3).

Solution:

$Q_1 = £68·25$, $Q_2 = P_{50} = £79·06$, $Q_3 = £90·75$, $P_{10} = D_1 = £58·12$, $P_{90} = D_9 = £101·00$.

(a) Quartile coefficient of skewness $= \dfrac{Q_3 - 2Q_2 + Q_1}{Q_3 - Q_1} = \dfrac{£90·75 - 2(£79·06) + £68·25}{£90·75 - £68·25} = 0·0391$

(b) Percentile coefficient of skewness $= \dfrac{P_{90} - 2P_{50} + P_{10}}{P_{90} - P_{10}} = \dfrac{£101·00 - 2(£79·06) + £58·12}{£101·00 - £58·12} = 0·0233$

5.12. Find the moment coefficient of skewness, a_3, for (a) the mass distribution of students at XYZ University (see Problem 5.6), (b) the I.Q.'s of elementary school children (see Problem 5.7).

Solution:

(a) $m_2 = s^2 = 8·5275$, $m_3 = -2·6932$.

Then $a_3 = \dfrac{m_3}{s^3} = \dfrac{m_3}{(\sqrt{m_2})^3} = \dfrac{-2·6932}{(\sqrt{8·5275})^3} = 0·1413$, or $-0·14$.

If Sheppard's corrections for grouping are used (see Problem 5.9(a)), then

Corrected $a_3 = \dfrac{m_3}{(\sqrt{\text{corrected } m_2})^3} = \dfrac{-2·6932}{(\sqrt{7·7775})^3} = -0·1242$ or $-0·12$

(b) $a_3 = \dfrac{m_3}{s^3} = \dfrac{m_3}{(\sqrt{m_2})^3} = \dfrac{202·8158}{(\sqrt{109·5988})^3} = 0·1768$, or $0·18$

If Sheppard's corrections for grouping are used (see Problem 5.9(b)), then

Corrected $a_3 = \dfrac{m_3}{(\sqrt{\text{corrected } m_2})^3} = \dfrac{202·8158}{(\sqrt{108·2655})^3} = 0·1800$, or $0·18$

Note that both distributions are moderately skewed, (a) to the left (negatively) and (b) to the right (positively). The distribution (b) is more skewed than (a), i.e. (a) is more symmetric than (b), as is evidenced by the fact that the numerical value or absolute value of the skewness coefficient for (b) is greater than that for (a).

KURTOSIS

5.13. Find the moment coefficient of kurtosis, a_4 for the data of (a) Problem 5.6, (b) Problem 5.7.

Solution:

(a) $a_4 = \dfrac{m_4}{s^4} = \dfrac{m_4}{m_2^2} = \dfrac{199·3759}{(8·5275)^2} = 2·7418$, or $2·74$.

If Sheppard's corrections are used (see Problem 5.9(a)), then

Corrected $a_4 = \dfrac{\text{corrected } m_4}{(\text{corrected } m_2)^2} = \dfrac{163·3646}{(7·7775)^2} = 2·7007$, or $2·70$

(b) $a_4 = \dfrac{m_4}{s^4} = \dfrac{m_4}{m_2^2} = \dfrac{35\,627 \cdot 2853}{(109 \cdot 5988)^2} = 2 \cdot 9660$, or $2 \cdot 97$.

If Sheppard's corrections are used (see Problem 5.9(b)), then

$$\text{Corrected } a_4 = \frac{\text{corrected } m_4}{(\text{corrected } m_2)^2} = \frac{34\,757 \cdot 9616}{(108 \cdot 2655)^2} = 2 \cdot 9653, \text{ or } 2 \cdot 97$$

Since for a normal distribution $a_4 = 3$, it follows that both distributions (a) and (b) are *platykurtic* with respect to the normal distribution (i.e. less peaked than the normal distribution).

Insofar as peakedness is concerned, the distribution (b) approximates the normal distribution much better than distribution (a). However, from Problem 5.12 the distribution (a) is more symmetric than (b), so that as far as symmetry is concerned (a) approximates the normal distribution better than (b).

5.14.   (a) Calculate the percentile coefficient of kurtosis, $\kappa = Q/(P_{90} - P_{10})$, for the distribution of Problem 5.11.

(b) How well would it be approximated by a normal distribution?

Solution:

(a) $Q = \frac{1}{2}(Q_3 - Q_1) = \frac{1}{2}(£90 \cdot 75 - £68 \cdot 25) = £11 \cdot 25$, $P_{90} - P_{10} = £101 \cdot 00 - £58 \cdot 12 = £42 \cdot 88$.
Then $\kappa = Q/(P_{90} - P_{10}) = 0 \cdot 262$.

(b)   Since κ for the normal distribution is $0 \cdot 263$, it follows that the given distribution is *mesokurtic* (i.e. about as peaked as the normal). Thus the kurtosis of the distribution is about the same as it should be for a normal distribution and leads us to believe that it would be approximated well by a normal distribution insofar as kurtosis is concerned.

Supplementary Problems

MOMENTS

5.15.   Find the (a) first, (b) second, (c) third and (d) fourth moments for the set of numbers 4, 7, 5, 9, 8, 3, 6.
Ans. (a) 6, (b) 40, (c) 288, (d) 2188

5.16.   Find the (a) first, (b) second, (c) third and (d) fourth moments about the mean for the set of numbers in Prob. 5.15.
Ans. (a) 0, (b) 4, (c) 0, (d) 25·86

5.17.   Find the (a) first, (b) second, (c) third and (d) fourth moments about the number 7 for the set of numbers in Prob. 5.15.
Ans. (a) -1, (b) 5, (c) -91, (d) 53

5.18.   Using the results of Problems 5.16 and 5.17, verify the relations between the moments (a) $m_2 = m_2' - m_1'^2$, (b) $m_3 = m_3' - 3m_1' m_2' + 2m_1'^3$, (c) $m_4 = m_4' - 4m_1' m_3' + 6m_1'^2 m_2' - 3m_1'^4$.

5.19.   Find the first four moments about the mean of the set of numbers in the arithmetic progression 2, 5, 8, 11, 14, 17.
Ans. 0, 26·25, 0, 1193·1

5.20.   Prove that (a) $m_2' = m_2 + h^2$, (b) $m_3' = m_3 + 3hm_2 + h^3$, (c) $m_4' = m_4 + 4hm_3 + 6h^2m_2 + h^4$, where $h = m_1'$.

5.21.   If the first moment about the number 2 is equal to 5, what is the mean?   *Ans.* 7

5.22.   If the first four moments of a set of numbers about the number 3 are equal to -2, 10, -25 and 50, determine the corresponding moments (a) about the mean, (b) about the number 5, (c) about zero.
Ans. (a) 0, 6, 19, 42; (b) -4, 22, -117, 560; (c) 1, 7, 38, 74

5.23.   Find the first four moments about the mean of the numbers 0, 0, 0, 1, 1, 1, 1, 1.
Ans. 0, 0·2344, $-0·0586$, $-0·0696$

5.24.   (a) Prove that $m_5 = m_5' - 5m_1' m_4' + 10m_1'^2 m_3' - 10m_1'^3 m_2' + 4m_1'^5$. (b) Derive a similar formula for m_6.

5.25.   Of a total of N numbers, the fraction p are ones while the fraction $q = 1 - p$ are zeros. Find (a) m_1, (b) m_2, (c) m_3 and (d) m_4 for the set of numbers. Compare with Prob. 5.23.
Ans. (a) $m_1 = 0$, (b) $m_2 = pq$, (c) $pq(q - p)$, (d) $pq(p^2 - pq + q^2)$

5.26.   Prove that the first four moments about the mean of the arithmetic progression $a, a + d, a + 2d, \ldots, a + (n - 1)d$ are
$$m_1 = 0, \quad m_2 = \tfrac{1}{12}(n^2 - 1)d^2, \quad m_3 = 0, \quad m_4 = \tfrac{1}{240}(n^2 - 1)(3n^2 - 7)d^4$$
Compare with Prob. 5.19. See also Prob. 4.69, Chap. 4.
[Hint: $1^4 + 2^4 + 3^4 + \cdots + (n - 1)^4 = \tfrac{1}{30}n(n - 1)(2n - 1)(3n^2 - 3n - 1)$.]

MOMENTS FOR GROUPED DATA

5.27. Calculate the first four moments about the mean for the distribution of Table 5.4.
  Ans. $m_1 = 0$, $m_2 = 5.97$, $m_3 = -0.397$, $m_4 = 89.22$

X	f
12	1
14	4
16	6
18	10
20	7
22	2

Table 5.4

Total 30

5.28. Illustrate the use of Charlier's check for the computations in Problem 5.27.

5.29. Apply Sheppard's corrections to the moments obtained in Prob. 5.27.
  Ans. m_1(corrected) $= 0$, m_2(corrected) $= 5.440$, m_3(corrected) $= -0.5920$, m_4(corrected) $= 76.2332$

5.30. Calculate the first four moments about the mean for the distribution of Prob. 3.59 of Chap. 3:
  (a) without Sheppard's corrections. (b) with Sheppard's corrections.
  Ans. (a) $m_1 = 0$, $m_2 = 53.743$, $m_3 = 61.853$, $m_4 = 8491.4$; (b) m_2(corrected) $= 51.660$, m_4(corrected) $= 7837.8$

5.31. Find (a) m_1, (b) m_2, (c) m_3, (d) m_4, (e) $\bar{X}$, (f) s, (g) $\overline{X^2}$, (h) $\overline{X^3}$, (i) $\overline{X^4}$ and (j) $\overline{(X + 1)^3}$ for the distribution of Prob. 3.62, Chap. 3.   *Ans.* (a) 0, (b) 52.95, (c) 92.35, (d) 7158.20, (e) 26.2, (f) 7.28, (g) 739.58, (h) 22 247, (i) 706 428, (j) 24 545

SKEWNESS

5.32. Find the moment coefficient of skewness, a_3, for the distribution of Prob. 5.27 (a) without and (b) with Sheppard's corrections.   *Ans.* (a) -0.2464, (b) -0.2464

5.33. Find the moment coefficient of skewness, a_3, for the distribution of Prob. 3.59, Chap. 3. See Prob. 5.30.   *Ans.* 0.1570

5.34. The second moments about the mean of two distributions are 9 and 16, while the third moments about the mean are -8.1 and -12.8 respectively. Which distribution is more skewed to the left?   *Ans.* First distribution

5.35. Find Pearson's (a) first and (b) second coefficient of skewness for the distribution of Prob. 3.59, Chap. 3. Account for the difference.   *Ans.* (a) 0.040, (b) 0.074

5.36. Find (a) the quartile and (b) the percentile coefficients of skewness for the distribution of Prob. 3.59, Chap. 3. Compare your results with that of Prob. 5.35 and explain.   *Ans.* (a) -0.02, (b) -0.13

5.37. (a) Explain why Pearson's coefficients of skewness are not appropriate for the distribution of Prob. 2.31, Chap. 2. (b) Find the quartile coefficient of skewness for this distribution and interpret the result.   *Ans.* (b) -0.078

KURTOSIS

5.38. Find the moment coefficient of kurtosis, a_4, for the distribution of Prob. 5.27 (a) without and (b) with Sheppard's corrections.   *Ans.* (a) 2.62. (b) 2.58

5.39. Find the moment coefficient of kurtosis for the distribution of Prob. 3.59, Chap. 3, (a) without and (b) with Sheppard's corrections (see Prob. 5.30).   *Ans.* 2.94, (b) 2.94

5.40. The fourth moments about the mean of the two distributions of Prob. 5.34 are 230 and 780 respectively. Which distribution more nearly approximates the normal distribution from the viewpoint of (a) peakedness, (b) skewness?
  Ans. (a) second, (b) first

5.41. Which of the distributions in Prob. 5.40 is (a) leptokurtic, (b) mesokurtic, (c) platykurtic?
  Ans. (a) second, (b) neither, (c) first

5.42. The standard deviation of a symmetric distribution is 5. What must be the value of the fourth moment about the mean in order that the distribution be (a) leptokurtic, (b) mesokurtic, (c) platykurtic?
  Ans. (a) greater than 1875, (b) equal to 1875, (c) less than 1875

5.43. (a) Calculate the percentile coefficient of kurtosis, κ, for the distribution of Prob. 3.59, Chap. 3. (b) Compare your result with the theoretical value 0.263 for the normal distribution and interpret. (c) How do you reconcile this result with that of Prob. 5.39?   *Ans.* (a) 0.313

CHAPTER 6

Elementary Probability Theory

CLASSICAL DEFINITION OF PROBABILITY

Suppose an event E can happen in h ways out of a total of n possible equally likely ways. Then the probability of occurrence of the event (called its *success*) is denoted by

$$p = \Pr\{E\} = \frac{h}{n}$$

The probability of non-occurrence of the event (called its *failure*) is denoted by

$$q = \Pr\{\text{not } E\} = \frac{n-h}{n} = 1 - \frac{h}{n} = 1 - p = 1 - \Pr\{E\}$$

Thus $p + q = 1$, or $\Pr\{E\} + \Pr\{\text{not } E\} = 1$.

The event "not E" is sometimes denoted by $\bar{E}$, $\widetilde{E}$ or $\sim E$.

Example: Let E be the event that the numbers 3 or 4 turn up in a single toss of a die. There are six ways in which the die can fall, resulting in the numbers 1, 2, 3, 4, 5 or 6; and if the die is *fair* (i.e. not *loaded*) we can *assume* these six ways to be equally likely. Since E can occur in two of these ways, we have $p = \Pr\{E\} = \frac{2}{6} = \frac{1}{3}$.

The probability of not getting a 3 or 4 (i.e. getting a 1, 2, 5 or 6) is

$$q = \Pr\{\bar{E}\} = 1 - \tfrac{1}{3} = \tfrac{2}{3}.$$

Note that the probability of an event is a number between 0 and 1. If the event cannot occur, its probability is 0. If it must occur, i.e. its occurrence is *certain*, its probability is 1.

If p is the probability that an event will occur, the *odds* in favour of its happening are $p:q$ (read "p to q"); the odds against its happening are $q:p$. Thus the odds against a 3 or 4 in a single toss of fair die are $q:p = \tfrac{2}{3}:\tfrac{1}{3} = 2:1$, i.e. 2 to 1.

RELATIVE FREQUENCY DEFINITION OF PROBABILITY

The above definition of probability has a disadvantage in that the words "equally likely" are vague. In fact, since these words seem to be synonymous with "equally probable", the definition is *circular* because we are essentially defining probability in terms of itself. For this reason a statistical definition of probability has been advocated by some people. According to this the estimated probability, or *empirical probability*, of an event is taken as the *relative frequency* of occurrence of the event when the number of observations is very large. The probability itself is the *limit* of the relative frequency as the number of observations increase indefinitely.

Example: If 1000 tosses of a coin result in 529 heads, the relative, frequency of heads is $529/1000 = 0\cdot529$. If another 1000 tosses result in 493 heads, the relative frequency in the total of 2000 tosses is $(529 + 493)/2000 = 0\cdot511$. According to the statistical definition, by continuing in this manner we should ultimately get closer and closer to a number which we call the probability of a head in a single toss of the coin. From results so far presented this should be $0\cdot5$ to one significant figure. To obtain more significant figures, further observations must be made.

The statistical definition, although useful in practice, has difficulties from a mathematical point of view, since an actual limiting number may not really exist. For this reason, modern probability theory has been developed *axiomatically* in which probability is an undefined concept much the same as *point* and *line* are undefined in geometry.

CONDITIONAL PROBABILITY. INDEPENDENT AND DEPENDENT EVENTS

If E_1 and E_2 are two events, the probability that E_2 occurs given that E_1 has occurred is denoted by $\Pr\{E_2|E_1\}$ or $\Pr\{E_2 \text{ given } E_1\}$ and is called the *conditional probability* of E_2 given that E_1 has occurred.

If the occurrence or non-occurrence of E_1 does not affect the probability of occurrence of E_2 then $\Pr\{E_2|E_1\} = \Pr\{E_2\}$ and we say that E_1 and E_2 are *independent events*; otherwise they are *dependent events*.

If we denote by E_1E_2 the event that "both E_1 and E_2 occur", sometimes called a *compound event*, then

$$\Pr\{E_1E_2\} = \Pr\{E_1\}\Pr\{E_2|E_1\} \tag{1}$$

In particular,

$$\Pr\{E_1E_2\} = \Pr\{E_1\}\Pr\{E_2\} \quad \text{for independent events} \tag{2}$$

For three events E_1, E_2, E_3 we have

$$\Pr\{E_1E_2E_3\} = \Pr\{E_1\}\Pr\{E_2|E_1\}\Pr\{E_3|E_1E_2\} \tag{3}$$

i.e. the probability of occurrence of E_1, E_2 and E_3 is equal to the probability of E_1 times the probability of E_2 given that E_1 has occurred, times the probability of E_3 given that both E_1 and E_2 have occurred. In particular,

$$\Pr\{E_1E_2E_3\} = \Pr\{E_1\}\Pr\{E_2\}\Pr\{E_3\} \text{ for independent events} \tag{4}$$

In general if E_1, E_2, E_3, . . ., E_n are n independent events having respective probabilities $p_1, p_2, p_3, \ldots, p_n$, then the probability of occurrence of E_1 and E_2 and E_3 and . . ., E_n is $p_1p_2p_3 \cdots p_n$.

Example 1. Let E_1 and E_2 be the events "heads on fifth toss" and "heads on sixth toss" of a coin, respectively. Then E_1 and E_2 are independent events, so that the probability of heads on both the fifth and sixth tosses is, assuming the coin is "fair",
$$\Pr\{E_1E_2\} = \Pr\{E_1\}\Pr\{E_2\} = (\tfrac{1}{2})(\tfrac{1}{2}) = \tfrac{1}{4}$$

Example 2. If the probability that A will be alive in 20 years is $0\cdot7$ and the probability that B will be alive in 20 years is $0\cdot5$, then the probability that they will both be alive in 20 years is $(0\cdot7)(0\cdot5) = 0\cdot35$.

Example 3. Suppose a box contains 3 white balls and 2 black balls. Let E_1 be the event "first ball drawn is black" and E_2 the event "second ball drawn is black" where the balls are not replaced after being drawn. Here E_1 and E_2 are dependent events.
$\Pr\{E_1\} = \tfrac{2}{3+2} = \tfrac{2}{5}$ is the probability that the first ball drawn is black, while $\Pr\{E_2|E_1\} = \tfrac{1}{3+1} = \tfrac{1}{4}$ is the probability that the second ball drawn is black given that the first ball drawn was black. Then the probability that both balls drawn are black is
$$\Pr\{E_1E_2\} = \Pr\{E_1\}\Pr\{E_2|E_1\} = \tfrac{2}{5} \cdot \tfrac{1}{4} = \tfrac{1}{10}$$

MUTUALLY EXCLUSIVE EVENTS

Two or more events are called *mutually exclusive* if the occurrence of any one of them excludes the occurrence of the others. Thus if E_1 and E_2 are mutually exclusive events, $\Pr\{E_1E_2\} = 0$.

If $E_1 + E_2$ denotes the event that "either E_1 or E_2 or both occur", then

$$\Pr\{E_1 + E_2\} = \Pr\{E_1\} + \Pr\{E_2\} - \Pr\{E_1 E_2\} \tag{5}$$

In particular,

$$\Pr\{E_1 + E_2) = \Pr\{E_1\} + \Pr\{E_2\} \text{ for mutually exclusive events} \tag{6}$$

As an extension of this, if $E_1, E_2, \ldots, E_n$ are n mutually exclusive events having respective probabilities of occurrence $p_1, p_2, \ldots, p_n$, then the probability of occurrence of either E_1 or E_2 or $\ldots E_n$ is $p_1 + p_2 + \cdots + p_n$.

Example 1. If E_1 is the event "drawing an ace from a deck of cards" and E_2 is the event "drawing a king", then $\Pr\{E_1\} = \frac{4}{52} = \frac{1}{13}$ and $\Pr\{E_2\} = \frac{4}{52} = \frac{1}{13}$. The probability of drawing either an ace or a king in a single draw is

$$\Pr\{E_1 + E_2\} = \Pr\{E_1\} + \Pr\{E_2\} = \frac{1}{13} + \frac{1}{13} = \frac{2}{13}$$

since both ace and king cannot be drawn in a single draw and are thus mutually exclusive events.

Example 2. If E_1 is the event "drawing an ace" from a deck of cards and E_2 is the event "drawing a spade", then E_1 and E_2 are not mutually exclusive since the ace of spades can be drawn. Thus the probability of drawing either an ace or a spade or both is

$$\Pr\{E_1 + E_2\} = \Pr\{E_1\} + \Pr\{E_2\} - \Pr\{E_1 E_2\}$$
$$= \frac{4}{52} + \frac{13}{52} - \frac{1}{52} = \frac{16}{52} = \frac{4}{13}$$

DISCRETE PROBABILITY DISTRIBUTIONS

If a variable X can assume a discrete set of values $X_1, X_2, \ldots, X_K$ with respective probabilities $p_1, p_2, \ldots, p_K$ where $p_1 + p_2 + \ldots + p_K = 1$, we say that a *discrete probability distribution* for X has been defined. The function $p(X)$ which has the respective values $p_1, p_2, \ldots, p_K$ for $X = X_1, X_2, \ldots, X_K$, is called the *probability function* or *frequency function* of X. Because X can assume certain values with given probabilities, it is often called a *discrete random variable*. A random variable is also known as a *chance variable* or *stochastic variable*.

Example: Let a pair of fair dice be tossed and let X denote the sum of the points obtained. Then the probability distribution is given by the following table.

X	2	3	4	5	6	7	8	9	10	11	12
$p(X)$	1/36	2/36	3/36	4/36	5/36	6/36	5/36	4/36	3/36	2/36	1/36

For example, the probability of getting sum 5 is $\frac{4}{36} = \frac{1}{9}$. Thus in 900 tosses of the dice we would expect 100 tosses to give the sum 5.

Note that this is analogous to a relative frequency distribution with probabilities replacing relative frequencies. Thus we can think of probability distributions as theoretical or ideal limiting forms of relative frequency distributions when the number of observations is made very large. For this reason we can think of probability distributions as being distributions for *populations*, whereas relative frequency distributions are distributions of *samples* drawn from this population.

The probability distribution can be represented graphically by plotting $p(X)$ against X, as for relative frequency distributions. See Problem 6.11.

By cumulating probabilities we obtain *cumulative probability distributions* which are analogous to cumulative relative frequency distributions. The function associated with this distribution is sometimes called a *distribution function*.

CONTINUOUS PROBABILITY DISTRIBUTIONS

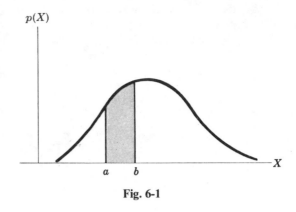

Fig. 6-1

The above ideas can be extended to the case where the variable X may assume a continuous set of values. The relative frequency polygon of a sample becomes, in the theoretical or limiting case of a population, a continuous curve such as shown in Fig. 6-1, whose equation is $Y = p(X)$. The total area under this curve bounded by the X axis is equal to one, and the area under the curve between lines $X = a$ and $X = b$ (shaded in the figure) gives the probability that X lies between a and b, which can be denoted by $\Pr\{a < X < b\}$.

We call $p(X)$ a *probability density function*, or briefly a *density function*, and when such a function is given we say that a *continuous probability distribution* for X has been defined. The variable X is then often called a *continuous random variable*.

As in the discrete case, we can define cumulative probability distributions and the associated distribution functions.

MATHEMATICAL EXPECTATION

If p is the probability that a person will receive a sum of money S, the *mathematical expectation*, or simply the *expectation*, is defined as pS.

Example: If the probability that a man wins a £10 prize is $\frac{1}{5}$, his expectation is $\frac{1}{5}(£10) = £2$.

The concept of expectation is easily extended. If X denotes a discrete random variable which can assume the values $X_1, X_2, \ldots, X_K$ with respective probabilities $p_1, p_2, \ldots, p_K$ where $p_1 + p_2 + \ldots p_K = 1$, the *mathematical expectation* of X or simply the *expectation* of X, denoted by $E(X)$, is defined as

$$E(X) = p_1 X_1 + p_2 X_2 + \ldots + p_K X_K = \sum_{j=1}^{K} p_j X_j = \Sigma p X \qquad (7)$$

If the probabilities p_j in this expectation are replaced by relative frequencies f_j/N where $N = \Sigma f_j$, the expectation reduces to $(\Sigma f X)/N$ which is the arithmetic mean $\bar{X}$ of a sample of size N in which $X_1, X_2, \ldots, X_K$ appear with these relative frequencies. As N gets larger and larger the relative frequencies f_j/N approach the probabilities p_j. This we are led to the interpretation that $E(X)$ represents the mean of the population from which the sample is drawn. If we call m the sample mean we can denote the population mean by the corresponding Greek letter μ (mu).

Expectation can also be defined for continuous random variables, but the definition requires use of the calculus.

RELATION BETWEEN POPULATION AND SAMPLE MEAN AND VARIANCE

If we select a sample of size N at random from a population (i.e. we assume all such samples are equally probable), then it is possible to show that the *expected value of the sample mean m is the population mean* μ.

It does not follow, however, that the expected value of any quantity computed from a sample is the corresponding population quantity. For example, the expected value of the sample variance as we have defined it is not the population variance but $(N - 1)/N$ times this variance. This is why some statisticians choose to define the sample variance as our variance multiplied by $N/(N - 1)$.

COMBINATORIAL ANALYSIS

In obtaining probabilities of complex events an enumeration of cases is often difficult, tedious, or both. To facilitate the labour involved, use is made of basic principles studied in a subject called *combinatorial analysis*.

FUNDAMENTAL PRINCIPLE

If an event can happen in any one of n_1 ways and if when this has occurred another event can happen in any one of n_2 ways, then the number of ways in which both events can happen in the specified order is $n_1 n_2$.

Example: If there are 3 candidates for governor and 5 for mayor, the two offices can be filled in $3 . 5 = 15$ ways.

FACTORIAL n

Factorial n, denoted by $n!$, is defined as

$$n! = n(n - 1)(n - 2) \ldots 1 \tag{8}$$

Thus $5! = 5.4.3.2.1 = 120$, $4! \, 3! = (4.3.2.1)(3.2.1) = 144$. It is convenient to define $0! = 1$.

PERMUTATIONS

A permutation of n different objects taken r at a time is an *arrangement* of r out of the n objects with attention given to the order of arrangement. The number of permutations of n objects taken r at a time is denoted by $_nP_r$, $P(n,r)$ or $P_{n,r}$ and is given by

$$_nP_r = n(n-1)(n-2) \ldots (n-r+1) = \frac{n!}{(n-r)!} \tag{9}$$

In particular, the number of permutations of n objects taken n at a time is

$$_nP_n = n(n-1)(n-2) \ldots 1 = n!$$

Example: The number of permutations of the letters, a, b, c taken two at a time is $_3P_2 = 3 . 2 = 6$. These are ab, ba, ac, ca, bc, cb.

The number of permutations of n objects consisting of groups of which n_1 are alike, n_2 are alike, $\ldots$ is

$$\frac{n!}{n_1! \, n_2! \ldots} \qquad \text{where} \quad n = n_1 + n_2 + \ldots \tag{10}$$

Example: The number of permutations of letters in the word *statistics* is $\dfrac{10!}{3! \, 3! \, 1! \, 2! \, 1!} = 50\,400$, since there are 3 s's, 3 t's, 1 a, 2 i's and 1 c.

COMBINATIONS

A combination of n different objects taken r at a time is a *selection* of r out of the n objects with no attention given to the order of arrangement. The number of combinations of n objects taken r at a time is denoted by $_nC_r$, $C(n,r)$, $C_{n,r}$ or $\binom{n}{r}$ and is given by

$$_nC_r = \frac{n(n-1) \cdots (n-r+1)}{r!} = \frac{n!}{r! \, (n-r)!} = \frac{_nP_r}{r!} \tag{11}$$

Example: The number of combinations of the letters a, b, c taken two at a time is $_3C_2 = \dfrac{3 \cdot 2}{2!} = 3$.

These are ab, ac, bc. Note that ab is the same combination as ba, but not the same permutation.

We have $_nC_r = {}_nC_{n-r}$. Thus $_{20}C_{17} = {}_{20}C_3 = \dfrac{20 \cdot 19 \cdot 18}{3!} = 1140$. The number of combinations of n objects taken 1 or 2, ..., or n at a time is $\quad {}_nC_1 + {}_nC_2 + \cdots + {}_nC_n = 2^n - 1$.

STIRLING'S APPROXIMATION TO $n!$

When n is large a direct evaluation of $n!$ is impractical. In such cases use is made of an approximate formula due to *Stirling*, namely

$$n! \quad \sim \quad \sqrt{2\pi n}\, n^n\, e^{-n} \tag{12}$$

where $e = 2 \cdot 718\ 28 \ldots$ is the natural base of logarithms. See Prob. 6.31.

RELATION OF PROBABILITY TO POINT SET THEORY

In modern probability theory, we think of all possible outcomes or results of an experiment, game, etc., as points in a space (which can be of one, two, three, etc., dimensions) called a *sample space* S. If S contains only a finite number of points then with each point we can associate a non-negative number, called a *probability*, such that the sum of all numbers corresponding to all points in S add to one. An event is a *set* or *collection* of points in S such as indicated by E_1 or E_2 in Fig. 6-2, called an *Euler diagram* or *Venn diagram*.

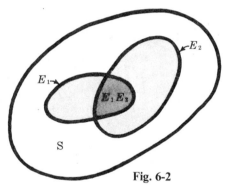

Fig. 6-2

The event $E_1 + E_2$ is the set of points which are *either in E_1 or E_2 or both*, while the event E_1E_2 is the set of points *common to both E_1 and E_2*. Then the probability of an event such as E_1 is the sum of the probabilities associated with all points contained in the set E_1. Similarly the probability of $E_1 + E_2$, denoted by $\Pr\{E_1 + E_2\}$, is the sum of the probabilities associated with all points contained in the set $E_1 + E_2$. If E_1 and E_2 have no points in common, i.e. the events are mutually exclusive, then $\Pr\{E_1 + E_2\} = \Pr\{E_1\} + \Pr\{E_2\}$. If they have points in common then $\Pr\{E_1 + E_2\} = \Pr\{E_1\} + \Pr\{E_2\} - \Pr\{E_1E_2\}$.

The set $E_1 + E_2$ is sometimes denoted by $E_1 \cup E_2$ and is called the *union* of the two sets. The set E_1E_2 is sometimes denoted by $E_1 \cap E_2$ and is called the *intersection* of the two sets. Extensions to more than two sets can be made. Thus instead of $E_1 + E_2 + E_3$ and $E_1E_2E_3$ we could use the notation $E_1 \cup E_2 \cup E_3$ and $E_1 \cap E_2 \cap E_3$ respectively.

A special symbol φ is sometimes used to denote a set with no points in it, called the *null set*. The probability associated with an event corresponding to this set is zero, i.e. $\Pr\{\phi\} = 0$. If E_1 and E_2 have no points in common, we can write $E_1E_2 = \phi$, which means that the corresponding events are mutually exclusive and $\Pr\{E_1E_2\} = 0$.

With this modern approach, a random variable is a function defined at each point of the sample space. For example, in Problem 6.37 the random variable is the sum of the co-ordinates of each point.

In the case where S has an infinite number of points the above ideas can be extended by means of concepts involving the calculus.

FUNDAMENTAL RULES OF PROBABILITY

6.1. Determine the probability p, or an estimate of it, for each of the following events.

(*a*) An odd number appears in a single toss of a fair die.

Out of 6 possible equally likely cases, 3 cases (where die comes up 1, 3 or 5) are favourable to the event. Then $p = \frac{3}{6} = \frac{1}{2}$.

(*b*) At least one head appears in two tosses of a fair coin.

If H denotes "head" and T denotes "tail", the two tosses can lead to four cases HH, HT, TH, TT all equally likely. Only the first three cases are favourable to the event. Then $p = \frac{3}{4}$.

(*c*) An ace, ten of diamonds or two of spades appears in drawing a single card from a well-shuffled ordinary deck of 52 cards.

The event can occur in 6 ways (ace of spades, ace of hearts, ace of clubs, ace of diamonds, ten of diamonds and two of spades) out of 52 equally likely cases. Then $p = \frac{6}{52} = \frac{3}{26}$.

(*d*) The sum 7 appears in a single toss of a pair of fair dice.

Each of the six faces of one die can be associated with each of the six faces of the other die, so that the total number of cases which can arise, all equally likely, is $6 \cdot 6 = 36$. These can be denoted by $(1,1)$, $(2,1)$, $(3,1)$, $\ldots$, $(6,6)$.

There are 6 ways of obtaining the sum 7, denoted by $(1,6)$, $(2,5)$, $(3,4)$, $(4,3)$, $(5,2)$, $(6,1)$ [see Prob. 6.37(*a*)]. Then $p = \frac{6}{36} = \frac{1}{6}$.

(*e*) A tail appears in the next toss of a coin if out of 100 tosses 56 were heads.

Since $(100 - 56) = 44$ tails were obtained in 100 tosses, the *estimated* or *empirical probability* of a tail is the relative frequency $44/100 = 0.44$.

6.2. An experiment consists of tossing a coin and a die. If E_1 is the event that "head" comes up in tossing the coin and E_2 is the event that "3 or 6" comes up in tossing the die, state in words the meaning of each of the following.

(*a*) $\bar{E}_1$ Tails on the coin and anything on the die.

(*b*) $\bar{E}_2$ 1, 2, 4 or 5 on the die and anything on the coin.

(*c*) $E_1 E_2$ Heads on the coin and 3 or 6 on the die.

(*d*) $\Pr\{E_1 \bar{E}_2\}$ Probability of heads on the coin and 1, 2, 4, or 5 on the die.

(*e*) $\Pr\{E_1 | E_2\}$ Probability of heads on the coin given that a 3 or 6 has come up on the die.

(*f*) $\Pr\{\bar{E}_1 + \bar{E}_2\}$ Probability of tails on the coin or 1, 2, 4 or 5 on the die, or both.

6.3. A ball is drawn at random from a box containing 6 red balls, 4 white balls and 5 blue balls. Determine the probability that it is (*a*) red, (*b*) white, (*c*) blue, (*d*) not red, (*e*) red or white.

Solution:
Let R, W and B denote the events of drawing a red ball, white ball and blue ball respectively. Then

(a) $\Pr\{R\} = \dfrac{\text{ways of choosing a red ball}}{\text{total ways of choosing a ball}} = \dfrac{6}{6+4+5} = \dfrac{6}{15} = \dfrac{2}{5}$

(b) $\Pr\{W\} = \dfrac{4}{6+4+5} = \dfrac{4}{15}$ (c) $\Pr\{B\} = \dfrac{6}{6+4+5} = \dfrac{5}{15} = \dfrac{1}{3}$

(d) $\Pr\{\bar{R}\} = 1 - \Pr\{R\} = 1 - \dfrac{2}{5} = \dfrac{3}{5}$ by (a).

(e) $\Pr\{R + W\} = \dfrac{\text{ways of choosing a red or white ball}}{\text{total ways of choosing a ball}} = \dfrac{6+4}{6+4+5} = \dfrac{10}{15} = \dfrac{2}{3}$

> **Another method:**
> $\Pr\{R + W\} = \Pr\{\bar{B}\} = 1 - \Pr\{B\} = 1 - \dfrac{1}{3} = \dfrac{2}{3}$ by part (c).
>   Note that $\Pr\{R + W\} = \Pr\{R\} + \Pr\{W\}$, i.e. $\dfrac{2}{3} = \dfrac{2}{5} + \dfrac{4}{15}$. This is an illustration of the general rule $\Pr\{E_1 + E_2\} = \Pr\{E_1\} + \Pr\{E_2\}$ which is true for *mutually exclusive* events E_1 and E_2.

6.4.   A fair die is tossed twice. Find the probability of getting a 4, 5 or 6 on the first toss and a 1, 2, 3 or 4 on the second toss.

> **Solution:**
>   Let $E_1 =$ event "4, 5 or 6" on first toss, and $E_2 =$ event "1, 2, 3 or 4" on second toss.
>   Each of the six ways in which the die can fall on the first toss can be associated with each of the six ways in which it can fall on the second toss, a total of $6 \cdot 6 = 36$ ways, all equally likely.
>   Each of the three ways in which E_1 can be associated with each of the four ways in which E_2 can occur, to give $3 \cdot 4 = 12$ ways in which both E_1 and E_2, or E_1E_2, occur.
>   Then $\Pr\{E_1E_2\} = 12/36 = 1/3$.
>   Note that $\Pr\{E_1E_2\} = \Pr\{E_1\}\Pr\{E_2\}$, i.e. $\dfrac{1}{3} = \dfrac{3}{6} \cdot \dfrac{4}{6}$, is valid for the *independent events* E_1 and E_2.

6.5.   Two cards are drawn from a well-shuffled ordinary deck of 52 cards. Find the probability that they are both aces if the first card is (a) replaced, (b) not replaced.

> **Solution:**
>   Let $E_1 =$ event "ace" on first draw, and $E_2 =$ event "ace" on second draw.
> (a)   If the first card is replaced, E_1 and E_2 are independent events. Then
>
> $$\Pr\{\text{both cards drawn are aces}\} = \Pr\{E_1E_2\} = \Pr\{E_1\}\Pr\{E_2\} = (4/52)(4/52) = 1/169$$
>
> (b)   The first card can be drawn in any one of 52 ways, and the second card can be drawn in any one of 51 ways since the first card is not replaced. Then both cards can be drawn in $52 \cdot 51$ ways, all equally likely.
>   There are 4 ways in which E_1 can occur and 3 ways in which E_2 can occur, so that both E_1 and E_2, or E_1E_2, can occur in $4 \cdot 3$ ways. Then $\Pr\{E_1E_2\} = \dfrac{4}{52} \cdot \dfrac{3}{51} = \dfrac{1}{221}$.
>   Note that $\Pr\{E_2 \mid E_1\} = \Pr\{\text{second card is an ace given that first card is an ace}\} = 3/51$. Thus our result is an illustration of the general rule $\Pr\{E_1E_2\} = \Pr\{E_1\}\Pr\{E_2 \mid E_1\}$ in case E_1 and E_2 are dependent events.

6.6.   Three balls are drawn successively from the box of Problem 6.3. Find the probability that they are drawn in the order red, white and blue if each ball is (a) replaced, (b) not replaced.

> **Solution:**
>   Let $R =$ event "red" on first draw, $W =$ event "white" on second draw, $B =$ event "blue" on third draw. We require $\Pr\{RWB\}$.
> (a)   If each ball is replaced, then R, W and B are independent events and
>
> $$\Pr\{RWB\} = \Pr\{R\}\,\Pr\{W\}\,\Pr\{B\} = \left(\frac{6}{6+4+5}\right)\left(\frac{4}{6+4+5}\right)\left(\frac{5}{6+4+5}\right) = \left(\frac{6}{15}\right)\left(\frac{4}{15}\right)\left(\frac{5}{15}\right) = \frac{8}{225}.$$

(b) If each ball is not replaced, then R, W and B are dependent events and

$$\Pr\{RWB\} = \Pr\{R\}\Pr\{W\,|\,R\}\Pr\{B\,|\,WR\} = \left(\frac{6}{6+4+5}\right)\left(\frac{4}{5+4+5}\right)\left(\frac{5}{5+3+5}\right)$$

$$= \left(\frac{6}{15}\right)\left(\frac{4}{14}\right)\left(\frac{5}{13}\right) = \frac{4}{91}$$

where $\Pr\{B\,|\,WR\}$ is the conditional probability of getting a blue ball if a white and red ball have already been chosen.

6.7. Find the probability of a 4 turning up at least once in two tosses of a fair die.

Solution:

Let $E_1 =$ event "4" on first toss and $E_2 =$ event "4" on second toss

$E_1 + E_2 =$ event "4" on first toss or "4" on second toss or both

$=$ event that at least one 4 turns up. We require $\Pr\{E_1 + E_2\}$.

Method 1:

Total number of equally likely ways in which both dice can fall $= 6 \cdot 6 = 36$.

Also, number of ways in which E_1 occurs but not $E_2 = 5$,

number of ways in which E_2 occurs but not $E_1 = 5$,

number of ways in which both E_1 and E_2 occur $= 1$.

Then the number of ways in which at least one of the events E_1 or E_2 occurs $= 5 + 5 + 1 = 11$, so that $\Pr\{E_1 + E_2\}$ $= 11/36$.

Method 2:

Since E_1 and E_2 are not mutually exclusive, $\Pr\{E_1 + E_2\} = \Pr\{E_1\} + \Pr\{E_2\} - \Pr\{E_1 E_2\}$.

Also, since E_1 and E_2 are independent, $\Pr\{E_1 E_2\} = \Pr\{E_1\}\Pr\{E_2\}$.

Then $\Pr\{E_1 + E_2\} = \Pr\{E_1\} + \Pr\{E_2\} - \Pr\{E_1\}\Pr\{E_2\} = \frac{1}{6} + \frac{1}{6} - (\frac{1}{6})(\frac{1}{6}) = \frac{11}{36}$.

Method 3:

$$\Pr\{\text{at least one 4 comes up}\} + \Pr\{\text{no 4 comes up}\} = 1$$

Then $\Pr$ {at least one 4 comes up} $= 1 - \Pr\{\text{no 4 comes up}\}$

$= 1 - \Pr\{\text{no 4 on 1st toss and no 4 on 2nd toss}\}$

$= 1 - \Pr\{\bar{E}_1 \bar{E}_2\} = 1 - \Pr\{\bar{E}_1\}\Pr\{\bar{E}_2\}$

$= 1 - (\frac{5}{6})(\frac{5}{6}) = \frac{11}{36}$.

6.8. One bag contains 4 white balls and 2 black balls; another contains 3 white balls and 5 black balls. If one ball is drawn from each bag, find the probability that (a) both are white, (b) both are black, (c) one is white and one black.

Solution:

Let $W_1 =$ event "white" ball from first bag, $W_2 =$ event "white" ball from second bag.

(a) $\Pr\{W_1 W_2\} = \Pr\{W_1\}\Pr\{W_2\} = (\frac{4}{4+2})(\frac{3}{3+5}) = \frac{1}{4}$

(b) $\Pr\{\bar{W}_1 \bar{W}_2\} = \Pr\{\bar{W}_1\}\Pr\{\bar{W}_2\} = (\frac{2}{4+2})(\frac{5}{3+5}) = \frac{5}{24}$

(c) The event "one is white and one is black" is the same as the event "either the first is white and the second is black or the first is black and the second is white", i.e. $W_1 \bar{W}_2 + \bar{W}_1 W_2$. Since events $W_1 \bar{W}_2$ and $\bar{W}_1 W_2$ are mutually exclusive, we have

$\Pr\{W_1 \bar{W}_2 + \bar{W}_1 W_2\} = \Pr\{W_1 \bar{W}_2\} + \Pr\{\bar{W}_1 W_2\}$

$= \Pr\{W_1\}\Pr\{\bar{W}_2\} + \Pr\{\bar{W}_1\}\Pr\{W_2\} = (\frac{4}{4+2})(\frac{5}{3+5}) + (\frac{2}{4+2})(\frac{3}{3+5}) = \frac{13}{24}$.

Another method:

The required probability is $1 - \Pr\{W_1 W_2\} - \Pr\{\bar{W}_1 \bar{W}_2\} = 1 - \frac{1}{4} - \frac{5}{24} = \frac{13}{24}$.

6.9. A and B play 12 games of chess of which 6 are won by A, 4 are won by B, and 2 end in a tie. They agree to play a tournament consisting of 3 games. Find the probability that (a) A wins all three games, (b) two games end in a tie, (c) A and B win alternately, (d) B wins at least one game.

Solution:

Let A_1, A_2, A_3 denote the events "A wins" in 1st, 2nd and 3rd games respectively
B_1, B_2, B_3 denote the events "B wins" in 1st, 2nd and 3rd games respectively
T_1, T_2, T_3 denote the events "there is a tie" in 1st, 2nd and 3rd games respectively.

On the basis of their past experience (empirical probability) we shall assume that
$$\Pr\{A \text{ wins any one game}\} = 6/12 = 1/2, \quad \Pr\{B \text{ wins any one game}\} = 4/12 = 1/3,$$
$$\Pr\{\text{any one game ends in a tie}\} = 2/12 = 1/6.$$

(a) $\Pr\{A \text{ wins all 3 games}\} = \Pr\{A_1 A_2 A_3\} = \Pr\{A_1\}\Pr\{A_2\}\Pr\{A_3\} = (\tfrac{1}{2})(\tfrac{1}{2})(\tfrac{1}{2}) = \tfrac{1}{8}$

assuming that the results of each game are independent of the results of any others, which appears to be justifiable (unless of course the players happen to be *psychologically influenced* by the other one's winning or losing).

(b) $\Pr\{2 \text{ games end in a tie}\} = \Pr\{\text{1st and 2nd } or \text{ 1st and 3rd } or \text{ 2nd and 3rd games end in a tie}\}$
$$= \Pr\{T_1 T_2 \bar{T}_3\} + \Pr\{T_1 \bar{T}_2 T_3\} + \Pr\{\bar{T}_1 T_2 T_3\}$$
$$= \Pr\{T_1\}\Pr\{T_2\}\Pr\{\bar{T}_3\} + \Pr\{T_1\}\Pr\{\bar{T}_2\}\Pr\{T_3\} + \Pr\{\bar{T}_1\}\Pr\{T_2\}\Pr\{T_3\}$$
$$= (\tfrac{1}{6})(\tfrac{1}{6})(\tfrac{5}{6}) + (\tfrac{1}{6})(\tfrac{5}{6})(\tfrac{1}{6}) + (\tfrac{5}{6})(\tfrac{1}{6})(\tfrac{1}{6}) = 15/216 = 5/72.$$

(c) $\Pr\{A \text{ and } B \text{ win alternately}\} = \Pr\{A \text{ wins then } B \text{ wins then } A \text{ wins } or \ B \text{ wins then } A \text{ wins then } B \text{ wins}\}$
$$= \Pr\{A_1 B_2 A_3 + B_1 A_2 B_3\} = \Pr\{A_1 B_2 A_3\} + \Pr\{B_1 A_2 B_3\}$$
$$= \Pr\{A_1\}\Pr\{B_2\}\Pr\{A_3\} + \Pr\{B_1\}\Pr\{A_2\}\Pr\{B_3\}$$
$$= (\tfrac{1}{2})(\tfrac{1}{3})(\tfrac{1}{2}) + (\tfrac{1}{3})(\tfrac{1}{2})(\tfrac{1}{3}) = 5/36.$$

(d) $\Pr\{B \text{ wins at least one game}\} = 1 - \Pr\{B \text{ wins no game}\}$
$$= 1 - \Pr\{\bar{B}_1 \bar{B}_2 \bar{B}_3\} = 1 - \Pr\{\bar{B}_1\}\Pr\{\bar{B}_2\}\Pr\{\bar{B}_3\}$$
$$= 1 - (\tfrac{2}{3})(\tfrac{2}{3})(\tfrac{2}{3}) = 19/27.$$

PROBABILITY DISTRIBUTIONS

6.10.   Find the probability of boys and girls in families with 3 children, assuming equal probabilities for boys and girls.

Solution:

Let B = event "boy in the family", and G = event "girl in the family". Then according to the assumption of equal probabilities, $\Pr\{B\} = \Pr\{G\} = \tfrac{1}{2}$. In families of 3 children the following mutually exclusive events can occur with the corresponding indicated probabilities.

(a)   3 boys (BBB). Then $\Pr\{BBB\} = \Pr\{B\}\Pr\{B\}\Pr\{B\} = \tfrac{1}{8}$.

Here we assume that the birth of a boy is not influenced in any manner by the fact that a previous child was also a boy, i.e. we assume the events are independent.

(b)   3 girls (GGG). Then as in part (a) or by symmetry, $\Pr\{GGG\} = \tfrac{1}{8}$.

(c)   2 boys and 1 girl ($BBG + BGB + GBB$). Then
$$\Pr\{BBG + BGB + GBB\} = \Pr\{BBG\} + \Pr\{BGB\} + \Pr\{GBB\}$$
$$= \Pr\{B\}\Pr\{B\}\Pr\{G\} + \Pr\{B\}\Pr\{G\}\Pr\{B\} + \Pr\{G\}\Pr\{B\}\Pr\{B\}$$
$$= \tfrac{1}{8} + \tfrac{1}{8} + \tfrac{1}{8} = \tfrac{3}{8}.$$

(d)   2 girls and 1 boy ($GGB + GBG + BGG$). As in (c) or by symmetry, probability is $\tfrac{3}{8}$.

If we call X the *random variable* showing the number of boys in families with 3 children, the probability distribution is as indicated in the table.

Number of boys X	0	1	2	3
Probability $p(X)$	1/8	3/8	3/8	1/8

6.11. Represent graphically the distribution of Problem 6.10.

Solution:

The graph can be represented either as in Fig. 6-3 or Fig. 6-4.

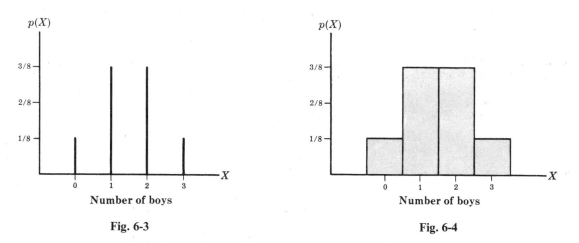

Fig. 6-3   Fig. 6-4

Note that the sum of the areas of the rectangles in Fig. 6-4 above is 1. In this figure, called a *probability histogram*, we are considering X as a continuous variable even though it is actually discrete, a procedure which is often found useful. Fig. 6-3 above, on the other hand, is used when one does not wish to consider the variable as continuous.

6.12. A continuous random variable X having values only between 0 and 4 has a density function given by $p(X) = \frac{1}{2} - aX$, where a is a constant. (*a*) Calculate a. (*b*) Find $\Pr\{1 < X < 2\}$.

Solution:

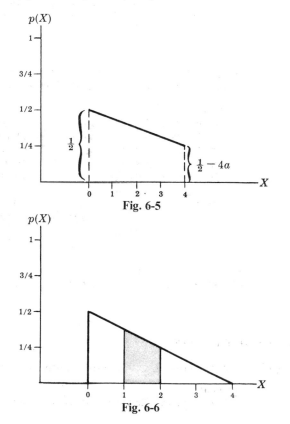

(*a*) The graph of $p(X) = \frac{1}{2} - aX$ is a straight line as indicated in Fig. 6-5.

To find a, we must realize that the total area under the line between $X = 0$ and $X = 4$ and above the X axis must be 1.

At $X = 0, p(X) = \frac{1}{2}$; at $X = 4, p(X) = \frac{1}{2} - 4a$.

Then we must choose a so that the trapezoidal area = 1.

Trapezoidal area = $\frac{1}{2}$(height)(sum of bases)
$= \frac{1}{2}(4)(\frac{1}{2} + \frac{1}{2} - 4a)$
$= 2(1 - 4a) = 1$

from which $(1 - 4a) = \frac{1}{2}$, $4a = \frac{1}{2}$, and $a = \frac{1}{8}$. This $(\frac{1}{2} - 4a)$ is actually equal to zero, and so the correct graphical representation is as given in Fig. 6-6.

(*b*) The required probability in the area between $X = 1$ and $X = 2$, shown shaded in Fig. 6-6.

From (*a*), $p(X) = \frac{1}{2} - \frac{1}{8}X$; then $p(1) = \frac{3}{8}$ and $p(2) = \frac{1}{4}$ are the ordinates at $X = 1$ and 2 respectively.

Required trapezoidal area is
$$\frac{1}{2}(1)(\frac{3}{8} + \frac{1}{4}) = \frac{5}{16}$$
which is the required probability.

MATHEMATICAL EXPECTATION

6.13. If a man purchases a raffle ticket, he can win a first prize of £5000 or a second prize of £2000 with probabilities 0·001 and 0·003. What should be a fair price to pay for the ticket?

Solution:
Expectation = (£5000)(0·001) + (£2000)(0·003) = £5 + £6 = £11, which is a fair price to pay.

6.14. In a given business venture a man can make a profit of £300 with probability 0·6 or take a loss of £100 with probability 0·4. Determine his expectation.

Solution:
Expectation = (£300)(0·6) + (£100)(0·4) = £180 − £40 = £140.

6.15. Find (a) $E(X)$, (b) $E(X^2)$, and (c) $E[(X − \bar{X})^2]$ for the following probability distribution.

X	8	12	16	20	24
$p(X)$	1/8	1/6	3/8	1/4	1/12

Solution:

(a) $E(X) = \Sigma X p(X) = (8)(1/8) + (12)(1/6) + (16)(3/8) + (20)(1/4) + (24)(1/12) = 16$
This represents the *mean* of the distribution.

(b) $E(X^2) = \Sigma X^2 p(X) = (8)^2(1/8) + (12)^2(1/6) + (16)^2(3/8) + (20)^2(1/4) + (24)^2(1/12) = 276$
This represents the *second moment* about the origin zero.

(c) $E[(X − \bar{X})^2] = \Sigma (X − \bar{X})^2 p(X)$
$= (8 − 16)^2(1/8) + (12 − 16)^2(1/6) + (16 − 16)^2(3/8) + (20 − 16)^2(1/4) + (24 − 16)^2(1/12) = 20$
This represents the *variance* of the distribution.

6.16. A bag contains 2 white balls and 3 black balls. Four persons A, B, C, D in the order named each draws one ball and does not replace it. The first to draw a white ball receives £20. Determine their expectations.

Solution:
Since only 3 black balls are present, one person must win on his first attempt. Denote by A, B, C and D the events "A wins", "B wins", "C wins" and "D wins" respectively.

$\Pr\{A \text{ wins}\} = \Pr\{A\} = \frac{2}{3+2} = \frac{2}{5}$. Then A's expectation $= \frac{2}{5}(£10) = £4$.

$\Pr\{A \text{ loses and } B \text{ wins}\} = \Pr\{\bar{A}B\} = \Pr\{\bar{A}\}\Pr\{B|\bar{A}\} = (\frac{3}{5})(\frac{2}{4}) = \frac{3}{10}$. Then B's expectation $= £3$.

$\Pr\{A \text{ loses and } B \text{ loses and } C \text{ wins}\} = \Pr\{\bar{A}\bar{B}C\} = \Pr\{\bar{A}\}\Pr\{\bar{B}|\bar{A}\}\Pr\{C|\bar{A}\bar{B}\} = (\frac{3}{5})(\frac{2}{4})(\frac{2}{3}) = \frac{1}{5}$.
Then C's expectation $= £2$.

$\Pr\{A \text{ loses and } B \text{ loses and } C \text{ loses and } D \text{ wins}\}$ is
$$\Pr\{\bar{A}\bar{B}\bar{C}D\} = \Pr\{\bar{A}\}\Pr\{\bar{B}|\bar{A}\}\Pr\{\bar{C}|\bar{A}\bar{B}\}\Pr\{D|\bar{A}\bar{B}\bar{C}\}$$
$$= (\tfrac{3}{5})(\tfrac{2}{4})(\tfrac{1}{3})(\tfrac{1}{1}) = \tfrac{1}{10}$$

Then D's expectation $= £1$.
Check: £4 + £3 + £2 + £1 = £10 and $\frac{2}{5} + \frac{3}{10} + \frac{1}{5} + \frac{1}{10} = 1$.

PERMUTATIONS

6.17. In how many ways can 5 differently coloured marbles be arranged in a row?

Solution:
We must arrange the 5 marbles in 5 positions thus: $- - - - -$.

The first position can be occupied by any one of 5 marbles, i.e. there are 5 ways of filling the first position. When this has been done there are 4 ways of filling the second position. Then there are 3 ways of filling the third position, 2 ways of filling the fourth position, and finally only 1 way of filling the last position. Therefore:

$$\text{Number of arrangements of 5 marbles in a row} = 5 \cdot 4 \cdot 3 \cdot 2 \cdot 1 = 5! = 120$$

In general,

$$\text{Number of arrangements of } n \text{ different objects in a row} = n(n-1)(n-2)\ldots 1 = n!$$

This is also called the number of *permutations* of n different objects taken n at a time and is denoted by $_nP_n$.

6.18. In how many ways can 10 people be seated on a bench if only 4 seats are available?

Solution:

The first seat can be filled in any one of 10 ways and when this has been done there are 9 ways of filling the second seat, 8 ways of filling the third seat and 7 ways of filling the fourth seat. Therefore:

$$\text{Number of arrangements of 10 people taken 4 at a time} = 10 \cdot 9 \cdot 8 \cdot 7 = 5040$$

In general,

$$\text{Number of arrangements of } n \text{ different objects taken } r \text{ at a time} = n(n-1)\ldots(n-r+1)$$

This is also called the number of *permutations* of n different objects taken r at a time and is denoted by $_nP_r$, $P(n,r)$ or $P_{n,r}$. Note that when $r = n$, $_nP_n = n!$ as in Problem 6.17.

6.19. Evaluate (a) $_8P_3$, (b) $_6P_4$, (c) $_{15}P_1$, (d) $_3P_3$.

Solution:

(a) $_8P_3 = 8 \cdot 7 \cdot 6 = 336$ (b) $_6P_4 = 6 \cdot 5 \cdot 4 \cdot 3 = 360$ (c) $_{15}P_1 = 15$ (d) $_3P_3 = 3 \cdot 2 \cdot 1 = 6$

6.20. It is required to seat 5 men and 4 women in a row so that the women occupy the even places. How many such arrangements are possible?

Solution:

The men may be seated in $_5P_5$ ways and the women in $_4P_4$ ways. Each arrangement of the men may be associated with each arrangement of the women.

Hence the required number of arrangements $= {}_5P_5 \cdot {}_4P_4 = 5! \, 4! = (120)(24) = 2880$.

6.21. How many 4 digit numbers can be formed with the 10 digits 0, 1, 2, 3, . . ., 9 if (a) repetitions are allowed, (b) repetitions are not allowed, (c) the last digit must be zero and repetitions are not allowed?

Solution:

(a) The first digit can be any one of 9 (since 0 is not allowed). The second, third and fourth digits can be any one of 10.
   Then $9 \cdot 10 \cdot 10 \cdot 10 = 9000$ numbers can be formed.

(b) The first digit can be any one of 9 (any one but 0).
   The second digit can be any one of 9 (any but that used for the first digit).
   The third digit can be any one of 8 (any but those used for the first two digits).
   The fourth digit can be any one of 7 (any but those used for the first three digits).

   Then $9 \cdot 9 \cdot 8 \cdot 7 = 4536$ numbers can be formed.

   Another method:
   The first digit can be any one of 9 and the remaining three can be chosen in $_9P_3$ ways. Then $9 \cdot {}_9P_3 = 9 \cdot 9 \cdot 8 \cdot 7$ $= 4536$ numbers can be formed.

(c) The first digit can be chosen in 9 ways, the second in 8 ways and the third in 7 ways.

   Then $9 \cdot 8 \cdot 7 = 504$ numbers can be formed.

   Another method:
   The first digit can be chosen in 9 ways and the next two digits in $_8P_2$ ways. Then $9 \cdot {}_8P_2 = 9 \cdot 8 \cdot 7 = 504$ numbers can be formed.

6.22. Four different mathematics books, six different physics books, and two different chemistry books are to be arranged on a shelf. How many different arrangements are possible if (*a*) the books in each particular subject must all stand together, (*b*) only the mathematics books must stand together?

Solution:

(*a*) The mathematics books can be arranged among themselves in $_4P_4 = 4!$ ways, the physics books in $_6P_6 = 6!$ ways, the chemistry books in $_2P_2 = 2!$ ways, and the three groups in $_3P_3 = 3!$ ways.

Then the required number of arrangements $= 4!\,6!\,2!\,3! = 207\,360$.

(*b*) Consider the four mathematics books as one big book. Then we have 9 books which can be arranged in $_9P_9 = 9!$ ways. In all of these ways the mathematics books are together. But the mathematics books can be arranged among themselves in $_4P_4 = 4!$ ways.

Then the required number of arrangements $= 9!\,4! = 8\,709\,120$.

6.23. Five red marbles, two white marbles, and three blue marbles are arranged in a row. If all the marbles of the same colour are not distinguishable from each other, how many different arrangements are possible?

Solution:

Assume that there are P different arrangements. Multiplying P by the number of ways of arranging (*a*) the five red marbles among themselves, (*b*) the two white marbles among themselves, and (*c*) the three blue marbles among themselves (i.e. multiplying P by $5!\,2!\,3!$), we obtain the number of ways of arranging the 10 marbles if they are distinguishable, i.e. $10!$.

Then $(5!\,2!\,3!)P = 10!$ and $P = 10!/(5!\,2!\,3!)$.

In general, the number of different arrangements of n objects of which n_1 are alike, n_2 are alike, . . ., n_k are alike is

$\dfrac{n!}{n_1!\,n_2!\ldots n_k!}$ where $n_1 + n_2 + \ldots + n_k = n$.

6.24. In how many ways can 7 people be seated at a round table if (*a*) they can sit anywhere, (*b*) 2 particular people must not sit next to each other?

Solution:

(*a*) Let 1 of them be seated anywhere. Then the remaining 6 people can be seated in $6! = 720$ ways, which is the total number of ways of arranging the 7 people in a circle.

(*b*) Consider the 2 particular people as one person. Then there are 6 people altogether and they can be arranged in $5!$ ways. But the 2 people considered as 1 can be arranged among themselves in $2!$ ways. Thus the number of ways of arranging 6 people at a round table with 2 particular people sitting together $= 5!\,2! = 240$.

Then using (*a*), the total number of ways in which 6 people can be seated at a round table so that the 2 people do not sit together $= 720 - 240 = 480$ ways.

COMBINATIONS

6.25. In how many ways can 10 objects be split into two groups containing 4 and 6 objects respectively?

Solution:

This is the same as the number of arrangements of 10 objects of which 4 objects are alike and 6 other objects are alike. By Problem 6.23 this is $\dfrac{10!}{4!\,6!} = \dfrac{10\cdot9\cdot8\cdot7}{4!} = 210$.

The problem is equivalent to finding the number of selections of 4 out of 10 objects (or 6 out of 10 objects), the order of selection being immaterial.

In general the number of selections of r out of n objects, called the number of *combinations* of n things taken r at a time is denoted by $_nC_r$, $C(n,r)$ or $\binom{n}{r}$ and is given by

$$_nC_r = \frac{n!}{r!\,(n-r)!} = \frac{n(n-1)\ldots(n-r+1)}{r!} = \frac{_nP_r}{r!}$$

6.26. Evaluate (a) $_7C_4$, (b) $_6C_5$, (c) $_4C_4$.

Solution:

(a) $_7C_4 = \dfrac{7!}{4!\,3!} = \dfrac{7 \cdot 6 \cdot 5 \cdot 4}{4!} = \dfrac{7 \cdot 6 \cdot 5}{3 \cdot 2 \cdot 1} = 35$.

(b) $_6C_5 = \dfrac{6!}{5!\,1!} = \dfrac{6 \cdot 5 \cdot 4 \cdot 3 \cdot 2}{5!} = 6$, or $_6C_5 = _6C_1 = 6$.

(c) $_4C_4$ is the number of selections of 4 objects taken all at a time, and there is only 1 such selection. Then $_4C_4 = 1$.

Note that formerly $_4C_4 = \dfrac{4!}{4!\,0!} = 1$ if we *define* $0! = 1$.

6.27. In how many ways can a committee of 5 people be chosen out of 9 people?

Solution:

$$_9C_5 = \frac{9!}{5!\,4!} = \frac{9 \cdot 8 \cdot 7 \cdot 6 \cdot 5}{5!} = 126$$

6.28. Out of 5 mathematicians and 7 physicists, a committee consisting of 2 mathematicians and 3 physicists is to be formed. In how many ways can this be done if (a) any mathematician and any physicist can be included, (b) one particular physicist must be on the committee, (c) two particular mathematicians cannot be on the committee?

Solution:

(a) 2 mathematicians out of 5 can be selected in $_5C_2$ ways.
3 physicists out of 7 can be selected in $_7C_3$ ways.

Total number of possible selections $= {}_5C_2 \cdot {}_7C_3 = 10 \cdot 35 = 250$.

(b) 2 mathematicians out of 5 can be selected in $_5C_2$ ways.
2 additional physicists out of 6 can be selected in $_6C_2$ ways.

Total number of possible selections $= {}_5C_2 \cdot {}_6C_2 = 10 \cdot 15 = 150$.

(c) 2 mathematicians out of 3 can be selected in $_3C_2$ ways.
3 physicists out of 7 can be selected in $_7C_3$ ways.

Total number of possible selections $= {}_3C_2 \cdot {}_7C_3 = 3 \cdot 35 = 105$.

6.29. A boy has five coins each of a different denomination. How many different sums of money can he form?

Solution:

Each coin can be dealt with in 2 ways, as it can be chosen or not chosen. Since each of the 2 ways of dealing with a coin is associated with 2 ways of dealing with each of the other coins, the number of ways of dealing with the five coins $= 2^5$ ways. But 2^5 ways includes the case in which no coin is chosen.

Hence the required number of sums of money $= 2^5 - 1 = 31$.

Another method:

He can select either 1 out of 5 coins, 2 out of 5 coins, . . ., 5 out of 5 coins. Then the required number of sums of money is

$$_5C_1 + {}_5C_2 + {}_5C_3 + {}_5C_4 + {}_5C_5 = 5 + 10 + 10 + 5 + 1 = 31$$

In general, for any positive integer n, $_nC_1 + {}_nC_2 + {}_nC_3 + \ldots + {}_nC_n = 2^n - 1$.

6.30. From 7 consonants and 5 vowels, how many words can be formed consisting of 4 different consonants and 3 different vowels? The words need not have meaning.

Solution:

The 4 different consonants can be selected in $_7C_4$ ways, the 3 different vowels can be selected in $_5C_3$ ways. and the resulting 7 different letters (4 consonants, 3 vowels) can then be arranged among themselves in $_7P_7 = 7!$ ways. Then:

$$\text{Number of words} = {}_7C_4 \cdot {}_5C_3 \cdot 7! = 35 \cdot 10 \cdot 5040 = 1\,764\,000$$

STIRLING'S APPROXIMATION TO $n!$

6.31. Evaluate 50!.

Solution:

For large n,
$$n! \sim \sqrt{2\pi n}\, n^n\, e^{-n}.$$

Thus
$$50! \sim \sqrt{2\pi(50)}\, 50^{50}\, e^{-50} = S.$$

To evaluate S use logarithms to the base 10. Then
$$\begin{aligned}
\log S = \log(\sqrt{100\pi}\ 50^{50}\, e^{-50}) &= \tfrac{1}{2}\log 100 + \tfrac{1}{2}\log \pi + 50\log 50 - 50\log e \\
&= \tfrac{1}{2}\log 100 + \tfrac{1}{2}\log 3\cdot142 + 50\log 50 - 50\log 2\cdot718 \\
&= \tfrac{1}{2}(2) + \tfrac{1}{2}(0\cdot4972) + 50(1\cdot6990) - 50(0\cdot4343) = 64\cdot4836
\end{aligned}$$

from which $S = 3\cdot04 \times 10^{64}$, a number which has 65 digits.

PROBABILITY AND COMBINATORIAL ANALYSIS

6.32. A box contains 8 red, 3 white and 9 blue balls. If 3 balls are drawn at random, determine the probability that (*a*) all 3 are red, (*b*) all 3 are white, (*c*) 2 are red and 1 is white, (*d*) at least 1 is white, (*e*) 1 of each colour is drawn, (*f*) the balls are drawn in the order red, white, blue.

Solution:

(*a*) **First method:**

Let R_1, R_2, R_3 denote the events "red ball on 1st draw", "red ball on 2nd draw", "red ball on 3rd draw" respectively. Then $R_1 R_2 R_3$ denotes the event that all 3 balls drawn are red.
$$\Pr\{R_1 R_2 R_3\} = \Pr\{R_1\}\Pr\{R_2 \mid R_1\}\Pr\{R_3 \mid R_1 R_2\} = (8/20)(7/19)(6/18) = 14/285$$

Second method:
$$\text{Required probability} = \frac{\text{number of selections of 3 out of 8 red balls}}{\text{number of selections of 3 out of 20 balls}} = \frac{{}_8C_3}{{}_{20}C_3} = \frac{14}{285}$$

(*b*) Using the method indicated in part (*a*), $\Pr\{\text{all 3 are white}\} = \dfrac{{}_3C_3}{{}_{20}C_3} = \dfrac{1}{1140}$.

The first method indicated in part (*a*) can also be used.

(*c*) $\Pr\{2 \text{ are red and } 1 \text{ is white}\}$

$$= \frac{(\text{selections of 2 out of 8 red balls})(\text{selections of 1 out of 3 white balls})}{\text{number of selections of 3 out of 20 balls}}$$

$$= \frac{{}_8C_2 \cdot {}_3C_1}{{}_{20}C_3} = \frac{7}{95}.$$

(*d*) $\Pr\{\text{none is white}\} = \dfrac{{}_{17}C_3}{{}_{20}C_3} = \dfrac{34}{57}.$ Then $\Pr\{\text{at least 1 is white}\} = 1 - \dfrac{34}{57} = \dfrac{23}{57}.$

(*e*) $\Pr\{1 \text{ of each colour is drawn}\} = \dfrac{{}_8C_1 \cdot {}_3C_1 \cdot {}_9C_1}{{}_{20}C_3} = \dfrac{18}{95}.$

(*f*) $\Pr\{\text{balls drawn in order red, white, blue}\} = \dfrac{1}{3!}\Pr\{1 \text{ of each colour is drawn}\}$

$$= \frac{1}{6}\left(\frac{18}{95}\right) = \frac{3}{95}, \text{ using } (e)$$

Another method: $\Pr\{R_1 W_2 B_3\} = \Pr\{R_1\}\Pr\{W_2 \mid R_1\}\Pr\{B_3 \mid R_1 W_2\} = (8/20)(3/19)(9/18) = 3/95.$

6.33. Five cards are drawn from a pack of 52 well-shuffled cards. Find the probability that (*a*) 4 are aces, (*b*) 4 are aces and 1 is a king, (*c*) 3 are tens and 2 are jacks, (*d*) a 9, 10, jack, queen, king are obtained in any order, (*e*) 3 are of any one suit and 2 are of another, (*f*) at least 1 ace is obtained.

Solution:

(*a*) $\Pr\{4 \text{ aces}\} = \dfrac{{}_4C_4 \cdot {}_{48}C_1}{{}_{52}C_5} = \dfrac{1}{54\,145}$.

(*b*) $\Pr\{4 \text{ aces and 1 king}\} = \dfrac{{}_4C_4 \cdot {}_4C_1}{{}_{52}C_5} = \dfrac{1}{649\,740}$

(*c*) $\Pr\{3 \text{ are tens and 2 are jacks}\} = \dfrac{{}_4C_3 \cdot {}_4C_2}{{}_{52}C_5} = \dfrac{1}{108\,290}$.

(*d*) $\Pr\{9, 10, \text{jack, queen, king in any order}\} = \dfrac{{}_4C_1 \cdot {}_4C_1 \cdot {}_4C_1 \cdot {}_4C_1 \cdot {}_4C_1}{{}_{52}C_5} = \dfrac{6}{162\,435}$.

(*e*) $\Pr\{3 \text{ of any one suit, 2 of another}\} = \dfrac{4 \, {}_{13}C_3 \cdot 3 \, {}_{13}C_2}{{}_{52}C_5} = \dfrac{429}{4165}$, since there are 4 ways of choosing the first suit and 3 ways of choosing the second suit.

(*f*) $\Pr\{\text{no ace}\} = \dfrac{{}_{48}C_5}{{}_{52}C_5} = \dfrac{35\,673}{54\,145}$. $\Pr\{\text{at least one ace}\} = 1 - \dfrac{35\,673}{54\,145} = \dfrac{18\,472}{54\,145}$.

6.34. Determine the probability of three 6's in 5 tosses of a fair die.

Solution:

Let the tosses of the die be represented by the 5 spaces $-\,-\,-\,-\,-$.

In each space we will have either the events 6 or non 6 ($\bar{6}$). For example, three 6's and two non 6's can occur as $6\,6\,\bar{6}\,6\,\bar{6}$ or $6\,\bar{6}\,6\,6\,\bar{6}$, etc.

Now the probability of an event such as $6\,6\,\bar{6}\,6\,\bar{6}$ is

$$\Pr\{6\,6\,\bar{6}\,6\,\bar{6}\} = \Pr\{6\} \Pr\{6\} \Pr\{\bar{6}\} \Pr\{6\} \Pr\{\bar{6}\} = \tfrac{1}{6} \cdot \tfrac{1}{6} \cdot \tfrac{5}{6} \cdot \tfrac{1}{6} \cdot \tfrac{5}{6} = (\tfrac{1}{6})^3(\tfrac{5}{6})^2$$

Similarly, $\Pr\{6\,\bar{6}\,6\,\bar{6}\,6\} = (\tfrac{1}{6})^3(\tfrac{5}{6})^2$, etc., for all events in which three 6's and two non 6's occur. But there are ${}_5C_3 = 10$ such events and these events are mutually exclusive. Hence the required probability is

$$\Pr\{6\,6\,\bar{6}\,6\,\bar{6} \text{ or } 6\,6\,6\,\bar{6}\,6 \text{ or etc.}\} = {}_5C_3(\tfrac{1}{6})^3(\tfrac{5}{6})^2 = \tfrac{125}{3888}.$$

In general, if $p = \Pr\{E\}$ and $q - \Pr\{\bar{E}\}$, then by using the same reasoning as given above the probability of getting exactly X E's in N trials is ${}_NC_X \, p^X \, q^{N-X}$.

6.35. A factory finds that on the average 20% of the bolts produced by a given machine will be defective for certain specified requirements. If 10 bolts are selected at random from the day's production of this machine, find the probability that (*a*) exactly 2 will be defective, (*b*) 2 or more will be defective, (*c*) more than 5 will be defective.

Solution:

(*a*) $\Pr\{2 \text{ defective bolts}\} = {}_{10}C_2(0.2)^2(0.8)^8 = 45(0.04)(0.1678) = 0.0302$, using reasoning similar to that of Problem 6.34.

(*b*) $\Pr\{2 \text{ or more defective bolts}\} = 1 - \Pr\{0 \text{ defective bolt}\} - \Pr\{1 \text{ defective bolt}\}$
$= 1 - {}_{10}C_0(0.2)^0(0.8)^{10} - {}_{10}C_1(0.2)^1(0.8)^9$
$= 1 - (0.8)^{10} - 10(0.2)(0.8)^9 = 1 - 0.1074 - 0.2684 = 0.6242.$

(*c*) $\Pr\{\text{more than 5 defec.}\} = \Pr\{6 \text{ defec.}\} + \Pr\{7 \text{ defec.}\} + \Pr\{8 \text{ defec.}\} + \Pr\{9 \text{ defec.}\} + \Pr\{10 \text{ defec.}\}$
$= {}_{10}C_6(0.2)^6(0.8)^4 + {}_{10}C_7(0.2)^7(0.8)^3 + {}_{10}C_8(0.2)^8(0.8)^2$
$+ {}_{10}C_9(0.2)^9(0.8) + {}_{10}C_{10}(0.2)^{10}$
$= 0.006\,37.$

6.36. If 1000 samples of 10 bolts each were taken in the preceding problem, in how many of these samples would we expect to find (*a*) exactly 2 defective bolts, (*b*) 2 or more defective bolts, (*c*) more than 5 defective bolts?

Solution:

(*a*) Expected number $= (1000)(0.0302) = 30$, by Problem 6.35(*a*).
(*b*) Expected number $= (1000)(0.6242) = 624$, by Problem 6.35(*b*).
(*c*) Expected number $= (1000)(0.006\,37) = 6$, by Problem 6.35(*c*).

SAMPLE SPACES AND EULER DIAGRAMS

6.37. (a) Set up a sample space for the single toss of a pair of fair dice. (b) From the sample space determine the probability that the sum in tossing a pair of dice is either 7 or 11.

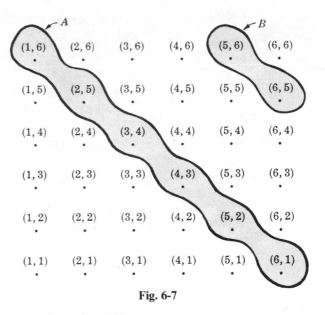

Fig. 6-7

Solution:

(a) The sample space consists of the set of points indicated in Fig. 6-7. The first co-ordinate of each point is the number on one die, and the second co-ordinate is the number on the other die.

There are 36 points in all, and to each point we assign a probability of 1/36. The sum of probabilities for all points in the space is 1.

(b) The sets of points corresponding to the events "sum 7" and "sum 11" are indicated by A and B respectively.

$\Pr\{A\}$ = sum of probabilities associated with each point in A = 6/36.
$\Pr\{B\}$ = sum of probabilities associated with each point in B = 2/36.
$\Pr\{A + B\}$ = sum of probabilities of points in A or in B or in both = $(6 + 2)/36 = 8/36 = 2/9$.
Note that in this case $\Pr\{A + B\} = \Pr\{A\} + \Pr\{B\}$. This is because A and B have no points in common, i.e. are mutually exclusive events.

6.38. Using a sample space, show that
(a) $\Pr\{A + B\} = \Pr\{A\} + \Pr\{B\} - \Pr\{AB\}$
(b) $\Pr\{A + B + C\} = \Pr\{A\} + \Pr\{B\} + \Pr\{C\} - \Pr\{AB\} - \Pr\{BC\} - \Pr\{AC\} + \Pr\{ABC\}$

Solution:

(a) Let A and B be two sets of points having points in common represented by AB, as in Fig. 6-8.

A is composed of $A\bar{B}$ and AB, while B is composed of $B\bar{A}$ and AB.

Totality of points in $A + B$ (either A or B or both)
= totality of points in A + totality of points in B − totality of points in AB.

Since the probability of an event or set = sum of probabilities associated with points of the set, we have

$$\Pr\{A + B\} = \Pr\{A\} + \Pr\{B\} - \Pr\{AB\}$$

Another method:

Let $A - AB$ denote the set of points in A but not in B (this is the same as $A\bar{B}$); then $A - AB$ and B are mutually exclusive (i.e. have no points in common). Also, $\Pr\{A - AB\} = \Pr\{A\} - \Pr\{AB\}$.

Thus,
$\Pr\{A + B\} = \Pr\{A - AB\} + \Pr\{B\} = \Pr\{A\} - \Pr\{AB\} + \Pr\{B\} = \Pr\{A\} + \Pr\{B\} - \Pr\{AB\}$

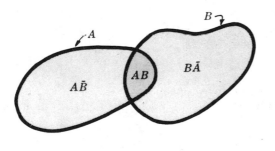

Fig. 6-8

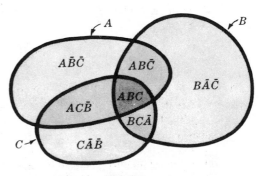

Fig. 6-9

(b) Let A, B and C be three sets of points as indicated in Fig. 6-9. The symbol $AB\overline{C}$ means the set of points in A and B but not in C, and the other symbols have similar meanings.

We can consider points which are either in A or B or C as points included in the 7 mutually exclusive sets of Fig. 6-9 above, 4 of which have been shown shaded and 3 unshaded. The required probability is given by

$$\Pr\{A + B + C\} = \Pr\{A\overline{B}\overline{C}\} + \Pr\{B\overline{C}\overline{A}\} + \Pr\{C\overline{A}\overline{B}\} + \Pr\{AB\overline{C}\} + \Pr\{BC\overline{A}\} + \Pr\{CA\overline{B}\} + \Pr\{ABC\}$$

Now to obtain $A\overline{B}\overline{C}$, for example, we remove points common to A and B and to A and C, but in so doing we have removed points common to A, B and C twice.

Hence $A\overline{B}\overline{C} = A - AB - AC + ABC$ and

$$\Pr\{A\overline{B}\overline{C}\} = \Pr\{A\} - \Pr\{AB\} - \Pr\{AC\} + \Pr\{ABC\}$$

Similarly, we find $\Pr\{B\overline{C}\overline{A}\} = \Pr\{B\} - \Pr\{BC\} - \Pr\{BA\} + \Pr\{BCA\}$
$$\Pr\{C\overline{A}\overline{B}\} = \Pr\{C\} - \Pr\{CA\} - \Pr\{CB\} + \Pr\{CAB\}$$
$$\Pr\{BC\overline{A}\} = \Pr\{BC\} - \Pr\{ABC\}$$
$$\Pr\{CA\overline{B}\} = \Pr\{CA\} - \Pr\{BCA\}$$
$$\Pr\{AB\overline{C}\} = \Pr\{AB\} - \Pr\{CAB\}$$
$$\Pr\{ABC\} = \Pr\{ABC\}$$

Adding these 7 equations and considering that $\Pr\{AB\} = \Pr\{BA\}$, etc., we obtain

$$\Pr\{A + B + C\} = \Pr\{A\} + \Pr\{B\} + \Pr\{C\} - \Pr\{AB\} - \Pr\{BC\} - \Pr\{AC\} + \Pr\{ABC\}$$

6.39. A survey of 500 students taking one or more courses in algebra, physics and statistics during one semester revealed the following numbers of students in the indicated subjects.

Algebra 329	Algebra and physics 83
Physics 186	Algebra and statistics 217
Statistics 295	Physics and statistics 63

How many students were taking (a) all 3 subjects, (b) algebra but not statistics, (c) physics but not algebra, (d) statistics but not physics, (e) algebra or statistics but not physics, (f) algebra but not physics or statistics?

Solution:

Let A denote the set of all students taking algebra, and denote by (A) the number of students belonging to this set. Similarly, let (B) denote the number taking physics and (C) the number taking statistics. Then $(A + B + C)$ denotes the number taking *either* algebra, physics or statistics or combinations, (AB) the number taking both algebra and physics, etc. As in the previous problem, it follows that

$$(A + B + C) = (A) + (B) + (C) - (AB) - (BC) - (AC) + (ABC)$$

(a) Substituting the given numbers in this expression we find

$$500 = 329 + 186 + 295 - 83 - 63 - 217 + (ABC)$$

or $(ABC) = 53$, which is the number of students taking algebra, physics *and* statistics. Note that the (empirical) probability of a student taking all 3 subjects is 53/500.

(b) To obtain the required information it is convenient to construct an Euler diagram showing the number of students belonging to each set.

Starting with the fact that 53 students are taking all 3 subjects, we deduce that the number taking both algebra and statistics but not physics is $217 - 53 = 164$, which is indicated in Fig. 6-10. From the given information the other numbers shown are obtained.

From the given data, the number taking algebra but not statistics $= 329 - 217$; or from Fig. 6-10, $82 + 30 = 112$.

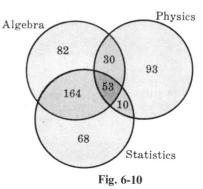

Fig. 6-10

(c) Number taking physics but not algebra $= 93 + 10 = 103$.

(d) Number taking statistics but not physics $= 68 + 164 = 232$.

(e) Number taking algebra or statistics but not physics $= 82 + 164 + 68 = 314$.

(f) Number taking algebra but not physics or statistics $= 82$.

Supplementary Problems

FUNDAMENTAL RULES OF PROBABILITY

6.40.   Determine the probability p, or an estimate of it, for each of the following events.

(a) A king, ace, jack of clubs or queen of diamonds appears in drawing a single card from a well-shuffled ordinary deck of cards.

(b) The sum 8 appears in a single toss of a pair of fair dice.

(c) A non-defective bolt will be found if out of 600 bolts already examined, 12 were defective.

(d) A 7 or 11 comes up in a single toss of a pair of fair dice.

(e) At least one head appears in three tosses of a fair coin.

Ans. (a) 5/26, (b) 5/36, (c) 0·98, (d) 2/9, (e) 7/8

6.41.   An experiment consists of drawing three cards in succession from a well-shuffled ordinary deck of cards. Let E_1 be the event "king" on first draw, E_2 the event "king" on second draw, and E_3 the event "king" on third draw. State in words the meaning of each of the following:

(a) $\Pr\{E_1\bar{E}_2\}$,   (b) $\Pr\{E_1 + E_2\}$,   (c) $\bar{E}_1 + \bar{E}_2$,   (d) $\Pr\{E_3 | E_1\bar{E}_2\}$,   (e) $\bar{E}_1\bar{E}_2\bar{E}_3$,   (f) $\Pr\{E_1E_2 + \bar{E}_2E_3\}$.

Ans. (a) Probability of king on the first draw and no king on the second draw.

(b) Probability of either a king on first draw or a king on second draw or both.

(c) No king on first draw or no king on second draw or both (no king on first and second draws).

(d) Probability of a king on third draw given that a king was drawn on first draw but not on second draw.

(e) No king on first, second and third draws.

(f) Probability of either king on first draw and king on second draw or no king on second draw and king on third draw.

6.42.   A ball is drawn at random from a box containing 10 red, 30 white, 20 blue and 15 orange marbles. Find the probability that it is (a) orange or red, (b) not red or blue, (c) not blue, (d) white , (e) red, white or blue.
Ans. (a) 1/3, (b) 3/5, (c) 11/15, (d) 2/5, (e) 4/5

6.43.   Two marbles are drawn in succession from the box of the preceding problem, replacement being made after each drawing. Find the probability that (a) both are white, (b) the first is red and the second is white, (c) neither is orange, (d) they are either red or white or both (red and white), (e) the second is not blue, (f) the first is orange, (g) at least one is blue, (h) at most one is red, (i) the first is white but the second is not, (j) only one is red.
Ans. (a) 4/25, (b) 4/75, (c) 16/25, (d) 64/225, (e) 11/15, (f) 1/5, (g) 104/225, (h) 221/225, (i) 6/25, (j) 52/225

6.44.   Work the preceding problem if there is no replacement after each drawing.
Ans (a) 29/185, (b) 2/37, (c) 118/185, (d) 52/185, (e) 11/15, (f) 1/5, (g) 86/185, (h) 182/185, (i) 9/37, (j) 26/111

6.45.   Find the probability of scoring a total of 7 points (a) once, (b) at least once, (c) twice in two tosses of a pair of fair dice.
Ans. (a) 5/18, (b) 11/36, (c) 1/36

6.46.   Two cards are drawn successively from an ordinary deck of 52 well-shuffled cards. Find the probability that (a) the first card is not a ten of clubs or an ace, (b) the first card is an ace but the second is not, (c) at least one card is a diamond, (d) the cards are not of the same suit, (e) not more than one card is a picture card (jack, queen, king), (f) the second card is not a picture card, (g) the second card is not a picture card given that the first was a picture card, (h) the cards are picture cards or spades or both.
Ans. (a) 47/52, (b) 16/221, (c) 15/34, (d) 13/17, (e) 210/221, (f) 10/13, (g) 40/51, (h) 77/442

6.47.   A box contains 9 tickets numbered from 1 to 9 inclusive. If 3 tickets are drawn from the box one at a time, find the probability that they are alternately either odd, even, odd or even, odd, even.   *Ans.* 5/18

6.48.   The odds in favour of A winning a game of chess against B are 3:2. If three games are to be played, what are the odds (a) in favour of A's winning at least two games out of three, (b) against A losing the first two games to B?
Ans. (a) 81 : 44, (b) 21 : 4

6.49. A purse contains 2 silver coins and 4 copper coins, and a second purse contains 4 silver coins and 3 copper coins. If a coin is selected at random from one of the two purses, what is the probability that it is a silver coin?   *Ans.* 19/42

6.50. The probability that a man will be alive in 25 years is $\frac{3}{5}$, and the probability that his wife will be alive in 25 years is $\frac{2}{3}$. Find the probability that (a) both will be alive, (b) only the man will be alive, (c) only the wife will be alive, (d) at least one will be alive.   *Ans.* (a) 2/5, (b) 1/5, (c) 4/15, (d) 13/15

6.51. Out of 800 families with 4 children each, what percentage would be expected to have (a) 2 boys and 2 girls, (b) at least one boy, (c) no girls, (d) at most 2 girls? Assume equal probabilities for boys and girls.
Ans. (a) 37·5%, (b) 93·75%, (c) 6·25%, (d) 68·75%

PROBABILITY DISTRIBUTIONS

6.52. If X is the random variable showing the number of boys in families with 4 children (see Prob. 6.51), (a) construct a table showing the probability distribution of X, (b) represent the distribution in (a) graphically.

Ans. (a)

X	0	1	2	3	4
$p(X)$	1/16	4/16	6/16	4/16	1/16

6.53. A continuous random variable X which can assume only values between $X = 2$ and 8 inclusive, has a density function given by $a(X + 3)$ where a is a constant. (a) Calculate a. Find (b) $\Pr\{3 < X < 5\}$, (c) $\Pr\{X \geqq 4\}$, (d) $\Pr\{|X - 5| < 0.5\}$.
Ans. (a) 1/48, (b) 7/24, (c) 3/4, (d) 1/6

6.54. Three marbles are drawn without replacement from an urn containing 4 red and 6 white marbles. If X is a random variable which denotes the total number of red marbles drawn, (a) construct a table showing the probability distribution of X, (b) represent the distribution graphically.

Ans. (a)

X	0	1	2	3
$p(X)$	1/6	1/2	3/10	1/30

6.55. For the preceding problem, find (a) $\Pr\{X = 2\}$, (b) $\Pr\{1 \leqq X \leqq 3\}$ and interpret.
Ans. (a) 3/10; this is the probability of drawing a total of 2 red marbles. (b) 5/6; this is the probability of drawing 1, 2 or 3 red marbles, i.e. drawing at least one red marble.

MATHEMATICAL EXPECTATION

6.56. What is a fair price to pay to enter a game in which one can win £25 with probability 0·2 and £10 with probability 0·4?
Ans. £9

6.57. If it rains, an umbrella salesman can earn £30 per day. If it is fair he can lose £6 per day. What is his expectation if the probability of rain is 0·3?   *Ans.* £4·80 per day

6.58. A and B play a game in which they toss a fair coin three times. The one obtaining heads first wins the game. If A tosses the coin first and if the total value of the stakes is £20, how much should be contributed by each in order that the game be considered fair?   *Ans.* A, £12·50; B, £7·50

6.59. Find (a) $E(X)$, (b) $E(X^2)$, (c) $E[(X - \bar{X})^2]$, and (d) $E(X^3)$ for the following probability distribution.

X	−10	−20	30
$p(X)$	1/5	3/10	1/2

Ans. (a) 7, (b) 590, (c) 541, (d) 10 900

6.60. Find the (a) mean, (b) variance, and (c) standard deviation of the distribution of X of Prob. 6.54 and interpret your results.   *Ans.* (a) 1·2, (b) 0·56, (c) $\sqrt{0·56} = 0·75$

6.61. A random variable assumes the value 1 with probability p, and 0 with probability $q = 1 - p$. Prove that (a) $E(X) = p$, (b) $E[(X - \bar{X})^2] = pq$.

6.62. Prove that (a) $E(2X + 3) = 2E(X) + 3$, (b) $E[(X - \bar{X})^2] = E(X^2) - [E(X)]^2$.

6.63. Let X and Y be two random variables having the same distribution. Show that $E(X + Y) = E(X) + E(Y)$.

PERMUTATIONS

6.64. Evaluate (a) $_4P_2$, (b) $_7P_5$, (c) $_{10}P_3$.   *Ans.* (a) 12, (b) 2520, (c) 720

6.65. For what value of n is $_{n+1}P_3 = \,_nP_4$?   *Ans.* $n = 5$

6.66. In how many ways can 5 people be seated on a sofa if there are only 3 seats available?   *Ans.* 60

6.67. In how many ways can 7 books be arranged on a shelf if (a) any arrangement is possible, (b) 3 particular books must always stand together, (c) two particular books must occupy the ends?   *Ans.* (a) 5040, (b) 720, (c) 240

6.68. How many numbers consisting of five different digits each can be made from the digits 1, 2, 3, . . ., 9 if (a) the numbers must be odd, (b) the first two digits of each number are even?   *Ans.* (a) 8400, (b) 2520

6.69. Solve the preceding problem if repetitions of the digits are allowed.   *Ans.* (a) 32 805, (b) 11 664

6.70. How may different three digit numbers can be made with 3 fours, 4 twos and 2 threes?   *Ans.* 20

6.71. In how many ways can 3 men and 3 women be seated at a round table if (a) no restriction is imposed, (b) two particular women must not sit together, (c) each woman is to be between two men?
Ans. (a) 120, (b) 72, (c) 12

COMBINATIONS

6.72. Evaluate (a) $_5C_3$, (b) $_8C_4$, (c) $_{10}C_8$.   *Ans.* (a) 10, (b) 70, (c) 45

6.73. For what value of n is $3 \cdot \,_{n+1}C_3 = 7 \cdot \,_nC_2$?   *Ans.* $n = 6$

6.74. In how many ways can 6 questions be selected out of 10?   *Ans.* 210

6.75. How many different committees of 3 men and 4 women can be formed from 8 men and 6 women?   *Ans.* 840

6.76. In how many ways can 2 men, 4 women, 3 boys and 3 girls be selected from 6 men, 8 women, 4 boys and 5 girls if (a) no restrictions are imposed, (b) a particular man and woman must be selected?   *Ans.* (a) 42 000, (b) 7000

6.77. In how many ways can a group of 10 people be divided into (a) two groups consisting of 7 and 3 people, (b) three groups consisting of 4, 3 and 2 people?   *Ans.* (a) 120, (b) 12 600

6.78. From 5 statisticians and 6 economists a committee consisting of 3 statisticians and 2 economists is to be formed. How many different committees can be formed if (a) no restrictions are imposed, (b) two particular statisticians must be on the committee, (c) one particular economist cannot be on the committee?   *Ans.* (a) 150, (b) 45, (c) 100

6.79. Find the number of (a) combinations and (b) permutations of four letters each that can be made from the letters of the word *Tennessee*?   *Ans.* (a) 17, (b) 163

6.80. Prove that $1 - \,_nC_1 + \,_nC_2 - \,_nC_3 + \cdots + (-1)^n \,_nC_n = 0$.

STIRLING'S APPROXIMATION TO n!

6.81. In how many ways can 30 individuals be selected out of 100?   *Ans.* $2 \cdot 95 \times 10^{25}$

6.82. Show that $_{2n}C_n = 2^{2n}/\sqrt{\pi n}$ approximately, for large values of n.

MISCELLANEOUS PROBLEMS

6.83. Three cards are drawn from a deck of 52 cards. Find the probability that (a) two are jacks and one is a king, (b) all cards are of one suit, (c) all cards of different suits, (d) at least two aces are drawn.
Ans. (a) 6/5525, (b) 22/425, (c) 169/425, (d) 73/5525

6.84. Find the probability of at least two 7's in four tosses of a pair of dice. *Ans.* 171/1296

6.85. If 10% of the rivets produced by a machine are defective, what is the probability that out of 5 rivets chosen at random (*a*) none will be defective, (*b*) one will be defective, (*c*) at least two will be defective? *Ans.* (*a*) 0·590 49, (*b*) 0·328 05, (*c*) 0·088 66

6.86. (*a*) Set up a sample space for the outcomes of 2 tosses of a fair coin using 1 to represent "heads" and 0 to represent "tails". (*b*) From this sample space determine the probability of at least one head. (*c*) Can you set up a sample space for the outcomes of 3 tosses of a coin? If so. determine with the aid of it the probability of at most two heads. *Ans.* (*b*) $\frac{3}{4}$, (*c*) $\frac{7}{8}$

6.87. A sample poll of 200 voters revealed the following information concerning three candidates A, B and C of a certain party who were running for three different offices.

28 in favour of both A and B	122 in favour of B or C but not A
98 in favour of A or B but not C	64 in favour of C but not A or B
42 in favour of B but not A or C	14 in favour of A and C but not B

How many of the voters were in favour of (*a*) all 3 candidates, (*b*) A irrespective of B or C, (*c*) B irrespective of A or C, (*d*) C irrespective of A or B, (*e*) A and B but not C, (*f*) only one of the candidates? *Ans.* (*a*) 8, (*b*) 78, (*c*) 86, (*d*) 102, (*e*) 20, (*f*) 142

6.88. (*a*) Prove that for any events E_1 and E_2, $\Pr\{E_1 + E_2\} \leqq \Pr\{E_1\} + \Pr\{E_2\}$. (*b*) Generalize the result in (*a*).

6.89. Let E_1, E_2, E_3 be 3 different events at least one of which is known to have occurred. Suppose that any of these events can result in another event A which is also known to have occurred. If all the probabilities $\Pr\{E_1\}$, $\Pr\{E_2\}$, $\Pr\{E_3\}$ and $\Pr\{A|E_1\}$, $\Pr\{A|E_2\}$, $\Pr\{A|E_3\}$ are assumed known prove that

$$\Pr\{E_1|A\} = \frac{\Pr\{E_1\}\,\Pr\{A|E_1\}}{\Pr\{E_1\}\,\Pr\{A|E_1\} + \Pr\{E_2\}\,\Pr\{A|E_2\} + \Pr\{E_3\}\,\Pr\{A|E_3\}}$$

with similar results for $\Pr\{E_2|A\}$ and $\Pr\{E_3|A\}$. This is known as *Bayes' rule or theorem*. It is useful in computing probabilities of various *hypotheses* E_1, E_2 or E_3 which have resulted in the event A. The result can be generalized.

6.90. Each of three identical jewellery boxes has two drawers. In each drawer of the first box there is a gold watch. In each drawer of the second box there is a silver watch. In one drawer of the third box there is a gold watch while in the other drawer there is a silver watch. If we select a box at random, open one of the drawers and find it to contain a silver watch, what is the probability that the other drawer has the gold watch? [Hint: Apply Problem 6.89.] *Ans.* $\frac{1}{3}$

6.91. A and B decide to meet between 3 and 4 p.m. but that each should wait no longer than 10 minutes for the other. Determine the probability that they meet. *Ans.* $\frac{11}{36}$

6.92. Two points are chosen at random on a line segment whose length is $a > 0$. Find the probability that the 3 line segments thus formed can be the sides of a triangle. *Ans.* $\frac{1}{4}$

CHAPTER 7

The Binomial, Normal and Poisson Distributions

THE BINOMIAL DISTRIBUTION

If p is the probability that an event will happen in any single trial (called the probability of a *success*) and $q = 1 - p$ is the probability that it will fail to happen in any single trial (calledthe probability of a *failure*) then the probability that the event will happen exactly X times in N trials (i.e. X successes and $N - X$ failures will occur) is given by

$$p(X) = {}_NC_X p^X q^{N-X} = \frac{N!}{X!(N-X)!} \, p^X q^{N-X} \tag{1}$$

where $X = 0, 1, 2, \ldots, N$ and $N! = N(N-1)(N-2)\ldots 1. \; 0! = 1$ by definition. (See Chap. 6, Prob. 6.34.)

Example 1. The probability of getting exactly 2 heads in 6 tosses of a fair coin is

$$_6C_2(\tfrac{1}{2})^2(\tfrac{1}{2})^{6-2} = \frac{6!}{2!\,4!} \, (\tfrac{1}{2})^6 = \tfrac{15}{64}$$

using (1) with $N = 6$, $X = 2$, and $p = q = \tfrac{1}{2}$.

Example 2. The probability of getting at least 4 heads in 6 tosses of a fair coin is

$$_6C_4(\tfrac{1}{2})^4(\tfrac{1}{2})^{6-4} + {}_6C_5(\tfrac{1}{2})^5(\tfrac{1}{2})^{6-5} + {}_6C_6(\tfrac{1}{2})^6 = \tfrac{15}{64} + \tfrac{6}{64} + \tfrac{1}{64} = \tfrac{11}{32}$$

The discrete probability distribution (*1*) is often called the *binomial distribution* since for $X = 0, 1, 2, \ldots,$ N it corresponds to successive terms in the *binomial formula* or *binomial expansion*

$$(q + p)^N = q^N + {}_NC_1 q^{N-1}p + {}_NC_2 q^{N-2}p^2 + \ldots + p^N \tag{2}$$

where $1, {}_NC_1, {}_NC_2, \ldots$ are called the *binomial coefficients*.

Example: $(q + p)^4 = q^4 + {}_4C_1 q^3 p + {}_4C_2 q^2 p^2 + {}_4C_3 q p^3 + p^4$
$= q^4 + 4q^3 p + 6q^2 p^2 + 4q p^3 + p^4$

The distribution (*1*) is also called the *Bernoulli distribution* after James Bernoulli who discovered it at the end of the 17th century.

SOME PROPERTIES OF THE BINOMIAL DISTRIBUTION are listed in the following table.

Table 7.1

Mean	$\mu = Np$
Variance	$\sigma^2 = Npq$
Standard deviation	$\sigma = \sqrt{Npq}$
Moment coefficient of skewness	$\alpha_3 = \dfrac{q - p}{\sqrt{Npq}}$
Moment coefficient of kurtosis	$\alpha_4 = 3 + \dfrac{1 - 6pq}{Npq}$

Example: In 100 tosses of a fair coin the mean number of heads is $\mu = Np = (100)(\frac{1}{2}) = 50$. This is the *expected* number of heads in 100 tosses of the coin.

The standard deviation is $\sigma = \sqrt{Npq} = \sqrt{(100)(\frac{1}{2})(\frac{1}{2})} = 5$.

THE NORMAL DISTRIBUTION

One of the most important examples of a continuous probability distribution is the *normal distribution, normal curve* or *Gaussian distribution* defined by the equation

$$Y = \frac{1}{\sigma\sqrt{2\pi}} e^{-\frac{1}{2}(X-\mu)^2/\sigma^2} \tag{3}$$

where $\mu =$ mean, $\sigma =$ standard deviation, $\pi = 3.141\,59\ldots$, $e = 2.718\,28\ldots$

The total area bounded by the curve (3) and the X axis is one; hence the area under the curve between two ordinates $X = a$ and $X = b$, where $a < b$, represents the probability that X lies between a and b, denoted by $\Pr\{a < X < b\}$.

When the variable X is expressed in terms of standard units, $z = (X - \mu)/\sigma$, equation (3) is replaced by the so-called *standard form*

$$Y = \frac{1}{\sqrt{2\pi}} e^{-\frac{1}{2}z^2} \tag{4}$$

In such case we say that z is *normally distributed with mean zero and variance one*.

A graph of this standardized normal curve is shown in Fig. 7-1. In this graph we have indicated the areas included between $z = -1$ and $+1$, $z = -2$ and $+2$, $z = -3$ and $+3$ as equal respectively to 68.27%, 95.45% and 99.73% of the total area which is one.

A table giving the areas under this curve bounded by the ordinates at $z = 0$ and any positive value of z, is given in Appendix II, Page 343. From this table the area between any two ordinates can be found by using the symmetry of the curve about $z = 0$.

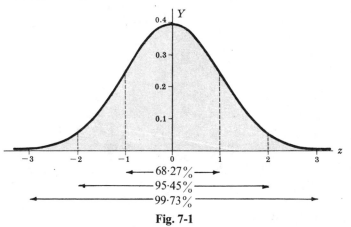

Fig. 7-1

Table 7.2

SOME PROPERTIES OF THE NORMAL DISTRIBUTION given by equation (3) are listed in Table 7.2.

Mean	μ
Variance	σ^2
Standard deviation	σ
Moment coefficient of skewness	$\alpha_3 = 0$
Moment coefficient of kurtosis	$\alpha_4 = 3$
Mean deviation	$\sigma\sqrt{2/\pi} = 0.7979\sigma$

RELATION BETWEEN BINOMIAL AND NORMAL DISTRIBUTIONS

If N is large and if neither p nor q is too close to zero, the binomial distribution can be closely approximated by a normal distribution with standardized variable given by $z = \dfrac{X - Np}{\sqrt{Npq}}$. The approximation becomes better with increasing N, and in the limiting case is exact. This is indicated in Tables 7.1 and 7.2 where it is clear that as N increases, the skewness and kurtosis for the binomial distribution approaches that of the normal distribution. In practice the approximation is very good if both Np and Nq are greater than 5.

THE POISSON DISTRIBUTION. The discrete probability distribution

$$p(X) \;=\; \frac{\lambda^X e^{-\lambda}}{X!} \qquad\qquad (X = 0, 1, 2, \ldots) \qquad\qquad (5)$$

where $e = 2\cdot718\,28\ldots$ and λ is a given constant, is called the *Poisson distribution*, after Poisson who discovered it in the early part of the 19th century.

The values of $p(X)$ can be computed by using the table on Page 348 which gives values of $e^{-\lambda}$ for various values of λ, or by using logarithms.

SOME PROPERTIES OF THE POISSON DISTRIBUTION

Some properties of the Poisson distribution are listed in the following table.

Table 7.3

Mean	$\mu = \lambda$
Variance	$\sigma^2 = \lambda$
Standard deviation	$\sigma = \sqrt{\lambda}$
Moment coefficient of skewness	$\alpha_3 = 1/\sqrt{\lambda}$
Moment coefficient of kurtosis	$\alpha_4 = 3 + 1/\lambda$

RELATION BETWEEN BINOMIAL AND POISSON DISTRIBUTIONS

In the binomial distribution (*1*), if N is large while the probability p of occurrence of an event is close to zero so that $q = (1 - p)$ is close to 1, the event is called a *rare event*. In practice we shall consider an event as rare if the number of trials is at least 50 ($N \geq 50$) while Np is less than 5. In such cases the binomial distribution (*1*) is very closely approximated by the Poisson distribution (*5*) with $\lambda = Np$. This is indicated by comparing Tables 7.1 and 7.3 above, for by placing $\lambda = Np$, $q \approx 1$ and $p \approx 0$ in Table 7.1 we get the results in Table 7.3.

Since there is a relation between the binomial and normal distributions, it follows that there also is a relation between the Poisson and normal distributions. It can in fact be shown that the Poisson distribution approaches a normal distribution with standardized variable $(X - \lambda)/\sqrt{\lambda}$ as λ increases indefinitely.

THE MULTINOMIAL DISTRIBUTION

If events $E_1, E_2, \ldots, E_K$ can occur with probabilities $p_1, p_2, \ldots, p_K$ respectively, then the probability that $E_1, E_2, \ldots, E_K$ will occur $X_1, X_2, \ldots, X_K$ times respectively is

$$\frac{N!}{X_1! X_2! \ldots X_K!} \, p_1^{X_1} p_2^{X_2} \cdots p_K^{X_K} \tag{6}$$

where $X_1 + X_2 + \ldots + X_K = N$.

This distribution, which is a generalization of the binomial distribution, is called the *multinomial distribution* since (6) is the general term in the *multinomial expansion* $(p_1 + p_2 + \ldots + p_K)^N$.

Example: If a fair die is tossed 12 times, the probability of getting 1, 2, 3, 4, 5 and 6 points exactly twice each is

$$\frac{12!}{2!\,2!\,2!\,2!\,2!\,2!} \, (\tfrac{1}{6})^2(\tfrac{1}{6})^2(\tfrac{1}{6})^2(\tfrac{1}{6})^2(\tfrac{1}{6})^2(\tfrac{1}{6})^2 = \frac{1925}{559\,872} = 0{\cdot}003\,44$$

The *expected* numbers of times that $E_1, E_2, \ldots, E_K$ will occur in N trials are $Np_1, Np_2, \ldots, Np_K$ respectively.

FITTING THEORETICAL DISTRIBUTIONS TO SAMPLE FREQUENCY DISTRIBUTIONS

When one has some indication of the distribution of a population by probabilistic reasoning or otherwise, it is often possible to fit such theoretical distributions (also called "model" or "expected" distributions) to frequency distributions obtained from a sample of the population. The method used in general consists of employing the mean and standard deviation of the sample to estimate the mean and sample of the population. See Problems 7.31, 7.33 and 7.34.

In order to test the *goodness of fit* of the theoretical distributions, use is made of the *chi-square test* which is given in Chapter 12.

In attempting to determine whether a normal distribution represents a good fit for given data, it is convenient to use *normal curve graph paper* or *probability graph paper* as it is sometimes called (see Problem 7.32).

Solved Problems

THE BINOMIAL DISTRIBUTION

7.1. Evaluate (*a*) 5!, (*b*) $\dfrac{6!}{2!\,4!}$, (*c*) $_8C_3$, (*d*) $_7C_5$, (*e*) $_4C_4$, (*f*) $_4C_0$.

Solution:

(*a*) $5! = 5\,.\,4\,.\,3\,.\,2\,.\,1 = 120$

(*b*) $\dfrac{6!}{2!\,4!} = \dfrac{6\,.\,5\,.\,5\,.\,3\,.\,2\,.\,1}{(2\,.\,1)(4\,.\,3\,.\,2\,.\,1)} = \dfrac{6\,.\,5}{2\,.\,1} = 15$

(*c*) $_8C_3 = \dfrac{8!}{3!\,(8-3)!} = \dfrac{8!}{3!\,5!} = \dfrac{8\,.\,7\,.\,6\,.\,5\,.\,4\,.\,3\,.\,2\,.\,1}{(3\,.\,2\,.\,1)(5\,.\,4\,.\,3\,.\,2\,.\,1)} = \dfrac{8\,.\,7\,.\,6}{3\,.\,2\,.\,1} = 56$

(*d*) $_7C_5 = \dfrac{7!}{5!\,2!} = \dfrac{7\,.\,6\,.\,5\,.\,4\,.\,3\,.\,2\,.\,1}{(5\,.\,4\,.\,3\,.\,2\,.\,1)(2\,.\,1)} = \dfrac{7\,.\,6}{2\,.\,1} = 21$

(*e*) $_4C_4 = \dfrac{4!}{4!\,0!} = 1$ (since $0! = 1$ by definition)   (*f*) $_4C_0 = \dfrac{4!}{0!\,4!} = 1$

7.2. Find the probability that in tossing a fair coin three times there will appear (*a*) 3 heads, (*b*) 2 heads and 1 tail, (*c*) 2 tails and 1 head, (*d*) 3 tails.

Solution:

Method 1:

Let H denote "heads" and T denote "tails" and suppose that we designate HTH, for example, to mean head on first toss, tail on second toss and then head on third toss.

Since 2 possibilities (head or tail) can occur on each toss, there are a total of $(2)(2)(2) = 8$ possible outcomes. These are

$$\text{HHH, HHT, HTH, HTT, TTH, THH, THT, TTT}$$

As each of these possibilities is equally likely, the probability of each is $\frac{1}{8}$.

(*a*) 3 heads (HHH) occur only once; hence the probability of three heads is $\frac{1}{8}$.

(*b*) 2 heads and 1 tail occur 3 times (HHT, HTH and THH); hence $\Pr\{2 \text{ heads and } 1 \text{ tail}\} = \frac{3}{8}$.

(*c*) 2 tails and 1 head occur 3 times (HTT, TTH and THT); then $\Pr\{2 \text{ tails and } 1 \text{ head}\} = \frac{3}{8}$.

(*d*) 3 tails (TTT) occur only once; then $\Pr\{\text{TTT}\} = \Pr\{3 \text{ tails}\} = \frac{1}{8}$.

Method 2: (using formula).

(*a*) $\Pr\{3 \text{ heads}\} \qquad\qquad = {}_3C_3(\frac{1}{2})^3(\frac{1}{2})^0 = (1)(\frac{1}{8})(1) = \frac{1}{8}$

(*b*) $\Pr\{2 \text{ heads and } 1 \text{ tail}\} = {}_3C_2(\frac{1}{2})^2(\frac{1}{2})^1 = (3)(\frac{1}{4})(\frac{1}{2}) = \frac{3}{8}$

(*c*) $\Pr\{1 \text{ head and } 2 \text{ tails}\} = {}_3C_1(\frac{1}{2})^1(\frac{1}{2})^2 = (3)(\frac{1}{2})(\frac{1}{4}) = \frac{3}{8}$

(*d*) $\Pr\{3 \text{ tails}\} \qquad\qquad = {}_3C_0(\frac{1}{2})^0(\frac{1}{2})^3 = (1)(1)(\frac{1}{8}) = \frac{1}{8}$

We can also proceed as in Chap. 6, Prob. 6.10.

7.3. Find the probability that in five tosses of a fair die a 3 appears (*a*) at no time, (*b*) once, (*c*) twice, (*d*) three times, (*e*) four times, (*f*) five times.

Solution:

Probability of 3 in a single toss $= p = \frac{1}{6}$.

Probability of no 3 in a single toss $= q = 1 - p = \frac{5}{6}$. Then,

(*a*) $\Pr\{3 \text{ occurs zero times}\} = {}_5C_0(\frac{1}{6})^0(\frac{5}{6})^5 = (1)(1)(\frac{5}{6})^5 = \frac{3125}{7776}$

(*b*) $\Pr\{3 \text{ occurs one time}\} = {}_5C_1(\frac{1}{6})^1(\frac{5}{6})^4 = (5)(\frac{1}{6})(\frac{5}{6})^4 = \frac{3125}{7776}$

(*c*) $\Pr\{3 \text{ occurs two times}\} = {}_5C_2(\frac{1}{6})^2(\frac{5}{6})^3 = (10)(\frac{1}{36})(\frac{125}{216}) = \frac{625}{3888}$

(*d*) $\Pr\{3 \text{ occurs three times}\} = {}_5C_3(\frac{1}{6})^3(\frac{5}{6})^2 = (10)(\frac{1}{216})(\frac{25}{36}) = \frac{125}{3888}$

(*e*) $\Pr\{3 \text{ occurs four times}\} = {}_5C_4(\frac{1}{6})^4(\frac{5}{6})^1 = (5)(\frac{1}{1296})(\frac{5}{6}) = \frac{25}{7776}$

(*f*) $\Pr\{3 \text{ occurs five times}\} = {}_5C_5(\frac{1}{6})^5(\frac{5}{6})^0 = (1)(\frac{1}{7776})(1) = \frac{1}{7776}$

Note that these probabilities represent the terms in the binomial expansion

$$(\tfrac{5}{6} + \tfrac{1}{6})^5 = (\tfrac{5}{6})^5 + {}_5C_1(\tfrac{5}{6})^4(\tfrac{1}{6}) + {}_5C_2(\tfrac{5}{6})^3(\tfrac{1}{6})^2 + {}_5C_3(\tfrac{5}{6})^2(\tfrac{1}{6})^3 + {}_5C_4(\tfrac{5}{6})(\tfrac{1}{6})^4 + (\tfrac{1}{6})^5$$

7.4. Write the binomial expansion for (*a*) $(q + p)^4$, (*b*) $(q + p)^6$.

Solution:

(*a*) $(q + p)^4 = q^4 + {}_4C_1 q^3 p + {}_4C_2 q^2 p^2 + {}_4C_3 q p^3 + p^4$

           $= q^4 + 4q^3 p + 6q^2 p^2 + 4q p^3 + p^4$

(*b*) $(q + p)^6 = q^6 + {}_6C_1 q^5 p + {}_6C_2 q^4 p^2 + {}_6C_3 q^3 p^3 + {}_6C_4 q^2 p^4 + {}_6C_5 q p^5 + p^6$

           $= q^6 + 6q^5 p + 15q^4 p^2 + 20q^3 p^3 + 15q^2 p^4 + 6q p^5 + p^6$

The coefficients 1, 4, 6, 4, 1 and 1, 6, 15, 20, 15, 6, 1 are called the *binomial coefficients* corresponding to $N - 4$ and $N = 6$ respectively. By writing these coefficients for $N = 0, 1, 2, 3, \ldots$ as shown at the right, we obtain an arrangement called *Pascal's triangle*. Note that the first and last numbers in each row are 1 and any other number can be obtained by adding the two numbers to the right and left of it in the preceding row.

```
            1
          1   1
        1   2   1
      1   3   3   1
    1   4   6   4   1
  1   5  10  10   5   1
1   6  15  20  15   6   1
```

7.5. Find the probability that in a family of 4 children will be (a) at least 1 boy, (b) at least 1 boy and 1 girl. Assume that the probability of a male birth is $\frac{1}{2}$.

Solution:

(a) $\Pr\{1 \text{ boy}\} = {}_4C_1(\frac{1}{2})^1(\frac{1}{2})^3 = \frac{1}{4}$ $\Pr\{3 \text{ boys}\} = {}_4C_3(\frac{1}{2})^3(\frac{1}{2})^1 = \frac{1}{4}$

$\Pr\{2 \text{ boys}\} = {}_4C_2(\frac{1}{2})^2(\frac{1}{2})^2 = \frac{3}{8}$ $\Pr\{4 \text{ boys}\} = {}_4C_4(\frac{1}{2})^4(\frac{1}{2})^0 = \frac{1}{16}$

Then $\Pr\{\text{at least 1 boy}\} = \Pr\{1 \text{ boy}\} + \Pr\{2 \text{ boys}\} + \Pr\{3 \text{ boys}\} + \Pr\{4 \text{ boys}\}$
$$= \frac{1}{4} + \frac{3}{8} + \frac{1}{4} + \frac{1}{16} = \frac{15}{16}$$

Another method: $\Pr\{\text{at least 1 boy}\} = 1 - \Pr\{\text{no boy}\} = 1 - (\frac{1}{2})^4 = 1 - \frac{1}{16} = \frac{15}{16}$

(b) $\Pr\{\text{at least 1 boy and 1 girl}\} = 1 - \Pr\{\text{no boy}\} - \Pr\{\text{no girl}\} = 1 - \frac{1}{16} - \frac{1}{16} = \frac{7}{8}$

7.6. Out of 2000 families with 4 children each, how many would you expect to have (a) at least 1 boy, (b) 2 boys, (c) 1 or 2 girls, (d) no girls? Refer to Problem 7.5(a).

Solution:

(a) Expected number of families with at least 1 boy $= 2000(\frac{15}{16}) = 1875$

(b) Expected number of families with 2 boys $= 2000 \cdot \Pr\{2 \text{ boys}\} = 2000(\frac{3}{8}) = 750$

(c) $\Pr\{1 \text{ or 2 girls}\} = \Pr\{1 \text{ girl}\} + \Pr\{2 \text{ girls}\} = \Pr\{1 \text{ boy}\} + \Pr\{2 \text{ boys}\} = \frac{1}{4} + \frac{3}{8} = \frac{5}{8}$

Expected number of families with 1 or 2 girls $= 2000(\frac{5}{8}) = 1250$.

(d) Expected number of families with no girls $= 2000(\frac{1}{16}) = 125$

7.7. If 20% of the bolts produced by a machine are defective, determine the probability that out of 4 bolts chosen at random (a) 1, (b) 0, (c) at most 2 bolts will be defective.

Solution:

The probability of a defective bolt is $p = 0.2$, of a non-defective bolt is $q = 1 - p = 0.8$.

(a) $\Pr\{1 \text{ defective bolt out of 4}\} = {}_4C_1(0.2)^1(0.8)^3 = 0.4096$

(b) $\Pr\{0 \text{ defective bolts}\} = {}_4C_0(0.2)^0(0.8)^4 = 0.4096$

(c) $\Pr\{2 \text{ defective bolts}\} = {}_4C_2(0.2)^2(0.8)^2 = 0.1536$. Then

$\Pr\{\text{at most 2 def. bolts}\} = \Pr\{0 \text{ def. bolts}\} + \Pr\{1 \text{ def. bolt}\} + \Pr\{2 \text{ def. bolts}\}$
$$= 0.4096 + 0.4096 + 0.1536 + 0.9728$$

7.8. The probability that an entering college student will graduate is 0.4. Determine the probability that out of 5 students (a) none, (b) one, (c) at least one will graduate.

Solution:

(a) $\Pr\{\text{none will graduate}\} = {}_5C_0(0.4)^0(0.6)^5 = 0.07776$, or about 0.08

(b) $\Pr\{\text{one will graduate}\} = {}_5C_1(0.4)^1(0.6)^4 = 0.2592$. or about 0.26

(c) $\Pr\{\text{at least one will graduate}\} = 1 - \Pr\{\text{none will graduate}\} = 0.92224$, or about 0.92

7.9. What is the probability of getting a total of 9 (a) twice, (b) at least twice in 6 tosses of a pair of dice?

Solution:

Each of the 6 ways in which the first die can fall can be associated with each of the 6 ways in which the second die can fall, so there are $6 \cdot 6 = 36$ ways in which both dice can fall. These are: 1 on the first die and 1 on the second die, 1 on the first die and 2 on the second die, etc., denoted by (1, 1), (1, 2), etc.

Of these 36 ways, all equally likely if the dice are fair, a total of 9 occurs in 4 cases: (3, 6), (4,5), (5, 4), (6, 3). Then the probability of a total of 9 in a single toss of a pair of dice is $p = 4/36 = 1/9$ and the probability of not getting a total of 9 in a single toss of a pair of dice is $q = 1 - p = 8/9$.

(a) $\Pr\{\text{two 9's in six tosses}\} = {}_6C_2(\frac{1}{9})^2(\frac{8}{9})^{6-2} = \dfrac{61\,440}{531\,441}$

(b) Pr{at least two 9's} = Pr{two 9's} + Pr{three 9's} + Pr{four 9's} + Pr{five 9's} + Pr{six 9's}

$$= {_6}C_2(\tfrac{1}{9})^2(\tfrac{8}{9})^4 + {_6}C_3(\tfrac{1}{9})^3(\tfrac{8}{9})^3 + {_6}C_4(\tfrac{1}{9})^4(\tfrac{8}{9})^2 + {_6}C_5(\tfrac{1}{9})^5\tfrac{8}{9} + {_6}C_6(\tfrac{1}{9})^6$$

$$= \frac{61\,440}{531\,441} + \frac{10\,240}{531\,441} + \frac{960}{531\,441} + \frac{48}{531\,441} + \frac{1}{531\,441} = \frac{72\,689}{531\,441}$$

Another method: Pr{at least two 9's} = 1 − Pr{zero 9's} − Pr{one 9}

$$= 1 - {_6}C_0(\tfrac{1}{9})^0(\tfrac{8}{9})^6 - {_6}C_0(\tfrac{1}{9})^1(\tfrac{8}{9})^5 = \frac{72\,689}{531\,441}$$

7.10. Evaluate   (a) $\displaystyle\sum_{X=0}^{N} Xp(X)$,   (b) $\displaystyle\sum_{X=0}^{N} X^2 p(X)$ where $p(X) = {_N}C_X\, p^X q^{N-X}$.

Solution:

(a) $\displaystyle\sum_{X=0}^{N} Xp(X) = \sum_{X=1}^{N} X \frac{N!}{X!\,(N-X)!} p^X q^{N-X} = Np\sum_{X=1}^{N} \frac{(N-1)!}{(X-1)!\,(N-X)!} p^{X-1} q^{N-X}$

$$= Np(q+p)^{N-1} = Np \qquad \text{since} \quad q + p = 1$$

(b) $\displaystyle\sum_{X=0}^{N} X^2 p(X) = \sum_{X=1}^{N} X^2 \frac{N!}{X!\,(N-X)!} p^X q^{N-X} = \sum_{X=1}^{N} [X(X-1) + X] \frac{N!}{X!\,(N-X)!} p^X q^{N-X}$

$$= \sum_{X=2}^{N} X(X-1) \frac{N!}{X!\,(N-X)!} p^X q^{N-X} + \sum_{X=1}^{N} X \frac{N!}{X!\,(N-X)!} p^X q^{N-X}$$

$$= N(N-1)p^2 \sum_{X=2}^{N} \frac{(N-2)!}{(X-2)!\,(N-X)!} p^{X-2} q^{N-X} + Np = N(N-1)p^2(q+p)^{N-2} + Np$$

$$= N(N-1)p^2 + Np$$

Note. The results in (a) and (b) are the *expectations* of X and X^2, denoted by $E(X)$ and $E(X^2)$ respectively (see Chapter 6).

7.11. If a variable is binomially distributed, determine its (a) mean μ, (b) variance σ^2.

Solution:

(a) μ = expectation of variable = $\displaystyle\sum_{X=0}^{N} Xp(X)$ = Np, by Problem 7.10(a).

(b) $\sigma^2 = \displaystyle\sum_{X=0}^{N} (X-\mu)^2 p(X) = \sum_{X=0}^{N} (X^2 - 2\mu X + \mu^2)\, p(X) = \sum_{X=0}^{N} X^2 p(X) - 2\mu \sum_{X=0}^{N} Xp(X) + \mu^2 \sum_{X=0}^{N} p(X)$

$= N(N-1)p^2 + Np - 2(Np)(Np) + (Np)^2(1) = Np - Np^2 = Np(1-p) = Npq$

using $\mu = Np$ and the result of Problem 7.10. It follows that the standard deviation of a variable which is binomially distributed is $\sigma = \sqrt{Npq}$.

Another method: $E[(X - \bar{X})]^2 = E(X^2) - [E(X)]^2 = N(N-1)p^2 + Np - N^2p^2 = Np - Np^2 = Npq$, by Prob. 6.62(b) of Chap. 6.

7.12. If the probability of a defective bolt is 0.1, find (a) the mean and (b) the standard deviation for the distribution of defective bolts in a total of 400.

Solution:

(a) Mean = Np = $400(0.1)$ = 40, i.e. we can *expect* 40 bolts to be defective.

(b) Variance = Npq = $400(0.1)(0.9)$ = 36. Hence the standard deviation = $\sqrt{36}$ = 6.

7.13. Find the moment coefficients of (a) skewness and (b) kurtosis of the distribution in Prob. 7.12.

Solution:

(a) Moment coefficient of skewness = $\dfrac{q - p}{\sqrt{Npq}} = \dfrac{0.9 - 0.1}{6} = 0.133$

Since this is positive the distribution is skewed to the right.

(b) Moment coefficient of kurtosis = $3 + \dfrac{1 - 6pq}{Npq} = 3 + \dfrac{1 - 6(0.1)(0.9)}{36} = 3.01$

The distribution is slightly *leptokurtic* with respect to the normal distribution (i.e. slightly more peaked, see Chap. 5).

THE NORMAL DISTRIBUTION

7.14. On a final examination in mathematics the mean was 72 and the standard deviation was 15. Determine the standard scores (i.e. marks in standard deviation units) of students receiving marks (*a*) 60, (*b*) 93, (*c*) 72.

Solution:

(*a*) $z = \dfrac{X - \bar{X}}{s} = \dfrac{60 - 72}{15} = -0.8$ (*b*) $z = \dfrac{X - \bar{X}}{s} = \dfrac{93 - 72}{15} = 1.4$

(*c*) $z = \dfrac{X - \bar{X}}{s} = \dfrac{72 - 72}{15} = 0$

7.15. Referring to Problem 7.14, find the marks corresponding to the standard scores (*a*) -1, (*b*) 1.6.

Solution:

(*a*) $X = \bar{X} + zs = 72 + (-1)(15) = 57$ (*b*) $X = \bar{X} + zs = 72 + (1.6)(15) = 96$

7.16. Two students were informed that they received standard scores of 0.8 and -0.4 respectively on a multiple choice examination in English. If their marks were 88 and 64 respectively, find the mean and the standard deviation of the examination marks.

Using the equation $X = \bar{X} + zs$, for first student   (1) $88 = \bar{X} + 0.8s$
for second student   (2) $64 = \bar{X} - 0.4s$.

Solving (1) and (2) simultaneously, mean $\bar{X} = 72$ and standard deviation $s = 20$.

7.17. Find the area under the normal curve in each of the cases (*a*) to (*g*) below. Use table on Page 343.

(*a*) Between $z = 0$ and $z = 1.2$.

In table on Page 343 proceed downwards under column marked z until entry 1.2 is reached. Then proceed right to column marked 0.

The result 0.3849 is the required area and represents the probability that z is between 0 and 1.2, denoted by $\Pr\{0 \leqq z \leqq 1.2\}$.

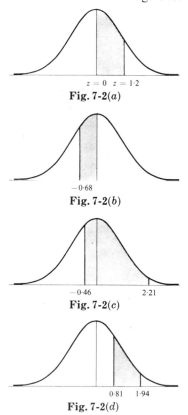

$z = 0 \quad z = 1.2$

Fig. 7-2(*a*)

(*b*) Between $z = -0.68$ and $z = 0$.

Required area = area between $z = 0$ and $z = 0.68$ (by symmetry).

To find the area between $z = 0$ and $z = 0.68$, proceed downwards under column marked z until entry 0.6 is reached. Then proceed right to column marked 8.

The result 0.2517 is the required area and represents the probability that z is between -0.68 and 0, denoted by $\Pr\{-0.68 \leqq z \leqq 0\}$.

-0.68

Fig. 7-2(*b*)

(*c*) Between $z = -0.46$ and $z = 2.21$.

Required area = (area between $z = 0.46$ and $z = 0$)
       + (area between $z = 0$ and $z = 2.21$)
= (area between $z = 0$ and $z = 0.46$)
       + (area between $z = 0$ and $z = 2.21$)
= $0.1772 + 0.4864 = 0.6636$

$-0.46 \qquad 2.21$

Fig. 7-2(*c*)

(*d*) Between $z = 0.81$ and $z = 1.94$.

Required area = (area between $z = 0$ and $z = 1.94$)
       − (area between $z = 0$ and $z = 0.81$)
= $0.4738 - 0.2910 = 0.1828$

$0.81 \quad 1.94$

Fig. 7-2(*d*)

(e) To the left of $z = 0.6$.

$$\text{Required area} = \text{(area to left of } z = 0)$$
$$\div \text{(area between } z = -0.6 \text{ and } z = 0)$$
$$= \text{(area to left of } z = 0)$$
$$- \text{(area between } z = 0 \text{ and } z = 0.6)$$
$$= 0.5 - 0.2258 = 0.2742$$

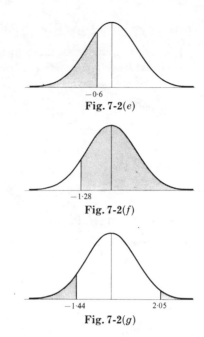

Fig. 7-2(e)

(f) To the right of $z = 1.28$.

$$\text{Required area} = \text{(area between } z = -1.28 \text{ and } z = 0)$$
$$+ \text{(area to right of } z = 0)$$
$$= 0.3997 + 0.5 = 0.8997.$$

This is the same as $\Pr\{z \geqq -1.28\}$.

Fig. 7-2(f)

(g) To the right of $z = 2.05$ and to the left of $z = -1.44$.

$$\text{Required area} = \text{total area}$$
$$- \text{(area between } z = -1.44 \text{ and } z = 0)$$
$$- \text{(area between } z = 0 \text{ and } z = 2.05)$$
$$= 1 - 0.4251 - 0.4798$$
$$= 1 - 0.9049 = 0.0951$$

Fig. 7-2(g)

7.18. Determine the value or values of z in each of the cases (a) to (c), where area refers to that under the normal curve.

(a) Area between 0 and z is 0.3770.

In Appendix II on Page 343, the entry 0.3770 is located to the right of the row marked 1.1 and under the column marked 6. Then the required $z = 1.16$.

By symmetry, $z = 1.16$ is another value of z. Thus $z = \pm 1.16$.

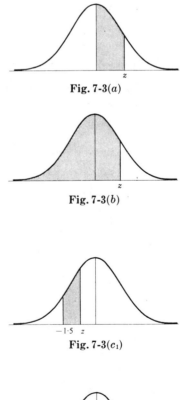

Fig. 7-3(a)

(b) Area to the left of z is 0.8621.

Since the area is greater than 0.5, z must be positive.

Area between 0 and $z = 0.8621 - 0.5 = 0.3621$, from which $z = 1.09$.

Fig. 7-3(b)

(c) Area between -1.5 and z is 0.0217.

If z were positive the area would be greater than the area between -1.5 and 0, which is 0.4332; hence z must be negative.

Case 1: z is negative but to the right of -1.5.

Area between -1.5 and z

$$= \text{(area between } -1.5 \text{ and } 0)$$
$$- \text{(area between } 0 \text{ and } z)$$
$$0.0217 = 0.4322 - \text{(area between } 0 \text{ and } z).$$

Then the area between 0 and $z = 0.4332 - 0.0217 = 0.4115$

from which $z = -1.35$.

Fig. 7-3(c₁)

Case 2: z is negative but to the left of -1.5.

Area between z and -1.5

$$= \text{(area between } z \text{ and } 0)$$
$$- \text{(area between } -1.5 \text{ and } 0)$$
$$0.0217 = \text{(area between } 0 \text{ and } z) - 0.4332$$

Then the area between 0 and $z = 0.0217 + 0.4332$
$$= 0.4549$$

and $z = -1.694$ by using linear interpolation; or, with slightly less precision, $z = -1.69$.

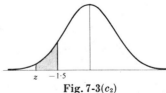

Fig. 7-3(c₂)

7.19. Find the ordinates of the normal curve at (*a*) $z = 0.84$, (*b*) $z = -1.27$, (*c*) $z = -0.05$.

Solution:

(*a*) In Appendix I on Page 342, proceed downwards under the column marked z until the entry 0.8 is reached. Then proceed right to the column marked 4. The entry 0.2803 is the required ordinate.

(*b*) By symmetry, (ordinate at $z = -1.27$) = (ordinate at $z = 1.27$) = 0.1781.

(*c*) (Ordinate at $z = -0.05$) = (ordinate at $z = 0.05$) = 0.3984.

7.20. The mean length of 500 laurel leaves from a certain bush is 151 mm and the standard deviation is 15 mm. Assuming that the lengths are normally distributed, find how many leaves measure (*a*) between 120 and 155 mm, (*b*) more than 185 mm.

Solution:

(*a*) Lengths recorded as being between 120 and 155 mm can actually have any value from 119·5 to 155·5 mm, assuming they are recorded to the nearest millimetre.

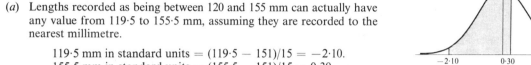

119·5 mm in standard units = $(119.5 - 151)/15 = -2.10$.
155·5 mm in standard units = $(155.5 - 151)/15 = 0.30$.
Required proportion of leaves
= (area between $z = -2.10$ and $z = 0.30$)
= (area between $z = -2.10$ and $z = 0$) + (area between $z = 0$ and $z = 0.30$)
= $0.4821 + 0.1179 = 0.6000$

Fig. 7-4(*a*)

Then the number of leaves having lengths between 120 and 155 mm = $500(0.6000) = 300$.

(*b*) Leaves measuring more than 185 mm must measure at least 185·5 mm.
185·5 mm in standard units = $(185.5 - 151)/15 = 2.30$.
Required proportion of leaves
= (area to right of $z = 2.30$)
= (area to right of $z = 0$)
    − (area between $z = 0$ and $z = 2.30$)
= $0.5 - 0.4893 = 0.0107$

Fig. 7-4(*b*)

Then the number of leaves having lengths greater than 185 mm = $500(0.0107) = 5$.

If L denotes the length of a leaf chosen at random, we can summarize the above results in terms of probability by writing

$$\Pr\{119.5 \leq L \leq 155.5\} = 0.6000 \quad \text{and} \quad \Pr\{L \geq 185.5\} = 0.0107$$

7.21. Determine how many leaves in the preceding problem measure (*a*) less than 128 mm, (*b*) 128 mm, (*c*) less than or equal to 128 mm.

Solution:

(*a*) Leaves measuring less than 128 mm must measure less than 127·5 mm.
127·5 mm in standard units = $(127.5 - 151)/15 = -1.57$.
Required proportion of leaves
= (area to left of $z = -1.57$)
= (area to left of $z = 0$)
    − (area between $z = -1.57$ and $z = 0$)
= $0.5 - 0.4418 = 0.0582$

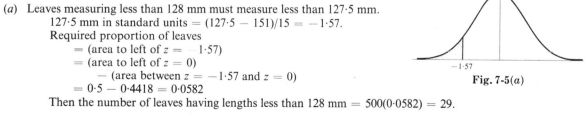

Fig. 7-5(*a*)

Then the number of leaves having lengths less than 128 mm = $500(0.0582) = 29$.

(*b*) Leaves measuring 128 mm have lengths between 127·5 and 128·5 mm. See Fig. 7-5(*b*) below.
127·5 mm in standard units = $(127.5 - 151)/15 = -1.57$.
128·5 mm in standard units = $(128.5 - 151)/15 = -1.50$.
Required proportion of leaves = (area between $z = -1.57$ and $z = -1.50$)
= (area between $z = -1.57$ and $z = 0$) − (area between $z = -1.50$ and $z = 0$)
= $0.4418 - 0.4332 = 0.0086$

Then the number of leaves having a length of 128 mm = $500(0.0086) = 4$.

Fig. 7-5(b)   **Fig. 7-5(c)**

(c) Leaves measuring less than or equal to 128 mm must measure less than 128·5 mm. See Fig. 7-5(c).

128·5 mm in standard units $= (128·5 - 151)/15 = -1·50$.

Required proportion of leaves $=$ (area to left of $z = -1·50$.
$=$ (area to left of $z = 0$) $-$ (area between $z = -1·50$ and $z = 0$)
$= 0·5 - 0·4332 = 0·0668$

Then the number of leaves having lengths of 128 mm or less $= 500(0·0668) = 33$.

Another method, using parts (a) and (b).
Number of leaves having lengths less than or equal to 128 mm
$=$ (number measuring less than 128 mm) $+$ (number measuring 128 mm) $= 29 + 4 = 33$.

7.22. The grades on a short quiz in biology were 0, 1, 2, . . ., 10 points, depending on the number answered correctly out of 10 questions. The mean grade was 6·7 and the standard deviation was 1·2. Assuming the grades to be normally distributed, determine (a) the percentage of students scoring 6 points, (b) the maximum grade of the lowest 10% of the class, (c) the minimum grade of the highest 10% of the class.

Solution:

(a) To apply the normal distribution to discrete data, it is necessary to treat the data as if they were continuous. Thus a score of 6 points is considered as 5·5 to 6·5 points. See Fig. 7-6(a).

5·5 in standard units $= (5·5 - 6·7)/1·2 = -1·0$.

6·5 in standard units $= (6·5 - 6·7)/1·2 = -0·17$.

Required proportion $=$ (area between $z = -1$ and $z = -0·17$)
$=$ (area between $z = -1$ and $z = 0$) $-$ (area between $z = -0·17$ and $z = 0$)
$= 0·3413 - 0·0675 = 0·2738 = 27\%$

Fig. 7-6(a)   **Fig. 7-6(b)**

(b) Let X_1 be the required maximum grade and z_1 the grade in standard units.

From Fig. 7-6(b) the area to the left of z_1 is $10\% = 0·10$; hence (area between z_1 and 0) $= 0·40$, and $z_1 = -1·28$ (very closely).

Then $z_1 = (X_1 - 6·7)/1·2 = 1·28$, and $X_1 = 5·2$ or 5 to the nearest integer.

(c) Let X_2 be the required minimum grade and z_2 the grade in standard units.

From (b), by symmetry, $z_2 = 1·28$. Then $(X_2 - 6·7)/1·2 = 1·28$, and $X_2 = 8·2$ or 8 to the nearest integer.

7.23. The mean inside diameter of a sample of 200 washers produced by a machine is 5·02 mm and the standard deviation is 0·05 mm. The purpose for which these washers are intended allows a maximum tolerance in the diameter of 4·96 to 5·08 mm, otherwise the washers are considered defective. Determine the percentage of defective washers produced by the machine, assuming the diameters are normally distributed.

Solution:
4·96 in standard units $= (4·96 - 5·02)/0·05 = -1·2$.
5·08 in standard units $= (5·08 - 5·02)/0·05 = 1·2$.

Proportion of non-defective washers
= (area under normal curve between $z = -1·2$ and $z = 1·2$)
= (twice the area between $z = 0$ and $z = 1·2$)
$= 2(0·3849) = 0·7698$, or 77%

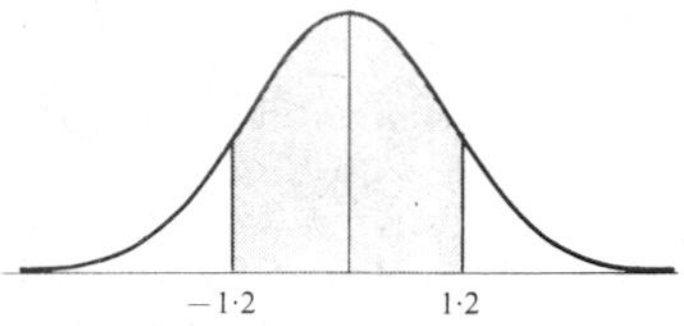

Thus the percentage of defective washers $= 100\% - 77\% = 23\%$.

Fig. 7-7

Note that if we think of the interval 4·96 to 5·08 mm as actually representing diameters of from 4·955 to 5·085 mm the above result is modified slightly. To two significant figures, however, the results are the same.

NORMAL APPROXIMATION TO BINOMIAL DISTRIBUTION

7.24. Find the probability of getting between 3 and 6 heads inclusive in 10 tosses of a fair coin by using (a) the binomial distribution, (b) the normal approximation to the binomial distribution.

Solution:
(a) $\Pr\{3 \text{ heads}\} = {}_{10}C_3(\tfrac{1}{2})^3(\tfrac{1}{2})^7 = \tfrac{15}{128}$ $\Pr\{5 \text{ heads}\} = {}_{10}C_5(\tfrac{1}{2})^5(\tfrac{1}{2})^5 = \tfrac{63}{256}$
$\Pr\{4 \text{ heads}\} = {}_{10}C_4(\tfrac{1}{2})^4(\tfrac{1}{2})^6 = \tfrac{105}{512}$ $\Pr\{6 \text{ heads}\} = {}_{10}C_6(\tfrac{1}{2})^6(\tfrac{1}{2})^4 = \tfrac{105}{512}$

Then $\Pr\{\text{between 3 and 6 heads inclusive}\} = \tfrac{15}{128} + \tfrac{105}{512} + \tfrac{63}{256} + \tfrac{105}{512} = \tfrac{99}{128} = 0·7734$.

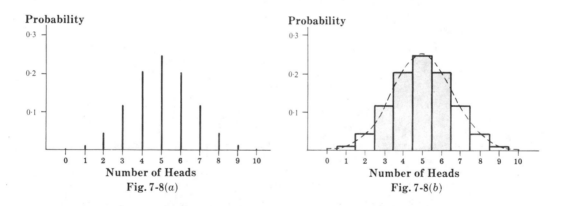

Fig. 7-8(a)   **Fig. 7-8(b)**

(b) The probability distribution for the number of heads in 10 tosses of the coin is shown graphically in Figs. 7-8(a) and 7-8(b) above, where Fig. 7-8(b) treats the data as if they were continuous. The required probability is the sum of the areas of the shaded rectangles in Fig. 7-8(b) and can be approximated by the area under the corresponding normal curve shown dashed.

Considering the data as continuous, it follows that 3 to 6 heads can be considered as 2·5 to 6·5 heads. Also, the mean and variance for the binomial distribution are given by $\mu = Np = 10(\tfrac{1}{2}) = 5$ and $\sigma = \sqrt{Npq} = \sqrt{(10)(\tfrac{1}{2})(\tfrac{1}{2})} = 1·58$.

Now 2·5 in standard units $= (2·5 - 5)/1·58 = -1·58$, and 6·5 in standard units $(6·5 - 5)/1·58 = 0·95$.

Required probability
= (area between $z = -1·58$ and $z = 0·95$)
= (area between $z = -1·58$ and $z = 0$)
    + (area between $z = 0$ and $z = 0·95$)
$= 0·4429 + 0·3289 = 0·7718$

which compares very well with the true value 0·7734 obtained in part (a). The accuracy is even better for larger values of N.

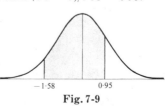

Fig. 7-9

7.25. A fair coin is tossed 500 times. Find the probability that the number of heads will not differ from 250 by (a) more than 10, (b) more than 30.

Solution:

$$\mu = Np = (500)(\tfrac{1}{2}) = 250 \qquad \sigma = \sqrt{Npq} = \sqrt{(500)(\tfrac{1}{2})(\tfrac{1}{2})} = 11\cdot18$$

(a) We require the probability that the number of heads will lie between 240 and 260 or, considering the data as continuous, between 239·5 and 260·5.

239·5 in standard units $= (239\cdot5 - 250)/11\cdot18 = -0\cdot94$, 260·5 in standard units $= 0\cdot94$.

Required probability $=$ (area under normal curve between $z = -0\cdot94$ and $z = 0\cdot94$)

$=$ (twice area between $z = 0$ and $z = 0\cdot94$) $= 2(0\cdot3264) = 0\cdot6528$

(b) We require the probability that the number of heads will lie between 220 and 280 or, considering the data as continuous, between 219·5 and 280·5.

219·5 in standard units $= (219\cdot5 - 250)/11\cdot18 = -2\cdot73$, 280·5 in standard units $= 2\cdot73$.

Required probability $=$ (twice area under normal curve between $z = 0$ and $z = -2\cdot73$)

$= 2(0\cdot4968) = 0\cdot9936$.

It follows that we can be very confident that the number of heads will not differ from that expected (250) by more than 30. Thus if it turned out that the *actual* number of heads was 280, we would strongly believe that the coin was not fair, i.e. was *loaded*.

7.26. A die is tossed 120 times. Find the probability that the face 4 will turn up (a) 18 times or less and (b) 14 times or less, assuming the die is fair.

Solution:

The face 4 has probability $p = \tfrac{1}{6}$ of turning up and probability $q = \tfrac{5}{6}$ of not turning up.

(a) We want the probability of the number of 4's being between 0 and 18. This is exactly equal to

$$_{120}C_{18}(\tfrac{1}{6})^{18}(\tfrac{5}{6})^{102} + {}_{120}C_{17}(\tfrac{1}{6})^{17}(\tfrac{5}{6})^{103} + \cdots + {}_{120}C_0(\tfrac{1}{6})^0(\tfrac{5}{6})^{120}$$

but since the labour involved in the computation is overwhelming, we use the normal approximation.

Considering the data as continuous, it follows that 0 to 18 4's can be considered as $-0\cdot5$ to 18·5 4's. Also,

$$\mu = Np = 120(\tfrac{1}{6}) = 20 \quad \text{and} \quad \sigma = \sqrt{Npq} = \sqrt{(120)(\tfrac{1}{6})(\tfrac{5}{6})} = 4\cdot08$$

Then $-0\cdot5$ in standard units $= (-0\cdot5 - 20)/4\cdot08 = -5\cdot02$, 18·5 in standard units $= -0\cdot37$.

Required probability $=$ (area under normal curve between $z = -5\cdot02$ and $z = -0\cdot37$)

$=$ (area between $z = 0$ and $z = -5\cdot02$)

$-$ (area between $z = 0$ and $z = -0\cdot37$)

$= 0\cdot5 - 0\cdot1443 = 0\cdot3557$

(b) We proceed as in part (a), replacing 18 by 14.

Then $-0\cdot5$ in standard units $= 5\cdot02$, 14·5 in standard units $= (14\cdot5 - 20)/4\cdot08 = 1\cdot35$.

Required probability $=$ (area under normal curve between $z = -5\cdot02$ and $z = -1\cdot35$)

$=$ (area between $z = 0$ and $z = -5\cdot02$)

$-$ (area between $z = 0$ and $z = -1\cdot35$)

$= 0\cdot5 - 0\cdot4115 = 0\cdot0885$

It follows that if we were to take repeated samples of 120 tosses of a die, a 4 should turn up 14 times or less in about one tenth of these samples.

THE POISSON DISTRIBUTION

7.27. Ten per cent of the tools produced in a certain manufacturing process turn out to be defective. Find the probability that in a sample of 10 tools chosen at random, exactly two will be defective by using (a) the binomial distribution, (b) the Poisson approximation to the binomial distribution.

Solution:

Probability of a defective tool $= p = 0\cdot1$.

(a) Pr {2 defective tools in 10} $= {}_{10}C_2(0\cdot1)^2(0\cdot9)^8 = 0\cdot1937$ or $0\cdot19$.

(b) $\lambda = Np = 10(0\cdot1) = 1$.

$$\text{Pr \{2 defective tools in 10\}} = \frac{\lambda^X e^{-\lambda}}{X!} = \frac{(1)^2 e^{-1}}{2!} = \frac{e^{-1}}{2} = \frac{1}{2e} = 0\cdot1839 \text{ or } 0\cdot18, \text{ using } e = 2\cdot718.$$

In general the approximation is good if $p \leq 0\cdot1$ and $\lambda = Np \leq 5$.

7.28. If the probability that an individual suffers a bad reaction from injection of a given serum is 0.001, determine the probability that out of 2000 individuals (*a*) exactly 3, (*b*) more than 2 individuals will suffer a bad reaction.

Solution:

$$\Pr\{X \text{ individuals suffer bad reaction}\} = \frac{\lambda^X e^{-\lambda}}{X!} = \frac{2^X e^{-2}}{X!}, \text{ where } \lambda = Np \; (2000)(0.001) = 2$$

(*a*) $\Pr\{3 \text{ individuals suffer bad reaction}) = \dfrac{2^3 e^{-2}}{3!} = \dfrac{4}{3e^2} = 0.180$

(*b*) $\Pr\{0 \text{ suffers}\} = \dfrac{2^0 e^{-2}}{0!} = \dfrac{1}{e^2}$ \qquad $\Pr\{1 \text{ suffers}\} = \dfrac{2^1 e^{-2}}{1!} = \dfrac{2}{e^2},$ \qquad $\Pr\{2 \text{ suffer}\} = \dfrac{2^2 e^{-2}}{2!} = \dfrac{2}{e^2}.$

$$\Pr\{\text{more than 2 suffer}\} = 1 - \Pr\{0 \text{ or } 1 \text{ or } 2 \text{ suffer}\}$$
$$= 1 - (1/e^2 + 2/e^2 + 2/e^2) = 1 - 5/e^2 = 0.323.$$

Note that according to the binomial distribution the required probabilities are

(*a*) ${}_{2000}C_3(0.001)^3 (0.999)^{1997}$

(*b*) $1 - \{ {}_{2000}C_0 (0.001)^0 (0.999)^{2000} + {}_{2000}C_1(0.001)^1 (0.999)^{1999} + {}_{2000}C_2 (0.001)^2 (0.999)^{1998} \}$

which would be very difficult to evaluate directly.

7.29. A Poisson distribution is given by $p(X) = \dfrac{(0.72)^X e^{-0.72}}{X!}$. Find (*a*) $p(0)$, (*b*) $p(1)$, (*c*) $p(2)$, (*d*) $p(3)$.

Solution:

(*a*) $p(0) = \dfrac{(0.72)^0 e^{-0.72}}{0!} = \dfrac{(1)e^{-0.72}}{1} = e^{-0.72} = 0.4868$, using Table of Appendix VI, Page 348.

(*b*) $p(1) = \dfrac{(0.72)^1 e^{-0.72}}{1!} = 0.72 e^{-0.72} = (0.72)(0.4868) = 0.3505$

(*c*) $p(2) = \dfrac{(0.72)^2 e^{-0.72}}{2!} = \dfrac{(0.5184)e^{-0.72}}{2} = (0.2592)(0.4868) = 0.1262$

Another method: $p(2) = \dfrac{0.72}{2} p(1) = (0.36)(0.3505) = 0.1262$

(*d*) $p(3) = \dfrac{(0.72)^3 e^{-0.72}}{3!} = \dfrac{0.72}{3} p(2) = (0.24)(0.1262) = 0.0303$

MULTINOMIAL DISTRIBUTION

7.30. A box contains 5 red balls, 4 white balls and 3 blue balls. A ball is selected at random from the box, its colour is noted, and then the ball is replaced. Find the probability that out of 6 balls selected in this manner 3 are red, 2 are white, and 1 is blue.

Solution:

Pt{red at any drawing} $= \frac{5}{12}$, Pr{white at any drawing} $= \frac{4}{12}$, Pr{blue at any drawing} $= \frac{3}{12}$.

Then Pr{3 are red, 2 are white, 1 is blue} $= \dfrac{6!}{3!\,2!\,1!}(\frac{5}{12})^3(\frac{4}{12})^2(\frac{3}{12})^1 = \dfrac{625}{5184}$

FITTING OF DATA BY THEORETICAL DISTRIBUTIONS

7.31. Fit a binomial distribution to the data of Problem 2.17, Chapter 2.

Solution:

We have $\Pr\{X \text{ heads in a toss of 5 pennies}\} = p(X) = {}_5C_X p^X q^{5-X}$, where p and q are the respective probabilities of a head and tail on a single toss of a penny.

By Problem 7.11(*a*) the mean number of heads is $\mu = Np = 5p$.

For the actual or observed frequency distribution, the mean number of heads is

$$\frac{\Sigma fX}{\Sigma f} = \frac{(38)(0) + (144)(1) + (342)(2) + (287)(3) + (164)(4) + (25)(5)}{1000} = \frac{2470}{1000} = 2.47$$

Equating the theoretical and actual means, $5p = 2\cdot47$ or $p = 0\cdot494$. Thus the fitted binomial distribution is given by $p(X) = {}_5C_X(0\cdot494)^X(0\cdot506)^{5-X}$.

In Table 7.4 these probabilities have been listed as well as the expected (theoretical) and actual frequencies. The fit is seen to be fair. The goodness of fit is investigated in Prob. 12.12 of Chap. 12.

Table 7.4

No. of heads, X	Pr{X heads}	Expected frequency	Observed frequency
0	0·0332	33·2 or 33	38
1	0·1619	161·9 or 162	144
2	0·3162	316·2 or 316	342
3	0·3087	308·7 or 309	287
4	0·1507	150·7 or 151	164
5	0·0294	29·4 or 29	25

7.32. Use probability graph paper to determine whether the frequency distribution of Table 2.1, Page 27, can be closely approximated by a normal distribution.

Solution:

Table 7.5

Mass (kg)	Cumulative relative frequency (%)
Less than 62·5	5·0
Less than 65·5	23·0
Less than 68·5	65·0
Less than 71·5	92·0
Less than 74·5	100·0

First the given frequency distribution is converted into a cumulative relative frequency distribution, as shown in Table 7.5. Then the cumulative relative frequencies expressed as percentages are plotted against upper class boundaries on special probability graph paper as shown in Fig. 7-10. The degree to which all plotted points lie on a line determines the closeness of fit of the given distribution to a normal distribution. From the above it is seen that there is a normal distribution which fits the data closely. See Problem 7.33.

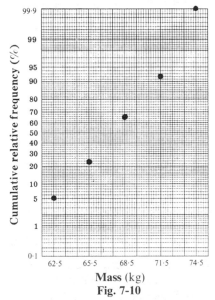

Mass (kg)

Fig. 7-10

7.33. Fit a normal curve to the data of Table 2.1, Page 27.

Solution:

Table 7.6

Mass (kg)	Class boundaries, X	z for class boundaries	Area under normal curve from 0 to z	Area for each class	Expected frequency	Observed frequency
60–62	59·5	−2·72	0·4967			5
	62·5	−1·70	0·4554	0·0413	4·13 or 4	
63–65				0·2068	20·68 or 21	18
	65·5	−0·67	0·2486 ⎫			
66–68			⎬ Add →	0·3892	38·92 or 39	42
	68·5	0·36	0·1406 ⎭			
69–71				0·2771	27·71 or 28	27
	71·5	1·39	0·4177			
72–74				0·0743	7·43 or 7	8
	74·5	2·41	0·4920			

$\overline{X} = 67\cdot45$ kg, $s = 2\cdot92$ kg

The work may be organized as in Table 7.6. In calculating z for the class boundaries, we use $z = (X - \bar{X})/s$ where the mean $\bar{X}$ and standard deviation s have been obtained respectively in Prob. 3.22 of Chap. 3 and Prob. 4.17 of Chap. 4.

In the fourth column, the areas under the normal curve from 0 to z have been obtained by using table of Appendix II, Page 343. From this we find the areas under the normal curve between successive values of z as in the fifth column. These are obtained by subtracting the successive areas in the fourth column when the corresponding z's have the same sign, and adding them when the z's have opposite signs (which occurs only once in the table). The reason for this is at once clear from a diagram.

Multiplying the entries in the fifth column (which represent relative frequencies) by the total frequency N (in this case $N = 100$) yields the expected frequencies as in the sixth column. It is seen that they agree well with the actual or observed frequencies of the last column.

If desired, the standard deviation modified by use of Sheppard's correction may be used (see Prob. 4.21(a), Chap. 4).

The "goodness of fit" of the distribution is considered in Prob. 12.13, Chap. 12.

7.34. Table 7.7 shows the number of days f in a 50 day period during which X automobile accidents occurred in a city. Fit a Poisson distribution to the data.

Solution:

Mean number of accidents is

$$\lambda = \frac{\Sigma f X}{\Sigma f} = \frac{(21)(0) + (18)(1) + (7)(2) + (3)(3) + (1)(4)}{50} = \frac{45}{50} = 0.90$$

Then, according to the Poisson distribution,

$$\Pr\{X \text{ accidents}\} = \frac{(0.90)^X e^{-0.90}}{X!}$$

Table 7.7

No. of Accidents, X	No. of Days, f
0	21
1	18
2	7
3	3
4	1

Total 50

In Table 7.8 are listed the probabilities for 0, 1, 2, 3 and 4 accidents as obtained from this Poisson distribution, as well as the expected or theoretical number of days during which X accidents take place (obtained by multiplying the respective probabilities by 50). For convenience of comparison, the fourth column giving the actual number of days has been repeated.

Table 7.8

No. of Accidents, X	$\Pr\{X \text{ accidents}\}$	Expected No. of Days	Actual No. of Days
0	0.4066	20.33 or 20	21
1	0.3659	18.30 or 18	18
2	0.1647	8.24 or 8	7
3	0.0494	2.47 or 2	3
4	0.0111	0.56 or 1	1

Note that the fit of the Poisson distribution to the given data is good.

For a true Poisson distribution, the variance $\sigma^2 = \lambda$. Computation of the variance of the given distribution gives 0.97. This compares favourably with the value 0.90 for λ, and this can be taken as further evidence for the suitability of the Poisson distribution in approximating the sample data.

Supplementary Problems

THE BINOMIAL DISTRIBUTION

7.35. Evaluate (a) $7!$, (b) $10!/(6!\ 4!)$, (c) $_9C_5$, (d) $_{11}C_8$, (e) $_6C_1$.
 Ans. (a) 5040, (b) 210, (c) 126, (d) 165, (e) 6

7.36. Expand (a) $(q + p)^7$, (b) $(q + p)^{10}$.
 Ans. (a) $q^7 + 7q^6p + 21q^5p^2 + 35q^4p^3 + 35q^3p^4 + 21q^2p^5 + 7qp^6 + p^7$
 (b) $q^{10} + 10q^9p + 45q^8p^2 + 120q^7p^3 + 210q^6p^4 + 252q^5p^5 + 210\ q^4p^6 + 120q^3p^7 + 45q^2p^8 + 10qp^9 + p^{10}$

7.37. Find the probability that in tossing a fair coin 6 times there will appear (a) 0, (b) 1, (c) 2, (d) 3, (e) 4, (f) 5, (g) 6 heads.
 Ans. (a) $\frac{1}{64}$, (b) $\frac{3}{32}$, (c) $\frac{15}{64}$, (d) $\frac{5}{16}$, (e) $\frac{15}{64}$, (f) $\frac{3}{32}$, (g) $\frac{1}{64}$

7.38. Find the probability of (a) 2 or more heads, (b) fewer than 4 heads in a single toss of 6 fair coins.
 Ans. (a) $\frac{57}{64}$, (b) $\frac{21}{32}$

7.39. If X denotes the number of heads in a single toss of 4 fair coins, find (a) $\Pr\{ = 3\}$, (b) $\Pr\{X < 2\}$, (c) $\Pr\{X \leq 2\}$, (d) $\Pr\{1 < X \leq 3\}$.   *Ans.* (a) $\frac{1}{4}$, (b) $\frac{5}{16}$, (c) $\frac{11}{16}$, (d) $\frac{5}{8}$

7.40. Out of 800 families with 5 children each, how many would you expect to have (a) 3 boys, (b) 5 girls, (c) either 2 or 3 boys. Assume equal probabilities for boys and girls.   *Ans.* 250, (b) 25, (c) 500

7.41. Find the probability of getting a total of 11 (a) once, (b) twice in two tosses of a pair of fair dice.
 Ans. (a) 17/162, (b) 1/324

7.42. What is the probability of getting a 9 exactly once in 3 throws with a pair of dice?   *Ans.* 64/243

7.43. Find the probability of guessing correctly at least 6 of the 10 answers on a true-false examination.   *Ans.* 193/512

7.44. An insurance salesman sells policies to 5 men, all of identical age and in good health. According to the actuarial tables the probability that a man of this particular age will be alive in 30 years hence is $\frac{2}{3}$. Find the probability that in 30 years (a) all 5 men, (b) at least 3 men, (c) only 2 men, (d) at least 1 man will be alive.
 Ans. (a) 32/243, (b) 192/243, (c) 40.243, (d) 242/243

7.45. Compute the (a) mean, (b) standard deviation, (c) moment coefficient of skewness, (d) moment coefficient of kurtosis for a binomial distribution in which $p = 0.7$ and $N = 60$. Interpret the results.
 Ans. (a) 42, (b) 3.550, (c) -0.1127, (d) 2.927

7.46. Show that if a binomial distribution with $N = 100$ is symmetric, its moment coefficient of kurtosis is 2.98.

7.47. Evaluate (a) $\Sigma\ (X - \mu)^3 p(X)$ and (b) $\Sigma\ (X - \mu)^4 p(X)$ for the binomial distribution.
 Ans. (a) $Npq(q - p)$, (b) $Npq(1 - 6pq) + 3N^2p^2q^2$

7.48. Prove the formulae on Page 122 for the moment coefficients of skewness and kurtosis.

THE NORMAL DISTRIBUTION

7.49. On a statistics examination the mean was 78 and the standard deviation was 10. (a) Determine the standard scores of two students whose marks were 93 and 62 respectively, (b) Determine the marks of two students whose standard scores were -0.6 and 1.2 respectively.   *Ans.* (a) 1.5, -1.6; (b) 72, 90

7.50. Find (a) the mean and (b) the standard deviation on an examination in which marks of 70 and 88 correspond to standard scores of -0.6 and 1.4 respectively.   *Ans.* (a) 75.4, (b) 9

7.51. Find the area under the normal curve between (a) $z = -1.20$ and $z = 2.40$, (b) $z = 1.23$ and $z = 1.87$, (c) $z = -2.35$ and $z = -0.50$.   *Ans.* (a) 0.8767, (b) 0.0786, (c) 0.2991

7.52. Find the area under the normal curve (a) to the left of $z = -1.78$, (b) to the left of $z = 0.56$, (c) to the right of $z = -1.45$, (d) corresponding to $z \geq 2.16$, (e) corresponding to $-0.80 \leq z \leq 1.53$, (f) to the left of $z = -2.52$ and to the right of $z = 1.83$.   *Ans.* (a) 0.0375, (b) 0.7123, (c) 0.9265, (d) 0.0154, (e) 0.7251, (f) 0.0395

7.53. If z is normally distributed with mean 0 and variance 1, find:
(a) $\Pr\{z \geqq -1.64\}$, (b) $\Pr\{-1.96 \leqq z \leqq 1.96\}$, (c) $\Pr\{|z| \geqq 1\}$.   *Ans.* (a) 0.9495, (b) 0.9500, (c) 0.6826

7.54. Find the value of z such that (a) the area to the right of z is 0.2266, (b) the area to the left of z is 0.0314, (c) the area between -0.23 and z is 0.5722, (d) the area between 1.15 and z is 0.0730, (e) the area between $-z$ and z is 0.9000.
Ans. (a) 0.75, (b) -1.86, (c) 2.08, (d) 1.625 or 0.849, (e) ± 1.645

7.55. Find z_1 if $\Pr\{z \geqq z_1\} = 0.84$, where z is normally distributed with mean 0 and variance 1.
Ans. -0.995

7.56. Find the ordinates of the normal curve at (a) $z = 2.25$, (b) $z = -0.32$, (c) $z = -1.18$.
Ans. (a) 0.0317, (b) 0.3790, (c) 0.1989

7.57. If the masses of 300 students are normally distributed with mean 68.0 kg and standard deviation 3.0 kg, how many students have masses (a) greater than 72 kg, (b) less than or equal to 64 kg, (c) between 65 and 71 kg inclusive, (d) equal to 68 kg. Assume the measurements to be recorded to the nearest kilogramme.   *Ans.* (a) 20, (b) 36, (c) 227, (d) 40

7.58. If the weights of ball bearings are normally distributed with mean 0.6140 newtons and standard deviation 0.0025 newtons, determine the percentage of ball bearings with weights (a) between 0.610 and 0.618 newtons inclusive, (b) greater than 0.617 newtons, (c) less than 0.608 newtons, (d) equal to 0.615 newtons.
Ans. (a) 93%, (b) 8.1%, (c) 0.47%, (d) 15%

7.59. The mean mark on a final examination was 72 and the standard deviation was 9. The top 10% of the students are to receive A's. What is the minimum mark a student must get in order to receive an A?   *Ans.* 84

7.60. If a set of measurements are normally distributed, what percentage of these differ from the mean by (a) more than half the standard deviation, (b) less than three quarters of the standard deviation?   *Ans.* (a) 61.7%, (b) 54.7%

7.61. If $\bar{X}$ is the mean and s is the standard deviation of a set of measurements which are normally distributed, what percentage of the measurement are (a) within the range $(\bar{X} \pm 2s)$, (b) outside the range $(\bar{X} \pm 1.2s)$, (c) greater than $(\bar{X} - 1.5s)$?
Ans. (a) 95.4%, (b) 23.0%, (c) 93.3%

7.62. In the preceding problem find the constant a such that the percentage of the cases (a) within the range $(\bar{X} \pm as)$ is 75%, (b) less than $(\bar{X} - as)$ is 22%.   *Ans.* (a) 1.15, (b) 0.77

NORMAL APPROXIMATION TO BINOMIAL DISTRIBUTION

7.63. Find the probability that 200 tosses of a coin will result in (a) between 80 and 120 heads inclusive, (b) less than 90 heads, (c) less than 85 or more than 115 heads, (d) exactly 100 heads.   *Ans.* (a) 0.9962, (b) 0.0687, (c) 0.0286, (d) 0.0558

7.64. Find the probability that a student can guess correctly the answers to (a) 12 or more out of 20, (b) 24 or more out of 40 questions on a true-false examination.   *Ans.* (a) 0.2511, (b) 0.1342

7.65. A machine produces bolts which are 10% defective. Find the probability that in a random sample of 400 bolts produced by this machine (a) at most 30, (b) between 30 and 50, (c) between 35 and 45, (d) 55 or more of the bolts will be defective.
Ans. (a) 0.0567, (b) 0.9198, (c) 0.6404, (d) 0.0079

7.66. Find the probability of getting more than 25 "sevens" in 100 tosses of a pair of fair dice.   *Ans.* 0.0089

THE POISSON DISTRIBUTION

7.67. If 3% of the electric bulbs manufactured by a company are defective, find the probability that in a sample of 100 bulbs (a) 0, (b) 1, (c) 2, (d) 3, (e) 4, (f) 5 bulbs will be defective.
Ans. (a) 0.049 79, (b) 0.1494, (c) 0.2241, (d) 0.2241, (e) 0.1680, (f) 0.1008

7.68. In the previous problem, find the probability that (a) more than 5, (b) between 1 and 3, (c) less than or equal to 2 bulbs will be defective.   *Ans.* (a) 0.0838, (b) 0.5976, (c) 0.4232

7.69. A bag contains one red and seven white marbles. A marble is drawn from the bag and its colour is observed. Then the marble is put back into the bag and the contents are thoroughly mixed. Using (a) the binomial distribution and (b) the Poisson approximation to the binomial distribution, find the probability that in 8 such drawings a red ball is selected exactly 3 times.   *Ans.* (a) 0.056 10, (b) 0.061 31

7.70. According to the National Office of Vital Statistics of the U.S. Department of Health, Education and Welfare, the average number of accidental drownings per year in the United States is 3·0 per 100 000 population. Find the probability that in a city of population 200 000 there will be (a) 0, (b) 2, (c) 6, (d) 8, (e) between 4 and 8, (f) fewer than 3 accidental drownings per year.   *Ans.* (a) 0·002 48, (b) 0·044 62, (c) 0·1607, (d) 0·1033, (e) 0·6964, (f) 0·0620

7.71. Between the hours of 2 and 4 p.m. the average number of phone calls per minute coming into the switchboard of a company is 2·5. Find the probability that during one particular minute there will be (a) 0, (b) 1, (c) 2, (d) 3, (e) 4 or fewer, (f) more than 6 phone calls.   *Ans.* (a) 0·082 08, (b) 0·2052, (c) 0·2565, (d) 0·2138, (e) 0·8911, (f) 0·0142

MULTINOMIAL DISTRIBUTION

7.72. A fair die is tossed 6 times. Find the probability that: (a) 1 "one", 2 "twos" and 3 "threes" turn up; (b) each side turns up only once.   *Ans.* (a) 5/3888, (b) 5/324

7.73. A box contains a very large number of red, white, blue and yellow marbles in the ratio 4 : 3 : 2 : 1 respectively. Find the probability that in 10 drawings: (a) 4 red, 3 white, 2 blue and 1 yellow marble will be drawn; (b) 8 red and 2 yellow marbles will be drawn.   *Ans.* (a) 0·000 348, (b) 0·000 295

7.74. Find the probability of not getting a 1, 2 or 3 in four tosses of a fair die.   *Ans.* $\frac{3}{8}$

FITTING OF DATA BY THEORETICAL DISTRIBUTIONS

7.75. Fit a binomial distribution to the following data.
Ans. $p(X) = {}_4C_X (0·32)^X (0·68)^{4-X}$. Expected frequencies are 32, 60, 43, 13 and 2 respectively.

X	0	1	2	3	4
f	30	62	46	10	2

7.76. Using probability graph paper to determine whether the data of Prob. 3.59 of Chap. 3 can be closely approximated by a normal distribution.

7.77. Fit a normal distribution to the data of Prob. 3.59, Chap. 3.
Ans. Expected frequencies are 1·7, 5·5, 12·0, 15·9, 13·7, 7·6, 2·7 and 0·6 respectively

7.78. Fit a normal distribution to the data of Prob. 3.61, Chap. 3.
Ans. Expected frequencies are 1·1, 4·0, 11·1, 23·9, 39·5, 50·2, 49·0, 36·6, 21·1, 9·4, 3·1 and 1·0 respectively

7.79. Fit a Poisson distribution to the data of Prob. 7.75 and compare with the fit obtained by using the binomial distribution.
Ans. Expected frequencies are 41·7, 53·4, 34·2, 14·6 and 4·7 respectively.

7.80. In 10 Prussian army corps units over a period of 20 years from 1875–1894 the number of deaths per army corps per year resulting from the kick of a horse are given in the following table. Fit a Poisson distribution to the data.

X	0	1	2	3	4
f	109	65	22	3	1

Ans. $p(X) = \dfrac{(0·61)^X e^{-0·61}}{X!}$, Expected frequencies are 108·7, 66·3, 20·2, 4·1 and 0·7 respectively.

CHAPTER 8

Elementary Sampling Theory

SAMPLING THEORY

Sampling theory is a study of relationships existing between a population and samples drawn from the population. It is of great value in many connections. For example it is useful in *estimation* of unknown population quantities (such as population mean, variance, etc.), often called *population parameters* or briefly *parameters*, from a knowledge of corresponding sample quantities (such as sample, mean, variance, etc.), often called *sample statistics* or briefly *statistics*. Estimation problems are considered in Chap. 9.

Sampling theory is also useful in determining whether observed differences between two samples are actually due to chance variation or whether they are really significant. Such questions arise, for example, in testing a new serum for use in treatment of a disease or in deciding whether one production process is better than another. Their answers involve use of so-called *tests of significance and hypotheses* which are important in the *theory of decisions*. These are considered in Chap. 10.

In general, a study of inferences made concerning a population by use of samples drawn from it, together with indications of the accuracy of such inferences using probability theory, is called *statistical inference*.

RANDOM SAMPLES. RANDOM NUMBERS

In order that conclusions of sampling theory and statistical inference be valid, samples must be chosen so as to be *representative* of a population. A study of methods of sampling and the related problems which arise is called the *design of the experiment*.

One way in which a representative sample may be obtained is by a process called *random sampling*, according to which each member of a population has an equal chance of being included in the sample. One technique for obtaining a random sample is to assign numbers to each member of the population, write these numbers on small pieces of paper, place them in an urn and then draw numbers from the urn, being careful to mix thoroughly before each drawing. This can be replaced by using a table of *random numbers* (see Page 349) specially constructed for such purposes. See Prob. 8.6.

SAMPLING WITH AND WITHOUT REPLACEMENT

If we draw a number from an urn, we have the choice of replacing or not replacing the number into the urn before a second drawing. In the first case the number can come up again and again, whereas in the second it can only come up once. Sampling where each member of a population may be chosen more than once is called *sampling with replacement*, while if each member cannot be chosen more than once it is called *sampling without replacement*.

Populations are either finite or infinite. If, for example, we draw 10 balls successively without replacement from an urn containing 100 balls, we are sampling from a finite population; while if we toss a coin 50 times and count the number of heads, we are sampling from an infinite population.

A finite population in which sampling is with replacement can theoretically be considered infinite, since any number of samples can be drawn without exhausting the population. For many practical purposes, sampling from a finite population which is very large can be considered as sampling from an infinite population.

SAMPLING DISTRIBUTIONS

Consider all possible samples of size N which can be drawn from a given population (either with or without replacement). For each sample we can compute a statistic, such as the mean, standard deviation, etc., which will vary from sample to sample. In this manner we obtain a distribution of the statistic which is called its *sampling distribution*.

If, for example, the particular statistic used is the sample mean, the distribution is called the *sampling distribution of means* or the *sampling distribution of the mean*. Similarly we could have sampling distributions of standard deviations, variances, medians, proportions, etc.

For each sampling distribution, we can compute the mean, standard deviation, etc. Thus we can speak of the mean and standard deviation of the sampling distribution of means, etc.

SAMPLING DISTRIBUTION OF MEANS

Suppose that all possible samples of size N are drawn without replacement from a finite population of size $N_p > N$. If we denote the mean and standard deviation of the sampling distribution of means by $\mu_{\bar{x}}$ and $\sigma_{\bar{x}}$ and the population mean and standard deviation by μ and σ respectively, then

$$\mu_{\bar{x}} = \mu \qquad \text{and} \qquad \sigma_{\bar{x}} = \frac{\sigma}{\sqrt{N}} \sqrt{\frac{N_p - N}{N_p - 1}} \qquad (1)$$

If the population is infinite or if sampling is with replacement, the above results reduce to

$$\mu_{\bar{x}} = \mu \qquad \text{and} \qquad \sigma_{\bar{x}} = \frac{\sigma}{\sqrt{N}} \qquad (2)$$

For large values of N ($N \geq 30$) the sampling distribution of means is approximately a normal distribution with mean $\mu_{\bar{x}}$ and standard deviation $\sigma_{\bar{x}}$ irrespective of the population (so long as the population mean and variance are finite and the population size is at least twice the sample size). This result for an infinite population is a special case of the *central limit theorem* of advanced probability theory which shows that the accuracy of the approximation improves as N gets larger. This is sometimes indicated by saying that the sampling distribution is *asymptotically normal*.

In case the population is normally distributed, the sampling distribution of means is also normally distributed even for small values of N (i.e. $N < 30$).

SAMPLING DISTRIBUTION OF PROPORTIONS

Suppose that a population is infinite and that the probability of occurrence of an event (called its success) is p while the probability of non-occurrence of the event is $q = 1 - p$. For example, the population may be all possible tosses of a fair coin in which the probability of the event "heads" is $p = \frac{1}{2}$.

Consider all possible samples of size N drawn from this population, and for each sample determine the proportion P of successes. In the case of the coin, P would be the proportion of heads turning up in N tosses. Then we obtain a *sampling distribution of proportions* whose mean μ_P and standard deviation σ_P are given by

$$\mu_P = p \qquad \text{and} \qquad \sigma_P = \sqrt{\frac{pq}{N}} = \sqrt{\frac{p(1-p)}{N}} \qquad (3)$$

which can be obtained from (2) by placing $\mu = p$ and $\sigma = \sqrt{pq}$.

For large values of N ($N \geq 30$) the sampling distribution is very closely normally distributed. Note that the population is *binomially distributed*.

The equations (3) are also valid for a finite population in which sampling is with replacement.

For finite populations in which sampling is without replacement, equations (3) are replaced by equations (1) with $\mu = p$ and $\sigma = \sqrt{pq}$.

Note that equations (*3*) are obtained most easily by dividing the mean and standard deviation (Np and $\sqrt{Npq}$) of the binomial distribution by N (see Chapter 7).

SAMPLING DISTRIBUTION OF DIFFERENCES AND SUMS

Suppose that we are given two populations. For each sample of size N_1 drawn from the first population let us compute a statistic S_1. This yields a sampling distribution for the statistic S_1 whose mean and standard deviation we denote by μ_{S_1} and σ_{S_1} respectively. Similarly, for each sample of size N_2 drawn from the second population let us compute a statistic S_2. This yields a sampling distribution for the statistic S_2 whose mean and standard deviation are denoted by μ_{S_2} and σ_{S_2}. From all possible combinations of these samples from the two populations we can obtain a distribution of the differences, $S_1 - S_2$, which is called the *sampling distribution of differences of the statistics*. The mean and standard deviation of this sampling distribution, denoted respectively by $\mu_{S_2-S_2}$ and $\sigma_{S_1-S_2}$, are given by

$$\mu_{S_1-S_2} = \mu_{S_1} - \mu_{S_2} \quad \text{and} \quad \sigma_{S_1-S_2} = \sqrt{\sigma_{S_1}^2 + \sigma_{S_2}^2} \tag{4}$$

provided that the samples chosen do not in any way depend on each other, i.e. the samples are *independent*.

If S_1 and S_2 are the sample means from the two populations, which we denote by $\bar{X}_1$ and $\bar{X}_2$, then the sampling distribution of the differences of means is given for infinite populations with mean and standard deviations μ_1, σ_2 and μ_2, σ_2 respectively by

$$\mu_{\bar{X}_1-\bar{X}_2} = \mu_{\bar{X}_1} - \mu_{\bar{X}_2} = \mu_1 - \mu_2 \quad \text{and} \quad \sigma_{\bar{X}_1-\bar{X}_2} = \sqrt{\sigma_{\bar{X}_1}^2 + \sigma_{\bar{X}_2}^2} = \sqrt{\frac{\sigma_1^2}{N_1} + \frac{\sigma_2^2}{N_2}} \tag{5}$$

using equations (*2*). The result also holds for finite populations if sampling is with replacement. Similar results can be obtained for finite populations in which sampling is without replacement by using equations (*1*).

Corresponding results can be obtained for the sampling distributions of differences of proportions from two binomially distributed populations with parameters p_1, q_1 and p_2, q_2 respectively. In this case S_1 and S_2 correspond to the proportion of successes, P_1 and P_2, and equations (*4*) yield the results

$$\mu_{P_1-P_2} = \mu_{P_1} - \mu_{P_2} = p_1 - p_2 \quad \text{and} \quad \sigma_{P_1-P_2} = \sqrt{\sigma_{P_1}^2 + \sigma_{P_2}^2} = \sqrt{\frac{p_1 q_1}{N_1} + \frac{p_2 q_2}{N_2}} \tag{6}$$

If N_1 and N_2 are large ($N_1, N_2 \geq 30$) the sampling distributions of differences of means or proportions are very closely normally distributed.

It is sometimes useful to speak of the *sampling distribution of the sum of statistics*. The mean and standard deviation of this distribution are given by

$$\mu_{S_1+S_2} = \mu_{S_1} + \mu_{S_2} \quad \text{and} \quad \sigma_{S_1+S_2} = \sqrt{\sigma_{S_1}^2 + \sigma_{S_2}^2} \tag{7}$$

assuming the samples are *independent*.

STANDARD ERRORS

The standard deviation of a sampling distribution of a statistic is often called its *standard error*. In Table 8.1 we have listed standard errors of sampling distributions for various statistics under the conditions of random sampling from an infinite (or very large) population or sampling with replacement from a finite population. Also listed are special remarks giving conditions under which results are valid and other pertinent statements.

The quantities μ, σ, p, μ_r and $\bar{X}, s, P, m_r$ denote respectively the population and sample means, standard deviations, proportions and rth moments about the mean.

It is noted that if the sample size N is large enough, the sampling distributions are normal or nearly normal. For this reason the methods are known as *large sampling methods*. When $N < 30$, samples are called *small*. The theory of *small samples*, or *exact sampling theory* as it is sometimes called, is treated in Chap. 11.

When population parameters such as σ, p or μ_r are unknown they may be estimated closely by their corresponding sample statistics, namely $\hat{s}$ (or $s = \sqrt{N/(g-1)s}$), P and m_r, if samples are large enough.

Table 8.1

Standard Errors for Some Sampling Distributions

Sampling Distribution	Standard Error	Special Remarks
Means	$$\sigma_{\bar{X}} = \frac{\sigma}{\sqrt{N}}$$	This is true for large or small samples. The sampling distribution of means is very nearly normal for $N \geqq 30$ even when the population is non-normal. $\mu_{\bar{X}} = \mu$, the population mean in all cases.
Proportions	$$\sigma_P = \sqrt{\frac{p(1-p)}{N}} = \sqrt{\frac{pq}{N}}$$	The remarks made for means apply here as well. $\mu_P = p$ in all cases.
Standard Deviations	$$(1) \quad \sigma_s = \frac{\sigma}{\sqrt{2N}}$$ $$(2) \quad \sigma_s = \sqrt{\frac{\mu_4 - \mu_2^2}{4N\mu_2}}$$	For $N \geqq 100$, the sampling distribution of s is very nearly normal. σ_s is given by (1) only if the population is normal (or approximately normal). If the population is non-normal, (2) can be used. Note that (2) reduces to (1) when $\mu_2 = \sigma^2$ and $\mu_4 = 3\sigma^4$, which is true for normal populations. For $N \geqq 100$, $\mu_s = \sigma$ very nearly.
Medians	$$\sigma_{\text{med.}} = \sigma \sqrt{\frac{\pi}{2N}} = \frac{1 \cdot 2533\sigma}{\sqrt{N}}$$	For $N \geqq 30$, the sampling distribution of the median is very nearly normal. The given result holds only if the population is normal (or approximately normal). $\mu_{\text{med.}} = \mu$.
First and Third Quartiles	$$\sigma_{Q_1} = \sigma_{Q_3} = \frac{1 \cdot 3626\sigma}{\sqrt{N}}$$	The remarks made for medians apply here as well. μ_{Q_1} and μ_{Q_3} are very nearly equal to the first and third quartiles of the population. Note that $\sigma_{Q_2} = \sigma_{\text{med.}}$.
Deciles	$$\sigma_{D_1} = \sigma_{D_9} = \frac{1 \cdot 7094\sigma}{\sqrt{N}}$$ $$\sigma_{D_2} = \sigma_{D_8} = \frac{1 \cdot 4288\sigma}{\sqrt{N}}$$ $$\sigma_{D_3} = \sigma_{D_7} = \frac{1 \cdot 3180\sigma}{\sqrt{N}}$$ $$\sigma_{D_4} = \sigma_{D_6} = \frac{1 \cdot 2680\sigma}{\sqrt{N}}$$	The remarks made for medians apply here as well. $\mu_{D_1}, \mu_{D_2}, \ldots$ are very nearly equal to the first, second, $\ldots$ deciles of the population. Note that $\sigma_{D_5} = \sigma_{\text{med.}}$.
Semi-interquartile Ranges	$$\sigma_Q = \frac{0 \cdot 7867\sigma}{\sqrt{N}}$$	The remarks made for medians apply here as well. μ_Q is very nearly equal to the population semi-interquartile range.
Variances	$$(1) \quad \sigma_{s^2} = \sigma^2 \sqrt{\frac{2}{N}}$$ $$(2) \quad \sigma_{s^2} = \sqrt{\frac{\mu_4 - \mu_2^2}{N}}$$	The remarks made for standard deviation apply here as well. Note that (2) yields (1) in case the population is normal. $\mu_{s^2} = \sigma^2(N-1)/N$, which is very nearly σ^2 for large N.
Coefficients of Variation	$$\sigma_V = \frac{v}{\sqrt{2N}} \sqrt{1 + 2v^2}$$	Here $v = \sigma/\mu$ is the population coefficient of variation. The given result holds for normal (or nearly normal) populations and $N \geqq 100$.

Solved Problems

SAMPLING DISTRIBUTION OF MEANS

8.1. A population consists of the five numbers 2, 3, 6, 8, 11. Consider all possible samples of size two which can be drawn with replacement from this population. Find (*a*) the mean of the population, (*b*) the standard deviation of the population, (*c*) the mean of the sampling distribution of means, (*d*) the standard deviation of the sampling distribution of means, i.e. the standard error of means.

Solution:

(*a*) $\mu = \dfrac{2 + 3 + 6 + 8 + 11}{5} = \dfrac{30}{5} = 6.0$

(*b*) $\sigma^2 = \dfrac{(2 - 6)^2 + (3 - 6)^2 + (6 - 6)^2 + (8 - 6)^2 + (11 - 6)^2}{5} = \dfrac{16 + 9 + 0 + 4 + 25}{5} = 10.8$, and $\sigma = 3.29$.

(*c*) There are 5(5) = 25 samples of size two which can be drawn with replacement (since any one of five numbers on the first draw can be associated with any one of the five numbers on the second draw). These are

$$
\begin{array}{ccccc}
(2, 2) & (2, 3) & (2, 6) & (2, 8) & (2, 11) \\
(3, 2) & (3, 3) & (3, 6) & (3, 8) & (3, 11) \\
(6, 2) & (6, 3) & (6, 6) & (6, 8) & (6, 11) \\
(8, 2) & (8, 3) & (8, 6) & (8, 8) & (8, 11) \\
(11, 2) & (11, 3) & (11, 6) & (11, 8) & (11, 11)
\end{array}
$$

The corresponding sample means are

(*1*)
$$
\begin{array}{ccccc}
2.0 & 2.5 & 4.0 & 5.0 & 6.5 \\
2.5 & 3.0 & 4.5 & 5.5 & 7.0 \\
4.0 & 4.5 & 6.0 & 7.0 & 8.5 \\
5.0 & 5.5 & 7.0 & 8.0 & 9.5 \\
6.5 & 7.0 & 8.5 & 9.5 & 11.0
\end{array}
$$

and the mean of sampling distribution of means is

$$\mu_{\bar{X}} = \frac{\text{sum of all sample means in } (1) \text{ above}}{25} = \frac{150}{25} = 6.0$$

illustrating the fact that $\mu_{\bar{X}} = \mu$.

(*d*) The variance $\sigma_{\bar{X}}^2$ of the sampling distribution of means is obtained by subtracting the mean 6 from each number in (*1*), squaring the result, adding all 25 numbers thus obtained and dividing by 25. The final result is

$$\sigma_{\bar{X}} = 135/25 = 5.40 \text{ so that } \sigma_{\bar{X}} = \sqrt{5.40} = 2.32$$

This illustrates the fact that for finite populations involving sampling with replacement (or infinite populations), $\sigma_{\bar{X}}^2 = \sigma^2/N$ since the right hand side is $10.8/2 = 5.40$, agreeing with the above value.

8.2. Solve Problem 8.1 in case sampling is without replacement.

Solution:

As in (*a*) and (*b*) of Problem 8.1, $\mu = 6$ and $\sigma = 3.29$.

(*c*) There are $_5C_2 = 10$ samples of size two which can be drawn without replacement (this means that we draw one number and then another number different from the first) from the population, namely

$$(2, 3), (2, 6), (2, 8), (2, 11), (3, 6), (3, 8), (3, 11), (6, 8), (6, 11), (8, 11)$$

The selection (2, 3), for example, is considered the same as (3, 2).

The corresponding sample means are

$$2 \cdot 5, 4 \cdot 0, 5 \cdot 0, 6 \cdot 5, 4 \cdot 5, 5 \cdot 5, 7 \cdot 0, 7 \cdot 0, 8 \cdot 5, 9 \cdot 5$$

and the mean of sampling distribution of means is

$$\mu_X = \frac{2 \cdot 5 + 4 \cdot 0 + 5 \cdot 0 + 6 \cdot 5 + 4 \cdot 5 + 5 \cdot 5 + 7 \cdot 0 + 7 \cdot 0 + 8 \cdot 5 + 9 \cdot 5}{10} = 6 \cdot 0$$

illustrating the fact that $\mu_X = \mu$.

(d) Variance of sampling distribution of means is

$$\sigma_{\bar{X}}^2 = \frac{(2 \cdot 5 - 6 \cdot 0)^2 + (4 \cdot 0 - 6 \cdot 0)^2 + (5 \cdot 0 - 6 \cdot 0)^2 + \ldots + (9 \cdot 5 - 6 \cdot 0)^2}{10} = 4 \cdot 05, \text{ and } \sigma_{\bar{X}} = 2 \cdot 01.$$

This illustrates $\sigma_{\bar{X}}^2 = \frac{\sigma^2}{N}\left(\frac{N_p - N}{N_p - 1}\right)$, since the right side equals $\frac{10 \cdot 8}{2}\left(\frac{5-2}{5-1}\right) = 4 \cdot 05$, as obtained above.

8.3.   Assume that the masses of 3000 male students at a university are normally distributed with mean $68 \cdot 0$ kg and standard deviation $3 \cdot 0$ kg. If 80 samples consisting of 25 students each are obtained, what would be the expected mean and standard deviation of the resulting sampling distribution of means if sampling were done (a) with replacement, (b) without replacement?

Solution:
   The numbers of samples of size 25 which could be obtained theoretically from a group of 3000 students with and without replacement are $(3000)^{25}$ and $_{3000}C_{25}$, which are much larger than 80. Hence we do not get a true sampling distribution of means but only an *experimental* sampling distribution. Nevertheless, since the number of samples is large, there should be close agreement between the two sampling distributions. Hence the expected mean and standard deviation would be close to those of the theoretical distribution . Thus we have

(a) $\qquad \mu_{\bar{X}} = \mu = 68 \cdot 0$ kg   and $\sigma_{\bar{X}} = \sigma/\sqrt{N} = 3/\sqrt{25} = 0 \cdot 6$ kg

(b) $\qquad \mu_{\bar{X}} = \mu = 68 \cdot 0$ kg   and $\sigma_{\bar{X}} = \dfrac{\sigma}{\sqrt{N}}\sqrt{\dfrac{N_p - N}{N_p - 1}} = \dfrac{3}{\sqrt{25}}\sqrt{\dfrac{3000 - 25}{3000 - 1}}$

which is only very slightly less than $0 \cdot 6$ kg and can therefore for all practical purposes be considered the same as in sampling with replacement.
   Thus we would expect the experimental sampling distribution of means to be approximately normally distributed with mean $68 \cdot 0$ kg and standard deviation $0 \cdot 6$ kg.

8.4.   In how many samples of Problem 8.3 would you expect to find the mean (a) between $66 \cdot 8$ and $68 \cdot 3$ kg, (b) less than $66 \cdot 4$ kg?

Solution:
   The mean $\bar{X}$ of a sample in standard units is here given by $z = \dfrac{\bar{X} - \mu_{\bar{X}}}{\sigma_{\bar{X}}} = \dfrac{\bar{X} - 68 \cdot 0}{0 \cdot 6}$.

(a) $66 \cdot 8$ in standard units $= (66 \cdot 8 - 68 \cdot 0)/0 \cdot 6 = -2 \cdot 0$
     $68 \cdot 3$ in standard units $= (68 \cdot 3 - 68 \cdot 0)/0 \cdot 6 = 0 \cdot 5$

Proportion of samples with means between $66 \cdot 8$ and $68 \cdot 3$ kg
   $=$ (area under normal curve between $z = -2 \cdot 0$ and $z = 0 \cdot 5$)
   $=$ (area bet. $z = -2$ and $z = 0$) $+$ (area bet. $z = 0$ and $z = 0 \cdot 5$)
   $= 0 \cdot 4772 + 0 \cdot 1915 = 0 \cdot 6687$

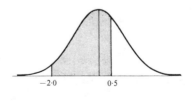

Then the expected number of samples $= (80)(0 \cdot 6687)$ or 53.

(b) $66 \cdot 4$ in standard units $= (66 \cdot 4 - 68 \cdot 0)/0 \cdot 6 = -2 \cdot 67$

Proportion of samples with means less than $66 \cdot 4$ kg
   $=$ (area under normal curve to left of $z = -2 \cdot 67$)
   $=$ (area to left of $z = 0$) $-$ (area bet. $z = -2 \cdot 67$ and $z = 0$)
   $= 0 \cdot 5 - 0 \cdot 4962 = 0 \cdot 0038$

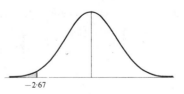

Then the expected number of samples $= (80)(0 \cdot 0038) = 0 \cdot 304$ or zero.

8.5. Five hundred castings have a mean weight of 5·02 N and a standard deviation of 0·30 N. Find the probability that a random sample of 100 castings chosen from this group will have a combined weight of (*a*) between 496 and 500 N, (*b*) more than 510 N.

Solution:

For the sampling distribution of means, $\mu_{\bar{X}} = \mu = 5.02$ N,

$$\sigma_{\bar{X}} = \frac{\sigma}{\sqrt{N}} \sqrt{\frac{N_p - N}{N_p - 1}} = \frac{0.30}{\sqrt{100}} \sqrt{\frac{500 - 100}{500 - 1}} = 0.027.$$

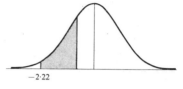

(*a*) The combined weight will lie between 496 and 500 N if the mean weight of the 100 castings lies between 4·96 and 500 N.

4·96 in standard units $= (4.96 - 5.02)/0.027 = -2.22$
5·00 in standard units $= (5.00 - 5.02)/0.027 = -0.74$

Required probability
= (area between $z = -2.22$ and $z = -0.74$)
= (area between $z = -2.22$ and $z = 0$)
   − (area between $z = -0.74$ and $z = 0$)
= $0.4868 - 0.2704 = 0.2164$

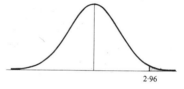

(*b*) The combined weight will exceed 510 N if the mean weight of the 100 castings exceeds 5·10 N.

5·10 in standard units $= (5.10 - 5.02)/0.227 = 2.96$

Required probability
= (area to right of $z = 2.96$)
= (area to right of $z = 0$)
   − (area between $z = 0$ and $z = 2.96$)
= $0.5 - 0.4985 = 0.0015$

Thus there are only 3 chances in 2000 of picking a sample of 100 castings with a combined weight exceeding 510 N.

RANDOM NUMBERS

8.6. (*a*) Show how to select 30 random samples of 4 students each (with replacement) from the table of masses on Page 27 by using random numbers. (*b*) Find the mean and standard deviation of the sampling distribution of means in (*a*). (*c*) Compare the results of (*b*) with theoretical values, explaining any discrepancies.

Solution:

(*a*) Use two digits to number each of the 100 students: 00, 01, 02, . . ., 99 (see Table 8.2). Thus the 5 students with masses 60–62 kg are numbered 00–04, the 18 students with masses 63–65 kg are numbered 05–22, etc. Each student number is called a *sampling number*.

We now draw sampling numbers from the random number table (Page 349). From the first line we find the sequence 51, 77, 27, 46, 40, etc., which we take as random sampling numbers, each of which yields the mass of a particular student. Thus 51 corresponds to a student having mass 66–68 kg, which we take as 67 kg (the class mark). Similarly 77, 27, 46 yield masses of 70, 67, 67 kg respectively.

Table 8.2

Mass (kg)	Frequency	Sampling Number
60–62	5	00–04
63–65	18	05–22
66–68	42	23–64
69–71	27	65–91
72–74	8	92–99

By this process we obtain Table 8.3 which shows the sample numbers drawn, the corresponding masses and the mean mass for each of 30 samples. It should be mentioned that although we have entered the random number table on the first line we could have started *anywhere* and chosen any specified pattern.

Table 8.3

Sample Number Drawn	Corresponding Masses	Mean Mass	Sample Number Drawn	Corresponding Masses	Mean Mass
1. 51, 77, 27, 46	67, 70, 67, 67	67·75	**16.** 11, 64, 55, 58	64, 67, 67, 67	66·25
2. 40, 42, 33, 12	67, 67, 67, 64	66·25	**17.** 70, 56, 97, 43	70, 67, 73, 67	69·25
3. 90, 44, 46, 62	70, 67, 67, 67	67·75	**18.** 74, 28, 93, 50	70, 67, 73, 67	69·25
4. 16, 28, 98, 93	64, 67, 73, 73	69·25	**19.** 79, 42, 71, 30	70, 67, 70, 67	68·50
5. 58, 20, 41, 86	67, 64, 67, 70	67·00	**20.** 58, 60, 21, 33	67, 67, 64, 67	66·25
6. 19, 64, 08, 70	64, 67, 64, 70	66·25	**21.** 75, 79, 74, 54	70, 70, 70, 67	69·25
7. 56, 24, 03, 32	67, 67, 61, 67	65·50	**22.** 06, 31, 04, 18	64, 67, 61, 64	64·00
8. 34, 91, 83, 58	67, 70, 70, 67	68·50	**23.** 67, 07, 12, 97	70, 64, 64, 73	67·75
9. 70, 65, 68, 21	70, 70, 70, 64	68·50	**24.** 31, 71, 69, 88	67, 70, 70, 70	69·25
10. 96, 02, 13, 87	73, 61, 64, 70	67·00	**25.** 11, 64, 21, 87	64, 67, 64, 70	66·25
11. 76, 10, 51, 08	70, 64, 67, 64	66·25	**26.** 03, 58, 57, 93	61, 67, 67, 73	67·00
12. 63, 97, 45, 39	67, 73, 67, 67	68·50	**27.** 53, 81, 93, 88	67, 70, 73, 70	70·00
13. 05, 81, 45, 93	64, 70, 67, 73	68·50	**28.** 23, 22, 96, 79	67, 64, 73, 70	68·50
14. 96, 01, 73, 52	73, 61, 70, 67	67·75	**29.** 98, 56, 59, 36	73, 67, 67, 67	68·50
15. 07, 82, 54, 24	64, 70, 67, 67	67·00	**30.** 08, 15, 08, 84	64, 64, 64, 70	65·50

Table 8.4

Sample Mean	Tally	f	u	fu	fu^2
64·00	/	1	−4	−4	16
64·75		0	−3	0	0
65·50	//	2	−2	−4	8
66·25	IIII /	6	−1	−6	6
$A \rightarrow$ 67·00	////	4	0	0	0
67·75	////	4	1	4	4
68·50	IIII //	7	2	14	28
69·25	IIII	5	3	15	45
70·00	/	1	4	4	16
		$\Sigma f = N = 30$		$\Sigma fu = 23$	$\Sigma fu^2 = 123$

(b) Table 8.4 gives the frequency distribution of sample mean masses obtained in (a). This is a *sampling distribution of means*. The mean and the standard deviation are obtained as usual by the coding methods of Chapters 3 and 4.

$$\text{Mean} = A + c\bar{u} = A + \frac{c\,\Sigma fu}{N} = 67\cdot00 + \frac{(0\cdot75)(23)}{30} = 67\cdot58 \text{ kg}$$

$$\text{Standard deviation} = c\sqrt{\overline{u^2} - \bar{u}^2} = c\sqrt{\frac{\Sigma fu^2}{N} - \left(\frac{\Sigma fu}{N}\right)^2} = 0\cdot75\sqrt{\frac{123}{30} - \left(\frac{23}{30}\right)^2} = 1\cdot41 \text{ kg}$$

(c) The theoretical mean of the sampling distribution of means, given by $\mu_{\bar{x}}$, should equal the population mean μ which is 67·45 kg (see Prob. 3.22, Chap. 3) in agreement with the value 67·58 kg of part (b).

　　The theoretical standard deviation (standard error) of the sampling distribution of means, given by $\sigma_{\bar{x}}$, should equal $\sigma/\sqrt{N}$, where the population standard deviation $\sigma = 2\cdot92$ kg (see Prob. 4.17, Chap. 4) and the sample size $N = 4$. Since $\sigma/\sqrt{N} = 2\cdot92/\sqrt{4} = 1\cdot46$ kg, we have agreement with the value 1·41 kg of part (b). Discrepancies are due to the fact that only 30 samples were selected and the sample size was small.

SAMPLING DISTRIBUTION OF PROPORTIONS

8.7.　Find the probability that in 120 tosses of a fair coin (a) between 40% and 60% will be heads, (b) $\frac{5}{8}$ or more will be heads.

Solution:
　　We consider the 120 tosses of the coin as a sample from the infinite population of all possible tosses of the coin. In this population the probability of heads is $p = \frac{1}{2}$ and the probability of tails is $q = 1 - p = \frac{1}{2}$.

(a) We require the probability that the number of heads in 120 tosses will be between (40% of 120) = 48 and (60% of 120) = 72. We proceed as in Chapter 7, using the normal approximation to the binomial. Since the number of heads is a discrete variable, we ask for the probability that the number of heads lies between 47·5 and 72·5.

μ = expected number of heads = Np = $120(\frac{1}{2})$ = 60, and σ = $\sqrt{Nqp}$ = $\sqrt{(120)(\frac{1}{2})(\frac{1}{2})}$ = 5·48.

47·5 in standard units = (47·5 − 60)/5·48 = −2·28.
72·5 in standard units = (72·5 − 60)/5·48 = 2·28.

Required probability
= (area under normal curve bet. z = −2·28 and z = 2·28)
= 2(area between z = 0 and z = 2·28)
= 2(0·4887) = 0·9774

Another method:

μ_P = p = $\frac{1}{2}$ = 0·50, σ_P = $\sqrt{pq/N}$ = $\sqrt{\frac{1}{2}(\frac{1}{2})/120}$ = 0·0456.

40% in standard units = (0·40 − 0·50)/0·0456 = −2·19.
60% in standard units = (0·60 − 0·50)/0·0456 = 2·19.

Thus the required probability is the area under the normal curve between (z = −2·19 and z = 2·19) = 2(0·4857) = 0·9714.

Although this result is accurate to two significant figures, it does not agree exactly since we have not used the fact that the proportion is actually a discrete variable. To account for this we subtract $\frac{1}{2N} = \frac{1}{2(120)}$ from 0·40 and add $\frac{1}{2N} = \frac{1}{2(120)}$ to 0·60. Thus the required proportions in standard units are, since 1/240 = 0·004 17,

$$\frac{0\cdot40 - 0\cdot004\ 17 - 0\cdot50}{0\cdot0456} = -2\cdot28 \quad \text{and} \quad \frac{0\cdot60 + 0\cdot004\ 17 - 0\cdot50}{0\cdot0456} = 2\cdot28$$

so that agreement with the first method is obtained.

Note that (0·40 − 0·004 17) and (0·60 + 0·004 17) corresponds to the proportions 47·5/120 and 72·5/120 in the first method above.

(b) Using the second method of (a), we find that since $\frac{5}{8}$ = 0·6250,

$$(0\cdot6250 - 0\cdot004\ 17) \text{ in standard units} = \frac{0\cdot6250 - 0\cdot004\ 17 - 0\cdot50}{-0\cdot0456} = 2\cdot65$$

Required probability = (area under normal curve to right of z = 2·65)
= (area to right of z = 0) − (area between z = 0 and z = 2·65)
= 0·5 − 0·4960 = 0·0040

8.8. Each person of a group of 500 people tosses a fair coin 120 times. How many people should be expected to report that (a) between 40% and 60% of their tosses resulted in heads, (b) $\frac{5}{8}$ or more of their tosses resulted in heads?

Solution:

This problem is closely related to the preceding problem. Here we consider 500 samples, of size 120 each, from the infinite population of all possible tosses of a coin.

(a) Part (a) of Prob. 8.7 states that of all possible samples, each consisting of 120 tosses of a coin, we can expect to find 97·74% with a percentage of heads between 40% and 60%. In 500 samples we can thus expect to find about (97·74% of 500) or 489 samples with this property. It follows that about 489 people would be expected to report that their experiment resulted in between 40% and 60% heads.

It is interesting to note that 500 − 489 = 11 people would be expected to report that the percentage of heads was not between 40% and 60%. Such people might reasonably conclude that their coins were loaded even though they were fair. This type of error is a *risk* which is always present whenever we deal with probability.

(b) By reasoning as in (a), we conclude that about (500)(0·0040) = 2 persons would report that $\frac{5}{8}$ or more of their tosses resulted in heads.

8.9. It has been found that 2% of the tools produced by a certain machine are defective. What is the probability that in a shipment of 400 such tools (a) 3% or more, (b) 2% or less will prove defective?

Solution:
$$\mu_P = p = 0.02 \quad \text{and} \quad \sigma_P = \sqrt{pq/N} = \sqrt{0.02(0.98)/400} = 0.14/20 = 0.007$$

(a) Using the correction for discrete variables, $1/2N = 1/800 = 0.00125$, we have

$$(0.03 - 0.00125) \text{ in standard units} = \frac{0.03 - 0.00125 - 0.02}{0.007} = 1.25$$

Required probability = (area under normal curve to right of $z = 1.25$) = 0.1056.
If we had not used the correction we would have obtained 0.0764.

Another Method:
(3% of 400) = 12 defective tools. On a continuous basis, 12 or more tools means 11.5 or more. $\bar{X} = $ (2% of 400) = 8, and $\sigma = \sqrt{Npq} = \sqrt{(400)(0.02)(0.98)} = 2.8$.
Then, 11.5 in standard units = $(11.5 - 8)/2.8 = 1.25$, and as before the required probability is 0.1056.

(b)
$$(0.02 + 0.00125) \text{ in standard units} = \frac{0.02 + 0.00125 - 0.02}{0.007} = 0.18$$

Required probability = (area under normal curve to left of $z = 0.18$) = 0.5000 + 0.0714 = 0.5714.
If we had not used the correction, we would have obtained 0.5000.
The second method of part (a) can also be used.

8.10. The election return showed that a certain candidate received 46% of the votes. Determine the probability that a poll of (a) 200, (b) 1000 people selected at random from the voting population would have shown a majority of votes in favour of the candidate.

Solution:
(a) $\mu_P = p = 0.46 \quad \text{and} \quad \sigma_P = \sqrt{pq/N} = \sqrt{0.46(0.54)/200} = 0.0352$

Since $1/2N = 1/400 = 0.0025$, a majority is indicated in the sample if the proportion in favour of the candidate is $(0.50 + 0.0025) = 0.5025$ or more. (This proportion can also be obtained by realizing that 101 or more indicates a majority but this as a continuous variable is 100.5, and so the proportion is $100.5/200 = 0.5025$.)

Then, 0.5025 in standard units = $(0.5025 - 0.46)/0.0352 = 1.21$.

Required probability = (area under normal curve to right of $z = 1.21$) = 0.5000 − 0.3869 = 0.1131

(b) $\mu_P = p = 0.46, \sigma_P = \sqrt{pq/N} = \sqrt{0.46(0.54)/1000} = $
0.5025 in standard units = $(0.5025 - 0.46)/0.0158 - 2.69 = 2.69$

Required probability = (area under normal curve to right of $z = 2.69$) = 0.5000 − 0.4964 = 0.0036.

SAMPLING DISTRIBUTIONS OF DIFFERENCES AND SUMS

8.11. Let U_1 be a variable which stands for any of the elements of the population 3, 7, 8 and U_2 a variable which stands for any of the elements of the population 2, 4. Compute (a) μ_{U_1}, (b) μ_{U_2}, (c) $\mu_{U_1-U_2}$, (d) σ_{U_1}, (e) σ_{U_2}, (f) $\sigma_{U_1-U_2}$.

Solution:
(a) $\mu_{U_1} = $ mean of population $U_1 = \frac{1}{3}(3 + 7 + 8) = 6$
(b) $\mu_{U_2} = $ mean of population $U_2 = \frac{1}{2}(2 + 4) = 3$
(c) The population consisting of the differences of any member of U_1 and any member of U_2 is

$$
\begin{array}{ccccccc}
3-2 & 7-2 & 8-2 & & 1 & 5 & 6 \\
3-4 & 7-4 & 8-4 & \text{or} & -1 & 3 & 4
\end{array}
$$

Then $\mu_{U_1-U_2} = $ mean of $(U_1 - U_2) = \dfrac{1 + 5 + 6 + (-1) + 3 + 4}{6} = 3$.

This illustrates the general result $\mu_{U_1-U_2} = \mu_{U_1} - \mu_{U_2}$, as seen from (a) and (b).

(d) $\sigma_{U_1}^2$ = variance of population $U_1 = \dfrac{(3-6)^2 + (7-6)^2 + (8-6)^2}{3} = \dfrac{14}{3}$, or $\sigma_{U_1} = \sqrt{\dfrac{14}{3}}$.

(e) $\sigma_{U_2}^2$ = variance of population $U_2 = \dfrac{(2-3)^2 + (4-3)^2}{2} = 1$, or $\sigma_{U_2} = 1$.

(f) $\sigma_{U_1-U_2}$ = variance of population $(U_1 - U_2)$

$$= \dfrac{(1-3)^2 + (5-3)^2 + (6-3)^2 + (-1-3)^2 + (3-3)^2 + (4-3)^2}{6} = \dfrac{17}{3}, \quad \text{or } \sigma_{U_1-U_2} = \sqrt{\dfrac{17}{3}}.$$

This illustrates the general result for independent samples, $\sigma_{U_1-U_2} = \sqrt{\sigma_{U_1}^2 + \sigma_{U_2}^2}$, as seen from (d) and (e).

8.12. The electric light bulbs of manufacturer A have a mean lifetime of 1400 hours with a standard deviation of 200 hours, while those of manufacturer B have a mean lifetime of 1200 hours with a standard deviation of 100 hours. If random samples of 125 bulbs of each brand are tested, what is the probability that the brand A bulbs will have a mean lifetime which is at least (a) 160 hours, (b) 250 hours more than the brand B bulbs?

Solution:

Let $\bar{X}_A$ and $\bar{X}_B$ denote the mean lifetimes of samples A and B respectively. Then

$$\mu_{\bar{X}_A - \bar{X}_B} = \mu_{\bar{X}_A} - \mu_{\bar{X}_B} = 1400 - 1200 = 200 \text{ h}$$

and

$$\sigma_{\bar{X}_A - \bar{X}_B} = \sqrt{\dfrac{\sigma_A^2}{N_A} + \dfrac{\sigma_B^2}{N_L}} = \sqrt{\dfrac{(100)^2}{125} + \dfrac{(200)^2}{125}} = 20 \text{ h}$$

The standardized variable for the difference in means is

$$z = \dfrac{(\bar{X}_A - \bar{X}_B) - (\mu_{\bar{X}_A - \bar{X}_B})}{\sigma_{\bar{X}_A - \bar{X}_B}} = \dfrac{(\bar{X}_A - \bar{X}_B) - 200}{20}$$

and is very closely normally distributed.

(a) The difference 160 hours in standard units $= (160 - 200)/20 = -2$.

     Required probability = (area under normal curve to right of $z = -2$)
             $= 0.5000 + 0.4772 = 0.9772$

(b) The difference 250 hours in standard units $= (250 - 200)/20 = 2.50$.

     Required probability = (area under normal curve to right of $z = 2.50$)
             $= 0.5000 - 0.4938 = 0.0062$

8.13. Zinc castings of a given brand weigh 0·50 N with a standard deviation of 0·02 N. What is the probability that two lots, of 1000 zinc castings each, will differ in weight by more than 2 newtons.

Solution:

Let $\bar{X}_1$ and $\bar{X}_2$ denote the mean weights of ball bearings in the two lots. Then

$$\mu_{\bar{X}_1 - \bar{X}_2} = \mu_{\bar{X}_1} - \mu_{\bar{X}_2} = 0.50 - 0.50 = 0$$

$$\sigma_{\bar{X}_1 - \bar{X}_2} = \sqrt{\dfrac{\sigma_1^2}{N_1} + \dfrac{\sigma_2^2}{N_2}} = \sqrt{\dfrac{(0.02)^2}{1000} + \dfrac{(0.02)^2}{1000}} = 0.000\,895$$

The standardized variable for the difference in mean is $z = \dfrac{(\bar{X}_1 - \bar{X}_2) - 0}{0.000\,895}$ and is very closely normally distributed.

A difference of 2 N in the lots is equivalent to a difference of $2/1000 = 0.002$ N in the means. This can occur either if $\bar{X}_1 - \bar{X}_2 \geq 0.002$ or $\bar{X}_1 - \bar{X}_2 \leq 60.002$, i.e. $z \geq \dfrac{0.002 - 0}{0.000\,895} = 2.23$ or $z \leq \dfrac{-0.002 - 0}{0.000\,895} = -2.23$.

Then $\Pr\{z \geq 2.23 \text{ or } z \leq -2.23\} = \Pr\{z \geq 2.23\} + \Pr\{z \leq -2.23\} = 2(0.5000 - 0.4871) = 0.0258$.

8.14. *A* and *B* play a game of "heads and tails", each tossing 50 coins. *A* will win the game if he tosses 5 or more heads than *B*, otherwise *B* wins. Determine the odds against *A* winning any particular game.

Solution:
Let P_A and P_B denote the proportion of heads obtained by *A* and *B*.
If we assume the coins are all fair, the probability p of heads is $\frac{1}{2}$. Then

$$\mu_{P_A-P_B} = \mu_{P_A} - \mu_{P_B} = 0 \quad \text{and} \quad \sigma_{P_A-P_B} = \sqrt{\sigma_{P_A}^2 + \sigma_{P_B}^2} = \sqrt{\frac{pq}{N_A} + \frac{pq}{N_B}} = \sqrt{\frac{2(\frac{1}{2})(\frac{1}{2})}{50}} = 0\cdot10$$

The standardized variable for the difference in proportions is $z = (P_A - P_B - 0)/0\cdot10$.

On a continuous variable basis, 5 or more heads means 4·5 or more heads, so that the difference in proportions should be $4\cdot5/50 = 0\cdot09$ or more, i.e. z is greater than or equal to $(0\cdot09 - 0)/0\cdot10 = 0\cdot9$ (or $z \geq 0\cdot9$). The probability of this is the area under the normal curve to the right of $z = 0\cdot9$, which is $(0\cdot5000 - 0\cdot3159) = 0\cdot1841$.

Thus the odds against *A* winning are $(1 - 0\cdot1841):0\cdot1841 = 0\cdot8159:0\cdot1841$, or 4·43 to 1.

8.15. Two distances are measured as 27·3 mm and 15·6 mm, with standard deviations (standard errors) of 0·16 mm and 0·08 mm respectively. Determine the mean and standard deviation of (*a*) the sum, (*b*) the difference of the distances.

Solution:
If the distances are denoted by D_1 and D_2 then

(*a*)
$$\mu_{D_1+D_2} = \mu_{D_1} + \mu_{D_2} = 27\cdot3 + 15\cdot6 = 42\cdot9 \text{ mm}$$
$$\sigma_{D_1+D_2} = \sqrt{\sigma_{D_1}^2 + \sigma_{D_2}^2} = \sqrt{(0\cdot16)^2 + (0\cdot08)^2} = 0\cdot18 \text{ mm}$$

(*b*)
$$\mu_{D_1-D_2} = \mu_{D_1} - \mu_{D_2} = 27\cdot3 - 15\cdot6 = 11\cdot7 \text{ mm}$$
$$\sigma_{D_1-D_2} = \sqrt{\sigma_{D_1}^2 + \sigma_{D_2}^2} = \sqrt{(0\cdot16)^2 + (0\cdot08)^2} = 0\cdot18 \text{ mm}$$

8.16. A certain type of electric light bulb has a mean lifetime of 1500 hours and a standard deviation of 150 hours. Three bulbs are connected so that when one burns out, another will go on. Assuming the lifetimes are normally distributed, what is the probability that lighting will take place for (*a*) at least 5000 hours, (*b*) at most 4200 hours?

Solution:
Assume the lifetimes to be L_1, L_2 and L_3. Then
$$\mu_{L_1+L_2+L_3} = \mu_{L_1} + \mu_{L_2} + \mu_{L_3} = 1500 + 1500 + 1500 = 4500 \text{ hours}$$
$$\sigma_{L_1+L_2+L_3} = \sqrt{\sigma_{L_1}^2 + \sigma_{L_2}^2 + \sigma_{L_3}^2} = \sqrt{3(150)^2} = 260 \text{ hours}$$

(*a*) 5000 hours in standard units $= (5000 - 4500)/260 = 1\cdot92$.
Required probability $=$ (area under normal curve to right of $z = 1\cdot92$)
$= 0\cdot5000 - 0\cdot4726 = 0\cdot0274$

(*b*) 4200 hours in standard units $= (4200 - 4500)/260 = -1\cdot15$.
Required probability $=$ (area under normal curve to left of $z = -1\cdot15$)
$= 0\cdot5000 - 0\cdot3749 = 0\cdot1251$

MISCELLANEOUS PROBLEMS

8.17. With reference to Problem 8.1, find (*a*) the mean of the sampling distribution of variances, (*b*) the standard deviation of the sampling distribution of variances, i.e. the standard error of variances.

Solution:
(*a*) The sample variances corresponding to each of the 25 samples in Problem 8.1 are

0	0·25	4·00	9·00	20·25
0·25	0	2·25	6·25	16·00
4·00	2·25	0	1·00	6·25
9·00	6·25	1·00	0	2·25
20·25	16·00	6·25	2·25	0

Mean of sampling distribution of variances is

$$\mu_{s^2} = \frac{\text{sum of all variances in the table above}}{25} = \frac{135}{25} = 5.40$$

This illustrates the fact that $\mu_{s^2} = (N-1)(\sigma^2)/N$, since for $N = 2$ and $\sigma^2 = 10.8$ [see Problem 8.1(b)], the right hand side is $\frac{1}{2}(10.8) = 5.4$.

The result shows the desirability of defining a corrected variance for samples as $\hat{s}^2 = \dfrac{N}{N-1} s^2$.

It would then follow that $\mu_{\hat{s}^2} = \sigma^2$ (see also the remarks on Page 70). It should be noted that population variances would be defined the same as before and only sample variances would be corrected.

(b) The variance of the sampling distribution of variances $\sigma_{s^2}^2$ is obtained by subtracting the mean 5.40 from each of the 25 numbers in the above table, squaring these numbers, adding them, and then dividing the result by 25. Thus, $\sigma_{s^2}^2 = 575.75/25 = 23.03$ or $\sigma_{s^2} = 4.80$.

8.18. Work the previous problem if sampling is without replacement.

Solution:

(a) There are 10 samples whose variances are given by the numbers above (or below) the diagonal of zeros in the table of Problem 8.17(a). Then

$$\mu_{s^2} = \frac{0.25 + 4.00 + 9.00 + 20.25 + 2.25 + 6.25 + 16.00 + 1.00 + 6.25 + 2.25}{10} = 6.75.$$

This is a special case of the general result $\mu_{s^2} = \left(\dfrac{N_p}{N_p - 1}\right)\left(\dfrac{N-1}{N}\right)\sigma^2$, as is verified by putting $N_P = 5$, $N = 2$ and $\sigma^2 = 10.8$ on the right hand side to obtain $\mu_{s^2} = (\frac{5}{4})(\frac{1}{2})(10.8) = 6.75$.

(b) Subtracting 6.75 from each of the 10 numbers above the diagonal of zeros in the table of Problem 8.17(a), squaring these numbers, adding them and dividing by 10, we find $\sigma_{s^2}^2 = 39.675$ or $\sigma_{s^2} = 6.30$.

8.19. The standard deviation of the masses of a very large population of males is 10.0 kg. Samples of 200 males each are drawn from this population, and the standard deviations of the masses in each sample are computed. Find (a) the mean and (b) the standard deviation of the sampling distribution of standard deviations.

Solution:

We can consider that sampling is either from an infinite population or with replacement from a finite population. From Page 144 we have:
(a) Means of sampling distribution of standard deviations $= \mu_s = \sigma = 10.0$ kg.
(b) Standard deviation of sampling distribution of standard deviations $= \sigma_s = \sigma/\sqrt{2N} = 10/\sqrt{400} = 0.50$ kg.

8.20. What percentage of the samples in the preceding problem would have standard deviations (a) greater than 11.0 kg, (b) less than 8.8 kg?

Solution:

The sampling distribution of standard deviations is approximately normally distributed with mean 10.0 kg and standard deviation 0.50 kg.

(a) 11.0 kg in standard units $= (11.0 - 10.0)/0.50 = 2.0$. Area under normal curve to right of $z = 2.0$ is $(0.5 - 0.4772) = 0.0228$; hence the required percentage is 2.3%.

(b) 8.8 kg in standard units $= (8.8 - 10.0)/0.50 = -2.4$. Area under normal curve to left of $z = -2.4$ is $(0.5 - 0.4918) = 0.0082$; hence the required percentage is 0.8%.

Supplementary Problems

SAMPLING DISTRIBUTION OF MEANS

8.21. A population consists of the four numbers 3, 7, 11, 15. Consider all possible samples of size two which can be drawn with replacement from this population. Find (a) the population mean, (b) the population standard deviation, (c) the mean of the sampling distribution of means, (d) the standard deviation of the sampling distribution of means. Verify (c) and (d) directly from (a) and (b) by use of suitable formulae.   *Ans.* (a) 9·0, (b) 4·47, (c) 9·0, (d) 3·16

8.22. Solve Problem 8.21 if sampling is without replacement.   *Ans.* (a) 9·0, (b) 4·47, (c) 9·0, (d) 2·58

8.23. The weights of 1500 iron castings are normally distributed with a mean of 22·40 newtons and a standard deviation of 0·048 newtons. If 300 random samples of size 36 are drawn from this population, determine the expected mean and standard deviation of the sampling distribution of means if sampling is done (a) with replacement, (b) without replacement.   *Ans.* (a) $\mu_{\bar{X}} = 22\cdot40$ N, $\sigma_{\bar{X}} = 0\cdot008$ N   (b) $\mu_{\bar{X}} = 22\cdot40$ N, $\sigma_{\bar{X}} = $ slightly less than 0·008 N

8.24. Solve Problem 8.23 if the population consists of 72 iron castings.
Ans. (a) $\mu_{\bar{X}} = 22\cdot4$ N, $\sigma_{\bar{X}} = 0\cdot008$ N,   (b) $\mu_{\bar{X}} = 22\cdot4$ N, $\sigma_{\bar{X}} = 0\cdot0057$ N

8.25. In Prob. 8.23, how many of the random samples would have their means (a) between 22·39 and 22·41 N, (b) greater than 22·42 N, (c) less than 22·37, (d) less than 22·38 or more than 22·41 N?   *Ans.* (a) 237, (b) 2, (c) none, (d) 34

8.26. Certain tubes manufactured by a company have a mean lifetime of 800 h and a standard deviation of 60 h. Find the probability that a random sample of 16 tubes taken from the group will have a mean lifetime of (a) between 790 and 810 h, (b) less than 785 h, (c) more than 820 h, (d) between 770 and 830 h.
Ans. (a) 0·4972, (b) 0·1587, (c) 0·0918, (d) 0·9544

8.27. Work Prob. 8.26 if a random sample of 64 tubes is taken. Explain the difference.
Ans. (a) 0·8164, (b) 0·0228, (c) 0·0038, (d) 1·0000

8.28. The weights of packages received by a department store have a mean of 300 newtons and a standard deviation of 50 newtons. What is the probability that 25 packages received at random and loaded on an elevator will exceed the specified safety limit of the elevator, listed as 8200 newtons?   *Ans.* 0·0026

RANDOM NUMBERS

8.29. Work Prob. 8.6 by using a different set of random numbers and selecting (a) 15, (b) 30, (c) 45, (d) 60 samples of size 4 with replacement. Compare with the theoretical results in each case.

8.30. Work Prob. 8.29 selecting samples of size (a) 2 and (b) 8 with replacement instead of 4.

8.31. Work Prob. 8.6 if sampling is without replacement. Compare with theoretical results.

8.32. (a) Show how to select 30 samples of size 2 from the distribution in Prob. 3.61, Chap. 3. (b) Compute the mean and standard deviation of the resulting sampling distribution of means and compare with theory.

8.33. Work the preceding problem using samples of size 4.

SAMPLING DISTRIBUTION OF PROPORTIONS

8.34. Find the probability that of the next 200 children born (a) less than 40% will be boys, (b) between 43% and 57% will be girls, (c) more than 54% will be boys. Assume equal probabilities for births of boys and girls.
Ans. (a) 0·0019, (b) 0·9596, (c) 0·1151

8.35. Out of 1000 samples of 200 children each, in how many would you expect to find that (a) less than 40% are boys, (b) between 40% and 60% are girls, (c) 53% or more are girls?   *Ans.* (a) 2, (b) 996, (c) 218

8.36. Work Prob. 8.34 if 100 instead of 200 children are considered and explain the differences in results.
Ans. (a) 0·0179, (b) 0·8664, (c) 0·1841

8.37. An urn contains 80 marbles of which 60% are red and 40% are white. Out of 50 samples of 20 marbles each selected with replacement from the urn, how many samples can be expected to consist of (a) equal numbers of red and white marbles, (b) 12 red and 8 white marbles, (c) 8 red and 12 white marbles, (d) 10 or more white marbles? *Ans.* (a) 6, (b) 9, (c) 2, (d) 12

8.38. Design an experiment intended to illustrate the results of Prob. 8.37. Instead of red and white marbles you may use slips of paper on which R or W are written in the correct proportions. What errors might you introduce by using two different sets of coins?

8.39. A manufacturer sends out 1000 lots, each consisting of 100 electric bulbs. If 5% of the bulbs are normally defective, in how many of the lots should we expect (a) fewer than 90 good bulbs, (b) 98 or more good bulbs?   *Ans.* (a) 6, (b) 125

SAMPLING DISTRIBUTIONS OF DIFFERENCES AND SUMS

8.40. A and B manufacture two types of cables, having mean breaking strengths of 4000 and 4500 N and standard deviations of 300 and 200 N respectively. If 100 cables of brand A and 50 cables of brand B are tested, what is the probability that the mean breaking strength of B will be (a) at least 600 N more than A, (b) at least 450 N more than A? *Ans.* (a) 0·0077, (b) 0·8869

8.41. What are the probabilities in Prob. 8.40 if 100 cables of both brands are tested? Account for the differences. *Ans.* (a) 0·0028, (b) 0·9172

8.42. The mean score of students on an aptitude test is 72 points with a standard deviation of 8 points. What is the probability that two groups of students, consisting of 28 and 36 students respectively, will differ in their mean scores by (a) 3 or more points, (b) 6 or more points, (c) between 2 and 5 points?   *Ans.* (a) 0·2150, (b) 0·0064, (c) 0·4504

8.43. An urn consists of 60 red marbles and 40 white marbles. Two sets of 30 marbles each are drawn with replacement from the urn and their colours are noted. What is the probability that the two sets differ by 8 or more red marbles? *Ans.* 0·0482

8.44. Solve Problem 8.43 if sampling is without replacement in obtaining each set.   *Ans.* 0·0136

8.45. The election returns showed that a certain candidate received 65% of the votes. Find the probability that two random samples, each consisting of 200 voters, indicated more than a 10% difference in the proportions that voted for the candidate.   *Ans.* 0·0316

8.46. If U_1 and U_2 are the sets of numbers in Prob. 8.11 verify that (a) $\mu_{U_1 + U_2} = \mu_{U_1} + \mu_{U_2}$, (b) $\sigma_{U_1 + U_2} = \sqrt{\sigma_{U_1}^2 + \sigma_{U_2}^2}$.

8.47. Three masses are measured as 20·48, 35·97 and 62·34 kg with standard deviations of 0·21, 0·46 and 0·54 kg respectively. Find the (a) mean and (b) standard deviation of the sum of the masses.   *Ans.* (a) 118·79 kg, (b) 0·74 kg

8.48. The mean voltage of a battery is 15·0 volts and the standard deviation is 0·2 volts. What is the probability that four such batteries connected in series will have a combined voltage of 60·8 or more volts?   *Ans.* 0·0228

MISCELLANEOUS PROBLEMS

8.49. A population of 7 numbers has a mean of 40 and a standard deviation of 3. If samples of size 5 are drawn from this population and the variance of each sample is computed, find the mean of the sampling distribution of variances if sampling is (a) with replacement, (b) without replacement.   *Ans.* (a) 7·2, (b) 8·4

8.50. Certain tubes produced by a company have a mean lifetime of 900 h, and a standard deviation of 80 h. The company sends out 1000 lots of 100 tubes each. In how many lots can we expect (a) the mean lifetime to exceed 910 h, (b) the standard deviations of the lifetimes to exceed 95 h? What assumptions must be made?   *Ans.* (a) 106, (b) 4

8.51. In Prob. 8.50 if the median lifetime is 900 h, in how many lots can we expect the median lifetimes to exceed 910 h? Compare your answer with Prob. 8.50(a) and explain the results.   *Ans.* 159

8.52. On a city-wide examination the marks were normally distributed with mean 72 and standard deviation 8. (a) Find the minimum mark of the top 20% of the students. (b) Find the probability that in a random sample of 100 students the minimum mark of the top 20% will be less than 76.   *Ans.* (a) 78·7, (b) 0·0090

CHAPTER 9

Statistical Estimation Theory

ESTIMATION OF PARAMETERS

In the last chapter we saw how sampling theory can be employed to obtain information about samples drawn at random from a known population. From a practical viewpoint, however, it is often more important to be able to infer information about a population by use of samples drawn from it. Such problems are dealt with in *statistical inference*, which uses principles of sampling theory.

One important problem of statistical inference is the estimation of *population parameters* or briefly *parameters* (such as population mean, variance, etc.) from the corresponding *sample statistics* or briefly *statistics* (i.e. sample mean, variance, etc.). We consider this problem in this chapter.

UNBIASED ESTIMATES

If the mean of the sampling distribution of a statistic equals the corresponding population parameter, the statistic is called an *unbiased estimator* of the parameter, otherwise it is called a *biased estimator*. The corresponding values of such statistics are called *unbiased* or *biased estimates* respectively.

Example 1. The mean of the sampling distribution of means $\mu_{\bar{X}} = \mu$, the population mean (see Page 142). Hence the sample mean $\bar{X}$ is an unbiased estimate of the population mean μ.

Example 2. The mean of the sampling distribution of variances $\mu_{s2} = \dfrac{N-1}{N} \sigma^2$, where σ^2 is the population variance and N is the sample size (see Page 144). Thus the sample variance s^2 is a biased estimate of the population variance σ^2. By use of the modified variance $\hat{s}^2 = \dfrac{N}{N-1} s^2$, we find $\mu_{\hat{s}2} = \sigma^2$ so that $\hat{s}^2$ is an unbiased estimate of σ^2. However, $\hat{s}$ is a biased estimate of σ.

In the language of expectation (see Chapter 6) we could say that a statistic is unbiased if its expectation equals the corresponding population parameter. Thus $\bar{X}$ and $\hat{s}^2$ are unbiased since $E\{\bar{X}\} = \mu$ and $E\{\hat{s}^2\} = \sigma^2$.

EFFICIENT ESTIMATES

If the sampling distributions of two statistics have the same mean (or expectation), the statistic with the smaller variance is called an *efficient estimator* of the mean while the other statistic is called an *inefficient estimator*. The corresponding values of the statistics are called *efficient* or *inefficient estimates* respectively.

If we consider all possible statistics whose sampling distributions have the same mean, the one with the smallest variance is sometimes called the *most efficient* or *best estimator* of this mean.

Example: The sampling distributions of the mean and median both have the same mean, namely the population mean. However, the variance of the sampling distribution of means is smaller than that for the sampling distribution of medians (see Page 144). Hence the sample mean gives an efficient estimate of the population mean, while the sample median gives an inefficient estimate of it.

Of all statistics estimating the population mean, the sample mean provides the best or most efficient estimate.

In practice inefficient estimates are often used because of the relative ease with which some of them can be obtained.

POINT ESTIMATES AND INTERVAL ESTIMATES. RELIABILITY

An estimate of a population parameter given by a single number is called a *point estimate* of the parameter. An estimate of a population parameter given by two numbers between which the parameter may be considered to lie is called an *interval estimate of* the parameter.

Interval estimates indicate the precision or accuracy of an estimate and are therefore preferable to point estimates.

Example : If we say that a distance is measured as 5·28 mm, we are giving a point estimate. If on the other hand we say that the distance is 5·28 ± 0·03 mm, i.e. the distance lies between 5·25 and 5·31 mm, we are giving an interval estimate.

A statement of the error or precision of an estimate is often called its *reliability*.

CONFIDENCE INTERVAL ESTIMATES OF POPULATION PARAMETERS

Let μ_S and σ_S be the mean and standard deviation (standard error) of the sampling distribution of a statistic S. Then if the sampling distribution of S is approximately normal (which as we have seen is true for many statistics if the sample size $N \geq 30$) we can expect to find an actual sample statistic S lying in the intervals $\mu_S - \sigma_S$ to $\mu_S + \sigma_S$, $\mu_S - 2\sigma_S$ to $\mu_S + 2\sigma_S$ or $\mu_S - 3\sigma_S$ to $\mu_S + 3\sigma_S$ about 68·27%, 95·45% and 99·73% of the time respectively.

Equivalently we can expect to find, or we can be *confident* of finding, μ_S in the intervals $S - \sigma_S$ to $S + \sigma_S$, $S - 2\sigma_S$ to $S + 2\sigma_S$ or $S - 3\sigma_S$ to $S + 3\sigma_S$ about 68·27%, 95·45% and 99·73% of the time respectively. Because of this we call these respective intervals the 68·27%, 95·45% and 99·73% *confidence intervals* for estimating μ_S. The end numbers of these intervals ($S \pm \sigma_S$, $S \pm 2\sigma_S$, $S \pm 3\sigma_S$) are then called the 68·27%, 95·45% and 99·73% *confidence limits* or, as they are sometimes called, *fiducial limits*.

Similarly, $S \pm 1·96\sigma_S$ and $S \pm 2·58\sigma_S$ are 95% and 99% (or 0·95 and 0·99) confidence limits for S. The percentage confidence is often called the *confidence level*. The numbers 1·96, 2·58, etc., in the confidence limits are called *confidence coefficients* or *critical values* and are denoted by z_c. From confidence levels we can find confidence coefficients, and conversely.

In Table 9.1 we give values of z_c corresponding to various confidence levels used in practice. For confidence levels not presented in the table, the values of z_c can be found from the normal curve area tables (see Prob. 9.7).

Table 9.1

Confidence level	99·73%	99%	98%	96%	95·45%	95%	90%	80%	68·27%	50%
z_c	3·00	2·58	2·33	2·05	2·00	1·96	1·645	1·28	1·00	0·6745

CONFIDENCE INTERVAL ESTIMATES FOR MEANS

If the statistic S is the sample mean $\bar{X}$, then 95% and 99% confidence limits for estimation of the population mean μ are given by $\bar{X} \pm 1·96\sigma_{\bar{X}}$ and $\bar{X} \pm 2·58\sigma_{\bar{X}}$ respectively. More generally, the confidence limits are given by $\bar{X} \pm z_c \sigma_{\bar{X}}$ where z_c, which depends on the particular level of confidence desired, can be read from the above table. Using the values of $\sigma_{\bar{X}}$ obtained in Chapter 8, we see that the confidence limits for the population mean are given by

$$\bar{X} \;\pm\; z_c \frac{\sigma}{\sqrt{N}} \tag{1}$$

in case sampling is from an infinite population or if sampling is with replacement from a finite population, and by

$$\bar{X} \;\pm\; z_c \frac{\sigma}{\sqrt{N}} \sqrt{\frac{N_p - N}{N_p - 1}} \tag{2}$$

if sampling is without replacement from a population of finite size N_p.

In general the population standard deviation σ is unknown, so that to obtain the above confidence limits we use the sample estimate $\hat{s}$ or s. This will prove satisfactory provided $N \geq 30$. For $N < 30$, the approximation is poor and small sampling theory must be employed (see Chapter 11).

CONFIDENCE INTERVALS FOR PROPORTIONS

If the statistic S is the proportion of "successes" in a sample of size N drawn from a binomial population in which p is the proportion of successes (i.e. the probability of success), the confidence limits for p are given by $P \pm z_c \sigma_P$, where P is the proportion of successes in the sample of size N. Using the values of σ_P obtained in Chapter 8, we see that the confidence limits for the population proportion are given by

$$P \pm z_c \sqrt{\frac{pq}{N}} \;=\; P \pm z_c \sqrt{\frac{p(1-p)}{N}} \tag{3}$$

in case sampling is from an infinite population or if sampling is with replacement from a finite population, and by

$$P \;\pm\; z_c \sqrt{\frac{pq}{N}} \sqrt{\frac{N_p - N}{N_p - 1}} \tag{4}$$

if sampling is without replacement from a population of finite size N_p.

To compute these confidence limits we can use the sample estimate P for p, which will generally prove satisfactory if $N \geq 30$. A more exact method for obtaining these confidence limits is given in Problem 9.12.

CONFIDENCE INTERVALS FOR DIFFERENCES AND SUMS

If S_1 and S_2 are two sample statistics with approximately normal sampling distributions, confidence limits for the differences of the population parameters corresponding to S_1 and S_2 are given by

$$S_1 - S_2 \pm z_c \sigma_{S_1 - S_2} \;=\; S_1 - S_2 \pm z_c \sqrt{\sigma_{S_1}^2 + \sigma_{S_2}^2} \tag{5}$$

while confidence limits for the sum of the population parameters are

$$S_1 + S_2 \pm z_c \sigma_{S_1 + S_2} \;=\; S_1 + S_2 \pm z_c \sqrt{\sigma_{S_1}^2 + \sigma_{S_2}^2} \tag{6}$$

provided that the samples are independent (see Chapter 8).

For example, confidence limits for the difference of two population means, in the case where the populations are infinite, are given by

$$\bar{X}_1 - \bar{X}_2 \pm z_c \sigma_{\bar{X}_1 - \bar{X}_2} = \bar{X}_1 - \bar{X}_2 \pm z_c \sqrt{\frac{\sigma_1^2}{N_1} + \frac{\sigma_2^2}{N_2}} \qquad (7)$$

where $\bar{X}_1$, σ_1, N_1 and $\bar{X}_2$, σ_2, N_2 are the respective means, standard deviations and sizes of the two samples drawn from the populations.

Similarly, confidence limits for the difference of two population proportions, where the populations are infinite, are given by

$$P_1 - P_2 \pm z_c \sigma_{P_1 - P_2} = P_1 - P_2 \pm z_c \sqrt{\frac{p_1(1 - p_1)}{N_1} + \frac{p_2(1 - p_2)}{N_2}} \qquad (8)$$

where P_1 and P_2 are the two sample proportions, N_1 and N_2 are the sizes of the two samples drawn from the populations, and p_1 and p_2 are the proportions in the two populations (estimated by P_1 and P_2).

CONFIDENCE INTERVALS FOR STANDARD DEVIATIONS

The confidence limits for the standard deviation σ of a normally distributed population as estimated from a sample with standard deviation s, are given by

$$s \pm z_c \sigma_s = s \pm z_c \frac{\sigma}{\sqrt{2N}} \qquad (9)$$

using Table 8.1, Page 144. In computing these confidence limits we use s or $\hat{s}$ to estimate σ.

PROBABLE ERROR

The 50% confidence limits of the population parameters corresponding to a statistic S are given by $S \pm 0.6745\sigma_S$. The quantity $0.6745\sigma_S$ is known as the *probable error* of the estimate.

Solved Problems

UNBIASED AND EFFICIENT ESTIMATES

9.1. Give an example of estimators (or estimates) which are (a) unbiased and efficient, (b) unbiased and inefficient, (c) biased and inefficient.

Solution:

(a) The sample mean $\bar{X}$ and the modified sample variance $\hat{s}^2 = \frac{N}{N - 1} s^2$ are two such examples.

(b) The sample median and the sample statistic $\frac{1}{2}(Q_1 + Q_3)$, where Q_1 and Q_3 are the lower and upper sample quartiles, are two such examples. Both statistics are unbiased estimates of the population mean, since the mean of their sampling distributions is the population mean.

(c) The sample standard deviation s, the modified standard deviation $\hat{s}$, the mean deviation and the semi-interquartile range are four such examples.

9.2. A sample of five measurements of the diameter of a sphere were recorded by a scientist as 6·33, 6·37, 6·36, 6·32 and 6·37 mm. Determine unbiased and efficient estimates of (*a*) the true mean, (*b*) the true variance.

Solution:

(*a*) Unbiased and efficient estimates of the true mean (i.e. the population mean)

$$= \bar{X} = \frac{\Sigma X}{N} = \frac{6\cdot33 + 6\cdot37 + 6\cdot36 + 6\cdot32 + 6\cdot37}{5} = 6\cdot35 \text{ mm}$$

(*b*) Unbiased and efficient estimates of the true variance (i.e. the population variance)

$$= \hat{s}^2 = \frac{N}{N-1} s^2 = \frac{\Sigma(X - \bar{X})^2}{N-1}$$

$$= \frac{(6\cdot33 - 6\cdot35)^2 + (6\cdot37 - 6\cdot35)^2 + (6\cdot36 - 6\cdot35)^2 + (6\cdot32 - 6\cdot35)^2 + (6\cdot37 - 6\cdot35)^2}{5-1}$$

$$= 0\cdot000\,55 \text{ mm}^2$$

        Note that $\hat{s} = \sqrt{0\cdot000\,55} = 0\cdot023$ is an estimate of the true standard deviation but this estimate is neither unbiased nor efficient.

9.3. Suppose that the masses of 100 male students at XYZ University represent a random sample of the masses of all 1546 students at the university. Determine unbiased and efficient estimates of (*a*) the true mean, (*b*) the true variance.

Solution:

(*a*) From Prob. 3.22 of Chap. 3:
      Unbiased and efficient estimate of true mean mass $= \bar{X} = 67\cdot45$ kg.

(*b*) From Prob. 4.17 of Chap. 4:
      Unbiased and efficient estimate of true variance $= \hat{s}^2 = \frac{N}{N-1} s^2 = \frac{100}{99} (8\cdot5275) = 8\cdot6136$.

        Thus $\hat{s} = \sqrt{8\cdot6136} = 2\cdot93$. Note that since N is large there is essentially no difference between s^2 and $\hat{s}^2$ or between s and $\hat{s}$.

        Note that we have not used Sheppard's corrections for grouping. To take these into account we would use $s = 2\cdot79$ above (see Prob. 4.21, Chap. 4).

9.4. Give an unbiased and efficient estimate of the true mean diameter of the sphere of Problem 9.2.

Solution:

       The median is one example of an unbiased and inefficient estimate of the population mean. For the five measurements arranged in order of magnitude the median is 6·36.

CONFIDENCE INTERVAL ESTIMATES FOR POPULATION MEANS

9.5. Find (*a*) 95% and (*b*) 99% confidence intervals for estimating the mean mass of the XYZ University students in Problem 9.3.

Solution:

(*a*) The 95% confidence limits are $\bar{X} \pm 1\cdot96\sigma/\sqrt{N}$.

        Using $\bar{X} = 67\cdot45$ kg, and $\hat{s} = 2\cdot93$ kg as an estimate of σ (see Problem 9.3), the confidence limits are $67\cdot45 \pm 1\cdot96(2\cdot93/\sqrt{100})$ or $67\cdot45 \pm 0\cdot57$ kg. Thus the 95% confidence interval for the population mean μ is 66·88 to 68·02 kg, which can be denoted by $66\cdot88 < \mu < 68\cdot02$.

        We can therefore say that the probability that the population mean mass lies between 66·88 and 68·02 kg is about 95% or 0·95. In symbols we write $\Pr\{66\cdot88 < \mu < 68\cdot02\} = 0\cdot95$. This is equivalent to saying that we are 95% *confident* that the population (or true mean) lies between 66·88 and 68·02 kg.

(*b*) The 99% confidence limits are $\bar{X} \pm 2\cdot58\sigma/\sqrt{N} = \bar{X} \pm 2\cdot58\hat{s}/\sqrt{N} = 67\cdot45 \pm 2\cdot58(2\cdot93/\sqrt{100}) = 67\cdot45 \pm 0\cdot76$ kg.

        Thus the 99% confidence interval for the population mean μ is 66·69 to 68·21 kg, which can be denoted by $66\cdot69 < \mu < 68\cdot21$.

In obtaining the above confidence intervals, we assumed that the population was infinite or so large that we could consider conditions to be the same as sampling with replacement. For finite populations where sampling is without replacement, we should use $\dfrac{\sigma}{\sqrt{N}}\sqrt{\dfrac{N_p - N}{N_p - 1}}$ in place of $\dfrac{\sigma}{\sqrt{N}}$. However, we can consider the factor

$$\sqrt{\frac{N_p - N}{N_p - 1}} = \sqrt{\frac{1546 - 100}{1546 - 1}} = 0.967$$ as essentially 1.0, so that it need not be used. If it is used, the above confidence limits become 67.45 ± 0.56 and 67.45 ± 0.73 kg respectively.

9.6. Measurements of the weights of a random sample of 200 ball bearings made by a certain machine during one week showed a mean of 0·824 newtons and a standard deviation of 0·042 newtons. Find (a) 95% and (b) 99% confidence limits for the mean weight of all the ball bearings.

Solution:
(a) The 95% confidence limits are
$$\bar{X} \pm 1.96\sigma/\sqrt{N} = \bar{X} \pm 1.96\hat{s}/\sqrt{N} = 0.824 \pm 1.96(0.042/\sqrt{200}) = 0.824 \pm 0.0058 \text{ N, or } 0.824 \pm 0.006 \text{ N.}$$

(b) The 99% confidence limits are
$$\bar{X} \pm 2.58\sigma/\sqrt{N} = \bar{X} \pm 2.58\hat{s}/\sqrt{N} = 0.824 \pm 2.58(0.042/\sqrt{200}) = 0.824 \pm 0.0077 \text{ N, or } 0.824 \pm 0.008 \text{ N.}$$

Note that we have assumed the reported standard deviation to be the *modified* standard deviation $\hat{s}$. If the standard deviation had been s, we would have used $\hat{s} = \sqrt{N/(N-1)}\,s = \sqrt{200/199}\,s$ which can be taken as s for all practical purposes. In general, for $N \geqq 30$ we may assume s and $\hat{s}$ as practically equal.

9.7. Find (a) 98%, (b) 90% and (c) 99·73% confidence limits for the mean weight of the ball bearings in Problem 9.6.

Solution:
(a) Let $z = z_c$ be such that the area under the normal curve to the right is 1%. Then by symmetry the area to the left of $z = -z_c$ is also 1%, so that the shaded area is 98% of the total area.

Since the total area under the curve is one, the area from $z = 0$ to $z = z_c$ is 0·49; hence $z_c = 2.33$.

Thus 98% confidence limits are $\bar{X} \pm 2.33\sigma/\sqrt{N} = 0.824 \pm 2.33(0.042/\sqrt{200}) = 0.824 \pm 0.0069$ N.

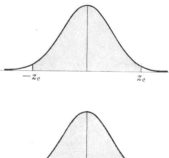

(b) We require z_c such that the area from $z = 0$ to $z = z_c$ is 0·45; then $z_c = 1.645$.

Thus 90% confidence limits are $\bar{X} \pm 1.645\sigma/\sqrt{N} = 0.824 \pm 1.645(0.042/\sqrt{200}) = 0.824 \pm 0.0049$ N.

(c) The 99·73% confidence limits are
$$\bar{X} \pm 3\sigma/\sqrt{N} = 0.824 \pm 3(0.042/\sqrt{200}) = 0.824 \pm 0.0089 \text{ N.}$$

9.8. In measuring reaction time, a psychologist estimates that the standard deviation is 0·05 seconds. How large a sample of measurements must he take in order to be (a) 95% and (b) 99% confident that the error of his estimate will not exceed 0·01 seconds?

Solution:
(a) The 95% confidence limits are $\bar{X} \pm 1.96\sigma/\sqrt{N}$, the error of the estimate being $1.96\sigma/\sqrt{N}$. Taking $\sigma = s = 0.05$ seconds, we see that this error will be equal to 0·01 seconds if $(1.96)(0.05)/\sqrt{N} = 0.01$, i.e. $\sqrt{N} = (1.96)(0.05)/0.01 = 9.8$ or $N = 96.04$. Thus we can be 95% confident that the error of the estimate will be less than 0·01 if N is 97 or larger.

Another method: $\dfrac{(1.96)(0.05)}{\sqrt{N}} \leqq 0.01$ if $\dfrac{\sqrt{N}}{(1.96)(0.05)} \geqq \dfrac{1}{0.01}$ or $\sqrt{N} \geqq \dfrac{(1.96)(0.05)}{0.01} = 9.8$.

Then $N \geqq 96.04$, or $N \geqq 97$.

(b) The 99% confidence limits are $\bar{X} \pm 2.58\sigma/\sqrt{N}$. Then $(2.58)(0.05)/\sqrt{N} = 0.01$, or $N = 166.4$. Thus we can be 99% confident that the error of the estimate will be less than 0·01 only if N is 167 or larger.

9.9.   A random sample of 50 mathematics marks out of a total of 200 showed a mean of 75 and a standard deviation of 10. (a) What are the 95% confidence limits for estimates of the mean of the 200 marks? (b) With what degree of confidence could we say that the mean of all 200 marks is 75 ± 1?

Solution:
(a)   Since the population size is not very large compared with the sample size, we must adjust for it. Then the 95% confidence limits are

$$\bar{X} \pm 1.96\sigma_{\bar{X}} = \bar{X} \pm 1.96 \frac{\sigma}{\sqrt{N}} \sqrt{\frac{N_p - N}{N_p - 1}} = 75 \pm 1.96 \frac{(10)}{\sqrt{50}} \sqrt{\frac{200 - 50}{200 - 1}} = 75 \pm 2.4$$

(b)   The confidence limits can be represented by

$$\bar{X} = z_c\sigma_{\bar{X}} = \bar{X} \pm z_c \frac{\sigma}{\sqrt{N}} \sqrt{\frac{N_p - N}{N_p - 1}} = 75 \pm z_c \frac{10}{\sqrt{50}} \sqrt{\frac{200 - 50}{200 - 1}} = 75 \pm 1.23z_c$$

Since this must equal 75 ± 1, we have $1.23z_c = 1$ or $z_c = 0.81$. The area under the normal curve from $z = 0$ to $z = 0.81$ is 0.2910; hence the required degree of confidence is $2(0.2910) = 0.582$ or 58.2%.

CONFIDENCE INTERVAL ESTIMATES FOR PROPORTIONS

9.10.   A sample poll of 100 voters chosen at random from all voters in a given district indicated that 55% of them were in favour of a particular candidate. Find (a) 95%, (b) 99% and (c) 99.73% confidence limits for the proportion of all the voters in favour of this candidate.

Solution:
(a)   The 95% confidence limits for the population p are
$$P \pm 1.96\sigma_P = P \pm 1.96\sqrt{p(1 - p)/N} = 0.55 \pm 1.96\sqrt{(0.55)(0.45)/100} = 0.55 \pm 0.10$$
where we have used the sample proportion P to estimate p.

(b)   The 99% confidence limits for p are $0.55 \pm 2.58\sqrt{(0.55)(0.45)/100} = 0.55 \pm 0.13$.

(c)   The 99.73% confidence limits for p are $0.55 \pm 3\sqrt{(0.55)(0.45)/100} = 0.55 \pm 0.15$.

    For a more exact method of working this problem, see Problem 9.12.

9.11.   How large a sample of voters should we take in Problem 9.10 in order to be (a) 95% and (b) 99.73% confident that the given candidate will be elected from two candidates?

Solution:
    The confidence limits for p are $P \pm z_c\sqrt{p(1 - p)/N} = 0.55 \pm z_c\sqrt{(0.55)(0.45)/N} = 0.55 \pm 0.50z_c/\sqrt{N}$, where we have used the estimate $P = p = 0.55$ on the basis of Problem 9.10.

    Since the candidate will win only if he receives more than 50% of the population votes, we require that $0.50z_c/\sqrt{N}$ be less than 0.05.

(a)   For 95% confidence, $0.50z_c/\sqrt{N} = 0.50(1.96)/\sqrt{N} = 0.05$ when $N = 384.2$. Thus N should be at least 385.

(b)   For 99.73% confidence, $0.50z_c/\sqrt{N} = 0.50(3)/\sqrt{N} = 0.05$ when $N = 900$. Thus N should be at least 901.

    Another method:
    $1.50/\sqrt{N} < 0.05$ when $\sqrt{N}/1.50 > 1/0.05$ or $\sqrt{N} > 1.50/0.05$. Then $\sqrt{N} > 30$ or $N > 900$, so that N should be at least 901.

9.12.   (a) If P is the observed proportion of successes in a sample of size N, show that the confidence limits for estimating the population proportion of successes p at the level of confidence determined by z_c are given by

$$p = \frac{P + \dfrac{z_c^2}{2N} \pm z_c \sqrt{\dfrac{P(1 - P)}{N} + \dfrac{z_c^2}{4N^2}}}{1 + \dfrac{z_c^2}{N}}$$

(b) Use the formula derived in (a) to obtain the 99.73% confidence limits of Prob. 9.10.

(c) Show that for large N the formula in (a) reduces to $p = P \pm z_c\sqrt{P(1-P)/N}$, as used in Prob. 9.10.

Solution:

(a) The sample proportion P in standard units $= \dfrac{P-p}{\sigma_P} = \dfrac{P-p}{\sqrt{p(1-p)/N}}$.

The largest and smallest values of this standardized variable are $\pm z_c$, where z_c determines the level of confidence. At these extreme values we must therefore have

$$P - p = \pm z_c\sqrt{p(1-p)/N}$$

Squaring both sides,

$$P^2 - 2pP + p^2 = z_c^2 p(1-p)/N$$

Multiplying both sizes by N and simplifying, we find

$$(N + z_c^2)p^2 - (2NP + z_c^2)p + NP^2 = 0$$

If $a = N + z_c^2$, $b = -(2NP + z_c^2)$ and $c = NP^2$, this equation becomes $ap^2 + bp + c = 0$ whose solution for p is given by the quadratic formula

$$p = \frac{-b \pm \sqrt{b^2 - 4ac}}{2a} = \frac{2NP + z_c^2 \pm \sqrt{(2NP + z_c^2)^2 - 4(N + z_c^2)(NP^2)}}{2(N + z_c^2)}$$

$$= \frac{2NP + z_c^2 \pm z_c\sqrt{4NP(1-P) + z_c^2}}{2(N + z_c^2)}$$

Dividing the numerator and denominator by $2N$, this becomes

$$p = \frac{P + \dfrac{z_c^2}{2N} \pm z_c\sqrt{\dfrac{P(1-P)}{N} + \dfrac{z_c^2}{4N^2}}}{1 + \dfrac{z_c^2}{N}}$$

(b) For 99·73% confidence limits, $z_c = 3$. Then using $P = 0.55$ and $N = 100$ in the formula derived in (a), we find $p = 0.40$ and 0.69, agreeing with Problem 9.10(c).

(c) If N is large, then $z_c^2/(2N)$, $z_c^2/(4N^2)$ and z_c^2/N are all negligibly small and can essentially be replaced by zero, so that the required result is obtained.

9.13. In 40 tosses of a coin, 24 heads were obtained. Find (a) 95% and (b) 99·73% confidence limits for the proportion of heads which would be obtained in an unlimited number of tosses of the coin.

Solution:

(a) At the 95% level, $z_c = 1.96$. Putting $P = 24/40 = 0.6$ and $N = 40$ in the formula of Problem 9.12(a), we find $p = 0.45$ and 0.74. Thus we can say with 95% confidence that p lies between 0·45 and 0·74.

Using the approximate formula $p = P \pm z_c\sqrt{P(1-P)/N}$, we find $p = 0.60 \pm 0.15$, yielding the interval 0·45 to 0·75.

(b) At the 99·73% level, $z_c = 3$. Using the formula of Problem 9.12(a), we find $p = 0.37$ and 0.79.

Using the approximate formula $p = P \pm z_c\sqrt{P(1-P)/N}$, we find $p = 0.60 \pm 0.23$, yielding the interval 0·37 to 0·83.

CONFIDENCE INTERVALS FOR DIFFERENCES AND SUMS

9.14. A sample of 150 brand A light bulbs showed a mean lifetime of 1400 hours and a standard deviation of 120 hours. A sample of 200 brand B light bulbs showed a mean lifetime of 1200 hours and a standard deviation of 80 hours. Find (a) 95% and (b) 99% confidence limits for the difference of the mean lifetimes of the populations of brands A and B.

Solution:

Confidence limits for the difference in means of brands A and B are given by

$$\bar{X}_A - \bar{X}_B \pm z_c\sqrt{\sigma_A^2/N_A + \sigma_B^2/N_B}$$

(a) The 95% confidence limits are: $1400 - 1200 \pm 1.96\sqrt{(120)^2/150 + (80)^2/100} = 200 \pm 24.8$.

Thus we can be 95% confident that the difference of population means lies between 175 and 225 h.

(b)   The 99% confidence limits are: $1400 - 1200 \pm 2 \cdot 58 \sqrt{(120)^2/150 + (80)^2/100} = 200 \pm 32 \cdot 6$.
    Thus we can be 99% confident that the difference of population means lies between 167 and 233 h.

9.15.   In a random sample of 400 adults and 600 teenagers who watched a certain television programme, 100 adults and 300 teenagers indicated that they liked it. Construct (a) 95% and (b) 99% confidence limits for the difference in proportions of all adults and all teenagers who watched the programme and liked it.

Solution:
    Confidence limits for the differences in proportions of the two groups are given by

$$P_1 - P_2 \pm z_c \sqrt{p_1 q_1/N_1 + p_2 q_2/N_2}$$

where subscripts 1 and 2 refer to teenagers and adults respectively. Here $P_1 = 300/600 = 0 \cdot 50$ and $P_2 = 100/400 = 0 \cdot 25$ are respectively the proportions of teenagers and adults who liked the programme.

(a)   95% confidence limits: $0 \cdot 50 - 0 \cdot 25 \pm 1 \cdot 96 \sqrt{(0 \cdot 50)(0 \cdot 50)/600 + (0 \cdot 25)(0 \cdot 75)/400} = 0 \cdot 25 \pm 0 \cdot 06$.
    Thus we can be 95% confident that the true difference in proportion lies between 0·19 and 0·31.

(b)   99% confidence limits: $0 \cdot 50 - 0 \cdot 25 \pm 2 \cdot 58 \sqrt{(0 \cdot 50)(0 \cdot 50)/600 + (0 \cdot 25)(0 \cdot 75)/400} = 0 \cdot 25 \pm 0 \cdot 08$.
    Thus we can be 99% confident that the true difference in proportion lies between 0·17 and 0·33.

9.16.   The mean e.m.f. of batteries produced by a company is 45·1 volts and the standard deviation is 0·04 volts. If four such batteries are connected in series, find (a) 95%, (b) 99%, (c) 99·73% and (d) 50% confidence limits for the total e.m.f.

Solution:
    If E_1, E_2, E_3, and E_4 represent the e.m.f.'s of the four batteries, we have

$$\mu_{E_1+E_2+E_3+E_4} = \mu_{E_1} + \mu_{E_2} + \mu_{E_3} + \mu_{E_4} \quad \text{and} \quad \sigma_{E_1+E_2+E_3+E_4} = \sqrt{\sigma_{E_1}^2 + \sigma_{E_2}^2 + \sigma_{E_3}^2 + \sigma_{E_4}^2}$$

Then since $\mu_{E_1} = \mu_{E_2} = \mu_{E_3} = \mu_{E_4} = 45 \cdot 1$ volts and $\sigma_{E_1} = \sigma_{E_2} = \sigma_{E_3} = \sigma_{E_4} = 0 \cdot 04$ volts,

$$\mu_{E_1+E_2+E_3+E_4} = 4(45 \cdot 1) = 180 \cdot 4 \quad \text{and} \quad \sigma_{E_1+E_2+E_3+E_4} = \sqrt{4(0 \cdot 04)^2} = 0 \cdot 08$$

(a)   95% confidence limits are: $180 \cdot 4 \pm 1 \cdot 96(0 \cdot 08) \quad = 180 \cdot 4 \pm 0 \cdot 16$ volts.

(b)   99% confidence limits are: $180 \cdot 4 \pm 2 \cdot 58(0 \cdot 08) \quad = 180 \cdot 4 \pm 0 \cdot 21$ volts.

(c)   99·73% confidence limits are: $180 \cdot 4 \pm 3(0 \cdot 08) \quad\quad = 180 \cdot 4 \pm 0 \cdot 24$ volts.

(d)   50% confidence limits are: $180 \cdot 4 \pm 0 \cdot 6745(0 \cdot 08) = 180 \cdot 4 \pm 0 \cdot 054$ volts.

    The value 0·054 volts is called the *probable error*.

CONFIDENCE INTERVALS FOR STANDARD DEVIATIONS

9.17.   The standard deviation of the lifetimes of a sample of 200 electric light bulbs was computed to be 100 hours. Find (a) 95% and (b) 99% confidence limits for the standard deviation of all such electric light bulbs.

Solution:
    Confidence limits for the population standard deviation σ are given by $s \pm z_c \sigma/\sqrt{2N}$, where z_c indicates the level of confidence. We use the sample standard deviation to estimate σ.

(a)   The 95% confidence limits are: $100 \pm 1 \cdot 96(100)/\sqrt{400} = 100 \pm 9 \cdot 8$.
    Thus we can be 95% confident that the population standard deviation will lie between 90·2 and 109·8 h.

(b)   The 99% confidence limits are: $100 \pm 2 \cdot 58(100)/\sqrt{400} = 100 \pm 12 \cdot 9$.
    Thus we can be 99% confident that the population standard deviation will lie between 87·1 and 112·9 h.

9.18.   How large a sample of the light bulbs in the previous problem must we take in order to be 99·73% confident that the true population standard deviation will not differ from the sample standard deviation by more than (a) 5%, (b) 10%?

Solution:

99·73% confidence limits for σ are $s \pm 3\sigma/\sqrt{2N} = s \pm 3s/\sqrt{2N}$ using s as an estimate of σ.

Then the percentage error in the standard deviation $= \dfrac{3s/\sqrt{2N}}{s} = \dfrac{300}{\sqrt{2N}}$ %.

(a) If $300/\sqrt{2N} = 5$, then $N = 1800$. Thus the sample size should be 1800 or more.

(b) If $300/\sqrt{2N} = 10$, then $N = 450$. Thus the sample size should be 450 or more.

PROBABLE ERROR

9.19.   The voltages of 50 batteries of the same type have a mean of 18·0 volts and a standard deviation of 0·5 volts. Find (a) the probable error of the mean, (b) the 50% confidence limits.

Solution:

(a) Probable error of the mean $= 0.6745 \sigma_{\bar{X}} = 0.6745\,\dfrac{\sigma}{\sqrt{N}} = 0.6745\,\dfrac{\hat{s}}{\sqrt{N}} = 0.6745\,\dfrac{s}{\sqrt{N-1}}$

$$= 0.6745(0.5)/\sqrt{49} = 0.048 \text{ volts}$$

Note that if the standard deviation of 0·5 volts is computed as $\hat{s}$, the probable error is $0.6745(0.5/\sqrt{50})$ $= 0.048$ also, so that either estimate can be used if N is large enough.

(b) 50% confidence limits are 18 ± 0.048 volts.

9.20.   A measurement was recorded as 216·480 grammes, with a probable error of 0·272 gramme. What are the 95% confidence limits for the measurement?

Solution:

Probable error $= 0.272 = 0.6745\sigma_{\bar{X}}$, or $\sigma_{\bar{X}} = 0.272/0.6745$.
95% confidence limits $= \bar{X} \pm 1.96\sigma_{\bar{X}} = 216.480 \pm 1.96(0.272/0.6745) = 216.480 \pm 0.790$ grammes.

Supplementary Problems

UNBIASED AND EFFICIENT ESTIMATES

9.21.   Measurements of a sample of masses were determined as 8·3, 10·6, 9·7, 8·8, 10·2 and 9·4 kg respectively. Determine unbiased and efficient estimates of (a) the population mean and (b) the population variance. (c) Compare the sample standard deviation with the estimated population standard deviation.
Ans. (a) 9·5 kg, (b) 0·74 kg², (c) 0·78 and 0·86 kg respectively

9.22.   A sample of 10 television tubes produced by a company showed a mean lifetime of 1200 h and a standard deviation of 100 h. Estimate (a) the mean and (b) the standard deviation of the population of all television tubes produced by this company.   *Ans.* (a) 1200 h, (b) 105·4 h

9.23.   (a) Work Problem 9.22 if the same results are obtained for 30, 50 and 100 television tubes. (b) What can you conclude about the relation between sample standard deviations and estimates of population standard deviations for different sample sizes?
Ans. (a) Estimates of population standard deviations for sample sizes of 30, 50 and 100 tubes are respectively 101·7, 101·0 and 100·5 h. Estimates of population means are 1200 h in all cases.

CONFIDENCE INTERVAL ESTIMATES FOR POPULATION MEANS

9.24.   The mean and standard deviation of the maximum loads supported by 60 cables (see Prob. 3.59, Chap. 3) are given by 11·09 kN and 0·73 kN respectively. Find (a) 95% and (b) 99% confidence limits for the mean of the maximum loads of all cables produced by the company.   *Ans.* (a) 11·09 $\pm$ 0·18 kN, (b) 11·09 $\pm$ 0·24 kN

9.25.   The mean and standard deviation of the diameters of a sample of 250 rivet heads manufactured by a company are 7·2642 mm and 0·0058 mm respectively (see Prob. 3.61, Chap. 3). Find (a) 99%, (b) 98%, (c) 95% and (d) 90% confidence limits for the mean diameter of all the rivet heads manufactured by the company.
Ans. (a) 7·2642 $\pm$ 0·000 95 mm, (b) 7·2642 $\pm$ 0·000 85 mm, (c) 7·2642 $\pm$ 0·000 72 mm, (d) 7·2642 $\pm$ 0·000 60 mm

9.26.   Find (a) the 50% confidence limits and (b) the probable error for the mean diameters in Prob. 9.25.
Ans. (a) $7 \cdot 2642 \pm 0 \cdot 00025$ mm, (b) $0 \cdot 00025$ mm

9.27.   If the standard deviation of the lifetimes of television tubes is estimated as 100 hours, how large a sample must we take in order to be (a) 95%, (b) 90%, (c) 99% and (d) 99·73% confident that the error in the estimated mean lifetime will not exceed 20 hours?   *Ans.* (a) at least 96, (b) at least 68, (c) at least 167, (d) at least 225

9.28.   What are the sample sizes in the previous problem if the error in the estimated mean lifetime must not exceed 10 hours?   *Ans.* (a) at least 384, (b) at least 271, (c) at least 666, (d) at least 900

9.29.   A company has 500 cables. A test of 40 cables selected at random showed a mean breaking strength of 2400 newtons and a standard deviation of 150 newtons. (a) What are the 95% and 99% confidence limits for estimating the mean breaking strength of the remaining 460 cables? (b) With what degree of confidence could we say that the mean breaking strength of the remaining 460 cables is 2400 ± 35 newtons?   *Ans.* (a) 2400 ± 45 N, 2400 ± 59 N; (b) 87·6%

CONFIDENCE INTERVAL ESTIMATES FOR PROPORTIONS

9.30.   An urn contains an unknown proportion of red and white marbles. A random sample of 60 marbles selected with replacement from the urn showed that 70% were red. Find (a) 95%, (b) 99% and (c) 99·73% confidence limits for the actual proportion of red marbles in the urn. Present the results using both the approximate formula and the more exact formula of Prob. 9.12.
Ans. (a) $0 \cdot 70 \pm 0 \cdot 12$, $0 \cdot 69 \pm 0 \cdot 11$; (b) $0 \cdot 70 \pm 0 \cdot 15$, $0 \cdot 68 \pm 0 \cdot 15$; (c) $0 \cdot 70 \pm 0 \cdot 18$, $0 \cdot 67 \pm 0 \cdot 17$

9.31.   How large a sample of marbles should one take in the preceding problem in order to be (a) 95%, (b) 99% and (c) 99·73% confident that the true proportion does not differ from the sample proportion by more than 5%?
Ans. (a) at least 323, (b) at least 560, (c) at least 756

9.32.   It is believed than an election will result in a very close vote between two candidates. What is the least number of voters which one should poll in order to be (a) 80%, (b) 90%, (c) 95% and (d) 99% confident of a decision in favour of either one of the candidates?   *Ans.* (a) 16 400, (b) 27 100, (c) 38 420, (d) 66 600

CONFIDENCE INTERVALS FOR DIFFERENCES AND SUMS

9.33.   Of two similar groups of patients, A and B consisting of 50 and 100 individuals respectively, the first was given a new type of sleeping pill and the second was given a conventional type. For patients in group A the mean number of hours sleep was 7·82 with a standard deviation of 0·24 hours. For patients in group B the mean number of hours of sleep was 6·75 with a standard deviation of 0·30 hours. Find (a) 95% and (b) 99% confidence limits for the difference in the mean number of hours of sleep induced by the two types of sleeping pills.
Ans. (a) $1 \cdot 07 \pm 0 \cdot 09$ h, (b) $1 \cdot 07 \pm 0 \cdot 12$ h

9.34.   A sample of 200 bolts from one machine showed that 15 were defective, while a sample of 100 bolts from another machine showed that 12 were defective. Find (a) 95%, (b) 99% and (c) 99·73% confidence limits for the difference in proportions of defective bolts from the two machines. Discuss the results obtained.
Ans. (a) $0 \cdot 045 \pm 0 \cdot 073$, (b) $0 \cdot 045 \pm 0 \cdot 097$, (c) $0 \cdot 045 \pm 0 \cdot 112$

9.35.   A company manufactures ball bearings having a mean mass of 0·638 kg and a standard deviation of 0·012 kg. Find (a) 95% and (b) 99% confidence limits for the masses of lots consisting of 100 ball bearings each.
Ans. (a) $63 \cdot 8 \pm 0 \cdot 24$ kg, (b) $63 \cdot 8 \pm 0 \cdot 31$ kg

CONFIDENCE INTERVALS FOR STANDARD DEVIATIONS

9.36.   The standard deviation of the breaking strength of 100 cables tested by a company was 180 N. Find (a) 95%, (b) 99% and (c) 99·73% confidence limits for the standard deviation of all cables produced by the company.
Ans. (a) $180 \pm 24 \cdot 9$ N, (b) $180 \pm 32 \cdot 8$ N, (c) $180 \pm 38 \cdot 2$ N

9.37.   Find the probable error of the standard deviation in the preceding problem.   *Ans.* 8·6 N

9.38.   How large a sample should one take in order to be (a) 95%, (b) 99% and (c) 99·73% confident that a population standard deviation will not differ from a sample standard deviation by more than 2%?
Ans. (a) at least 4802, (b) at least 8321, (c) at least 11 250

CHAPTER 10

Statistical Decision Theory,
Tests of Hypotheses and Significance

STATISTICAL DECISIONS

Very often in practice we are called upon to make decisions about populations on the basis of sample information. Such decisions are called *statistical decisions*. For example, we may wish to decide on the basis of sample data whether a new serum is really effective in curing a disease, whether one educational procedure is better than another, whether a given coin is loaded, etc.

STATISTICAL HYPOTHESES. NULL HYPOTHESES

In attempting to reach decisions, it is useful to make assumptions or guesses about the populations involved. Such assumptions, which may or may not be true, are called *statistical hypotheses* and in general are statements about the probability distributions of the populations.

In many instances we formulate a statistical hypothesis for the sole purpose of rejecting or nullifying it. For example, if we want to decide whether a given coin is loaded we formulate the hypothesis that the coin is fair, i.e. $p = 0.5$, where p is the probability of heads. Similarly, if we want to decide whether one procedure is better than another, we formulate the hypothesis that there is *no difference* between the procedures (i.e. any observed differences are merely due to fluctuations in sampling from the *same* population). Such hypotheses are often called *null hypotheses* and are denoted by H_0.

Any hypothesis which differs from a given hypothesis is called an *alternative hypothesis*. For example, if one hypothesis is $p = 0.5$, alternative hypotheses are $p = 0.7$, $p \neq 0.5$, or $p > 0.5$. A hypothesis alternative to the null hypothesis is denoted by H_1.

TESTS OF HYPOTHESES AND SIGNIFICANCE

If on the supposition that a particular hypothesis is true we find that results observed in a random sample differ markedly from those expected under the hypothesis on the basis of pure chance using sampling theory, we would say that the observed differences are *significant* and we would be inclined to reject the hypothesis (or at least not accept it on the basis of the evidence obtained). For example, if 20 tosses of a coin yield 16 heads we would be inclined to reject the hypothesis that the coin is fair, although it is conceivable that we might be wrong.

Procedures which enable us to decide whether to accept or reject hypotheses or to determine whether observed samples differ significantly from expected results are called *tests of hypotheses*, *tests of significance*, or *rules of decision*.

TYPE I AND TYPE II ERRORS

If we reject a hypothesis when it should be accepted, we say that a *Type I error* has been made. If, on the other hand, we accept a hypothesis when it should be rejected, we say that a *Type II error* has been made.

In either case a wrong decision or error in judgment has occurred.

In order for any tests of hypotheses or rules of decisions to be good, they must be designed so as to minimize errors of decision. This is not a simple matter since, for a given sample size, an attempt to decrease one type of error is accompanied in general by an increase in the other type of error. In practice one type of error may be more serious than the other, and so a compromise should be reached in favour of a limitation of the more serious error. The only way to reduce both types of error is to increase the sample size, which may or may not be possible.

LEVEL OF SIGNIFICANCE

In testing a given hypothesis, the maximum probability with which we would be willing to risk a Type I error is called the *level of significance* of the test. This probability, often denoted by α, is generally specified before any samples are drawn, so that results obtained will not influence our choice.

In practice a level of significance of 0·05 or 0·01 is customary, although other values are used. If for example a 0·05 or 5% level of significance is chosen in designing a test of hypothesis, then there are about 5 chances in 100 that we would reject the hypothesis when it should be accepted, i.e. we are about 95% *confident* that we have made the right decision. In such case we say that the hypothesis has been *rejected at a* 0·05 *level of significance*, which means that we could be wrong with probability 0·05.

TESTS INVOLVING THE NORMAL DISTRIBUTION

To illustrate the ideas presented above, suppose that under a given hypothesis the sampling distribution of a statistic S is a normal distribution with mean μ_S and standard deviation σ. Then the distribution of the standardized variable (or z score), given by $z = (S - \mu_S)/\sigma_S$, is the standardized normal distribution (mean 0, variance 1) and is shown in Fig. 10-1.

As indicated in the figure we can be 95% confident that, if the hypothesis is true, the z score of an actual sample statistic S will lie between −1·96 and 1·96 (since the area under the normal curve between these values is 0·95).

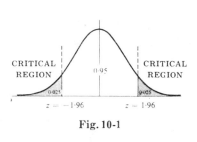

Fig. 10-1

However, if on choosing a single sample at random we find that the z score of its statistic lies *outside* the range −1·96 to 1·96, we would conclude that such an event could happen with probability of only 0·05 (total shaded area in the figure) if the given hypothesis were true. We would then say that this z score differed *significantly* from what would be expected under the hypothesis and would be inclined to reject the hypothesis.

The total shaded area 0·05 is the level of significance of the test. It represents the probability of our being wrong in rejecting the hypothesis, i.e. the probability of making a Type I error. Thus we say that the hypothesis is *rejected at a* 0·05 *level of significance* or that the z score of the given sample statistic is *significant at a* 0·05 *level of significance*.

The set of z scores outside the range −1·96 to 1·96 constitutes what is called the *critical region or region of rejection of the hypothesis* or the *region of significance*. The set of z scores inside the range −1·96 to 1·96 could then be called the *region of acceptance of the hypothesis* or the *region of non-significance*.

On the basis of the previous remarks we can formulate the following rule of decision or test of hypothesis or significance.

(*a*) Reject the hypothesis at a 0·05 level of significance if the *z* score of the statistic *S* lies outside the range −1·96 to 1·96 (i.e. either *z* > 1·96 or *z* < −1·96). This is equivalent to saying that the observed sample statistic is significant at the 0·05 level.

(*b*) Accept the hypothesis (or if desired make no decision at all) otherwise.

Because the *z* score plays such an important part in tests of hypotheses and significance, it is also called a *test statistic*.

It should be noted that other levels of significance could have been used. For example, if a 0·01 level were used we would replace 1·96 everywhere above by 2·58 (see Table 10.1 below). Table 9.1, Page 157, can also be used since the sum of the level of significance and level of confidence is 100%.

ONE-TAILED AND TWO-TAILED TESTS

In the above test we displayed interest in extreme values of the statistic *S* or its corresponding *z* score on both sides of the mean, i.e. in both "tails" of the distribution. For this reason such tests are called *two-tailed tests* or *two-sided tests*.

Often, however, we may be interested only in extreme values to one side of the mean, i.e. in one "tail" of the distribution, as for example when we are testing the hypothesis that one process is better than another (which is different from testing whether one process is better or worse than the other). Such tests are called *one-tailed tests* or *one-sided tests*. In such cases the critical region is a region to one side of the distribution, with area equal to the level of significance.

Table 10.1, which gives critical values of *z* for both one-tailed and two-tailed tests at various levels of significance, will be found useful for purposes of reference. Critical values of *z* for other levels of significance are found by use of the table of normal curve areas.

Table 10.1

Level of Significance α	0·10	0·05	0·01	0·005	0·002
Critical Values of *z* for One-Tailed Tests	−1·28 *or* 1·28	−1·645 *or* 1·645	−2·33 *or* 2·33	−2·58 *or* 2·58	−2·88 *or* 2·88
Critical Values of *z* for Two-Tailed Tests	−1·645 *and* 1·645	−1·96 *and* 1·96	−2·58 *and* 2·58	−2·81 *and* 2·81	−3·08 *and* 3·08

SPECIAL TESTS

For large samples the sampling distributions of many statistics are normal distributions (or at least nearly normal) with mean μ_S and standard deviation σ_S. In such cases we can use the above results to formulate decision rules or tests of hypotheses and significance. The following special cases, taken from Table 8.1, Page 144, are just a few of the statistics of practical interest. In each case the results hold for infinite populations or for sampling with replacement. For sampling without replacement from finite populations the results must be modified. See Page 142.

1. Means. Here $S = \bar{X}$, the sample mean; $\mu_S = \mu_{\bar{X}} = \mu$, the population mean; $\sigma_S = \sigma_{\bar{X}} = \sigma/\sqrt{N}$, where σ is the population standard deviation and N is the sample size. The *z* score is given by

$$z = \frac{\bar{X} - \mu}{\sigma/\sqrt{N}}$$

When necessary the sample deviation s or $\hat{s}$ is used to estimate σ.

2. Proportions. Here $S = P$, the proportion of "successes" in a sample; $\mu_S = \mu_P = p$, where p is the population proportion of successes and N is the sample size; $\sigma_S = \sigma_P = \sqrt{pq/N}$, where $q = 1 - p$. The z score is given by

$$z = \frac{P - p}{\sqrt{pq/N}}$$

In case $P = X/N$, where X is the actual number of successes in a sample, the z score becomes

$$z = \frac{X - Np}{\sqrt{Npq}}$$

i.e. $\mu_X = \mu = Np$, $\sigma_X = \sigma = \sqrt{Npq}$, and $S = X$.

Results for other statistics can similarly be obtained.

OPERATING CHARACTERISTIC CURVES. POWER OF A TEST

We have seen how the Type I error can be limited by properly choosing a level of significance. It is possible to avoid risking Type II errors altogether by simply not making them, which amounts to never accepting hypotheses. In many practical cases, however, this cannot be done. In such cases use is often made of *operating characteristic curves*, or *OC curves*, which are graphs showing the probabilities of Type II errors under various hypotheses. These provide indications of how well given tests will enable us to minimize Type II errors, i.e. they indicate the *power of a test* to avoid making wrong decisions. They are useful in designing experiments by showing, for instance, what sample sizes to use.

CONTROL CHARTS

It is often important in practice to know when a process has changed sufficiently so that steps may be taken to remedy the situation. Such problems arise, for example, in *quality control* where one must, often quickly, decide whether observed changes are due simply to chance fluctuations or to actual changes in a manufacturing process because of deterioration of machine parts, mistakes of employees, etc. *Control charts* provide a useful and simple method for dealing with such problems (see Prob. 10.16).

TESTS OF SIGNIFICANCE INVOLVING SAMPLE DIFFERENCES

1. Differences of Means.

Let $\bar{X}_1$ and $\bar{X}_2$ be the sample means obtained in large samples of sizes N_1 and N_2 drawn from respective populations having means μ_1 and μ_2 and standard deviations σ_1 and σ_2. Consider the null hypothesis that there is *no difference* between the population means, i.e. $\mu_1 = \mu_2$ or the samples are drawn from two populations having the same mean.

From Chapter 8, Page 143, Equation (5), placing $\mu_1 = \mu_2$ we see that the sampling distribution of differences in means is approximately normally distributed with mean and standard deviation given by

$$\mu_{\bar{X}_1 - \bar{X}_2} = 0 \quad \text{and} \quad \sigma_{\bar{X}_1 - \bar{X}_2} = \sqrt{(\sigma_1^2/N_1) + (\sigma_2^2/N_2)} \tag{1}$$

where we can, if necessary, use the sample standard deviations s_1 and s_2 (or $\hat{s}_1$ and $\hat{s}_2$) as estimates of σ_1 and σ_2.

By using the standardized variable or z score given by

$$z = \frac{\bar{X}_1 - \bar{X}_2 - 0}{\sigma_{\bar{X}_1 - \bar{X}_2}} = \frac{\bar{X}_1 - \bar{X}_2}{\sigma_{\bar{X}_1 - \bar{X}_2}}$$

we can test the null hypothesis against alternative hypotheses (or the significance of an observed difference) at an appropriate level of significance.

2. Differences of Proportions.

Let P_1 and P_2 be the sample proportions obtained in large samples of sizes N_1 and N_2 drawn from respective populations having proportions p_1 and p_2. Consider the null hypothesis that there is *no difference* between the population parameters, i.e. $p_1 = p_2$, and thus the samples are really drawn from the same population.

From Chapter 8, Page 143, Equation (6) placing $p_1 = p_2 = p$ we see that the sampling distribution of differences in proportions is approximately normally distributed with mean and standard deviation given by

$$\mu_{P_1-P_2} = 0 \quad \text{and} \quad \sigma_{P_1-P_2} = \sqrt{pq(1/N_1 + 1/N_2)} \tag{3}$$

where $p = \dfrac{N_1 P_1 + N_2 P_2}{N_1 + N_2}$ is used as an estimate of the population proportion, and $q = 1 - p$.

By using the standardized variable

$$z = \frac{P_1 - P_2 - 0}{\sigma_{P_1-P_2}} = \frac{P_1 - P_2}{\sigma_{P_1-P_2}} \tag{4}$$

we can test observed differences at an appropriate level of significance and thereby test the null hypothesis. Tests involving other statistics can similarly be designed.

TESTS INVOLVING THE BINOMIAL DISTRIBUTION

Tests involving the binomial distribution as well as other distributions can be designed in a manner analogous to those using the normal distribution, the basic principles being essentially the same. See Problems 10.23 to 10.28.

Solved Problems

TESTS OF MEANS AND PROPORTIONS USING NORMAL DISTRIBUTIONS

10.1. Find the probability of getting between 40 and 60 heads inclusive in 100 tosses of a fair coin.

Solution:

According to the binomial distribution the required probability is

$$_{100}C_{40}(\tfrac{1}{2})^{40}(\tfrac{1}{2})^{60} + {}_{100}C_{41}(\tfrac{1}{2})^{41}(\tfrac{1}{2})^{59} + \cdots + {}_{100}C_{60}(\tfrac{1}{2})^{60}(\tfrac{1}{2})^{40}$$

Since $Np = 100(\tfrac{1}{2})$ and $Nq = 100(\tfrac{1}{2})$ are both greater than 5, the normal approximation to the binomial distribution can be used in evaluating this sum.

The mean and standard deviation of the number of heads in 100 tosses are given by

$$\mu = Np = 100(\tfrac{1}{2}) = 50 \quad \text{and} \quad \sigma = \sqrt{Npq} = \sqrt{(100)(\tfrac{1}{2})(\tfrac{1}{2})} = 5$$

On a continuous scale, between 40 and 60 heads inclusive is the same as between 39·5 and 60·5 heads.

39·5 in standard units = $(39.5 - 50)/5 = -2.10$, 60·5 in standard units = $(60.5 - 50)/5 = 2.10$.

Required probability = area under normal curve between $z = -2.10$ and $z = 2.10$
= 2(area between $z = 0$ and $z = 2.10$) = 2(0.4821) = 0.9642

10.2. To test the hypothesis that a coin is fair, the following rule of decision is adopted: (*1*) accept the hypothesis if the number of heads in a single sample of 100 tosses is between 40 and 60 inclusive, (*2*) reject the hypothesis otherwise.

(*a*) Find the probability of rejecting the hypothesis when it is actually correct.

(*b*) Interpret graphically the decision rule and the results of part (*a*).

(*c*) What conclusions would you draw if the sample of 100 tosses yielded 53 heads? 60 heads?

(*d*) Could you be wrong in your conclusions to (*c*)? Explain.

Solution:

(*a*) By Problem 10.1, the probability of not getting between 40 and 60 heads inclusive if the coin is fair $= 1 - 0.9642 = 0.0358$. Then the probability of rejecting the hypothesis when it is correct $= 0.0358$.

(*b*) The decision rule is illustrated by Fig. 10-2 which shows the probability distribution of heads in 100 tosses of a fair coin.

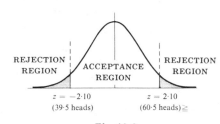

Fig. 10-2

If a single sample of 100 tosses yields a *z* score between -2.10 and 2.10, we accept the hypothesis; otherwise we reject the hypothesis and decide that the coin is not fair.

The error made in rejecting the hypothesis when it should be accepted is the *Type I error* of the decision rule: and the probability of making this error, equal to 0.0358 from part (*a*), is represented by the total shaded area of the figure.

If a single sample of 100 tosses yields a number of heads whose *z* score (or *z* statistic) lies in the shaded regions, we would say that this score differed *significantly* from what would be expected if the hypothesis were true. For this reason the total shaded area (i.e. probability of a Type I error) is called the *level of significance* of the decision rule and equals 0.0358 in this case. Thus we speak of rejecting the hypothesis at 0.0358 or 3.58% level of significance.

(*c*) According to the decision rule, we would have to accept the hypothesis that the coin is fair in both cases. One might argue that if only one more head had been obtained we would have rejected the hypothesis. This is what one must face when any sharp line of division is used in making decisions.

(*d*) Yes. We could accept the hypothesis when it actually should be rejected, as would be the case for example when the probability of heads is really 0.7 instead of 0.5.

The error made in accepting the hypothesis when it should be rejected is the *Type II error* of the decision. For further discussion see Problems 10.10 to 10.12.

10.3. Design a decision rule to test the hypothesis that a coin is fair if a sample of 64 tosses of the coin is taken and if a level of significance of (*a*) 0.05 and (*b*) 0.01 is used.

Solution:

(*a*) **First Method:** If the level of significance is 0.05, each shaded area in Fig. 10-3 is 0.025 by symmetry. Then the area between 0 and $z_1 = 0.5000 - 0.0250 = 0.4750$, and $z_1 = 1.96$.

Thus a possible decision rule is:

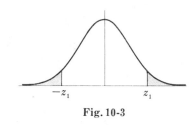

Fig. 10-3

(*1*) Accept the hypothesis that the coin is fair if *z* is between -1.96 and 1.96.

(*2*) Reject the hypothesis otherwise.

The critical values -1.96 and 1.96 can also be read from Table 10.1.

To express this decision rule in terms of the number of heads to be obtained in 64 tosses of the coin, note that the mean and standard deviation of the distribution of heads are given by

$$\mu = Np = 64(0.5) = 32, \quad \text{and} \quad \sigma = \sqrt{Npq} = \sqrt{64(0.5)(0.5)} = 4$$

under the hypothesis that the coin is fair. Then $z = (X - \mu)/\sigma = (X - 32)/4$.

If $z = 1.96$, $(X - 32)/4 = 1.96$ or $X = 39.84$. If $z = -1.96$, $(X - 32)/4 = -1.96$ or $X = 24.16$.

Thus the decision rule becomes:

(*1*) Accept the hypothesis that the coin is fair if the number of heads is between 24.16 and 39.84, i.e. between 25 and 39 inclusive.

(*2*) Reject the hypothesis otherwise.

Second method: With probability 0.95, the number of heads will lie between $\mu - 1.96\sigma$ and $\mu + 1.96\sigma$, i.e. $Np - 1.96\sqrt{Npq}$ and $Np + 1.96\sqrt{Npq}$ or between $32 - 1.96(4) = 24.16$ and $32 + 1.96(4) = 39.84$, which leads to the above decision rule.

Third method: $-1.96 < z < 1.96$ is equivalent to $-1.96 < \frac{1}{4}(X - 32) < 1.96$. Then $-1.96(4) < (X - 32) < 1.96(4)$ or $32 - 1.96(4) < X < 32 + 1.96(4)$, i.e. $24.16 < X < 39.84$, which also leads to the above decision rule.

(*b*) If the level of significance is 0.01, each shaded area in the above figure is 0.005. Then the area between 0 and z_1 $= 0.5000 - 0.0050 = 0.4950$, and $z_1 = 2.58$ (more exactly 2.575). This can also be read from Table 10.1.

Following the procedure in the second method of part (*a*), we see that with probability 0.99 the number of heads will lie between $\mu - 2.58\sigma$ and $\mu + 2.58\sigma$, i.e. $32 - 2.58(4) = 21.68$ and $32 + 2.58(4) = 42.32$.

Thus the decision rule becomes:

(*1*) Accept the hypothesis if the number of heads is between 22 and 42 inclusive.

(*2*) Reject the hypothesis otherwise.

10.4. How could you design a decision rule in Problem 10.3 so as to avoid a Type II error?

Solution:

A Type II error is made by accepting a hypothesis when it should be rejected. To avoid this error, instead of accepting the hypothesis we simply do not reject it, which could mean that we are withholding any decision in this case. Thus, for example, we could word the decision rule of Problem 10.3(*b*) as:

(*1*) Do not reject the hypothesis if the number of heads is between 22 and 42 inclusive.

(*2*) Reject the hypothesis otherwise.

In many practical instances, however, it is important to decide whether a hypothesis should be accepted or rejected. A complete discussion of such cases requires consideration of Type II errors (see Problems 10.10 to 10.12).

10.5. In an experiment on extra-sensory perception (E.S.P.) an individual (subject) in one room is asked to state the colour (red or blue) of a card chosen from a deck of 50 well-shuffled cards by an individual in another room. It is unknown to the subject how many red or blue cards are in the deck. If the subject identifies 32 cards correctly, determine whether the results are significant at the (*a*) 0.05 and (*b*) 0.01 level of significance.

Solution:

If p is the probability of the subject choosing the colour of a card correctly, then we have to decide between the following two hypotheses:

H_0: $p = 0.5$, and subject is simply guessing, i.e. results are due to chance.
H_1: $p > 0.5$, and the subject has powers of E.S.P.

We choose a one-tailed test, since we are not interested in ability to obtain extremely low scores but rather in ability to obtain high scores.

If the hypothesis H_0 is true, the mean and standard deviation of the number of cards identified correctly is given by

$$\mu = Np = 50(0.5) = 25 \quad \text{and} \quad \sigma = \sqrt{Npq} = \sqrt{50(0.5)(0.5)} = \sqrt{12.5} = 3.54$$

(a) For a one-tailed test at a level of significance of 0·05 we must choose z_1 in Fig. 10-4 so that the shaded area in the critical region of high scores is 0·05. Then the area between 0 and $z_1 = 0·4500$, and $z_1 = 1·645$. This can also be read from Table 10.1.

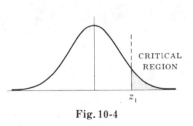

Fig. 10-4

Thus our decision rule or test of significance is:

(1) If the z score observed is greater than 1·645, the results are significant at the 0·05 level and the individual has powers of E.S.P.

(2) If the z score is less than 1·645 the results are due to chance, i.e. not significant at the 0·05 level.

Since 32 in standard units $= (32 - 25)/3·54 = 1·98$, which is greater than 1·645, decision (1) holds, i.e. we conclude at the 0·05 level that the individual has powers of E.S.P.

Note that we should really apply a continuity correction, since 32 on a continuous scale is between 31·5 and 32·5. However, 31·5 has a standard score of $(31·5 - 25)/3·54 = 1·84$, and so the same conclusion is reached.

(b) If the level of significance is 0·01, then the area between 0 and $z_1 = 0·4900$, and $z_1 = 2·33$. Since 32 (or 31·5) in standard units is 1·98 (or 1·84) which is less than 2·33, we conclude that the results are *not significant* at the 0·01 level.

Some statisticians adopt the terminology that results significant at the 0·01 level are *highly significant*, results significant at the 0·05 level but not at the 0·01 level are *probably significant*, while results significant at levels larger than 0·05 are *not significant*.

According to this terminology, we would conclude that the above experimental results are *probably significant*, so that further investigations of the phenomena are probably warranted.

Since significance levels serve as guides in making decisions, some statisticians quote actual probabilities involved. For instance in this problem, since $\Pr\{z \geq 1·84\} = 0·0322$, the statistician could say that on the basis of the experiment the chances of being wrong in concluding that the individual has powers of E.P.S. is about 3 in 100. The quoted probability, in this case 0·0322, is sometimes called an *experimental* or *descriptive significance level*.

10.6. The manufacturer of a patent medicine claimed that it was 90% effective in relieving an allergy for a period of 8 hours. In a sample of 200 people who had the allergy, the medicine provided relief for 160 people. Determine whether the manufacturer's claim is legitimate.

Solution:

Let p denote the probability of obtaining relief from the allergy by using the medicine. Then we must decide between the two hypotheses:

$$H_0: p = 0·9, \text{ and the claim is correct}$$
$$H_1: p < 0·9, \text{ and the claim is false}$$

We choose a one-tailed test, since we are interested in determining whether the proportion of people relieved by the medicine is too low.

If the level of significance is taken as 0·01, i.e. if the shaded area in Fig. 10-5 is 0·01, then $z_1 = -2·33$ as can be seen from Problem 10.5(b) using the symmetry of the curve, or from Table 10.1.

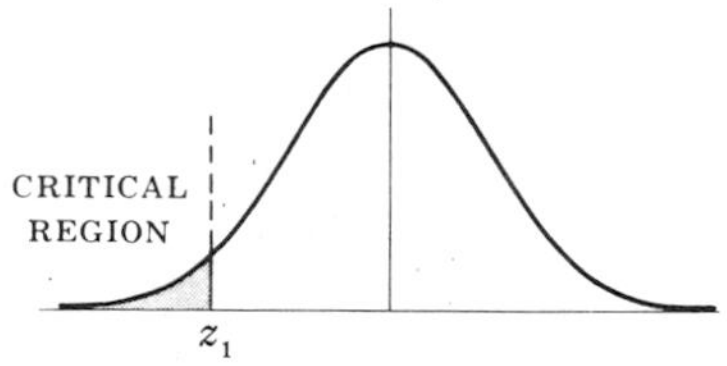

Fig. 10-5

We take as our decision rule:

(1) The claim is not legitimate if z is less than $-2·33$ (in which case we reject H_0).

(2) Otherwise, the claim is legitimate and the observed results are due to chance (in which case we accept H_0).

If H_0 is true, $\mu = Np = 200(0·9) = 180$ and $\sigma = \sqrt{Npq} = \sqrt{(200)(0·9)(0·1)} = 4·23$.

Now 160 in standard units $= (160 - 180)/4·23 = -4·73$, which is much less than $-2·33$. Thus by our decision rule we conclude that the claim is not legitimate and that the sample results are *highly significant* (see end of Problem 10.5).

10.7. The mean lifetime of a sample of 100 fluorescent light bulbs produced by a company is computed to be 1570 hours with a standard deviation of 120 hours. If μ is the mean lifetime of all the bulbs produced by the company, test the hypothesis $\mu = 1600$ hours against the alternative hypothesis $\mu \neq 1600$ hours, using a level of significance of (a) 0·05, (b) 0·01.

Solution:

We must decide between the two hypotheses:

$$H_0: \mu = 1600 \text{ hours}, \quad H_1: \mu \neq 1600 \text{ hours}.$$

A two-tailed test should be used here since $\mu \neq 1600$ includes values both larger and smaller than 1600.

(a) For a two-tailed test at a level of significance of 0·05, we have the following decision rule:

  (1) Reject H_0 if the z score of the sample mean is outside the range $-1·96$ to $1·96$.

  (2) Accept H_0 (or withhold any decision) otherwise.

  The statistic under consideration is the sample mean. $\bar{X}$. The sampling distribution of X has a mean $\mu_{\bar{X}} = \mu$ and standard deviation $\sigma_{\bar{X}} = \sigma/\sqrt{N}$, where μ and σ are the mean and standard deviation of the population of all bulbs produced by the company.

  Under the hypothesis H_0, we have $\mu = 1600$ and $\sigma_X = \sigma/\sqrt{N} = 120/\sqrt{100} = 12$, using the sample standard deviation as an estimate of σ. Since $z = (\bar{X} - 1600)/12 = (1570 - 1600)/12 = -2·50$ lies outside the range $-1·96$ to $1·96$, we reject H_0 at a 0·05 level of significance.

(b) If the level of significance is 0·01, the range $-1·96$ to $1·96$ in the decision rule of part (a) is replaced by $-2·58$ to $2·58$. Then since the z score of $-2·50$ lies inside this range, we accept H_0 (or withhold any decision) at a 0·01 level of significance.

10.8. In Problem 10.7, test the hypothesis $\mu = 1600$ hours against the alternative $\mu < 1600$ hours, using a level of significance of (a) 0·05 and (b) 0·01.

Solution:

We must decide between the two hypotheses:

$$H_0: \mu = 1600 \text{ hours}, \quad H_1: \mu < 1600 \text{ hours}$$

A one-tailed test should be used here, the corresponding figure being identical with that of Problem 10.6.

(a) If the level of significance is 0·05, the shaded region of Fig. 10-5 has an area of 0·05, and we find that $z_1 = -1·645$.

  We therefore adopt the decision rule:

  (1) Reject H_0 if z is less than $-1·645$.

  (2) Accept H_0 (or withhold any decision) otherwise.

  Since, as in Problem 10.7(a), the z score is $-2·50$ which is less than $-1·645$, we reject H_0 at a 0·05 level of significance. Note that this decision is identical with that reached in Problem 10.7(a) using a two-tailed test.

(b) If the level of significance is 0·01, the z_1 value in Fig. 10-5 is $-2·33$. Hence we adopt the decision rule:

  (1) Reject H_0 if z is less than $-2·33$.

  (2) Accept H_0 (or withhold any decision) otherwise.

  Since, as in Problem 10.7(a), the score is $-2·50$ which is less than $-2·33$, we reject H_0 at a 0·01 level of significance. Note that this decision is not the same as that reached in Problem 10.7(b) using a two-tailed test.

  It follows that decisions concerning a given hypothesis H_0 based on one-tailed and two-tailed tests are not always in agreement. This is, of course, to be expected since we are testing H_0 against a different alternative in each case.

10.9. The breaking strengths of ropes produced by a manufacturer have mean 1800 N and standard deviation of 100 N. By a new technique in the manufacturing process it is claimed that the breaking strength can be increased. To test this claim a sample of 50 ropes is tested and it is found that the mean breaking strength is 1850 N. Can we support the claim at a 0·01 level of significance?

Solution:

We have to decide between the two hypotheses:

H_0: $\mu = 1800$ N, and there is really no change in breaking strength.
H_1: $\mu > 1800$ N, and there is a change in breaking strength.

A one-tailed test should be used here. The diagram associated with this test is identical with that of Problem 10.5.

At a 0·01 level of significance the decision rule is:

(*1*) If the z score observed is greater than 2·33, the results are significant at the 0·01 level and H_0 is rejected.

(*2*) Otherwise H_0 is accepted (or the decision is withheld).

Under the hypothesis that H_0 is true, we find

$$z = \frac{\bar{X} - \mu}{\sigma/\sqrt{N}} = \frac{1850 - 1800}{100/\sqrt{50}} = 3.55$$

which is greater than 2·33. Hence we conclude that the results are *highly significant* and the claim should be supported.

OPERATING CHARACTERISTIC CURVES

10.10. Referring to Problem 10.2, what is the probability of accepting the hypothesis that the coin is fair when the actual probability of heads is $p = 0.7$?

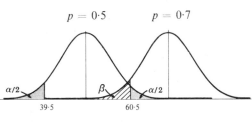

Fig. 10-6

Solution:
The hypothesis H_0 that the coin is fair, i.e. $p = 0.5$, is accepted when the number of heads in 100 tosses lies between 39·5 and 60·5. The probability of rejecting H_0 when it should be accepted (i.e. the probability of a Type I error) is represented by the total area α of the shaded region under the normal curve to the left in Fig. 10-6. As computed in Problem 10.2(*a*), this area α, which represents the level of significance of the test of H_0, is equal to 0·0358.

If the probability of heads is $p = 0.7$, then the distribution of heads in 100 tosses is represented by the normal curve to the right in Fig. 10-6. From the diagram it is clear that the probability of accepting H_0 when actually $p = 0.07$ (i.e. the probability of a Type II error) is given by the cross-hatched area β of the figure.

To compute this area we observe that the distribution under the hypothesis $p = 0.7$ has mean and standard deviation given by

$$\mu = Np = (100)(0.7) = 70 \quad \text{and} \quad \sigma = \sqrt{Npq} = \sqrt{(100)(0.7)(0.3)} = 4.58$$

60·5 in standard units $= (60.5 - 70)/4.58 = -2.07$
39·5 in standard units $= (39.5 - 70)/4.58 = -6.66$

Then $\beta = $ (area under normal curve between $z = -6.66$ and $z = -2.07$) $= 0.0192$

Thus with the given decision rule there is very little chance of accepting the hypothesis that the coin is fair when actually $p = 0.7$.

Note that in this problem we were given the decision rule from which we computed α and β. In practice two other possibilities may arise:

(*1*) We decide on α (such as 0·05 or 0·01), arrive at a decision rule and then compute β.

(*2*) We decide on α and β and then arrive at a decision rule.

10.11. Work the previous problem if (a) $p = 0.6$, (b) $p = 0.8$, (c) $p = 0.9$, (d) $p = 0.4$.

Solution:

(a) If $p = 0.6$, the distribution of heads has mean and standard deviation given by

$$\mu = Np = (100)(0.6) = 60 \quad \text{and} \quad \sigma = \sqrt{Npq} = \sqrt{(100)(0.6)(0.4)} = 4.90$$

$$60.5 \text{ in standard units} = (60.5 - 60)/4.90 = 0.0102$$
$$39.5 \text{ in standard units} = (39.5 - 60)/4.90 = -4.18$$

Then β = (area under normal curve between $z = -4.18$ and $z = 0.0102$) = 0.5040.

Thus with the given decision rule there is a large chance of accepting the hypothesis that the coin is fair when actually $p = 0.6$.

(b) If $p = 0.8$, then $\mu = Np = (100)(0.8) = 80$ and $\sigma = \sqrt{Npq} = \sqrt{(100)(0.8)(0.2)} = 4$.

$$60.5 \text{ in standard units} = (60.5 - 80)/4 = -4.88$$
$$39.5 \text{ in standard units} = (39.5 - 80)/4 = -10.12$$

Then β = (area under normal curve between $z = -10.12$ and $z = -4.88$) = 0.0000, very closely.

(c) From comparison with (b) or by calculation, we see that if $p = 0.9$, $\beta = 0$ for all practical purposes.

(d) By symmetry, $p = 0.4$ yields the same value of β as $p = 0.6$, i.e. $\beta = 0.5040$.

10.12. Represent graphically the results of Problems 10.10 and 10.11 by constructing a graph of (a) β vs. p, (b) $(1 - \beta)$ vs. p. Interpret the graphs obtained.

Solution:

Table 10.2 shows the values of β corresponding to given values of p as obtained in Problems 10.10 and 10.11.

Table 10.2

p	0.1	0.2	0.3	0.4	0.5	0.6	0.7	0.8	0.9
β	0.0000	0.0000	0.0192	0.5040	0.9642	0.5040	0.0192	0.0000	0.0000

Note that β represents the probability of accepting the hypothesis $p = 0.5$ when actually p is a value other than 0.5. However, if it is actually true that $p = 0.5$, we can interpret β as the probability of accepting $p = 0.5$ when it should be accepted. This probability equals $1 - 0.0358 = 0.9642$ and has been entered into Table 10.2.

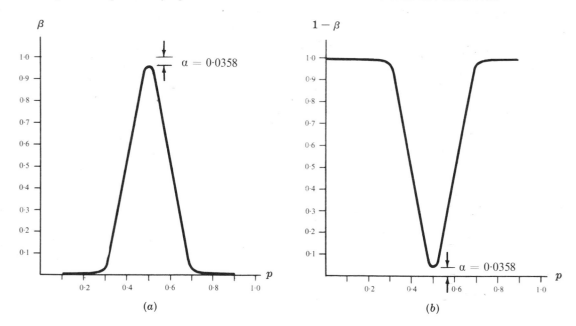

(a) (b)

Fig. 10-7

(a) The graph of β vs. p, shown in Fig. 10-7(a), is called the *operating characteristic curve* or *OC curve* of the decision rule or test of hypothesis.

    The distance from the maximum point of the *OC* curve to the line $\beta = 1$ is equal to $\alpha = 0 \cdot 0358$, the level of significance of the test.

    In general, the sharper the peak of the *OC* curve the better is the decision rule for rejecting hypotheses which are not valid.

(b) The graph of $(1 - \beta)$ vs. p, shown in Fig. 10-7(b), is called the *power curve* of the decision rule or test of hypothesis. This curve is obtained simply by inverting the *OC* curve, so that actually both graphs are equivalent.

    The quantity $(1 - \beta)$ is often called a *power function* since it indicates the ability or *power of a test* to reject hypotheses which are false, i.e. should be rejected. The quantity β is also called the *operating characteristic function* of a test.

10.13. A company manufactures cords whose breaking strengths have a mean of 300 N and standard deviation 24 N. It is believed that by a newly developed process the mean breaking strength can be increased.

(a) Design a decision rule for rejecting the old process at a $0 \cdot 01$ level of significance if it is agreed to test 64 cords.

(b) Under the decision rule adopted in (a), what is the probability of accepting the old process when in fact the new process has increased the mean breaking strength to 310 N? Assume the standard deviation is still 24 N.

Solution:

(a) If μ is the mean breaking strength, we wish to decide between the hypotheses:

$$H_0 : \mu = 300 \text{ N, and the new process is the same as the old one.}$$
$$H_1 : \mu > 300 \text{ N, and the new process is better than the old one.}$$

    For a one-tailed test at a $0 \cdot 01$ level of significance, we have the following decision rule [refer to Fig. 10-8(a)]:

(1) Reject H_0 if the z score of the sample mean breaking strength is greater than $2 \cdot 33$.

(2) Accept H_0 otherwise.

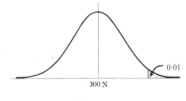

    Since $z = \dfrac{\bar{X} - \mu}{\sigma/\sqrt{N}} = \dfrac{\bar{X} - 300}{24/\sqrt{64}}$, $\bar{X} = 300 + 3z$. Then if $z > 2 \cdot 33$, $\bar{X} > 300 + 3(2 \cdot 33) = 307 \cdot 0$ N.

Fig. 10-8(a)

    Thus the above decision rule becomes:

(1) Reject H_0 if the mean breaking strength of 64 cords exceeds $307 \cdot 0$ N.

(2) Accept H_0 otherwise.

(b) Consider the two hypotheses $H_0 : \mu = 300$ N and $H_1 : \mu = 310$ N. The distributions of mean breaking strengths corresponding to these two hypotheses are represented respectively by the left and right normal distributions of Fig. 10-8(b).

    The probability of accepting the old process when the new mean breaking strength is actually 310 N is represented by the region of area β in Fig. 10-8(b). To find this, note that $307 \cdot 0$ N in standard units $= (307 \cdot 0 - 310)/3 = -1 \cdot 00$; hence

Fig. 10-8(b)

$$\beta = (\text{area under right-hand normal curve to left of } z = -1 \cdot 00) = 0 \cdot 1587$$

This is the probability of accepting $H_0 : \mu = 300$ N when actually $H_1 : \mu = 310$ N is true, i.e. it is the probability of making a Type II error.

10.14. Construct (a) an *OC* curve and (b) a power curve for Problem 10.13, assuming that the standard deviation of breaking strengths remains at 24 N.

Solution:

    By reasoning similar to that used in Problem 10.13(b), we can find β for the cases where the new process yields mean breaking strengths μ equal to 305 N, 315 N, etc. For example if μ = 305 N, then 307·0 N in standard units = (307·0 − 305)/3 = 0·67, and hence

$$\beta = (\text{area under right-hand normal curve to left of } z = 0·67) = 0·7486$$

In this manner Table 10.3 is obtained.

Table 10.3

μ	290	295	300	305	310	315	320
β	1·0000	1·0000	0·9900	0·7486	0·1587	0·0038	0·0000

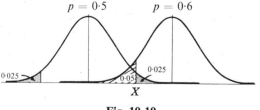

Fig. 10-9

(a)   (b)

(a) The *OC* curve is shown in Fig. 10-9(a). From this curve we see that the probability of keeping the old process if the new breaking strength is less than 300 N, is practically 1 (except for the level of significance of 0·01 when the new process gives a mean of 300 N). It then drops rather sharply to zero so that there is practically no chance of keeping the old process when the mean breaking strength is greater than 315 N.

(b) The power curve shown in Fig. 10-9(b) is capable of exactly the same interpretations as that for the *OC* curve. In fact the two curves are essentially equivalent.

10.15. To test the hypothesis that a coin is fair (i.e. $p = 0·5$) by a number of tosses of the coin, we wish to impose the following restrictions: (*A*) the probability of rejecting the hypothesis when it is actually correct must be 0·05 at most; (*B*) the probability of accepting the hypothesis when actually p differs from 0·5 by 0·1 or more (i.e. $p \geqq 0·6$ or $p \leqq 0·4$) must be 0·05 at most. Determine the minimum sample size which is necessary and state the resulting decision rule.

Solution:

Here we have placed limits on the risks of Type I and Type II errors. For example, the imposed restriction (*A*) requires that the probability of a Type I error $= \alpha = 0·05$ at most, while restriction (*B*) requires that the probability of a Type II error $= \beta = 0·05$ at most. The situation is illustrated graphically in Fig. 10-10.

Let N denote the required sample size and X the number of heads in N tosses, above which we reject the hypothesis $p = 0·5$. From Fig. 10-10:

Fig. 10-10

(*1*) Area under normal curve $p = 0·5$ to right of $\dfrac{X - Np}{\sqrt{Npq}} = \dfrac{X - 0·5N}{\sqrt{N(0·5)(0·5)}} = \dfrac{X - 0·5N}{0·5\sqrt{N}}$ is 0·025.

(*2*) Area under normal curve $p = 0·6$ to left of $\dfrac{X - Np}{\sqrt{Npq}} = \dfrac{X - 0·6N}{\sqrt{N(0·6)(0·4)}} = \dfrac{X - 0·6N}{0·49\sqrt{N}}$ is 0·05.

(Actually, Area between $(X - 0·6N)/0·49\sqrt{N}$ and $[(N - X) - 0·6N]/0·49\sqrt{N}$ is 0·05; (*2*) is a close approx.)

From (1), $\dfrac{X - 0\cdot5N}{0\cdot5\sqrt{N}} = 1\cdot96$ or   (3) $X = 0\cdot5N + 0\cdot980\sqrt{N}$.

From (2), $\dfrac{X - 0\cdot6N}{0\cdot49\sqrt{N}} = -1\cdot645$ or   (4) $X = 0\cdot6N - 0\cdot806\sqrt{N}$.

Then from (3) and (4), $N = 318\cdot98$. It follows that the sample size must be at least 319, i.e. we must toss the coin at least 319 times. Putting $N = 319$ in (3) or (4), $X = 177$.

For $p = 0\cdot5$, $X - Np = 177 - 159\cdot5 = 17\cdot5$. Thus we adopt the following decision rule:

(a) Accept the hypothesis $p = 0\cdot5$ if the number of heads in 319 tosses is in the range $159\cdot5 \pm 17\cdot5$, i.e. between 142 and 177 heads.

(b) Reject the hypothesis otherwise.

CONTROL CHARTS

10.16. A machine is constructed to produce ball bearings having a mean diameter of 5.74 mm and a standard deviation of 0·08 mm. To determine whether the machine is in proper working order, a sample of 6 ball bearings is taken every 2 hours, for example, and the mean diameter is computed from this sample.

(a) Design a decision rule whereby one can be fairly certain that the quality of the products are conforming to required standards.

(b) Show how to represent graphically the decision rule in (a).

Solution:

(a) With 99·73% confidence we can say that the sample mean $\bar{X}$ must lie in the range $(\mu_{\bar{X}} - 3\sigma_{\bar{X}})$ to $(\mu_{\bar{X}} + 3\sigma_{\bar{X}})$ or $(\mu - 3\sigma/\sqrt{N})$ to $(\mu + 3\sigma/\sqrt{N})$. Since $\mu = 0\cdot574$, $\sigma = 0\cdot08$ and $N = 6$, it follows that with 99·73% confidence the sample mean should lie between $(5\cdot74 - 0\cdot24/\sqrt{6})$ and $(5\cdot74 + 0\cdot24/\sqrt{6})$ or between 5·64 and 5·84 mm.

Hence our decision rule is as follows:

(1) If a sample mean falls inside the range 5·64 to 5·84 mm, assume the machine is in proper working order.

(2) Otherwise conclude that the machine is not in proper working order and seek to determine the reason.

(b) A record of the sample means can be kept by means of a chart such as shown in Fig. 10-11, called *a quality control chart*. Each time a sample mean is computed it is represented by a particular point. As long as the points lie between the lower limit 5·64 mm and upper limit 5·84 mm, the process is under control. When a point goes outside of these control limits (such as in the third sample taken on Thursday), there is a possibility that something is wrong and investigation is warranted.

The control limits specified above are called the 99·73% confidence limits or briefly 3σ limits. However, confidence limits, such as 99% or 95%, can be determined as well. The choice in each case depends on particular circumstances.

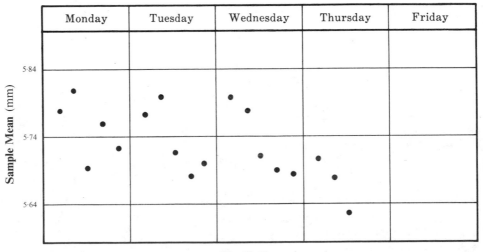

Fig. 10-11

TESTS INVOLVING DIFFERENCES OF MEANS AND PROPORTIONS

10.17. An examination was given to two classes consisting of 40 and 50 students respectively. In the first class the mean mark was 74 with a standard deviation of 8, while in the second class the mean mark was 78 with a standard deviation of 7. Is there a significant difference between the performance of the two classes at a level of significance of (*a*) 0·05, (*b*) 0·01?

Solution:

Suppose the two classes come from two populations having the respective means μ_1 and μ_2. Then we have to decide between the hypotheses:

H_0: $\mu_1 = \mu_2$, and the difference is merely due to chance.
H_1: $\mu_1 \neq \mu_2$, and there is a significant difference between classes.

Under the hypothesis H_0, both classes come from the same population. The mean and standard deviation of the difference in means are given by

$$\mu_{\bar{X}_1 - \bar{X}_2} = 0 \quad \text{and} \quad \sigma_{\bar{X}_1 - \bar{X}_2} = \sqrt{\sigma_1^2/N_1 + \sigma_2^2/N_2} = \sqrt{8^2/40 + 7^2/50} = 1\cdot606$$

where we have used the sample standard deviations as estimates of σ_1 and σ_2.

Then $z = (\bar{X}_1 - \bar{X}_2)/\sigma_{\bar{X} - \bar{X}_2} = (74 - 78)/1\cdot606 = -2\cdot49$.

(*a*) For a two-tailed test the results are significant at a 0·05 level if z lies outside the range $-1\cdot96$ to $1\cdot96$. Hence we conclude that a 0·05 level there is a significant difference in performance of the two classes and that the second class is probably better.

(*b*) For a two-tailed test the results are significant at a 0·01 level if z lies outside the range $-2\cdot58$ and $2\cdot58$. Hence we conclude that at a 0·01 level there is no significant difference between the classes.

Since the results are significant at the 0·05 level but not at the 0·01 level, we conclude that the results are *probably significant* according to the terminology used at the end of Problem 10.5.

10.18. The mean mass of 50 male students who showed above average participation in college athletics was 68·2 kg with a standard deviation of 2·5 kg, while 50 male students who showed no interest in such participation had a mean mass of 67·5 kg with a standard deviation of 2·8 kg. Test the hypothesis that male students who participate in college athletics are more massive than other male students.

Solution:

We must decide between the hypotheses:

H_0: $\mu_1 = \mu_2$, there is no difference between the mean masses.
H_1: $\mu_1 > \mu_2$, mean mass of first group is greater than that of second group.

Under the hypothesis H_0,

$$\mu_{\bar{X}_1 - \bar{X}_2} = 9 \quad \text{and} \quad \sigma_{\bar{X}_1 - \bar{X}_2} = \sqrt{\sigma_1^2/N_1 + \sigma_2^2/N_2} = \sqrt{(2\cdot5)^2/50 + (2\cdot8)^2/50} = 0\cdot53$$

where we have used the sample standard deviations as estimates of σ_1 and σ_2.

Then $z = (\bar{X}_1 - \bar{X}_2)/\sigma_{\bar{X}_1 - \bar{X}_2} = (68\cdot2 - 67\cdot5)/0\cdot53 = 1\cdot32$.

On the basis of a one-tailed test at a level of significance of 0·05, we would reject the hypothesis H_0 if the z score were greater than 1·645. Thus we cannot reject the hypothesis at this level of significance.

It should be noted, however, that the hypothesis can be rejected at a level of 0·10 if we are willing to take the risk of being wrong with a probability of 0·10, i.e. 1 chance in 10.

10.19. By how much should the sample size of each of the two groups in Problem 10.18 be increased in order that the observed difference of 0·7 kg in the mean masses be significant at the level of significance (*a*) 0·05, (*b*) 0·01?

Solution:

Suppose the sample size of each group is N and that the standard deviations for the two groups remain the same. Then under the hypothesis H_0, we have $\mu_{\bar{X}_1 - \bar{X}_2} = 0$ and

$$\sigma_{\bar{X}_1 - \bar{X}_2} = \sqrt{\sigma_1^2/N + \sigma_2^2/N} = \sqrt{[(2\cdot5)^2 + (2\cdot8)^2]/N} = \sqrt{14\cdot09/N} = 3\cdot75/\sqrt{N}$$

For an observed difference in mean masses of 0·7 kg, $z = \dfrac{\bar{X}_1 - \bar{X}_2}{\sigma_{\bar{X}_1 - \bar{X}_2}} = \dfrac{0\cdot7}{3\cdot75/\sqrt{N}} = \dfrac{0\cdot7\sqrt{N}}{3\cdot75}$.

(a) The observed difference will be significant at a 0·05 level if $0·7\sqrt{N}/3·75 = 1·645$ at least, so that N must be 78 at least. Thus we must increase the sample size in each group by at least $(78 - 50) = 28$.

Another method:

$$0·7\sqrt{N}/3·75 \geq 1·645, \quad \sqrt{N} \geq (3·75)(1·645)/0·7, \quad \sqrt{N} \geq 8·8, \quad N \geq 77·4 \text{ or } N \geq 78.$$

(b) The observed difference will be significant at a 0·01 level if

$$0·7\sqrt{N}/3·75 \geq 2·33, \quad \sqrt{N} \geq (3·75)(2·33)/0·7, \quad \sqrt{N} \geq 12·5, \quad N \geq 156·3 \text{ or } N \geq 157.$$

Hence we must increase the sample size in each group by at least $(157 - 50) = 107$.

10.20. Two groups, A and B, consist of 100 people each who have a disease. A serum is given to group A but not to group B (which is called the *control group*); otherwise, the two groups are treated identically. It is found that in groups A and B, 75 and 65 people, respectively, recover from the disease. Test the hypothesis that the serum helps to cure the disease using a level of significance of (a) 0·01, (b) 0·05, (c) 0·10.

Solution:

Let p_1 and p_2 denote respectively the population proportions cured by (1) using the serum, (2) not using the serum. We must decide between the two hypotheses:

$H_0: p_1 = p_2$, and observed differences are due to chance, i.e. serum is ineffective.

$H_0: p_1 > p_2$, and serum is effective.

Under the hypothesis H_0,

$$\mu_{P_1 - P_2} = 0 \quad \text{and} \quad \sigma_{P_1 - P_2} = \sqrt{pq(1/N_1 + 1/N_2)} = \sqrt{(0·70)(0·30)(1/100 + 1/100)} = 0·0648$$

where we have used as an estimate of p the average proportion of cures in the two sample groups given by $(75 + 65)/200 = 0·70$, and where $q = 1 - p = 0·30$.

Then $z = (P_1 - P_2)/\sigma_{P_1 - P_2} = (0·750 - 0·650)/0·0648 = 1·54$.

(a) On the basis of a one-tailed test at a 0·01 level of significance, we would reject the hypothesis H_0 only if the z score were greater than 2·33. Since the z score is only 1·54, we must conclude that the results are due to chance at this level of significance.

(b) On the basis of a one-tailed test at a 0·05 level of significance, we would reject H_0 only if the z score were greater than 1·645. Hence we must conclude that the results are due to chance at this level also.

(c) If a one-tailed test at a 0·10 level of significance were used, we would reject H_0 only if the z score were greater than 1·28. Since this condition is satisfied, we would conclude that the serum is effective at a 0·10 level of significance.

Note that our conclusions above depended on how much we were willing to risk being wrong. If results are actually due to chance and we conclude that they are due to the serum (Type I error), we might proceed to give the serum to large groups of people only to find then that it is actually ineffective. This is a risk which we are not always willing to assume.

On the other hand, we could conclude that the serum does not help when it actually does help (Type II error). Such a conclusion is very dangerous especially if human lives are at stake.

10.21. Work the preceding problem if each group consists of 300 people and if 225 people in group A and 195 people in group B are cured.

Solution:

Note that in this case the proportions of people cured in the two groups are $225/300 = 0·750$ and $195/300 = 0·650$ respectively, which are the same as in the previous problem. Under the hypothesis H_0,

$$\mu_{P_1 - P_2} = 0 \quad \text{and} \quad \sigma_{P_1 - P_2} = \sqrt{pq(1/N_1 + 1/N_2)} = \sqrt{(0·70)(0·30)(1/300 + 1/300)} = 0·0374$$

where $(225 + 195)/600 = 0·70$ is used as an estimate of p.

Then $z = (P_1 - P_2)/\sigma_{P_1 - P_2} = (0·750 - 0·650)/0·0374 = 2·67$.

Since this value of z is greater than 2·33, we can reject the hypothesis at a 0·01 level of significance, i.e. we can conclude that the serum is effective with only a 0·01 probability of being wrong.

This shows how increasing the sample size can increase the reliability of decisions. In many cases, however, it may be impractical to increase sample sizes. In such cases we are forced to make decisions on the basis of available information and so must contend with greater risks of incorrect decisions.

10.22. A sample poll of 300 voters from district A and 200 voters from district B showed that 56% and 48% respectively were in favour of a given candidate. At a level of significance of 0.05, test the hypothesis that (a) there is a difference between the districts, (b) the candidate is preferred in district A.

Solution:

Let p_1 and p_2 denote the proportions from all voters of districts A and B respectively who are in favour of the candidate.

Under the hypothesis $H_0: p_1 = p_2$, we have

$$\mu_{P_1-P_2} = 0 \quad \text{and} \quad \sigma_{P_1-P_2} = \sqrt{pq(1/N_1 + 1/N_2)} = \sqrt{(0.528)(0.472)(1/300 + 1/200)} = 0.0456$$

where we have used as estimates of p and q the values $[(0.56)(300) + (0.48)(200)]/500 = 0.528$ and $(1 - 0.528) = 0.472$.

Then $z = (P_1 - P_2)/\sigma_{P_1-P_2} = (0.560 - 0.480)/0.0456 = 1.75$.

(a) If we only wish to determine whether there is a difference between the districts, we must decide between the hypotheses ($H_0: p_1 = p_2$) and ($H_1: p_1 \neq p_2$), which involves a two-tailed test.

On the basis of a two-tailed test at a 0.05 level of significance, we would reject H_0 if z were outside the interval -1.96 to 1.96. Since $z = 1.75$ lies inside this interval, we cannot reject H_0 at this level, i.e. there is no significant difference between the districts.

(b) If we wish to determine whether the candidate is preferred in district A, we must decide between the hypotheses ($H_0: p_1 = p_2$) and ($H_1: p_1 > p_2$), which involves a one-tailed test.

On the basis of a one-tailed test at a 0.05 level of significance, we would reject H_0 if z were greater than 1.645. Since this is the case, we can reject H_0 at this level and conclude that the candidate is preferred in district A.

TESTS INVOLVING BINOMIAL DISTRIBUTIONS

10.23. An instructor gives a short quiz involving 10 true-false questions. To test the hypothesis that the student is guessing, the following decision rule is adopted: (i) if 7 or more are correct the student is not guessing; (ii) if less than 7 are correct the student is guessing. Find the probability of rejecting the hypothesis when it is correct.

Solution:

Let $p =$ probability that a question is answered correctly.

The probability of getting X problems out of 10 correct is $_{10}C_X p^X q^{10-X}$, where $q = 1 - p$.

Then under the hypothesis $p = 0.5$ (i.e. the student is guessing),

$$\Pr\{7 \text{ or more correct}\} = \Pr\{7 \text{ correct}\} + \Pr\{8 \text{ correct}\} + \Pr\{9 \text{ correct}\} + \Pr\{10 \text{ correct}\}$$
$$= {}_{10}C_7(\tfrac{1}{2})^7(\tfrac{1}{2})^3 + {}_{10}C_8(\tfrac{1}{2})^8(\tfrac{1}{2})^2 + {}_{10}C_9(\tfrac{1}{2})^9(\tfrac{1}{2}) + {}_{10}C_{10}(\tfrac{1}{2})^{10} = 0.1719$$

Thus the probability of concluding that the student is not guessing when in fact he is guessing is 0.1719. Note that this is the probability of a Type I error.

10.24. In the preceding problem, find the probability of accepting the hypothesis $p = 0.5$ when actually $p = 0.7$.

Solution:

Under the hypothesis $p = 0.7$,

$$\Pr\{\text{less than 7 correct}\} = 1 - \Pr\{7 \text{ or more correct}\}$$
$$= 1 - [{}_{10}C_7(0.7)^7(0.3)^3 + {}_{10}C_8(0.7)^8(0.3)^2 + {}_{10}C_9(0.7)^9(0.3) + {}_{10}C_{10}(0.3)^{10}] = 0.3504$$

10.25. In Problem 10.23, find the probability of accepting the hypothesis $p = 0.5$ when actually (a) $p = 0.6$, (b) $p = 0.8$, (c) $p = 0.9$, (d) $p = 0.4$, (e) $p = 0.3$, (f) $p = 0.2$, (g) $p = 0.1$.

Solution:

(a) If $p = 0.6$, the required probability

$$= 1 - [\Pr\{7 \text{ correct}\} + \Pr\{8 \text{ correct}\} + \Pr\{9 \text{ correct}\} + \Pr\{10 \text{ correct}\}]$$
$$= 1 - [{}_{10}C_7(0.6)^7(0.4)^3 + {}_{10}C_8(0.6)^8(0.4)^2 + {}_{10}C_9(0.6)^9(0.4) + {}_{10}C_{10}(0.6)^{10}] = 0.618$$

The results for (b), (c), . . ., (g) can be similarly found and are indicated in Table 10.4 together with the values corresponding to $p = 0.6$ and $p = 0.7$. Note that the probability is denoted by β (probability of a Type II error). We have also included the entry for $p = 0.5$, given by $\beta = 1 - 0.1719 = 0.828$ from Prob. 10.23, and that for $p = 0.7$, from Prob. 10.24.

Table 10.4

p	0·1	0·2	0·3	0·4	0·5	0·6	0·7	0·8	0·9
β	1·000	0·999	0·989	0·945	0·828	0·618	0·350	0·121	0·13

10.26. Use Problem 10.25 to construct the graph of β vs. p, thus obtaining the operating characteristic curve of the decision rule in Prob. 10.23.

Solution:

The required graph is shown in Fig. 10-12. Note the similarity with the *OC* curve of Problem 10.14.

If we had plotted $(1 - \beta)$ vs. p, the *power curve* of the decision rule would have been obtained.

The graph indicates that the given decision rule is *powerful* for rejecting $p = 0.5$ when actually $p \leq 0.4$ or $p \geq 0.8$.

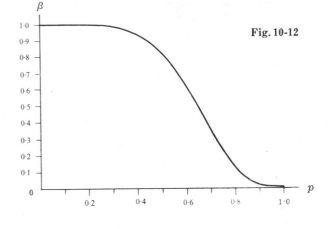

Fig. 10-12

10.27. A coin which is tossed 6 times comes up heads 6 times. Can we conclude at a (*a*) 0·05 and (*b*) 0·01 significance level that the coin is not fair? Consider both a one-tailed and two-tailed test.

Solution:

Let $p =$ probability of heads in a single toss of the coin.

Under the hypothesis H_0: $p = 0.5$ (i.e. the coin is fair),

$$p(X) = \Pr\{X \text{ heads in 6 tosses}\} = {}_6C_X(\tfrac{1}{2})^X(\tfrac{1}{2})^{6-X} = {}_6C_X/64$$

Then the probabilities of 0, 1, 2, 3, 4, 5 and 6 heads are given respectively by $\tfrac{1}{64}, \tfrac{6}{64}, \tfrac{15}{64}, \tfrac{20}{64}, \tfrac{15}{64}, \tfrac{6}{64}$ and $\tfrac{1}{64}$, as shown graphically in the probability distribution in Fig. 10-13.

One-tailed test.

Here we wish to decide between the hypotheses (H_0: $p = 0.5$) and (H_1: $p > 0.5$). Since $\Pr\{6 \text{ heads}\} = \tfrac{1}{64} = 0.01562$ and $\Pr\{5 \text{ or } 6 \text{ heads}\} = \tfrac{6}{64} + \tfrac{1}{64} = 0.1094$, we can reject H_0 at a 0·05 but not a 0·01 level (i.e. the result observed is significant at a 0·05 but not a 0·01 level).

Two-tailed test.

Here we wish to decide between the hypotheses (H_0: $p = 0.5$) and (H_1: $p \neq 0.5$). Since $\Pr\{0 \text{ or } 6 \text{ heads}\} = \tfrac{1}{64} + \tfrac{1}{64} = 0.03125$, we can reject H_0 at a 0·05 but not a 0·01 level.

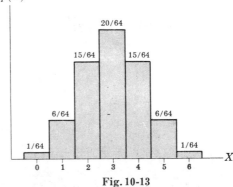

Fig. 10-13

10.28. Work Prob. 10.27 if the coin comes up heads 5 times.

Solution:

One-tailed test. Since $\Pr\{5 \text{ or } 6 \text{ heads}\} = \tfrac{6}{64} + \tfrac{1}{64} = \tfrac{7}{64} = 0.1094$, we cannot reject H_0 at a level of 0·05 or 0·01.

Two-tailed test. Since $\Pr\{0 \text{ or } 1 \text{ or } 5 \text{ or } 6 \text{ heads}\} = 2(\tfrac{7}{64}) = 0.2188$, we cannot reject H_0 at a level of 0·05 or 0·01.

Supplementary Problems

TESTS OF MEANS AND PROPORTIONS USING NORMAL DISTRIBUTIONS

10.29. An urn contains marbles which are either red or blue. To test the hypothesis of equal proportions of these colours, we agree to sample 64 marbles with replacement, noting the colours drawn and adopt the following decision rule: (*1*) accept the hypothesis if between 28 and 36 red marbles are drawn; (*2*) reject the hypothesis otherwise.
(*a*) Find the probability of rejecting the hypothesis when it is actually correct.
(*b*) Interpret graphically the decision rule and the result obtained in (*a*).
Ans. (*a*) 0·2606

10.30. (*a*) What decision rule would you adopt in Prob. 29 if you require the probability of rejecting the hypothesis when it is actually correct to be at most 0·01, i.e. you want a 0·01 level of significance?
(*b*) At what level of confidence would you accept the hypothesis?
(*c*) What would be the decision rule if a 0·05 level of significance were adopted?
Ans. (*a*) Accept the hypothesis if between 22 and 42 red marbles are drawn; reject it otherwise.
(*b*) 0·99. (*c*) Accept the hypothesis if between 24 and 40 red marbles are drawn; reject it otherwise.

10.31. Suppose that in Prob. 10.29 we wish to test the hypothesis that there is a *greater proportion* of red than blue marbles. (*a*) What would you take as the null hypothesis and what would be the alternative? (*b*) Should you use a one- or two-tailed test? Why? (*c*) What decision rule should you adopt if the level of significance is 0·05? (*d*) What is the decision rule if the level of significance is 0·01?
Ans. (*a*) $H_0: p = 0·5$, $H_1: p > 0·5$. (*b*) One-tailed test. (*c*) Reject H_0 if more than 39 red marbles are drawn, and accept it otherwise (or withhold decision). (*d*) Reject H_0 if more than 41 red marbles are drawn, and accept it otherwise (or withhold decision).

10.32. A pair of dice is tossed 100 times and it is observed that "sevens" appear 23 times. Test the hypothesis that the dice are fair, i.e. not loaded, using (*a*) a two-tailed test and (*b*) a one-tailed test and a significance level of 0·05. Discuss your reasons, if any, for preferring one of these tests over the other.
Ans. (*a*) We cannot reject the hypothesis at 0·05 level. (*b*) We can reject the hypothesis at 0·05 level.

10.33. Work Prob. 10.32 if the level of significance is 0·01.
Ans. We cannot reject the hypothesis at 0·01 level in either (*a*) or (*b*).

10.34. A manufacturer claimed that at least 95% of the equipment which he supplied to a factory conformed to specifications. An examination of a sample of 200 pieces of equipment revealed that 18 were faulty. Test his claim at a significance level of (*a*) 0·01, (*b*) 0·05.   *Ans.* We can reject the claim at both levels using a one-tailed test.

10.35. The percentage of *A*'s given in a physics course at a certain university over a long period of time was 10%. During one particular term there were 40 *A*'s in a group of 300 students. Test the significance of this result at level of (*a*) 0·05, (*b*) 0·01.
Ans. Using a one-tailed test, the result is significant at a 0·05 level but is not significant at a 0·01 level.

10.36. It has been found from experience that the mean breaking strength of a particular brand of thread is 9·72 N with a standard deviation of 1·40 N. Recently a sample of 36 pieces of thread showed a mean breaking strength of 8·93 N. Can one conclude at a significance level of (*a*) 0·05 and (*b*) 0·01 that the thread has become inferior?
Ans. Yes, at both levels, using a one-tailed test in each case.

10.37. On an examination given to students at a large number of different schools, the mean mark was 74·5 and the standard deviation was 8·0. At one particular school where 200 students took the examination, the mean mark was 75·9. Discuss the significance of this result at a 0·05 level from the viewpoint of (*a*) a one-tailed test and (*b*) a two-tailed test, explaining carefully your conclusions on the basis of these tests.
Ans. The result is significant at a 0·05 level in both a one-tailed and two-tailed test.

10.38. Answer Prob. 10.37 if the significance level is 0·01.
Ans. The result is significant at a 0·01 level in a one-tailed test but not in a two-tailed test.

OPERATING CHARACTERISTIC CURVES

10.39. Referring to Prob. 10.29, determine the probability of accepting the hypothesis that there are equal proportions of red and blue marbles when the actual proportion p of red marbles is (a) 0·6, (b) 0·7, (c) 0·8, (d) 0·9, (e) 0·3.
Ans. (a) 0·3112, (b) 0·0118, (c) 0, (d) 0, (e) 0·0118

10.40. Represent graphically the results of the preceding problem by constructing a graph of (a) β vs. p, (b) $1 - \beta$ vs. p. Compare these graphs with those of Prob. 10.12 by considering the analogy of red and blue marbles to heads and tails respectively.

10.41. (a) Work Problems 10.13 and 10.14 if it is agreed to test 400 cords. (b) What conclusions can you draw regarding risks of Type II errors when sample sizes are increased?

10.42. Construct (a) an OC curve and (b) a power curve corresponding to Prob. 10.31. Compare these curves with those of Prob. 10.14.

QUALITY CONTROL CHARTS

10.43. In the past a certain type of thread produced by a manufacturer has had a mean breaking strength of 8·64 N and a standard deviation of 1·28 N. To determine whether the product is conforming to standards, a sample of 16 pieces of thread is taken every 3 hours and the mean breaking strength is determined. Find the (a) 99·73% or 3σ, (b) 99% and (c) 95% control limits on a quality control chart and explain their applications.
Ans. The upper control limits are respectively (a) 6 and (b) 4 defective bolts

10.44. On the average about 3% of the bolts produced by a company are defective. To maintain this quality of performance, a sample of 200 bolts produced is examined every 4 hours. Determine (a) 99% and (b) 95% control limits for the number of defective bolts in each sample. Note that only *upper control limits* are needed in this case.
Ans. The upper control limits are respectively (a) 6 and (b) 4 defective bolts

TESTS INVOLVING DIFFERENCES OF MEANS AND PROPORTIONS

10.45. A sample of 100 electric light bulbs produced by manufacturer A showed a mean lifetime of 1190 hours and a standard deviation of 90 hours. A sample of 75 bulbs produced by manufacturer B showed a mean lifetime of 1230 hours with a standard deviation of 120 hours. Is there a difference between the mean lifetimes of the two brands at a significance level of (a) 0·05, (b) 0·01?   *Ans.* (a) Yes, (b) No

10.46. In the preceding problem test the hypothesis that the bulbs of manufacturer B are superior to those of manufacturer A using a significance level of (a) 0·05, (b) 0·01. Explain the differences between this and what was asked in the preceding problem. Do the results contradict those of the preceding problem?
Ans. A one-tailed test at both levels of significance shows that brand B is superior to brand A.

10.47. On an elementary examination in spelling, the mean mark of 32 boys was 72 with a standard deviation of 8, while the mean mark of 36 girls was 75 with a standard deviation of 6. Test the hypothesis at a (a) 0·05 and (b) 0·01 level of significance that the girls are better in spelling than the boys.
Ans. A one-tailed test shows that the difference is significant at a 0·05 level but not at a 0·01 level.

10.48. To test the effects of a new fertilizer on wheat production, a tract of land was divided into 60 squares of equal areas, all portions having identical qualities as to soil, exposure to sunlight, etc. The new fertilizer was applied to 30 squares and the old fertilizer was applied to the remaining squares. The mean number of bushels of wheat harvested per square of land using the new fertilizer was 18·2 litres with a standard deviation of 0·63 litres. The corresponding mean and standard deviation for the squares using the old fertilizer were 17·8 and 0·54 litres respectively. Using a significance level of (a) 0·05 and (b) 0·01, test the hypothesis that the new fertilizer is better than the old one.
Ans. A one-tailed test shows that the new fertilizer is superior at both levels of significance.

10.49. Random samples of 200 bolts manufactured by machine A and 100 bolts manufactured by machine B showed 19 and 5 defective bolts respectively. Test the hypothesis that (a) the two machines are showing different qualities of performance and (b) machine B is performing better than A. Use a 0·05 level of significance.
Ans. (a) A two-tailed test shows no difference in quality of performance at a 0·05 level.
(b) A one-tailed test shows that B is not performing better than A at a 0·05 level.

10.50. Two urns, A and B, contain equal numbers of marbles, but the proportion of red and white marbles in each of the urns is unknown. A sample of 50 marbles selected with replacement from each of the urns, revealed 32 red marbles from A and 23 red marbles from B. Using a significance level of 0·05, test the hypothesis that (a) the two urns have equal proportions of red marbles and (b) A has a greater proportion of red marbles than B.
Ans. (a) A two-tailed test at a 0·05 level fails to reject the hypothesis of equal proportions.
   (b) A one-tailed test at a 0·05 level indicates that A has a greater proportion of red marbles than B.

TESTS INVOLVING BINOMIAL DISTRIBUTIONS

10.51. Referring to Prob. 10.23, find the least number of questions a student must answer correctly before the instructor is sure at a significance level of (a) 0·05, (b) 0·01, (c) 0·001, (d) 0·06 that the student is not merely guessing. Discuss the results.
Ans. (a) 9, (b) 10, (c) 10, (d) 8

10.52. Construct graphs similar to those of Prob. 10.10 for Prob. 10.24.

10.53. Work Problems 10.23–10.25 if the 7 in the decision rule of Prob. 10.23 is changed to 8.

10.54. A coin which is tossed 8 times comes up heads 7 times. Can we reject the hypothesis that the coin is fair at a significance level of (a) 0·05, (b) 0·10, (c) 0·01? Use a two-tailed test. *Ans.* (a) No, (b) Yes, (c) No

10.55. Work Prob. 10.54 if a one-tailed test is used. *Ans.* (a) Yes, (b) Yes, (c) No

10.56. Work Prob. 10.54 if the coin comes up heads 8 times. *Ans.* (a) Yes, (b) Yes, (c) Yes

10.57. Work Prob. 10.54 if the coin comes up heads 6 times. *Ans.* (a) No, (b) No, (c) No

10.58. An urn contains a large number of red and white marbles. A random sample of 8 marbles revealed 6 white and 2 red. Using appropriate tests and levels of significance, discuss the proportions of red and white marbles in the urn.

10.59. Discuss how sampling theory can be used to investigate the proportions of different types of fish present in a lake.

CHAPTER 11

Small Sampling Theory

"STUDENT'S" t DISTRIBUTION and
THE CHI-SQUARE DISTRIBUTION

SMALL SAMPLES

In previous chapters we often made use of the fact that for samples of size $N > 30$, called *large samples*, the sampling distributions of many statistics were approximately normal, the approximation becoming better with increasing N. For samples of size $N < 30$, called *small samples*, this approximation is not good and becomes worse with decreasing N, so that appropriate modifications must be made.

A study of sampling distributions of statistics for small samples is called *small sampling theory*. However, a more suitable name would be *exact sampling theory*, since the results obtained hold for large as well as small samples. In this chapter we study two important distributions, called *"Student's" t distribution* and the *chi-square distribution*.

"STUDENT'S" t DISTRIBUTION

Let us define the statistic

$$t = \frac{\bar{X} - \mu}{s}\sqrt{N - 1} = \frac{\bar{X} - \mu}{\hat{s}/\sqrt{N}} \tag{1}$$

which is analogous to the z statistic given by $z = \dfrac{\bar{X} - \mu}{\sigma/\sqrt{N}}$ (see Page 169).

If we consider samples of size N drawn from a normal (or approximately normal) population with mean μ and if for each sample we compute t, using the sample mean $\bar{X}$ and sample standard deviation s or $\hat{s}$, the sampling distribution for t can be obtained. This distribution (see Fig. 11-1) is given by

$$Y = \frac{Y_0}{\left(1 + \dfrac{t^2}{N-1}\right)^{N/2}} = \frac{Y_0}{\left(1 + \dfrac{t^2}{\nu}\right)^{(\nu+1)/2}} \tag{2}$$

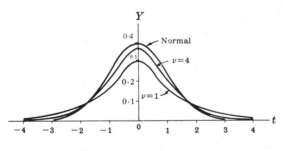

Student's t distributions for various values of ν

Fig. 11-1

where Y_0 is a constant depending on N such that the total area under the curve is one, and where the constant $\nu = (N - 1)$ is called the *number of degrees of freedom* (ν is the Greek letter *nu*). For a definition of degrees of freedom, see Page 191.

The distribution (2) is called *"Student's" t distribution* after its discoverer *Gosset*, who published his works under the pseudonym of "Student" during the early part of the twentieth century.

For large values of ν or N (certainly $N \geq 30$) the curves (2) closely approximate the standardized normal curve $Y = \dfrac{1}{\sqrt{2\pi}}e^{-\frac{1}{2}t^2}$, as indicated in Fig. 11-1.

CONFIDENCE INTERVALS

As done with normal distributions in Chapter 9, we can define 95%, 99% or other confidence intervals by using the table of the t distribution in the Appendix, Page 344. In this manner we can estimate within specified limits of confidence the population mean μ.

For example, if $-t_{0.975}$ and $t_{0.975}$ are the values of t for which 2·5% of the area lies in each "tail" of the t distribution, then a 95% confidence interval for t is

$$-t_{0.975} < \frac{\bar{X} - \mu}{s} \sqrt{N - 1} < t_{0.975} \tag{3}$$

from which we see that μ is estimated to lie in the interval

$$\bar{X} - t_{0.975} \frac{s}{\sqrt{N - 1}} < \mu < \bar{X} + t_{0.975} \frac{s}{\sqrt{N - 1}} \tag{4}$$

with 95% confidence (i.e. probability 0·95). Note that $t_{0.975}$ represents the 97·5 percentile value, while $t_{0.025} = -t_{0.975}$ represents the 2·5 percentile value.

In general, we can represent confidence limits for population means by

$$\bar{X} \pm t_c \frac{s}{\sqrt{N - 1}} \tag{5}$$

where the values $\pm t_c$, called *critical values* or *confidence coefficients*, depend on the level of confidence desired and the sample size. They can be read from the table on Page 344.

A comparison of (5) with the confidence limits ($\bar{X} \pm z_c \sigma/\sqrt{N}$) of Chapter 9, Page 158, shows that for samples we replace z_c (obtained from the normal distribution) by t_c (obtained from the t distribution) and we replace σ by $\sqrt{N/(N-1)}s = \hat{s}$, which is the sample estimate of σ. As N increases, both methods tend towards agreement.

TESTS OF HYPOTHESES AND SIGNIFICANCE

Tests of hypotheses and significance as discussed in Chapter 10 are easily extended to problems involving small samples, the only difference being that the z *score* or z *statistic* is replaced by a suitable t *score* or t *statistic*.

1. **Means.**

   To test the hypothesis H_0 that a normal population has mean μ, we use the t score or t statistic

$$t = \frac{\bar{X} - \mu}{s} \sqrt{N - 1} = \frac{\bar{X} - \mu}{\hat{s}} \sqrt{N} \tag{6}$$

   where $\bar{X}$ is the mean of a sample of size N.

   This is analogous to using the z score, $z = \dfrac{\bar{X} - \mu}{\sigma/\sqrt{N}}$, for large N except that $\hat{s} = \sqrt{N/(N-1)}s$ is used in place of σ. The difference is that while z is normally distributed, t follows Student's distribution. As N increases, these tend towards agreement.

2. **Differences of Means.**

   Suppose that two random samples of sizes N_1 and N_2 are drawn from normal populations whose standard deviations are equal ($\sigma_1 = \sigma_2$). Suppose further that these two samples have means and standard deviations given by $\bar{X}_1$, $\bar{X}_2$ and s_1, s_2 respectively. To test the hypothesis H_0 that the samples come from the same population (i.e. $\mu_1 = \mu_2$ as well as $\sigma_1 = \sigma_2$) we use the t score given by

$$t = \frac{\bar{X}_1 - \bar{X}_2}{\sigma\sqrt{1/N_1 + 1/N_2}} \qquad \text{where } \sigma = \sqrt{\frac{N_1 s_1^2 + N_2 s_2^2}{N_1 + N_2 - 2}} \tag{7}$$

The distribution of t is Student's distribution with $v = N_1 + N_2 - 2$ degrees of freedom.

Use of (7) is made plausible on placing $\sigma_1 = \sigma_2 = \sigma$ in the z score of Equation (2), Page 170 and then using as an estimate of σ^2 the weighted mean $\dfrac{N_1 - 1)\hat{s}_1^2 + (N_2 - 1)\hat{s}_2^2}{(N_1 - 1) + (N_2 - 1)} = \dfrac{N_1 s_1^2 + N_2 s_2^2}{N_1 + N_2 - 2}$ where $\hat{s}_1^2$ and $\hat{s}_2^2$ are the unbiased estimates of σ_1^2 and σ_2^2 (see Property 3, Page 72).

THE CHI-SQUARE DISTRIBUTION

Let us define the statistic

$$\chi^2 = \frac{Ns^2}{\sigma^2} = \frac{(X_1 - \bar{X})^2 + (X_2 - \bar{X})^2 + \ldots + (X_N - \bar{X})^2}{\sigma^2} \tag{8}$$

where χ is the Greek letter *chi* and χ^2 is read *chi-square*.

If we consider samples of size N drawn from a normal population with standard deviation σ, and if for each sample we compute χ^2, a sampling distribution for χ^2 can be obtained. This distribution, called the *chi-square distribution*, is given by

$$Y = Y_0(\chi^2)^{\frac{1}{2}(v-2)} e^{-\frac{1}{2}\chi^2} = Y_0\chi^{v-2} e^{-\frac{1}{2}\chi^2} \tag{9}$$

where $v = N - 1$ is the *number of degrees of freedom*, and Y_0 is a constant depending on v such that the total area under the curve is one. The chi-square distributions corresponding to various values of v are shown in Fig. 11-2. The maximum value of Y occurs at $\chi^2 = v - 2$ for $v \geqq 2$.

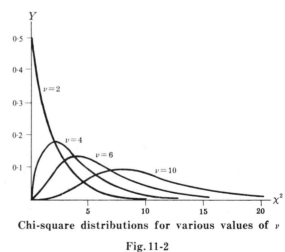

Chi-square distributions for various values of v

Fig. 11-2

CONFIDENCE INTERVALS FOR χ^2

As done with the normal and t distributions, we can define 95%, 99% or other confidence limits and intervals for χ^2 by use of the table of the χ^2 distribution in the Appendix, Page 345. In this manner we can estimate within specified limits of confidence the population standard deviation σ in terms of a sample standard deviation s.

For example, if $\chi^2_{0.025}$ and $\chi^2_{0.975}$ are the values of χ^2 (called *critical values*) for which 2.5% of the area lies in each "tail" of the distribution, then the 95% confidence interval is

$$\chi^2_{0.025} < \frac{Ns^2}{\sigma^2} < \chi^2_{0.975} \qquad (10)$$

from which we see that σ is estimated to lie in the interval

$$\frac{s\sqrt{N}}{\chi_{0.975}} < \sigma < \frac{s\sqrt{N}}{\chi_{0.025}} \qquad (11)$$

with 95% confidence. Similarly other confidence intervals can be found. The values $\chi_{0.025}$ and $\chi_{0.975}$ represent respectively the 2.5 and 97.5 percentile values.

The table of Appendix IV, Page 345 gives percentile values corresponding to the number of degrees of freedom v. For large values of v ($v \geq 30$) we can use the fact that $(\sqrt{2\chi^2} - \sqrt{2v-1})$ is very nearly normally distributed with mean zero and standard deviation one, so that normal distribution tables can be used if $v \geq 30$. Then if χ^2_P and z_P are the pth percentiles of the chi-square and normal distributions respectively, we have

$$\chi^2_P = \tfrac{1}{2}(z_P + \sqrt{2v-1})^2 \qquad (12)$$

In these cases agreement is close to results obtained in Chapters 8 and 9.

For further applications of the chi-square distribution see Chapter 12.

DEGREES OF FREEDOM

In order to compute a statistic such as (1) or (8), it is necessary to use observations obtained from a sample as well as certain population parameters. If these parameters are unknown they must be estimated from the sample.

The *number of degrees of freedom* of a statistic generally denoted by v is defined as the number N of independent observations in the sample (i.e. the sample size) minus the number k of population parameters which must be estimated from sample observations. In symbols, $v = N - k$.

In the case of the statistic (1) the number of independent observations in the sample is N, from which we can compute $\bar{X}$ and s. However, since we must estimate μ, $k = 1$ and so $v = N - 1$.

In the case of the statistic (8), the number of independent observations in the sample is N, from which we can compute s. However, since we must estimate σ, $k = 1$ and so $v = N - 1$.

Solved Problems

STUDENT'S t DISTRIBUTION

11.1. The graph of Student's t distribution with 9 degrees of freedom is shown in Fig. 11-3. Find the value of t_1 for which

(a) the shaded area on the right = 0·05,

(b) the total shaded area = 0·05,

(c) the total unshaded area = 0·99,

(d) the shaded area on the left = 0·01,

(e) the area to the left of t_1 is 0·90,

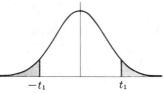

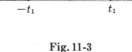

Fig. 11-3

Solution:

(a) If the shaded area on the right is 0·05, then the area to the left of t_1 is $(1 - 0\cdot05) = 0\cdot95$ and t_1 represents the 95th percentile, $t_{0.95}$.

Referring to table of Appendix III on Page 344, proceed downward under column headed v until entry 9 is reached. Then proceed right to column headed $t_{0.95}$. The result 1·83 is the required value of t.

(b) If the total shaded area is 0·05, then the shaded area on the right is 0·025 by symmetry. Thus the area to the left of t_1 is $(1 - 0\cdot025) = 0\cdot975$ and t_1 represents the 97·5th percentile $t_{0.975}$. From table of Appendix III, Page 344, we find 2·26 as the required value of t.

(c) If the total unshaded area is 0·99, then the total shaded area is $(1 - 0\cdot99) = 0\cdot01$ and the shaded area to the right is $0\cdot01/2 = 0\cdot005$. From the table we find $t_{0.995} = 3\cdot25$.

(d) If the shaded area on the left is 0·01, then by symmetry the shaded area on the right is 0·01. From the table, $t_{0.99} = 2\cdot82$. Thus the critical value of t for which the shaded area on the left is 0·01 equals $-2\cdot82$.

(e) If the area to the left of t_1 is 0·90, then t_1 corresponds to the 90th percentile, $t_{0.90}$, which from the table equals 1·38.

11.2. Find the critical values of t for which the area of the right-hand tail of the t distribution is 0·05 if the number of degrees of freedom v is equal to (a) 16, (b) 27, (c) 200.

Solution:

Using table of Appendix III, Page 344, we find in the column headed $t_{0.95}$ the values: (a) 1·75 corresponding to $v = 16$; (b) 1·70 corresponding to $v = 27$; (c) 1·645 corresponding to $v = 200$. (The latter is the value which would be obtained by using the normal curve. In table of Appendix III, Page 344, it corresponds to the entry in the last row marked ∞, i.e. *infinity*.)

11.3. The 95% confidence coefficients ("two-tailed") for the normal distribution are given by $\pm 1\cdot96$. What are the corresponding coefficients for the t distribution if (a) $v = 9$, (b) $v = 20$, (c) $v = 30$, (d) $v = 60$?

Solution:

For 95% confidence coefficients ("two-tailed") the total shaded area in Fig. 11-3 must be 0·05. Thus the shaded area in the right tail is 0·025 and the corresponding critical value of t is $t_{0.975}$. Then the required confidence coefficients are $\pm t_{0.975}$. For the given values of v these are: (a) $\pm 2\cdot26$, (b) $\pm2\cdot09$, (c) $\pm 2\cdot04$, (d) $\pm 2\cdot00$.

11.4. A sample of 10 measurements of the diameter of a marble gave a mean $\bar{X} = 4 \cdot 38$ mm and a standard deviation $s = 0 \cdot 06$ mm. Find (a) 95% and (b) 99% confidence limits for the actual diameter.

Solution:

(a) The 95% confidence limits are given by $\bar{X} \pm t_{0.975}(s/\sqrt{N-1})$.

Since $v = N - 1 = 10 - 1 = 9$, we find $t_{0.975} = 2 \cdot 26$ (see also Problem 11.3(a)). Then using $\bar{X} = 4 \cdot 38$ and $s = 0 \cdot 06$, the required 95% confidence limits are $4 \cdot 38 \pm 2 \cdot 26(0 \cdot 06)/\sqrt{10-1} = 4 \cdot 38 \pm 0 \cdot 0452$ mm.

Thus we can be 95% confident that the true mean lies between $(4 \cdot 38 - 0 \cdot 045) = 4 \cdot 335$ mm and $(4 \cdot 38 + 0 \cdot 045) = 4 \cdot 425$ mm.

(b) The 99% confidence limits are given by $\bar{X} \pm t_{0.995}(s/\sqrt{N-1})$.

For $v = 9$, $t_{0.995} = 3 \cdot 25$. Then the 99% confidence limits are $4 \cdot 38 \pm 3 \cdot 25(0 \cdot 06/\sqrt{10-1}) = 4 \cdot 38 \pm 0 \cdot 0650$ mm, and the 99% confidence interval is $4 \cdot 315$ to $4 \cdot 445$ mm.

11.5. (a) Work the previous problem assuming that the methods of large sampling theory are valid. (b) Compare the results of the two methods.

Solution:

(a) Using large sampling theory methods, the 95% confidence limits are

$$\bar{X} \pm 1 \cdot 96\sigma/\sqrt{N} = 4 \cdot 38 \pm 1 \cdot 96(0 \cdot 06/\sqrt{10}) = 4 \cdot 38 \pm 0 \cdot 037 \text{ mm}$$

where we have used the sample standard deviation $0 \cdot 06$, as estimate of σ.

Similarly, the 99% confidence limits are

$$\bar{X} \pm 2 \cdot 58\sigma/\sqrt{N} = 4 \cdot 38 \pm 2 \cdot 58(0 \cdot 06/\sqrt{10}) = 4 \cdot 38 \pm 0 \cdot 049 \text{ mm}$$

(b) In each case the confidence intervals using the small or exact sampling methods are wider than those obtained by using large sampling methods. This is to be expected since less precision is available with small samples than with large samples.

11.6. In the past a machine has produced washers having a thickness of $0 \cdot 50$ mm. To determine whether the machine is in proper working order, a sample of 10 washers is chosen for which the mean thickness is $0 \cdot 53$ mm and the standard deviation is $0 \cdot 03$ mm. Test the hypothesis that the machine is in proper working order using a level of significance of (a) $0 \cdot 05$, (b) $0 \cdot 01$.

Solution:

We wish to decide between the hypotheses

H_0: $\mu = 0 \cdot 50$, and the machine is in proper working order
H_1: $\mu \neq 0 \cdot 50$, and the machine is not in proper working order

so that a two-tailed test is required.

Under the hypothesis H_0, we have $t = \dfrac{\bar{X} - \mu}{s} \sqrt{N-1} = \dfrac{0 \cdot 53 - 0 \cdot 50}{0 \cdot 03} \sqrt{10-1} = 3 \cdot 00$.

(a) For a two-tailed test at a $0 \cdot 05$ level of significance, we adopt the decision rule:
  (1) Accept H_0 if t lies inside the interval $-t_{0.975}$ to $t_{0.975}$, which for $10 - 1 = 9$ degrees of freedom is the interval $-2 \cdot 26$ to $2 \cdot 26$.
  (2) Reject H_0 otherwise.
     Since $t = 3 \cdot 00$, we reject H_0 at the $0 \cdot 05$ level.

(b) For a two-tailed test at a $0 \cdot 01$ level of significance, we adopt the decision rule:
  (1) Accept H_0 if t lies inside the interval $-t_{0.995}$ to t_{0995}, which for $10 - 1 = 9$ degrees of freedom is the interval $-3 \cdot 25$ to $3 \cdot 25$.
  (2) Reject H_0 otherwise.
     Since $t = 3 \cdot 00$, we accept H_0 at the $0 \cdot 01$ level.

Because we can reject H_0 at the $0 \cdot 05$ level but not at the $0 \cdot 01$ level, we can say that the sample result is *probably significant* (see terminology at the end of Problem 10.5 of Chapter 10). It would be advisable to check the machine or at least to take another sample.

11.7. A test of the breaking strengths of 6 ropes manufactured by a company showed a mean breaking strength of 7750 N and a standard deviation of 145 N, whereas the manufacturer claimed a mean breaking strength of 8000 N. Can we support the manufacturer's claim at a level of significance of (*a*) 0·05, (*b*) 0·01?

Solution:

We must decide between the hypotheses

$$H_0: \ \mu = 8000 \text{ N, and the manufacturer's claim is justified}$$
$$H_1: \ \mu < 8000 \text{ N, and the manufacturer's claim is not justified}$$

so that a one-tailed test is required.

Under the hypothesis H_0, we have $t = \dfrac{\bar{X} - \mu}{s}\sqrt{N-1} = \dfrac{7750 - 8000}{145}\sqrt{6-1} = 3{\cdot}86$.

(*a*) For a one-tailed test at a 0·05 level of significance, we adopt the decision rule:

    (*1*) Accept H_0 if t is greater than $-t_{0.95}$, which for $6 - 1 = 5$ degrees of freedom means $t > -2{\cdot}01$.
    (*2*) Reject H_0 otherwise.
        Since $t = -3{\cdot}86$, we reject H_0.

(*b*) For a one-tailed test at a 0·01 level of significance, we adopt the decision rule:

    (*1*) Accept H_0 if t is greater than $-t_{0.99}$, which for 5 degrees of freedom means $t > -3{\cdot}36$.
    (*2*) Reject H_0 otherwise.
        Since $t = 3{\cdot}86$, we reject H_0.
        We conclude that it is extremely unlikely that the manufacturer's claim is justified.

11.8. The I.Q.'s (intelligence quotients) of 16 students from one area of a city showed a mean of 107 with a standard deviation of 10, while the I.Q.'s of 14 students from another area of the city showed a mean of 112 with a standard deviation of 8. Is there a significant difference between the I.Q.'s of the two groups at a (*a*) 0·01 and (*b*) 0·05 level of significance?

Solution:

If μ_1 and μ_2 denote population mean I.Q.'s of students from the two areas, we have to decide between the hypotheses

$$H_0: \ \mu_1 = \mu_2, \text{ and there is essentially no difference between the}$$
$$H_1: \ \mu_1 \neq \mu_2, \text{ and there is a significant difference between the groups.}$$

Under the hypothesis H_0, $t = \dfrac{\bar{X}_1 - \bar{X}_2}{\sigma\sqrt{1/N_1 + 1/N_2}}$ where $\sigma = \sqrt{\dfrac{N_1 s_1^2 + N_2 s_2^2}{N_1 + N_2 - 2}}$.

Then $\sigma = \sqrt{\dfrac{16(10)^2 + 14(8)^2}{16 + 14 - 2}} = 9{\cdot}44$ and $t = \dfrac{112 - 107}{9{\cdot}44\sqrt{1/16 + 1/14}} = 1{\cdot}45$.

(*a*) On the basis of a two-tailed test at a 0·01 level of significance, we would reject H_0 if t were outside the range $-t_{0.995}$ to $t_{0.995}$, which for $(N_1 + N_2 - 2) = (16 + 14 - 2) = 28$ degrees of freedom is the range $-2{\cdot}76$ to $2{\cdot}76$.

    Thus we cannot reject H_0 at a 0·01 level of significance.

(*b*) On the basis of a two-tailed test at a 0·05 level of significance, we would reject H_0 if t were outside the range $-t_{0.975}$ to $t_{0.975}$, which for 28 degrees of freedom is the range $-2{\cdot}05$ to $2{\cdot}05$.

    Thus we cannot reject H_0 at a 0·05 level of significance.

    We conclude that there is no significant difference between the I.Q.'s of the two groups.

11.9. At an agricultural station it was desired to test the effect of a given fertilizer on wheat production. To accomplish this, 24 plots of land having equal areas were chosen; half of these were treated with the fertilizer and the other half were untreated (control group). Otherwise the conditions were the same. The mean yield of wheat on the untreated plots was 4·8 litres with a standard deviation of 4 litres, while the mean yield on the treated plots was 5·1 litres with a standard deviation of 3·6 litres. Can we conclude that there is a significant improvement in wheat production because of the fertilizer if a significance level of (a) 1% and (b) 5% is used?

Solution:

If μ_1 and μ_2 denote population mean yields of wheat on treated and untreated land respectively, we have to decide between the hypotheses

$$H_0: \mu_1 = \mu_2, \text{ and the difference is due to chance.}$$
$$H_1: \mu_1 > \mu_2, \text{ and the fertilizer improves the yield.}$$

Under the hypothesis H_0, $t = \dfrac{\bar{X}_1 - \bar{X}_2}{\sigma \sqrt{1/N_1 + 1/N_2}}$ where $\sigma = \sqrt{\dfrac{N_1 s_1^2 + N_2 s_2^2}{N_1 + N_2 - 2}}$.

Then $\sigma = \sqrt{\dfrac{12(4)^2 + 12(3\cdot6)^2}{12 + 12 - 2}} = 3\cdot97$ and $t = \dfrac{5\cdot1 - 4\cdot8}{3\cdot97\sqrt{1/12 + 1/12}} = 1\cdot85$.

(a) On the basis of a one-tailed test at a 0·01 level of significance, we would reject H_0 if t were greater than $t_{0.99}$, which for $(N_1 + N_2 - 2) = (12 + 12 - 2) = 22$ degrees of freedom is 2·51.

Thus we cannot reject H_0 at a 0·01 level of significance.

(b) On the basis of a one-tailed test at a 0·05 level of significance, we would reject H_0 if t were greater than $t_{0.95}$, which for 22 degrees of freedom is 1·72.

Thus we can reject H_0 at a 0·05 level of significance.

We conclude that the improvement in yield of wheat by use of the fertilizer is *probably significant*. However, before definite conclusions are drawn concerning the usefulness of the fertilizer it may be desirable to have some further evidence.

THE CHI-SQUARE DISTRIBUTION

11.10. The graph of the chi-square distribution with 5 degrees of freedom is shown in Fig. 11-4. Find the critical values of χ^2 for which

(a) the shaded area on the right = 0·05,

(b) the total shaded area = 0·05,

(c) the shaded area on the left = 0·10,

(d) the shaded area on the right = 0·01.

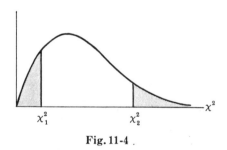

Fig. 11-4

Solution:

(a) If the shaded area on the right is 0·05, then the area to the left of χ_2^2 is $(1 - 0\cdot05) = 0\cdot95$ and χ_2^2 represents the 95th percentile, $\chi_{0.95}^2$.

Referring to table of Appendix IV, Page 345, proceed downwards under column headed ν until entry 5 is reached. Then proceed right to column headed $\chi_{0.95}^2$. The result 11.1 is the required critical value of χ^2.

(b) Since the distribution is not symmetric, there are many critical values for which the total shaded area = 0·05. For example, the right hand shaded area could be 0·04, while the left hand shaded area is 0·01. It is customary, however, unless otherwise specified, to choose the two areas equal. In this case, then, each area = 0·025.

If the shaded area on the right is 0·025, the area to the left of χ_2^2 is $1 - 0\cdot025 = 0\cdot975$ and χ_2^2 represents the 97·5th percentile, $\chi_{0.975}^2$, which from table of Appendix IV, Page 345, is 12·8.

Similarly, if the shaded area on the left is 0·025, the area to the left of χ_1^2 is 0·025 and χ_1^2 represents the 2·5th percentile, $\chi_{0.025}^2$, which equals 0·831.

Thus the critical values are 0·831 and 12·8.

(c) If the shaded area on the left is 0·10, χ_1^2 represents the 10th percentile, $\chi_{0.10}^2$, which equals 1·61.

(d) If the shaded area on the right is 0·01, the area to the left of χ_2^2 is 0·99 and χ_2^2 represents the 99th percentile, $\chi_{0.99}^2$, which equals 15·1.

11.11. Find the critical values of χ^2 for which the area of the right hand tail of the χ^2 distribution is 0·05, if the number of degrees of freedom ν is equal to (a) 15, (b) 21, (c) 50.

Solution:

Using table of Appendix IV, Page 345, we find in the column headed $\chi^2_{0.50}$ (since the median is the 50th percentile) to $\nu = 15$; (b) 32·7 corresponding to $\nu = 21$; (c) 67·5 corresponding to $\nu = 50$.

11.12. Find the median value of χ^2 corresponding to (a) 9, (b) 28 and (c) 40 degrees of freedom.

Solution:

Using table of Appendix IV, Page 345, we find in the column headed $\chi^2_{0.50}$ (since the median is the 50th percentile) the values: (a) 8·34 corresponding to $\nu = 9$; (b) 27·3 corresponding to $\nu = 28$; (c) 39·3 corresponding to $\nu = 40$.

It is of interest to note that the median values are very nearly equal to the number of degrees of freedom. In fact, for $\nu > 10$ the median values are equal to $(\nu - 0.7)$, as can be seen from the table.

11.13. The standard deviation of the masses of 16 male students chosen at random in a school of 1000 male students is 2·40 kg. Find (a) 95% and (b) 99% confidence limits of the standard deviation for all male students at the school.

Solution:

(a) 95% confidence limits are given by $s\sqrt{N}/\chi_{0.975}$ and $s\sqrt{N}/\chi_{0.025}$.

For $\nu = 16 - 1 = 15$ degrees of freedom, $\chi^2_{0.975} = 27.5$ or $\chi_{0.975} = 5.24$ and $\chi^2_{0.025} = 6.26$ or $\chi_{0.25} = 2.50$.

Then the 95% confidence limits are $2.40\sqrt{16}/5.24$ and $2.40\sqrt{16}/2.50$, i.e. 1·83 and 3·84 kg. Thus we can be 95% confident that the population standard deviation lies between 1·83 and 3·84 kg.

(b) 99% confidence limits are given by $s\sqrt{N}/\chi_{0.995}$ and $s\sqrt{N}/\chi_{0.005}$.

For $\nu = 16 - 1 = 15$ degrees of freedom, $\chi^2_{0.995} = 32.8$ or $\chi_{0.995} = 5.73$ and $\chi^2_{0.005} = 4.60$ or $\chi_{0.005} = 2.14$.

Then the 99% confidence limits are $2.40\sqrt{16}/5.73$ and $2.40\sqrt{16}/2.14$, i.e. 1·68 and 4·49 kg. Thus we can be 99% confident that the population standard deviation lies between 1·68 and 4·49 kg.

11.14. Find $\chi^2_{0.95}$ for (a) $\nu = 50$ and (b) $\nu = 100$ degrees of freedom.

Solution:

For ν greater than 30, we can use the fact that $(\sqrt{2\chi^2} - \sqrt{2\nu - 1})$ is very closely normally distributed with mean zero and standard deviation one. Then if z_p is the z score percentile of the standardized normal distribution, we can write, to a high degree of approximation,

$$\sqrt{2\chi_p^2} - \sqrt{2\nu - 1} = z_p \quad \text{or} \quad \sqrt{2\chi_p^2} = z_p + \sqrt{2\nu - 1}$$

from which

$$\chi_p^2 = \tfrac{1}{2}(z_p + \sqrt{2\nu - 1})^2$$

(a) If $\nu = 50$, $\chi^2_{0.95} = \tfrac{1}{2}(z_{0.95} + \sqrt{2(50) - 1})^2 = \tfrac{1}{2}(1.64 + \sqrt{99})^2 = 69.2$, which agrees very well with the value 67·5 given in the table on Page 345.

(b) If $\nu = 100$, $\chi^2_{0.95} = \tfrac{1}{2}(z_{0.95} + \sqrt{2(100) - 1})^2 = \tfrac{1}{2}(1.64 + \sqrt{199})^2 = 124.0$ (actual value = 124·3).

11.15. The standard deviation of the lifetimes of a sample of 200 electric light bulbs is 100 hours. Find (a) 95% and (b) 99% confidence limits for the standard deviation of all such electric light bulbs.

Solution:

(a) 95% confidence limits are given by $s\sqrt{N}/\chi_{0.995}$ and $s\sqrt{N}/\chi_{0.025}$.

For $v = 200 - 1 = 199$ degrees of freedom, we find as in Problem 11.14,

$$\chi_{0.975}^2 = \tfrac{1}{2}(z_{0.975} + \sqrt{2(199) - 1})^2 = \tfrac{1}{2}(1.96 + 19.92)^2 = 239$$
$$\chi_{0.025}^2 = \tfrac{1}{2}(z_{0.025} + \sqrt{2(199) - 1})^2 = \tfrac{1}{2}(-1.96 + 19.92)^2 = 161$$

from which $\chi_{0.975} = 15.5$ and $\chi_{0.025} = 12.7$.

Then the 95% confidence limits are $100\sqrt{200}/15.5 = 91.2$ and $100\sqrt{200}/12.7 = 111.3$ hours respectively. Thus we can be 95% confident that the population standard deviation will lie between 91.2 and 111.3 hours. This should be compared with Problem 9.17(a) of Chapter 9.

(b) 99% confidence limits are given by $s\sqrt{N}/\chi_{0.995}$ and $s\sqrt{N}/\chi_{0.005}$.

For $v = 200 - 1 = 199$ degrees of freedom,

$$\chi_{0.995}^2 = \tfrac{1}{2}(z_{0.995} + \sqrt{2(199) - 1})^2 = \tfrac{1}{2}(2.58 + 19.92)^2 = 253$$
$$\chi_{0.005}^2 = \tfrac{1}{2}(z_{0.005} + \sqrt{2(199) - 1})^2 = \tfrac{1}{2}(-2.58 + 19.92)^2 = 150$$

from which $\chi_{0.995} = 15.9$ and $\chi_{0.005} = 12.2$.

Then the 99% confidence limits are $100\sqrt{200}/15.9 = 88.9$ and $100\sqrt{200}/12.2 = 115.9$ hours respectively. Thus we can be 99% confident that the population standard deviation will lie between 88.9 and 115.9 hours. This should be compared with Problem 9.17(b) of Chapter 9.

11.16. Is it possible to obtain a 95% confidence interval for the population standard deviation whose width is smaller than that found in Problem 11.15(a)?

Solution:

The 95% confidence limits for the population standard deviation as found in Problem 11.15(a) were obtained by choosing critical values of χ^2 such that the area in each tail was 2.5%. It is possible to find other 95% confidence limits by choosing critical values of χ^2 for which the sum of the areas in the tails is 5%, or 0.05, but such that the areas in each tail are not equal.

In Table 11.1 several such critical values have been obtained (using methods of Problem 11.14) and the corresponding 95% confidence intervals shown.

Table 11.1

Critical Values	95% Confidence Interval	Width
$\chi_{0.01} = 12.44,\ \chi_{0.96} = 15.32$	92.3 to 113.7	21.4
$\chi_{0.02} = 12.64,\ \chi_{0.97} = 15.42$	91.7 to 111.9	20.2
$\chi_{0.03} = 12.76,\ \chi_{0.98} = 15.54$	91.0 to 110.8	19.8
$\chi_{0.04} = 12.85,\ \chi_{0.99} = 15.73$	88.9 to 110.0	20.1

From this table is it seen that a 95% interval of width only 19.8 is 91.0 to 110.8.

An interval with even smaller width can be found by continuing the same method of approach, using critical values such as $\chi_{0.031}$ and $\chi_{0.981}$, $\chi_{0.032}$ and $\chi_{0.982}$, etc.

In general, however, the decrease in the interval which is thereby obtained is usually negligible and is not worth the labour involved.

11.17. In the past the standard deviation of weights of certain 40.0 newton packages filled by a machine was 0.25 newtons. A random sample of 20 packages showed a standard deviation of 0.32 newtons. Is the apparent increase in variability significant at the (a) 0.05 and (b) 0.01 level of significance?

Solution:

We have to decide between the hypotheses

$$H_0:\ \sigma = 0.25, \text{ and the observed result is due to chance}$$
$$H_1:\ \sigma > 0.25, \text{ and the variability has increased.}$$

The value of χ^2 for the sample is $\chi^2 = Ns^2/\sigma^2 = 20(0.32)^2/(0.25)^2 = 32.8$.

(a) Using a one-tailed test, we would reject H_0 at a 0·05 level of significance if the sample value of χ^2 were greater than $\chi^2_{0.95}$, which equals 30·1 for $\nu = 20 - 1 = 19$ degrees of freedom. Thus we would reject H_0 at a 0·05 level of significance.

(b) Using a one-tailed test, we would reject H_0 at a 0·01 level of significance if the sample value of χ^2 were greater than $\chi^2_{0.99}$, which equals 36·2 for 19 degrees of freedom. Thus we would not reject H_0 at a 0·01 level of significance. We conclude that the variability has probably increased. An examination of the machine should be made.

Supplementary Problems

STUDENT'S t DISTRIBUTION

11.18. For a Student's distribution with 15 degrees of freedom, find the value of t_1 such that (a) the area to the right of t_1 is 0·01, (b) the area to the left of t_1 is 0·95, (c) the area to the right of t_1 is 0·10, (d) the combined area to the right of t_1 and to the left of $-t_1$ is 0·01, (e) the area between $-t_1$ and t_1 is 0·95.
Ans. (a) 2·60, (b) 1·75, (c) 1·34, (d) 2·95, (e) 2·13

11.19. Find the critical values of t for which the area of the right hand tail of the t distribution is 0·01, if the number of degrees of freedom ν is equal to (a) 4, (b) 12, (c) 25, (d) 60, (e) 150.
Ans. (a) 3·75, (b) 2·68, (c) 2·48, (d) 2·39, (e) 2·33

11.20. Find the values of t_1 for Student's distribution which satisfy each of the following conditions: (a) the area between $-t_1$ and t_1 is 0·9 and $\nu = 25$, (b) the area to the left of $-t_1$ is 0·025 and $\nu = 20$, (c) the combined area to the right of t_1 and the left of $-t_1$ is 0·01 and $\nu = 5$, (d) the area to the right of t_1 is 0·55 and $\nu = 16$.
Ans. (a) 1·71, (b) 2·09, (c) 4·03, (d) $-0·128$

11.21. If a variable U has a Student's distribution with $\nu = 10$, find the constant C such that (a) $\Pr\{U > C\} = 0·05$, (b) $\Pr\{-C \le U \le C\} = 0·98$, (c) $\Pr\{U \le C\} = 0·20$, (d) $\Pr\{U \ge C\} = 0·90$.
Ans. (a) 1·81, (b) 2·76, (c) $-8·79$, (d) $-1·37$

11.22. The 99% confidence coefficients ("two-tailed") for the normal distribution are given by $\pm 25·8$. What are the corresponding coefficients for the t distribution if (a) $\nu = 4$, (b) $\nu = 12$, (c) $\nu = 25$, (d) $\nu = 30$, (e) $\nu = 40$?
Ans. (a) $\pm 4·60$, (b) $\pm 3·06$, (c) $\pm 2·79$, (d) $\pm 2·75$, (e) $\pm 2·70$

11.23. A sample of 12 measurements of the breaking strengths of nylon threads gave a mean of 7·38 newtons and a standard deviation of 1·24 newtons. Find (a) 95% and (b) 99% confidence limits for the actual breaking strength.
Ans. (a) $7·38 \pm 0·82$, (b) $7·38 \pm 1·16$ N

11.24. Work the preceding problem assuming that the methods of large sampling theory are applicable and compare the results obtained. *Ans.* (a) $7·38 \pm 0·73$, (b) $7·38 \pm 0·96$ N

11.25. Five measurements of the reaction time of an individual to certain stimuli were recorded as 0·28, 0·30, 0·27, 0·33, 0·31 seconds. Find (a) 95% and (b) 99% confidence limits for the actual reaction time.
Ans. (a) $0·298 \pm 0·030$, (b) $0·298 \pm 0·049$ seconds

11.26. The mean lifetime of electric light bulbs produced by a company has in the past been 1120 hours with a standard deviation of 125 hours. A sample of 8 electric light bulbs recently chosen from a supply of newly produced bulbs showed a mean lifetime of 1070 hours. Test the hypothesis that the mean lifetime of the bulbs has not changed, using a level of significance of (a) 0·05 and (b) 0·01.
Ans. A two-tailed test shows that there is no evidence at either of the levels 0·05 or 0·01 to indicate that the mean lifetime has changed.

11.27. In the previous problem test the hypothesis $\mu = 1120$ hours against the alternative hypothesis $\mu < 1120$ hours, using a significance level of (a) 0·05 and (b) 0·01.
Ans. A one-tailed test indicates no decrease in the mean at either the 0·05 or 0·01 level.

11.28. The specifications for the production of a certain alloy call for 23·2% copper. A sample of 10 analyses of the product showed a mean copper content of 23·5% and a standard deviation of 0·24%. Can we conclude at a (a) 0·01 and (b) 0·05 significance level that the product meets the required specifications?
Ans. A two-tailed test at both levels shows that the product does not meet the required specifications.

11.29. In Problem 11.28 test the hypothesis that the mean copper content is higher than in required specifications, using a significance level of (a) 0·01, (b) 0·05.
Ans. A one-tailed test at both levels shows that the mean copper content is higher than specifications require.

11.30. An efficiency expert claims that by introducing a new type of machinery into a production process he can decrease substantially the time required for production. Because of the expense involved in maintenance of the machines, management feels that unless the production time can be decreased by at least 8·0% they cannot afford to introduce the process. Six resulting experiments show that the time for production is decreased by 8·4% with a standard deviation of 0·32%. Using a level of significance of (a) 0·01 and (b) 0·05, test the hypothesis that the process should be introduced.
Ans. A one-tailed test shows that the process should not be introduced if the significance level adopted is 0·01, but should be introduced if the significance level adopted is 0·05.

11.31. Using brand *A* petrol the mean number of kilometres per litre travelled by 5 similar motorcycles under identical conditions was 22·6 with a standard deviation of 0·48. Using brand *B*, the mean number was 21·4 with a standard deviation of 0·54. Choosing a significance level of 0·05, investigate whether brand *A* is really better than brand *B* in providing more kilometres per litre.
Ans. A one-tailed test shows that brand *A* is better than brand *B* at a 0·05 level of significance.

11.32. Two types of chemical solutions, *A* and *B*, were tested for their pH (degree of acidity of the solution). Analysis of 6 samples of *A* showed a mean pH of 7·52 with a standard deviation of 0·024. Analysis of 5 samples of *B* showed a mean pH of 7·49 with a standard deviation of 0·032. Using a 0·05 significance level, determine whether the two types of solutions have different pH values.
Ans. Using a two-tailed test at a 0·05 level of significance, we would not conclude on the basis of the samples that there is a difference in acidity between the two types of solutions.

11.33. On an examination in psychology 12 students in one class had a mean mark of 78 with a standard deviation of 6, while 15 students in another class had a mean mark of 74 with a standard deviation of 8. Using a significance level of 0·05, determine whether the first group is superior to the second group.
Ans. Using a one-tailed test at a 0·05 level of significance, we would conclude that the first group is not superior to the second.

THE CHI-SQUARE DISTRIBUTION

11.34. For a chi-square distribution with 12 degrees of freedom, find the value of χ_c^2 such that (a) the area to the right of χ_c^2 is 0·05, (b) the area to the left of χ_c^2 is 0·99, (c) the area to the right of χ_c^2 is 0·025.   *Ans.* (a) 21·0, (b) 26·2, (c) 23·3

11.35. Find the critical values of χ^2 for which the area of the right hand tail of the χ^2 distribution is 0·05, if the number of degrees of freedom v is equal to (a) 8, (b) 19, (c) 28, (d) 40.
Ans. (a) 15·5, (b) 30·1, (c) 41·3, (d) 55·8

11.36. Work Prob. 11.35 if the area of the right hand tail is 0·01.   *Ans.* (a) 20·1, (b) 36·2, (c) 48·3, (d) 63·7

11.37. (a) Find χ_1^2 and χ_2^2 such that the area under the χ^2 distribution corresponding to $v = 20$ between χ_1^2 and χ_2^2 is 0·95, assuming equal areas to the right of χ_2^2 and left of χ_1^2. (b) Show that if the assumption of equal areas in part (a) is not made, the values χ_1^2 and χ_2^2 are not unique.   *Ans.* (a) 9·59 and 34·2

11.38. If the variable *U* is chi-square distributed with $v = 7$, find χ_1^2 and χ_2^2 such that (a) $\Pr\{U > \chi_2^2\} = 0·025$, (b) $\Pr\{U < \chi_1^2\} = 0·50$, (c) $\Pr\{\chi_1^2 \leq U \leq \chi_2^2\} = 0·90$.
Ans. (a) 16·0, (b) 6·35, (c) assuming equal areas in the two tails, $\chi_1^2 = 2·17$ and $\chi_2^2 = 14·1$

11.39. The standard deviation of the lifetimes of 10 electric light bulbs manufactured by a company is 120 hours. Find (a) 95% and (b) 99% confidence limits for the standard deviation of all bulbs manufactured by the company.
Ans. (a) 87·0 to 230·9, (b) 78·1 to 288·5 hours

11.40. Work the preceding problem if 25 electric light bulbs show the same standard deviation of 120 hours.
Ans. (a) 95·6 to 170·4, (b) 88·9 to 190·8 hours

11.41. Find (a) $\chi^2_{0.05}$ and (b) $\chi^2_{0.95}$ for $\nu = 150$.   *Ans.* (a) 122·5, (b) 179·2

11.42. Find (a) $\chi^2_{0.025}$ and (b) $\chi^2_{0.975}$ for $\nu = 250$.   *Ans.* (a) 207·7, (b) 295·2

11.43. Show that for large values of ν a good approximation to χ^2 is given by $(\nu + z_p\sqrt{2\nu})$ where z_p is the pth percentile of the standard normal distribution.

11.44. Work Prob. 11.39 by using the χ^2 distribution if a sample of 100 electric bulbs show the same standard deviation of 120 hours. Compare the results with those obtained by the methods of Chap. 9.
Ans. (a) 106·1 to 140·5, (b) 102·1 to 148·1 hours

11.45. What is the 95% confidence interval of Prob. 11.44 which has the least width?   *Ans.* 105·5 to 139·6 hours

11.46. The standard deviation of the breaking strengths of certain cables produced by a company is given as 240 kN. After a change was introduced in the process of manufacture of these cables, the breaking strengths of a sample of 8 cables showed a standard deviation of 300 kN. Investigate the significance of the apparent increase in variability, using a significance level of (a) 0·05, (b) 0·01.
Ans. On the basis of the given sample the apparent increase in variability is not significant at either level.

11.47. The standard deviation of the annual temperatures of a city over a period of 100 years was 8° Celsius. Using the mean temperature on the 15th day of each month during the last 15 years, a standard deviation of annual temperatures was computed as 5° Celsius. Test the hypothesis that the temperatures in the city have become less variable than in the past, using a significance level of (a) 0·05, (b) 0·01.
Ans. The apparent decrease is significant at the 0·05 level but not at the 0·01 level.

CHAPTER 12

The Chi-Square Test

OBSERVED AND THEORETICAL FREQUENCIES

As we have already seen many times, results obtained in samples do not always agree exactly with theoretical results expected according to rules of probability. For example, although theoretical considerations lead us to expect 50 heads and 50 tails when we toss a fair coin 100 times, it is rare that these results are obtained exactly.

Suppose that in a particular sample a set of possible events E_1, E_2, E_3, ..., E_k (see Table 12.1) are observed to occur with frequencies o_1, o_2, o_3, ..., o_k, called *observed frequencies*, and that according to probability rules they are expected to occur with frequencies e_1, e_2, e_3, ..., e_k, called *expected* or *theoretical frequencies*.

Table 12.1

Event	E_1	E_2	E_3	...	E_k
Observed frequency	o_1	o_2	o_3	...	o_k
Expected frequency	e_1	e_2	e_3	...	e_k

Often we wish to know whether observed frequencies differ significantly from expected frequencies. For the case where only two events E_1 and E_2 are possible (sometimes called *a dichotomy* or *dichotomous classification*), as for example heads or tails, defective or non-defective bolts, etc., the problem is answered satisfactorily by methods of previous chapters. In this chapter the general problem is considered.

DEFINITION OF χ^2

A measure of the discrepancy existing between observed and expected frequencies is supplied by the statistic χ^2 (read *chi-square*) given by

$$\chi^2 = \frac{(o_1 - e_1)^2}{e_1} + \frac{(o_2 - e_2)^2}{e_2} + \cdots + \frac{(o_k - e_k)^2}{e_k} = \sum_{j=1}^{k} \frac{(o_j - e_j)^2}{e_j} \tag{1}$$

where if the total frequency is N,

$$\Sigma o_j = \Sigma e_j = N \tag{2}$$

An expression equivalent to (*1*) is (see Prob. 12.11)

$$\chi^2 = \Sigma \frac{o_j^2}{e_j} - N \tag{3}$$

If $\chi^2 = 0$, observed and theoretical frequencies agree exactly, while if $\chi^2 > 0$, they do not agree exactly. The larger the value of χ^2, the greater is the discrepancy between observed and expected frequencies.

The sampling distribution of χ^2 is approximated very closely by the chi-square distribution

$$Y = Y_0 (\chi^2)^{\frac{1}{2}(\nu-2)} e^{-\frac{1}{2}\chi^2} = Y_0 \chi^{\nu-2} e^{-\frac{1}{2}\chi^2} \tag{4}$$

(already considered in Chap. 11) if expected frequencies are at least equal to 5, the approximation improving for larger values.

The number of degrees of freedom v is given by

(a) $v = k - 1$ if expected frequencies can be computed without having to estimate population parameters from sample statistics. Note that we subtract 1 from k because of the constraint condition (2) which states that if we know $k - 1$ of the expected frequencies the remaining frequency can be determined.

(b) $v = k - 1 - m$ if the expected frequencies can be computed only by estimating m population parameters from sample statistics.

SIGNIFICANCE TESTS

In practice, expected frequencies are computed on the basis of a hypothesis H_0. If under this hypothesis the computed value of χ^2 given by (1) or (3) is greater than some critical value (such as $\chi^2_{0.95}$ or $\chi^2_{0.99}$, which are the critical values at the 0·05 and 0·01 significance levels respectively), we would conclude that observed frequencies differ *significantly* from expected frequencies and would reject H_0 at the corresponding level of significance. Otherwise we would accept it or at least not reject it. This procedure is called *the chi-square test* of hypothesis or significance.

It should be noted that we must look with suspicion upon circumstances where χ^2 is *too close to zero* since it is rare that observed frequencies agree *too well* with expected frequencies. To examine such situations, can determine whether the computed value of χ^2 is less than $\chi^2_{0.05}$ or $\chi^2_{0.01}$, in which cases we would decide that the agreement is *too good* at the 0·05 or 0·01 levels of significance respectively.

THE CHI-SQUARE TEST FOR GOODNESS OF FIT

The chi-square test can be used to determine how well theoretical distributions, such as the normal, binomial, etc., fit empirical distributions, i.e. those obtained from sample data. See Problems 12.12 and 12.13.

CONTINGENCY TABLES

Table 12.1 above, in which observed frequencies occupy a single row, is called a *one-way classification table*. Since the number of columns is k, this is also called a $1 \times k$ (read "1 by k") *table*. By extending these ideas we can arrive at *two-way classification tables* or $h + k$ *tables* in which the observed frequencies occupy h rows and k columns. Such tables are often called *contingency tables*.

Corresponding to each observed frequency in an $h \times k$ contingency table, there is an *expected* or *theoretical frequency* which is computed subject to some hypothesis according to rules of probability. These frequencies which occupy the *cells* of a contingency table are called *cell frequencies*. The total frequency in each row or each column is called the *marginal frequency*.

To investigate agreement between observed and expected frequencies, we compute the statistic

$$\chi^2 = \sum_j \frac{(o_j - e_j)^2}{e_j} \tag{5}$$

where the sum is taken over all cells in the contingency table, the symbols o_j and e_j representing respectively the observed and expected frequencies in the jth cell. This sum which is analogous to (1) contains hk terms. The sum of all observed frequencies is denoted by N and is equal to the sum of all expected frequencies (compare with equation (2)).

As before, the statistic (5) has a sampling distribution given very closely by (4), provided expected frequencies are not too small. The number of degrees of freedom v of this chi-square distribution is given for $h > 1, k > 1$ by

(a) $v = (h - 1)(k - 1)$ if the expected frequencies can be computed without having to estimate population parameters from sample statistics. For a proof of this see Prob. 12.18.

(b) $v = (h - 1)(k - 1) - m$ if the expected frequencies can be computed only by estimating m population parameters from sample statistics.

Significance tests for $h \times k$ tables are similar to those for $1 \times k$ tables. Expected frequencies are found subject to a particular hypothesis H_0. A hypothesis commonly assumed is that the two classifications are independent of each other.

Contingency tables can be extended to higher dimensions. Thus, for example, we can have $h \times k \times l$ tables where 3 classifications are present.

YATES' CORRECTION FOR CONTINUITY

When results for continuous distributions are applied to discrete data, certain corrections for continuity can be made as we have seen in previous chapters. A similar correction is available when the chi-square distribution is used. The correction consists in rewriting (1) as

$$\chi^2 \text{ (corrected)} = \frac{(|o_1 - e_1| - 0.5)^2}{e_1} + \frac{(|o_2 - e_2| - 0.5)^2}{e_2} + \cdots + \frac{(|o_k - e_k| - 0.5)^2}{e_k} \tag{6}$$

and is often referred to as *Yates' correction*. An analogous modification of (5) also exists.

In general the correction is made only when the number of degrees of freedom is $v = 1$. For large samples this yields practically the same results as the uncorrected χ^2, but difficulties can arise near critical values (see Prob. 12.8). For small samples where each expected frequency is between 5 and 10, it is perhaps best to compare both the corrected and uncorrected values of χ^2. If both values lead to the same conclusion regarding a hypothesis, such as rejection at the 0.05 level, difficulties are rarely encountered. If they lead to different conclusions one can either resort to increasing sample sizes or, if this proves impractical, one can employ exact methods of probability involving the *multinomial distribution* of Chapter 6.

SIMPLE FORMULAE FOR COMPUTING χ^2

Simple formulae for computing χ^2 which involve only the observed frequencies can be derived. In the following we give the results for 2×2 and 2×3 contingency tables.

2×2 Tables

$$\chi^2 = \frac{N(a_1 b_2 - a_2 b_1)^2}{(a_1 + b_1)(a_2 + b_2)(a_1 + a_2)(b_1 + b_2)} = \frac{N\Delta^2}{N_1 N_2 N_A N_B} \tag{7}$$

where $\Delta = a_1 b_2 - a_2 b_1$, $N = a_1 + a_2 + b_1 + b_2$, $N_1 = a_1 + b_1$, $N_2 = a_2 + b_2$, $N_A = a_1 + a_2$, $N_B = b_1 + b_2$. See Prob. 12.19.

	I	II	Totals
A	a_1	a_2	N_A
B	b_1	b_2	N_B
Totals	N_1	N_2	N

With *Yates' correction* this becomes

$$\chi^2 \text{ (corrected)} = \frac{N(|a_1 b_2 - a_2 b_1| - \frac{1}{2}N)^2}{(a_1 + b_1)(a_2 + b_2)(a_1 + a_2)(b_1 + b_2)}$$

$$= \frac{N(|\Delta| - \frac{1}{2}N)^2}{N_1 N_2 N_A N_B} \tag{8}$$

2 × 3 Tables

$$\chi^2 = \frac{N}{N_A}\left[\frac{a_1^2}{N_1} + \frac{a_2^2}{N_2} + \frac{a_3^2}{N_3}\right] + \frac{N}{N_B}\left[\frac{b_1^2}{N_1} + \frac{b_2^2}{N_2} + \frac{b_3^2}{N_3}\right] - N \qquad (9)$$

	I	II	III	Totals
A	a_1	a_2	a_3	N_A
B	b_1	b_2	b_3	N_B
Totals	N_1	N_2	N_3	N

where we have used the general result valid for all contingency tables,

$$\chi^2 = \sum \frac{o_j^2}{e_j} - N \qquad (10)$$

See Prob. 12.43. The result (9) for $2 \times k$ tables where $k > 3$, can be generalized. (See Prob. 12.46.)

COEFFICIENT OF CONTINGENCY

A measure of the degree of relationship, association or dependence of the classifications in a contingency table is given by

$$C = \sqrt{\frac{\chi^2}{\chi^2 + N}} \qquad (11)$$

which is called the *coefficient of contingency*. The larger the value of C, the greater is the degree of association. The number of rows and columns in the contingency table determines the maximum value of C, which is never greater than one. If the number of rows and columns of a contingency table is equal to k, the maximum value of C is given by $\sqrt{(k-1)/k}$. (See Problems 12.22, 12.52 and 12.53.)

CORRELATION OF ATTRIBUTES

Because classifications in a contingency table often describe characteristics of individuals or objects, they are often referred to as *attributes* and the degree of dependence, association or relationship is called *correlation* of attributes. For $k \times k$ tables we define

$$r = \sqrt{\frac{\chi^2}{N(k-1)}} \qquad (12)$$

as the correlation coefficient between attributes or classifications. This coefficient lies between 0 and 1 (see Prob. 12.24). For 2×2 tables in which $k = 2$, the correlation is often called *tetrachoric correlation*.

The general problem of correlation of numerical variables is considered in Chapter 14.

ADDITIVE PROPERTY OF χ^2

Suppose the results of repeated experiments yield sample values of χ^2 given by $\chi_1^2, \chi_2^2, \chi_3^2, \ldots$ with $\nu_1, \nu_2, \nu_3, \ldots$ degrees of freedom respectively. Then the result of all these experiments can be considered equivalent to a χ^2 value given by $\chi_1^2 + \chi_2^2 + \chi_3^2 + \ldots$ with $\nu_1 + \nu_2 + \nu_3 + \ldots$ degrees of freedom. See Prob. 12.25.

Solved Problems

THE CHI-SQUARE TEST

12.1. In 200 tosses of a coin, 115 heads and 85 tails were observed. Test the hypothesis that the coin is fair using a level of significance of (a) 0·05, (b) 0·01.

Solution:

Observed frequencies of heads and tails are respectively $o_1 = 115$, $o_2 = 85$.

Expected frequencies of heads and tails if the coin is fair are $e_1 = 100$, $e_2 = 100$ respectively. Then

$$\chi^2 = \frac{(o_1 - e_1)^2}{e_1} + \frac{(o_2 - e_2)^2}{e_2} = \frac{(115 - 100)^2}{100} + \frac{(85 - 100)^2}{100} = 4\cdot50$$

Since the number of categories or classes (heads, tails) is $k = 2$, $v = k - 1 = 2 - 1 = 1$.

(a) The critical value $\chi^2_{0.95}$ for 1 degree of freedom = 3·84. Then since 4·50 > 3·84, we reject the hypothesis that the coin is fair at a 0·05 level of significance.

(b) The critical value $\chi^2_{0.99}$ for 1 degree of freedom = 6·63. Then since 4·50 > 6·63, we cannot reject the hypothesis that the coin is fair at a 0·01 level of significance.

We conclude that the observed results are *probably significant* and the coin is *probably not fair*. For a comparison of this method with previous methods used, see Prob. 12.3.

12.2. Work Problem 12.1 using Yates' correction.

Solution:

$$\chi^2 \text{ (corrected)} = \frac{(|o_1 - e_1| - 0\cdot5)^2}{e_1} + \frac{(|o_2 - e_2| - 0\cdot5)^2}{e_2} = \frac{(|115 - 100| - 0\cdot5)^2}{100} + \frac{(|85 - 100| - 0\cdot5)^2}{100}$$

$$= \frac{(14\cdot5)^2}{100} + \frac{(14\cdot5)^2}{100} = 4\cdot205.$$

Since 4·205 > 3·84 and 4·205 < 6·63, the conclusions arrived at in Problem 12.1 are valid. For a comparison with previous methods, see Prob. 12.3.

12.3. Work Problem 12.1 by using the normal approximation to the binomial distribution.

Solution:

Under the hypothesis that the coin is fair, the mean and standard deviation of the number of heads expected in 200 tosses of a coin are $\mu = Np = (200)(0\cdot5) = 100$ and $\sigma = \sqrt{Npq} = \sqrt{(200)(0\cdot5)(0\cdot5)} = 7\cdot07$ respectively.

First method:

115 heads in standard units = $(115 - 100)/7\cdot07 = 2\cdot12$.

Using a 0·05 significance level and a two-tailed test, we would reject the hypothesis that the coin is fair if the z score were outside the interval $-1\cdot96$ to $1\cdot96$. With a 0·01 level the corresponding interval would be $-2\cdot58$ to $2\cdot58$. It follows as in Prob. 12.1 that we can reject the hypothesis at a 0·05 level but cannot reject it at a 0·01 level.

Note that the square of the above standard score $(2\cdot12)^2 = 4\cdot50$, is the same as the value of χ^2 obtained in Prob. 12.1. This is always the case for a chi-square test involving two categories. See Prob. 12.10.

Second method:

Using the correction for continuity, 115 or more heads is equivalent to 114·5 or more heads. Then 114·5 in standard units = $(114\cdot5 - 100)/7\cdot07 = 2\cdot05$. This leads to the same conclusions as in the first method.

Note that the square of this standard score is $(2\cdot05)^2 = 4\cdot20$, agreeing with the value of χ^2 corrected for continuity using Yates' correction of Prob. 12.2. This is always the case for a chi-square test involving two categories in which Yates' correction is applied.

12.4. Table 12.2 shows the observed and expected frequencies in tossing a die 120 times. Test the hypothesis that the die is fair, using a significance level of 0·05.

Table 12.2

Face	1	2	3	4	5	6
Observed frequency	25	17	15	23	24	16
Expected frequency	20	20	20	20	20	20

Solution:

$$\chi^2 = \frac{(o_1 - e_1)^2}{e_1} + \frac{(o_2 - e_2)^2}{e_2} + \frac{(o_3 - e_3)^2}{e_3} + \frac{(o_4 - e_4)^2}{e_4} + \frac{(o_5 - e_5)^2}{e_5} + \frac{(o_6 - e_6)^2}{e_6}$$

$$= \frac{(25 - 20)^2}{20} + \frac{(17 - 20)^2}{20} + \frac{(15 - 20)^2}{20} + \frac{(23 - 20)^2}{20} + \frac{(24 - 20)^2}{20} + \frac{(16 - 20)^2}{20} = 5·00$$

Since the number of categories or classes (faces 1, 2, 3, 4, 5, 6) is $k = 6$, $v = k - 1 = 6 - 1 = 5$.

The critical value $\chi^2_{0.95}$ for 5 degrees of freedom is 11·1. Then since $5·00 < 11·1$, we cannot reject the hypothesis that the die is fair.

For 5 degrees of freedom $\chi^2_{0.05} = 1·15$, so that $\chi^2 = 5·00 > 1·15$. It follows that the agreement is not so exceptionally good that we would look upon it with suspicion.

12.5. A random number table of 250 digits showed the following distribution of the digits 0, 1, 2, . . ., 9. Does the observed distribution differ significantly from the expected distribution?

Digit	0	1	2	3	4	5	6	7	8	9
Observed frequency	17	31	29	18	14	20	35	30	20	36
Expected frequency	25	25	25	25	25	25	25	25	25	25

Solution:

$$\chi^2 = \frac{(17 - 25)^2}{25} + \frac{(31 - 25)^2}{25} + \frac{(29 - 25)^2}{25} + \frac{(18 - 25)^2}{25} + \cdots + \frac{(36 - 25)^2}{25} = 23·3$$

The critical value $\chi^2_{0.99}$ for $v = k - 1 = 9$ degrees of freedom is 21·7, and $23·3 > 21·7$. Hence we conclude that the observed distribution differs significantly from the expected distribution at the 0·01 level of significance. It follows that some suspicion is cast upon the table of random numbers.

12.6. In Mendel's experiments with peas he observed 315 round and yellow, 108 round and green, 101 wrinkled and yellow, and 32 wrinkled and green. According to his theory of heredity the numbers should be in the proportion 9:3:3:1. Is there any evidence to doubt his theory at the (a) 0·01, (b) 0·05 level of significance?

Solution:

The total number of peas $= 315 + 108 + 101 + 32 = 556$. Since the expected numbers are in the proportion 9:3:3:1 (and $9 + 3 + 3 + 1 = 16$), we would expect

$\frac{9}{16}(556) = 312·75$ round and yellow $\frac{3}{16}(556) = 104·25$ wrinkled and yellow

$\frac{3}{16}(556) = 104·25$ round and green $\frac{1}{16}(556) = 34·75$ wrinkled and green

Then $$\chi^2 = \frac{(315 - 312·75)^2}{312·75} + \frac{(108 - 104·25)^2}{104·25} + \frac{(101 - 104·25)^2}{104·25} + \frac{(32 - 34·75)^2}{34·75} = 0·470$$

Since there are four categories, $k = 4$ and the number of degrees of freedom is $v = 4 - 1 = 3$.

(a) For $\nu = 3$, $\chi^2_{0.99} = 11.3$ so that we cannot reject the theory at the 0.01 level.

(b) For $\nu = 3$, $\chi^2_{0.95} = 7.81$ so that we cannot reject the theory at the 0.05 level.

We conclude the theory and experiment are in agreement.

Note that for 3 degrees of freedom, $\chi^2_{0.05} = 0.352$ and $\chi^2 = 0.470 > 0.352$. Thus although the agreement is good, the results obtained are subject to a reasonable amount of sampling error.

12.7. An urn consists of a very large number of marbles of four different colours: red, orange, yellow and green. A sample of 12 marbles drawn at random from the urn revealed 2 red, 5 orange, 4 yellow and 1 green marble. Test the hypothesis that the urn contains equal proportions of the differently coloured marbles.

Solution:

Under the hypothesis that the urn contains equal proportions of the differently coloured marbles, we would expect 3 of each kind in a sample of 12 marbles.

Since these expected numbers are less than 5, the chi-square approximation will be in error. To avoid this, we combine categories so that the expected number in each category is at least 5.

If we wish to reject the hypothesis, we should combine categories in such a way that the evidence against the hypothesis shows up best. This is achieved in our case by considering the categories "red or green" and "orange or yellow", for which the sample revealed 3 and 9 marbles respectively. Since the expected number in each category under the hypothesis of equal proportions is 6, we have

$$\chi^2 = \frac{(3-6)^2}{6} + \frac{(9-6)^2}{6} = 3$$

For $\nu = 2 - 1 = 1$, $\chi^2_{0.95} = 3.84$. Thus we cannot reject the hypothesis at the 0.05 level of significance (although we can at the 0.10 level). Conceivably the observed results could arise on the basis of chance even when equal proportions of the colours are present.

Another method: Using Yates' correction, we find

$$\chi^2 = \frac{(|3-6|-0.5)^2}{6} + \frac{(|9-6|-0.5)^2}{6} = \frac{(2.5)^2}{6} + \frac{(2.5)^2}{6} = 2.1$$

which leads to the same conclusion given above. This is of course to be expected since Yates' correction always *reduces* the value of χ^2.

It should be noted that if the χ^2 approximation is used despite the fact that the frequencies are too small, we would obtain

$$\chi^2 = \frac{(2-3)^2}{3} + \frac{(5-3)^2}{3} + \frac{(4-3)^2}{3} + \frac{(1-3)^2}{3} = 3.33$$

Since for $\nu = 4 - 1 = 3$, $\chi^2_{0.95} = 7.81$, we would arrive at the same conclusions as above. Unfortunately the χ^2 approximation for small frequencies is poor; hence when it is not advisable to combine frequencies we must resort to the exact probability methods of Chapter 6.

12.8. In 360 tosses of a pair of dice, 74 "sevens" and 24 "elevens" are observed. Using a 0.05 level of significance, test the hypothesis that the dice are fair.

Solution:

A pair of dice can fall in 36 ways. A "seven" can occur in 6 ways, an "eleven" in 2 ways.

Then $\Pr\{\text{"seven"}\} = \frac{6}{36} = \frac{1}{6}$ and $\Pr\{\text{"eleven"}\} = \frac{2}{36} = \frac{1}{18}$. Thus in 360 tosses we would expect $\frac{1}{6}(360) = 60$ sevens and $\frac{1}{18}(360) = 20$ elevens, so that

$$\chi^2 = \frac{(74-60)^2}{60} + \frac{(24-20)^2}{20} = 4.07$$

For $\nu = 2 - 1 = 1$, $\chi^2_{0.95} = 3.84$. Then since $4.07 > 3.84$, we would be inclined to reject the hypothesis that the dice are fair. Using Yates' correction, however, we find

$$\chi^2 \text{ (corrected)} = \frac{(|74-60|-0.5)^2}{60} + \frac{(|24-20|-0.5)^2}{20} = \frac{(13.5)^2}{60} + \frac{(3.5)^2}{20} = 3.65$$

Thus on the basis of the corrected χ^2, we could not reject the hypothesis at the 0.05 level.

In general for large samples such as we have here, results using Yates' correction prove to be more reliable than uncorrected results. However, since even the corrected value of χ^2 lies so close to the critical value, we are hesitant about making decisions one way or the other. In such cases it is perhaps best to increase the sample size by taking more observations if we are interested especially in the 0·05 level for some reason. Otherwise we could reject the hypothesis at some other level (such as 0·10) if this is satisfactory.

12.9. A survey of 320 families with 5 children revealed the distribution shown in Table 12.3. Is the result consistent with the hypothesis that male and female births are equally probable?

Table 12.3

Number of Boys and Girls	5 boys 0 girls	4 boys 1 girl	3 boys 2 girls	2 boys 3 girls	1 boy 4 girls	0 boys 5 girls	Total
Number of Families	18	56	110	88	40	8	320

Solution:

Let p = probability of a male birth, and $q = 1 - p$ = probability of a female birth. Then the probabilities of (5 boys), (4 boys and 1 girl), . . ., (5 girls) are given by the terms in the binomial expansion

$$(p + q)^5 = p^5 + 5p^4q + 10p^3q^2 + 10p^2q^3 + 5pq^4 + q^5$$

If $p = q = \frac{1}{2}$, we have

$$\Pr\{5 \text{ boys and } 0 \text{ girls}\} = (\tfrac{1}{2})^5 = \tfrac{1}{32} \qquad \Pr\{2 \text{ boys and } 3 \text{ girls}\} = 10(\tfrac{1}{2})^2(\tfrac{1}{2})^3 = \tfrac{10}{32}$$
$$\Pr\{4 \text{ boys and } 1 \text{ girl}\} = 5(\tfrac{1}{2})^4(\tfrac{1}{2}) = \tfrac{5}{32} \qquad \Pr\{1 \text{ boy and } 4 \text{ girls}\} = 5(\tfrac{1}{2})(\tfrac{1}{2})^4 = \tfrac{5}{32}$$
$$\Pr\{3 \text{ boys and } 2 \text{ girls}\} = 10(\tfrac{1}{2})^3(\tfrac{1}{2})^2 = \tfrac{10}{32} \qquad \Pr\{0 \text{ boys and } 5 \text{ girls}\} = (\tfrac{1}{2})^5 = \tfrac{1}{32}$$

Then the expected number of families with 5, 4, 3, 2, 1 and 0 boys are obtained respectively by multiplying the above probabilities by 320, and the results are 10, 50, 100, 100, 50, 10. Hence

$$\chi^2 = \frac{(18 - 10)^2}{10} + \frac{(56 - 50)^2}{50} + \frac{(110 - 100)^2}{100} + \frac{(88 - 100)^2}{100} + \frac{(40 - 50)^2}{50} + \frac{(8 - 10)^2}{10} = 12\cdot0$$

Since $\chi^2_{0.95} = 11\cdot1$ and $\chi^2_{0.99} = 15\cdot1$ for $\nu = 6 - 1 = 5$ degrees of freedom, we can reject the hypothesis at the 0·05 but not at the 0·01 significance level. Thus we conclude that the results are probably significant, and male and female births are not equally probable.

12.10. Show that a chi-square test involving only two categories is equivalent to the significance test at the top of Page 170, Chap. 10.

Solution:

If P is the sample proportion for category I, p is the population proportion, and N is the total frequency, we can describe the situations by means of the accompanying table. Then by definition,

	I	II	Total
Observed frequency	NP	$N(1 - P)$	N
Expected frequency	Np	$N(1 - p) = Nq$	N

$$\chi^2 = \frac{(NP - Np)^2}{Np} + \frac{[N(1 - P) - N(1 - p)]^2}{Nq}$$

$$= \frac{N^2(P - p)^2}{Np} + \frac{N^2(P - p)^2}{Nq} = N(P - p)^2\left(\frac{1}{p} + \frac{1}{q}\right) = \frac{N(P - p)^2}{pq} = \frac{(P - p)^2}{pq/N}$$

which is the square of the z statistic on Page 170.

12.11. (a) Prove that formula (1), Page 201, can be written $\chi^2 = \Sigma \dfrac{o_j^2}{e_j} - N$.

(b) Use the result of (a) to verify the value of χ^2 computed in Prob. 12.6.

Solution:

(a) By definition, $\chi^2 = \sum \dfrac{(o_j - e_j)^2}{e_j} = \sum \left(\dfrac{o_j^2 - 2o_j e_j + e_j^2}{e_j}\right)$

$$= \sum \dfrac{o_j^2}{e_j} - 2\sum o_j + \sum e_j = \sum \dfrac{o_j^2}{e_j} - 2N + N = \sum \dfrac{o_j^2}{e_j} - N$$

where we have used the results (2) of Page 201.

(b) $\chi^2 = \sum \dfrac{o_j^2}{e_j} - N = \dfrac{(315)^2}{312 \cdot 75} + \dfrac{(108)^2}{104 \cdot 25} + \dfrac{(101)^2}{104 \cdot 25} + \dfrac{(32)^2}{34 \cdot 75} - 556 = 0 \cdot 470$

GOODNESS OF FIT

12.12. Use the chi-square test to determine the goodness of fit of the data in Prob. 7.31 of Chap. 7.

Solution:

$$\chi^2 = \dfrac{(38 - 33 \cdot 2)^2}{33 \cdot 2} + \dfrac{(144 - 161 \cdot 9)^2}{161 \cdot 9} + \dfrac{(342 - 316 \cdot 2)^2}{316 \cdot 2} + \dfrac{(287 - 308 \cdot 7)^2}{308 \cdot 7} + \dfrac{(164 - 150 \cdot 7)^2}{150 \cdot 7} + \dfrac{(25 - 29 \cdot 4)^2}{29 \cdot 4}$$

$$= 7 \cdot 54.$$

Since the number of parameters used in estimating the expected frequencies in $m = 2$ (namely the parameter p of the binomial distribution), $\nu = k - 1 - m = 6 - 1 - 1 = 4$.

For $\nu = 4$, $\chi^2_{0 \cdot 95} = 9 \cdot 49$. Hence the fit of the data is good.
For $\nu = 4$, $\chi^2_{0 \cdot 05} = 0 \cdot 711$. Thus, since $\chi^2 = 7 \cdot 54 > 0 \cdot 711$, the fit is not so good as to be unbelievable.

12.13. Determine the goodness of fit of the data in Prob. 7.33 of Chap. 7.

Solution:

$$\chi^2 = \dfrac{(5 - 4 \cdot 13)^2}{4 \cdot 13} + \dfrac{(18 - 20 \cdot 68)^2}{20 \cdot 68} + \dfrac{(42 - 38 \cdot 92)^2}{38 \cdot 92} + \dfrac{(27 - 27 \cdot 71)^2}{27 \cdot 71} + \dfrac{(8 - 7 \cdot 43)^2}{7 \cdot 43} = 0 \cdot 959$$

Since the number of parameters used in estimating the expected frequencies is $m = 2$ (namely the mean μ and standard deviation σ of the normal distribution), $\nu = k - 1 - m = 5 - 1 - 2 = 2$.

For $\nu = 2$, $\chi^2_{0 \cdot 95} = 5 \cdot 99$. Thus we conclude that the fit of the data is very good.
For $\nu = 2$, $\chi^2_{0 \cdot 05} = 0 \cdot 103$. Then since $\chi^2 = 0 \cdot 959 > 0 \cdot 103$, the fit is not "too good".

CONTINGENCY TABLES

12.14. Work Problem 10.20, Chapter 10, by using the chi-square test.

Solution:

The conditions of the problem are presented in Table 12.4(a). Under the null hypothesis H_0 that the serum has no effect, we would expect 70 people in each of the groups to recover and 30 in each group not to recover, as indicated in Table 12.4(b). Note that H_0 is equivalent to the statement that recovery is *independent* of the use of the serum, i.e. the classifications are independent.

Table 12.4(a). Frequencies Observed

	Recover	Do Not recover	Total
Group A (using serum)	75	25	100
Group B (not using serum)	65	35	100
Total	140	60	200

Table 12.4(b). Frequencies Expected under H_0

	Recover	Do Not recover	Total
Group A (using serum)	70	30	100
Group B (not using serum)	70	30	100
Total	140	60	200

$$\chi^2 = \dfrac{(75 - 70)^2}{70} + \dfrac{(65 - 70)^2}{70} + \dfrac{(25 - 30)^2}{30} + \dfrac{(35 - 30)^2}{30} = 2 \cdot 38$$

To determine the number of degrees of freedom, consider Table 12.5 which is the same as those given above except that only totals are shown. It is clear that we have the freedom of placing only one number in any of the four empty cells, since once this is done the numbers in the remaining cells are uniquely determined from the indicated totals. Thus there is one degree of freedom.

Table 12.5

	Recover	Do Not Recover	Total
Group A			100
Group B			100
Total	140	60	200

Another method: By formula (see Prob. 12.18), $\nu = (h - 1)(k - 1) = (2 - 1)(2 - 1) = 1$.

Since $\chi^2_{0.95} = 3.84$ for 1 degree of freedom, and since $\chi^2 = 2.38 < 3.84$, we conclude that the results are *not significant* at a 0.05 level. We are thus unable to reject H_0 at this level, and we conclude either that the serum is not effective or else withhold decision pending further tests.

Note that $\chi^2 = 2.38$ is the square of the z score, $z = 1.54$, obtained in Prob. 10.20 of Chap. 10. In general, the chi-square test involving sample proportions in a 2×2 contingency table is equivalent to a test of significance of differences in proportions using the normal approximation as on Page 171 of Chap. 10. (See Prob. 10.20.)

Note also that a one-tailed test using χ^2 is equivalent to a two-tailed test using χ since, for ecample $\chi^2 > \chi_{0.95}$ corresponds to $(\chi > \chi_{0.95})$ or $(\chi < -\chi_{0.95})$. Since for 2×2 tables, χ^2 is the square of the z score, it follows that χ is the same as z for this case. Thus a rejection of a hypothesis at the 0.05 level using χ^2 is equivalent to a rejection in a one-tailed test at the 0.10 level using z.

12.15. Work the preceding problem by using Yates' correction.

Solution:

$$\chi^2 \text{ (corrected)} = \frac{(|75 - 70| - 0.5)^2}{70} + \frac{(|65 - 70| - 0.5)^2}{70} + \frac{(|25 - 30| - 0.5)^2}{30} + \frac{(|35 - 30| - 0.5)^2}{30} = 1.93$$

Thus the conclusions given in the preceding problem are valid. This could have been realized at once by noting that Yates' correction always decreases the value of χ^2.

12.16. In Table 12.6 are indicated the numbers of students passed and failed by three instructors: Mr. X, Mr. Y and Mr. Z. Test the hypothesis that the proportions of students failed by the three instructors are equal.

Table 12.6. Frequencies Observed

	Mr. X	Mr. Y	Mr. Z	Total
Passed	50	47	56	153
Failed	5	14	8	27
Total	55	61	64	180

Solution:

Under the hypothesis H_0 that the proportions of students failed by the three instructors are the same, they would have failed $27/180 = 15\%$ of the students and thus passed 85% of the students. In this case Mr. X, for example, would have failed 15% of 55 students and passed 85% of 55 students. The frequencies expected under H_0 are shown in Table 12.7.

Table 12.7. Frequencies Expected Under H_0

	Mr. X	Mr. Y	Mr. Z	Total
Passed	85% of 55 = 46.75	85% of 61 = 51.85	85% of 64 = 54.40	153
Failed	15% of 55 = 8.25	15% of 61 = 9.15	15% of 64 = 9.60	27
Total	55	61	64	180

Table 12.8

	Mr. X	Mr. Y	Mr. Z	Total
Passed				153
Failed				27
Total	55	61	64	180

Then

$$\chi^2 = \frac{(50 - 46.75)^2}{46.75} + \frac{(47 - 51.85)^2}{51.85} + \frac{(56 - 54.40)^2}{54.40} + \frac{(5 - 8.25)^2}{8.25} + \frac{(14 - 9.15)^2}{9.15} + \frac{(8 - 9.60)^2}{9.60} = 4.84.$$

To determine the number of degrees of freedom, consider Table 12.8 which is the same as those given above except that only totals are shown. It is clear that we have the freedom of placing only one number into an empty cell of the first column and one number into an empty cell of the second or third column, after which all numbers in the remaining cells will be uniquely determined from the indicated totals. Thus there are two degrees of freedom in this case.

Another method: By formula, $\nu = (h - 1)(k - 1) = (2 - 1)(3 - 1) = 2$.

Since $\chi^2_{0.95} = 5.99$, we cannot reject H_0 at the 0.05 level. Note, however, that since $\chi^2_{0.90} = 4.61$, we can reject H_0 at the 0.10 level if we are willing to take the risk of 1 chance in 10 of being wrong.

12.17. Use formula (9), Page 204, to compute the value of χ^2 for the preceding problem.

Solution:

We have $a_1 = 50$, $a_2 = 47$, $a_3 = 56$, $b_1 = 5$, $b_2 = 14$, $b_3 = 8$, $N_A = a_1 + a_2 + a_3 = 153$, $N_B = b_1 + b_2 + b_3 = 27$, $N_1 = a_1 + b_1 = 55$, $N_2 = a_2 + b_2 = 61$, $N_3 = a_3 + b_3 = 64$, $N = N_A + N_B = N_1 + N_2 + N_3 = 180$. Then

$$\chi^2 = \frac{N}{N_A}\left[\frac{a_1^2}{N_1} + \frac{a_2^2}{N_2} + \frac{a_3^2}{N_3}\right] + \frac{N}{N_B}\left[\frac{b_1^2}{N_1} + \frac{b_2^2}{N_2} + \frac{b_3^2}{N_3}\right] - N$$

$$= \frac{180}{153}\left[\frac{(50)^2}{55} + \frac{(47)^2}{61} + \frac{(56)^2}{64}\right] + \frac{180}{27}\left[\frac{(5)^2}{55} + \frac{(14)^2}{61} + \frac{(8)^2}{64}\right] - 180 = 4.84$$

12.18. Show that for an $h \times k$ contingency table the number of degrees of freedom is $(h - 1) \times (k - 1)$ where $h > 1, k > 1$.

Solution:

In a table with h rows and k columns, we can leave out a single number in each row and column, since such numbers can easily be restored from a knowledge of the totals of each column and row. It follows that we have the freedom of placing only $(h - 1)(k - 1)$ numbers into the table, the others being then automatically determined uniquely. Thus the number of degrees of freedom is $(h - 1)(k - 1)$. Note that this result holds provided the population parameters needed in obtaining expected frequencies are known.

12.19. (a) Prove that for the 2×2 contingency table shown in Table 12.9(a),

$$\chi^2 = \frac{N(a_1 b_2 - a_2 b_1)^2}{N_1 N_2 N_A N_B}$$

(b) Illustrate the result in (a) with reference to the data of Problem 12.14.

Table 12.9(a). Results Observed

	I	II	Total
A	a_1	a_2	N_A
B	b_1	b_2	N_B
Total	N_1	N_2	N

Table 12.9(b). Results Expected

	I	II	Total
A	$N_1 N_A/N$	$N_2 N_A/N$	N_A
B	$N_1 N_B/N$	$N_2 N_B/N$	N_B
Total	N_1	N_2	N

Solution:

As in Problem 12.14, the results expected under a null hypothesis are shown in Table 12.9(b). Then

$$\chi^2 = \frac{(a_1 - N_1 N_A/N)^2}{N_1 N_A/N} + \frac{(a_2 - N_2 N_A/N)^2}{N_2 N_A/N} + \frac{(b_1 - N_1 N_B/N)^2}{N_1 N_B/N} + \frac{(b_2 - N_2 N_B/N)^2}{N_2 N_B/N}$$

But

$$a_1 - \frac{N_1 N_A}{N} = a_1 - \frac{(a_1 + b_1)(a_1 + a_2)}{a_1 + b_1 + a_2 + b_2} = \frac{a_1 b_2 - a_2 b_1}{N}$$

Similarly $\left(a_2 - \dfrac{N_2 N_A}{N}\right)$, $\left(b_1 - \dfrac{N_1 N_B}{N}\right)$, and $\left(b_2 - \dfrac{N_2 N_B}{N}\right)$ are also equal to $\left(\dfrac{a_1 b_2 - a_2 b_1}{N}\right)$.

Thus we can write

$$\chi^2 = \frac{N}{N_1 N_A}\left(\frac{a_1 b_2 - a_2 b_1}{N}\right)^2 + \frac{N}{N_2 N_A}\left(\frac{a_1 b_2 - a_2 b_1}{N}\right)^2 + \frac{N}{N_1 N_B}\left(\frac{a_1 b_2 - a_2 b_1}{N}\right)^2 + \frac{N}{N_2 N_B}\left(\frac{a_1 b_2 - a_2 b_1}{N}\right)^2$$

which simplifies to

$$\chi^2 = \frac{N(a_1 b_2 - a_2 b_1)^2}{N_1 N_2 N_A N_B}$$

In Problem 12.14, $a_1 = 75$, $a_2 = 25$, $b_1 = 65$, $b_2 = 35$, $N_1 = 140$, $N_2 = 60$, $N_A = 100$, $N_B = 100$, and $N = 200$; then, as obtained before,

$$\chi^2 = \frac{200[(75)(35) - (25)(65)]^2}{(140)(60)(100)(100)} = 2.38$$

Using Yates' correction, the result is the same as in Problem 12.15:

$$\chi^2 \text{ (corrected)} = \frac{N(|a_1 b_2 - a_2 b_1| - \frac{1}{2}N)^2}{N_1 N_2 N_A N_B} = \frac{200[|(75)(35) - (25)(65)| - 100]^2}{(140)(60)(100)(100)} = 1.93$$

12.20. Show that a chi-square test involving two sample proportions is equivalent to a significant test of differences in proportions using the normal approximation (see Page 171).

Solution:

Let P_1 and P_2 denote the two sample proportions and p the population proportion. With reference to Prob. 12.19, we have

(1) $P_1 = a_1/N_1$, $P_2 = a_2/N_2$, $1 - P_1 = b_1/N_1$, $1 - P_2 = b_2/N_2$

(2) $p = N_A/N$, $1 - p = q = N_B/N$

so that

(3) $a_1 = N_1 P_1$, $a_2 = N_2 P_2$, $b_1 = N_1(1 - P_1)$, $b_2 = N_2(1 - P_2)$

(4) $N_A = Np$, $N_B = Nq$

Using (3) and (4), we have from Prob. 19,

$$\chi^2 = \frac{N(a_1 b_2 - a_2 b_1)^2}{N_1 N_2 N_A N_B} = \frac{N[N_1 P_1 N_2(1 - P_2) - N_2 P_2 N_1(1 - P_1)]^2}{N_1 N_2 Np Nq}$$

$$= \frac{N_1 N_2 (P_1 - P_2)^2}{Npq} = \frac{(P_1 - P_2)^2}{pq(1/N_1 + 1/N_2)} \quad (\text{since } N = N_1 + N_2)$$

which is the square of the z statistic given on Page 171.

COEFFICIENT OF CONTINGENCY

12.21. Find the coefficient of contingency for the data in the contingency table of Prob. 12.14.

Solution:

$$C = \sqrt{\frac{\chi^2}{\chi^2 + N}} = \sqrt{\frac{2.38}{2.38 + 200}} = \sqrt{0.01176} = 0.1084$$

12.22. Find the maximum value of C for the 2×2 table of Prob. 12.14.

Table 12.10

	Recover	Do Not Recover	Total
Group A (using serum)	100	0	100
Group B (not using serum)	0	100	100
Total	100	100	200

Solution:

The maximum value of C occurs when the two classifications are perfectly dependent or associated. In such case all those who take the serum will recover and all those who do not take the serum will not recover. The contingency table then appears as in Table 12.10.

Since the expected cell frequencies assuming complete independence, are all equal to 50.

$$\chi^2 = \frac{(100 - 50)^2}{50} + \frac{(0 - 50)^2}{50} + \frac{(0 - 50)^2}{50} + \frac{(100 - 50)^2}{50} = 200$$

Then the maximum value of $C = \sqrt{\chi^2/(\chi^2 + N)} = \sqrt{200/(200 + 200)} = 0.7071$.

In general for perfect dependence in a contingency table where the number of rows and columns are both equal to k, the only non-zero cell frequencies occur in the diagonal from upper left to lower right of the contingency table. For such cases, $C_{max} = \sqrt{(k - 1)/k}$. (See Problems 12.52, 12.53.)

CORRELATION OF ATTRIBUTES

12.23. For the table of Problem 12.14, find the correlation coefficient (*a*) without and (*b*) with Yates' correction.

Solution:

(*a*) Since $\chi^2 = 2 \cdot 38$, $N = 200$ and $k = 2$, we have $r = \sqrt{\dfrac{\chi^2}{N(k-1)}} = \sqrt{\dfrac{2.38}{200}} = 0 \cdot 1091$, indicating very little correlation between recovery and use of the serum.

(*b*) From Problem 12.15, r (corrected) $= \sqrt{1 \cdot 93/200} = 0 \cdot 0982$.

12.24. Prove that the correlation coefficient for contingency tables, as defined by Equation (*12*), Page 204, lies between 0 and 1.

Solution:

By Problem 12.53, the maximum value of $\sqrt{\chi^2/(\chi^2 + N)}$ is $\sqrt{(k-1)/k}$. Then

$$\frac{\chi^2}{\chi^2 + N} \leqq \frac{k-1}{k}, \quad k\chi^2 \leqq (k-1)(\chi^2 + N), \quad k\chi^2 \leqq k\chi^2 - \chi^2 + kN - N$$

$$\chi^2 \leqq (k-1)N, \quad \frac{\chi^2}{N(k-1)} \leqq 1, \quad \text{and} \quad r = \sqrt{\frac{\chi^2}{N(k-1)}} \leqq 1$$

Since $\chi^2 \geqq 0$, $r \geqq 0$. Then $0 \leqq r \leqq 1$, as required.

ADDITIVE PROPERTY OF χ^2

12.25. To test a hypothesis H_0, an experiment is performed three times. The resulting values of χ^2 are $2 \cdot 37$, $1 \cdot 86$ and $3 \cdot 54$, each of which correspond to one degree of freedom. Show that while H_0 cannot be rejected at the $0 \cdot 05$ level on the basis of any individual experiment, it can be rejected when the three experiments are combined.

Solution:

The value of χ^2 obtained by combining the results of the 3 experiments is, according to the *additive property*,

$$\chi^2 = 2 \cdot 37 + 2 \cdot 86 + 3 \cdot 54 = 9 \cdot 77 \text{ with } 1 + 1 + 1 = 3 \text{ degrees of freedom}$$

Since $\chi^2_{0.95}$ for 3 degrees of freedom is $7 \cdot 81$, we can reject H_0 at the $0 \cdot 05$ level of significance. But since $\chi^2_{0.95} = 3 \cdot 84$ for one degree of freedom, we cannot reject H_0 on the basis of any one experiment.

In combining experiments in which values of χ^2 corresponding to one degree of freedom are obtained, Yates' correction is omitted since it has a tendency to overcorrect.

Supplementary Problems

THE CHI-SQUARE TEST

12.26. In 60 tosses of a coin, 37 heads and 23 tails were observed. Test the hypothesis that the coin is fair using a significance level of (*a*) $0 \cdot 05$, (*b*) $0 \cdot 01$.
Ans. The hypothesis cannot be rejected at either level.

12.27. Work Problem 12.26 using Yates' correction. *Ans.* The conclusion is the same as before.

12.28. Over a long period of time the marks given by a group of instructors in a particular course have averaged 12% A's, 18% B's, 40% C's, 18% D's and 12% F's. A new instructor gives 22 A's, 34 B's, 66 C's, 16 D's and 12 F's during two terms. Determine at a $0 \cdot 05$ significance level whether the new instructor is following the mark pattern set by the others.
Ans. The new instructor is not following the mark pattern of the others. (The fact that the marks happen to be better than average *may* be due to better teaching ability or lower standards or both.)

12.29. Three coins were tossed a total of 240 times and each time the number of heads turning up was observed. The results are shown in Table 12.11 together with results expected under the hypothesis that the coins are fair. Test this hypothesis at a significance level of 0·05.

Ans. There is no reason to reject the hypothesis that the coins are fair.

Table 12.11

	0 heads	1 head	2 heads	3 heads
Observed frequency	24	108	95	23
Expected frequency	30	90	90	30

12.30. The number of books borrowed from a public library during a particular week is given in Table 12.12. Test the hypothesis that the number of books borrowed does not depend on the day of the week, using a significance level of (*a*) 0·05, (*b*) 0·01.

Ans. There is no reason to reject the hypothesis at either level.

Table 12.12

	Mon.	Tues.	Wed.	Thur.	Fri.
Number of books borrowed	135	108	120	114	146

12.31. An urn consists of 6 red marbles and 3 white ones. Two marbles are selected at random from the urn, their colours are noted and then the marbles are replaced in the urn. This process is performed a total of 120 times and the results obtained are shown in Table 12.13. (*a*) Determine the expected frequencies. (*b*) Determine at a level of significance of 0·05 whether the results are consistent with those expected.

Ans. (*a*) 10, 60, 50 respectively. (*b*) The hypothesis that the results are the same as those expected cannot be rejected at a 0·05 level of significance.

Table 12.13

	0 red 2 white	1 red 1 white	2 red 0 white
Number of drawings	6	53	61

12.32. Two hundred bolts were selected at random from the production of each of 4 machines. The numbers of defective bolts found were 2, 9, 10, 3. Determine whether there is a significant difference between the machines using a significance level of 0·05.   *Ans.* The difference is significant at the 0·05 level.

GOODNESS OF FIT

12.33. (*a*) Use the chi-square test to determine the goodness of fit of the data of Prob. 7.75, Chap 7. (*b*) Is the fit "too good"? Use a 0·05 level of significance.   *Ans.* (*a*) The fit is good. (*b*) No.

12.34. Use the chi-square test to determine the goodness of fit of the data referred to in (*a*) Prob. 7.77 of Chap. 7, (*b*) Prob. 7.78 of Chap. 7. Use a level of significance of 0·05 and in each case determine whether the fit is "too good".

Ans. (*a*) The fit is "too good". (*b*) The fit is poor at the 0·05 level.

12.35. Use the chi-square test to determine the goodness of fit of the data referred to in (*a*) Prob. 7.79 of Chap. 7, (*b*) Prob. 7.80 of Chap. 7. Is your result in (*a*) consistent with that of Prob. 12.33?

Ans. (*a*) The fit is very poor at the 0·05 level. Since the binomial distribution gives a good fit of the data, this is consistent with Prob. 12.33. (*b*) The fit is good but not "too good".

CONTINGENCY TABLES

12.36. Table 12.14 shows the result of an experiment to investigate the effect of vaccination of laboratory animals against a particular disease. Using a (*a*) 0·01 and (*b*) 0·05 significance level, test the hypothesis that there is no difference between the vaccinated and unvaccinated groups, i.e. vaccination and this disease are independent.

Ans. The hypothesis can be rejected at the 0·05 level but not at the 0·01 level.

Table 12.14

	Got disease	Did not get disease
Vaccinated	9	42
Not vaccinated	17	28

12.37. Work the preceding problem using Yates' correction.   *Ans.* Same conclusion.

12.38. Table 12.15 shows the numbers of students in each of two classes A and B who passed and failed an examination given to both groups. Using a (*a*) 0·05 and (*b*) 0·01 significance level, test the hypothesis that there is no difference between the two classes. Work the problem with and without Yates' correction.
Ans. The hypothesis cannot be rejected at either level.

Table 12.15

	Passed	Failed
Class A	72	17
Class B	64	23

12.39. Of a group of patients who complained that they did not sleep well, some were given sleeping pills while others were given sugar pills (although they all *thought* they were getting sleeping pills). They were later asked whether the pills helped them or not. The results of their responses are shown in Table 12.16. Assuming that all patients told the truth, test the hypothesis that there is no difference between sleeping pills and sugar pills at a significance level of 0·05.
Ans. The hypothesis cannot be rejected at the 0·05 level.

Table 12.16

	Slept well	Did not sleep well
Took sleeping pills	44	10
Took sugar pills	·81·	35

Table 12.17

	In Favour	Opposed	Undecided
Democrats	85	78	37
Republicans	118	61	25

12.40. On a particular proposal of national importance, Democrats and Republicans cast votes as indicated in Table 12.17. At a level of significance of (*a*) 0·01 and (*b*) 0·05, test the hypothesis that there is no difference between the two parties insofar as this proposal is concerned.
Ans. The hypothesis can be rejected at both levels.

12.41. Table 12.18 shows the relation between the performances of students in mathematics and physics. Test the hypothesis that performance in physics is independent of performance in mathematics, using a (*a*) 0·05 and (*b*) 0·01 significance level.

Table 12.18

		Mathematics		
		High marks	Medium marks	Low marks
Physics	High marks	56	71	12
	Medium marks	47	163	38
	Low marks	14	42	85

Ans. The hypothesis can be rejected at both levels.

12.42. The results of a survey made to determine whether the age of a driver 21 years of age and older has any effect on the number of automobile accidents in which he is involved (including all minor accidents) are indicated in Table 12.19. At a level of significance of (*a*) 0·05 and (*b*) 0·01, test the hypothesis that number of accidents is independent of the age of the driver. What possible sources of difficulty in sampling techniques as well as other considerations could affect your conclusions?

Table 12.19

		Age of Driver				
		21 — 30	31 — 40	41 — 50	51 — 60	61 — 70
Number of Accidents	0	748	821	786	720	672
	1	74	60	51	66	50
	2	31	25	22	16	15
	more than 2	9	10	6	5	7

Ans. The hypothesis cannot be rejected at either level.

12.43. (a) Prove that $\chi^2 = \Sigma(o_j^2/e_j) - N$ for all contingency tables, where N is the total frequency of all cells, (b) Usine the result in (a), work Prob. 12.41.

12.44. If N_i and N_j denote respectively the sum of the frequencies in the ith row and jth columns of a contingency table (the *marginal frequencies*), show that the expected frequency for the cell belonging to the ith and jth column is $N_i N_j/N$, where N is the total frequency of all cells.

12.45. Prove formula (9), Page 204. (*Hint*: Use Problems 12.43 and 12.44.)

12.46. Extend the result of formula (9), Page 204, to $2 \times k$ contingency tables where $k > 3$.

12.47. Prove formula (8), Page 203.

12.48. By analogy with the ideas developed for $h \times k$ contingency tables, discuss $h \times k \times l$ contingency tables, pointing out possible applications which they may have.

COEFFICIENT OF CONTINGENCY

12.49. Table 12.20 shows the relationship between hair and eye colour of a sample of 200 students. (a) Find the coefficient of contingency without and with Yates' correction. (b) Compare the result of (a) with the maximum coefficient of contingency.
Ans. (a) 0·3863, 0·3779 (with Yates' correction).

Table 12.20

		Hair Colour	
		Blonde	Not blonde
Eye Colour	Blue	49	25
	Not blue	30	96

12.50. Find the coefficient of contingency for the data of (a) Prob. 12.36 and (b) Prob. 12.38 without and with Yates' correction. *Ans.* (a) 0·2205, 0·1985 (corrected). (b) 0·0872, 0·0738 (corrected).

12.51. Find the coefficient of contingency for the data of Prob. 12.41.   *Ans.* 0·4651

12.52. Prove that the maximum coefficient of contingency for a 3×3 table is $\sqrt{\frac{2}{3}} = 0.8165$ approximately.

12.53. Prove that the maximum coefficient of contingency for a $k \times k$ table is $\sqrt{(k-1)/k}$.

CORRELATION OF ATTRIBUTES

12.54. Find the correlation coefficient for the data in the table of Prob. 12.49.
Ans. (a) 0·4188, 0·4082 with Yates' correction.

12.55. Find the correlation coefficient for the data in the tables of (a) Prob. 12.36 and (b) Prob. 12.38 without and with Yates' correction.   *Ans.* (a) 0·2261, 0·2026 (corrected). (b) 0·0875, 0·0740 (corrected).

12.56. Find the correlation coefficient between the mathematics and physics marks in the table of Prob. 12.41.
Ans. 0·3715

12.57. If C is the coefficient of contingency for a $k \times k$ table and r is the corresponding coefficient of correlation, prove that $r = C/\sqrt{(1 - C^2)(k - 1)}$.

ADDITIVE PROPERTY OF χ^2

12.58. To test a hypothesis H_0, an experiment is performed five times. The resulting values of χ^2, each corresponding to 4 degrees of freedom, are 8·3, 9·1, 8·9, 7·8, 8·6 respectively. Show that while H_0 cannot even be rejected at a 0·05 level on the basis of each experiment separately, it can be rejected at a 0·005 level on the basis of the combined experiments.

CHAPTER 13

Curve Fitting and
The Method of Least Squares

RELATIONSHIP BETWEEN VARIABLES

Very often in practice a relationship is found to exist between two (or more) variables. For example: weights of adult males depend to some degree on their heights; circumferences of circles depend on their radii; and the pressure of a given mass of gas depends on its temperature and volume.

It is frequently desirable to express this relationship in mathematical form by determining an equation connecting the variables.

CURVE FITTING

To aid in determining an equation connecting variables, a first step is the collection of data showing corresponding values of the variables under consideration.

For example, suppose X and Y denote respectively the height and weight of adult males. Then a sample of N individuals would reveal the heights $X_1, X_2, \ldots, X_N$ and the corresponding weights $Y_1, Y_2, \ldots, Y_N$.

A next step is to plot the points $(X_1, Y_1), (X_2, Y_2), \ldots, (X_N, Y_N)$ on a rectangular co-ordinate system The resulting set of points is sometimes called a *scatter diagram*.

From the scatter diagram it is often possible to visualize a smooth curve approximating the data. Such a curve is called an *approximating curve*. In Fig. 13-1, for example, the data appear to be approximated well by a straight line and we say that a *linear relationship* exists between the variables. In Fig. 13-2, however, although a relationship exists between the variables it is not a linear relationship and so we call it a *non-linear relationship*.

The general problem of finding equations of approximating curves which fit given sets of data is called *curve fitting*.

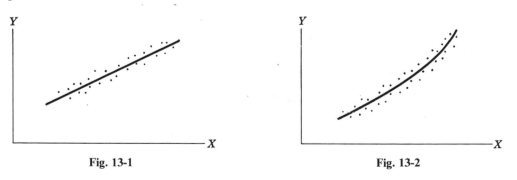

Fig. 13-1   Fig. 13-2

EQUATIONS OF APPROXIMATING CURVES

For purposes of reference we have listed below several common types of approximating curves and their equations. All letters other than X and Y represent constants. The variables X and Y are often referred to as *independent* and *dependent variables* respectively, although these roles can be interchanged.

(*1*) $Y = a_0 + a_1 X$ Straight line

(*2*) $Y = a_0 + a_1 X + a_2 X^2$ Parabola or Quadratic curve

(*3*) $Y = a_0 + a_1 X + a_2 X^2 + a_3 X^3$ Cubic curve

(*4*) $Y = a_0 + a_1 X + a_2 X^2 + a_3 X^3 + a_4 X^4$ Quartic curve

(*5*) $Y = a_0 + a_1 X + a_2 X^2 + \ldots + a_n X^n$ *n*th degree curve

The right sides of the above equations are called *polynomials* of the first, second, third, fourth and *n*th degrees respectively. The functions defined by the first four of these equations are sometimes called *linear*, *quadratic*, *cubic* and *quartic* functions respectively.

As other possible equations (among many) used in practice we mention the following.

(*6*) $Y = \dfrac{1}{a_0 + a_1 X}$ or $\dfrac{1}{Y} = a_0 + a_1 X$ Hyperbola

(*7*) $Y = ab^X$ or $\log Y = \log a + (\log b)X = a_0 + a_1 X$ Exponential curve

(*8*) $Y = aX^b$ or $\log Y = \log a + b \log X$ Geometric curve

(*9*) $Y = ab^X + g$ Modified exponential curve

(*10*) $Y = aX^b + g$ Modified geometric curve

(*11*) $Y = pq^{bX}$ or $\log Y = \log p + b^X \log q = ab^X + g$ Gompertz curve

(*12*) $Y = pq^{bX} + h$ Modified Gompertz curve

(*13*) $Y = \dfrac{1}{ab^X + g}$ or $\dfrac{1}{Y} = ab^X + g$ Logistic curve

(*14*) $Y = a_0 + a_1 (\log X) + a_2 (\log X)^2$

To decide which curve should be used, it is helpful to obtain scatter diagrams of transformed variables. For example, if a scatter diagram of log Y vs. X shows a linear relationship the equation has the form (*7*), while if log Y vs. log X shows a linear relationship the equation has the form (*8*). To facilitate this we frequently employ special graph paper for which one or both scales are calibrated logarithmically. These are referred to as *semi-log* or *log-log graph paper* respectively.

FREEHAND METHOD OF CURVE FITTING

Individual judgment can often be used to draw an approximating curve to fit a set of data. This is called a *freehand method of curve fitting*. If the type of equation of this curve is known, it is possible to obtain the constants in the equation by choosing as many points on the curve as there are constants in the equation. For example, if the curve is a straight line, two points are necessary; if it is a parabola, three points are necessary. The method has the disadvantage that different observers will obtain different curves and equations.

THE STRAIGHT LINE

The simplest type of approximating curve is a straight line, whose equation can be written

$$Y = a_0 + a_1 X \tag{15}$$

Given any two points (X_1, Y_1) and (X_2, Y_2) on the line, the constants a_0 and a_1 can be determined. The resulting equation of the line can be written

$$Y - Y_1 = \left(\frac{Y_2 - Y_1}{X_2 - X_1}\right) (X - X_1) \quad \text{or} \quad Y - Y_1 = m(X - X_1) \qquad (16)$$

where $m = \dfrac{Y_2 - Y_1}{X_2 - X_1}$ is called the *slope* of the line and represents the change in Y divided by the corresponding change in X.

When the equation is written in the form (15), the constant a_1 is the slope m. The constant a_0, which is the value of Y when $X = 0$, is called the Y *intercept*.

THE METHOD OF LEAST SQUARES

To avoid individual judgment in constructing lines, parabolas or other approximating curves to fit sets of data, it is necessary to agree on a definition of a "best fitting line", "best fitting parabola", etc.

To motivate a possible definition, consider Fig. 13-3 in which the data points are given by (X_1, Y_1), (X_2, Y_2), ..., (X_N, Y_N). For a given value of X, say X_1, there will be a difference between the value Y_1 and the corresponding value as determined from the curve C. As indicated in the figure we denote this difference by D_1, which is sometimes referred to as a *deviation*, *error* or *residual* and may be positive, negative or zero. Similarly, corresponding to the values X_2, ..., X_N we obtain the deviations D_2, ..., D_N.

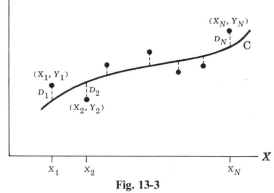

Fig. 13-3

A measure of the "goodness of fit" of the curve C to the given data is provided by the quantity $D_1^2 + D_2^2 + \ldots + D_N^2$. If this is small the fit is good, if it is large the fit is bad. We therefore make the following

Definition. Of all curves approximating a given set of data points, the curve having the property that

$$D_1^2 + D_2^2 + \ldots + D_N^2 \text{ is a minimum}$$

is called a *best fitting curve*.

A curve having this property is said to fit the data in the *least square sense* and is called a *least square curve*. Thus a line having this property is called a *least square line*, a parabola with this property is called a *least square parabola*, etc.

It is customary to employ the above definition when X is the independent variable and Y is the dependent variable. If X is the dependent variable the definition is modified by considering horizontal instead of vertical deviations, which amounts to an interchange of the X and Y axes. These two definitions in general lead to different least square curves. Unless otherwise specified we shall consider Y as the dependent and X as the independent variable.

It is possible to define another least square curve by considering perpendicular distances from each of the data points to the curve instead of either vertical or horizontal distances. However, this is not used too often.

THE LEAST SQUARE LINE

The least square line approximating the set of points $(X_1, Y_1), (X_2, Y_2), \ldots, (X_N, Y_N)$ has the equation

$$Y = a_0 + a_1 X \tag{18}$$

where the constants a_0 and a_1 are determined by solving simultaneously the equations

$$\begin{cases} \Sigma Y = a_0 N + a_1 \Sigma X \\ \Sigma XY = a_0 \Sigma X + a_1 \Sigma X^2 \end{cases} \tag{19}$$

which are called the *normal equations for the least square line* (*18*).

The constants a_0 and a_1 of (*19*) can, if desired, be found from the formulae

$$a_0 = \frac{(\Sigma Y)(\Sigma X^2) - (\Sigma X)(\Sigma XY)}{N \Sigma X^2 - (\Sigma X)^2} \qquad a_1 = \frac{N \Sigma XY - (\Sigma X)(\Sigma Y)}{N \Sigma X^2 - (\Sigma X)^2} \tag{20}$$

The normal equations (*19*) are easily remembered by observing that the first equation can be obtained formally by summing on both sides of (*18*), i.e. $\Sigma Y = \Sigma(a_0 + a_1 X) = a_0 N + a_1 \Sigma X$, while the second equation is obtained formally by first multiplying both sides of (*18*) by X and then summing, i.e. $\Sigma XY = \Sigma Y = (a_0 + a_1 X) = a_0 \Sigma X + a_1 \Sigma X^2$. Note that this is not a derivation of the normal equations but simply a means for remembering them. For a derivation using the calculus, see Appendix VIII, Page 350.

Note also that in (*19*) and (*20*) we have used the short notation $\Sigma X, \Sigma XY$, etc., in place of $\sum\limits_{j=1}^{N} X_j, \sum\limits_{j=1}^{N} X_j Y_j$, etc.

The labour involved in finding a least square line can sometimes be shortened by transforming the data so that $x = X - \bar{X}$ and $y = Y - \bar{Y}$. The equation of the least square line can then be written (see Problem 13.15)

$$y = \left(\frac{\Sigma xy}{\Sigma x^2}\right) x \quad \text{or} \quad y = \left(\frac{\Sigma xY}{\Sigma x^2}\right) x \tag{21}$$

In particular if X is such that $\Sigma X = 0$, i.e. $\bar{X} = 0$, this becomes

$$Y = \bar{Y} + \left(\frac{\Sigma XY}{\Sigma X^2}\right) X \tag{22}$$

From these equations it is at once evident that the least square line passes through the point $(\bar{X}, \bar{Y})$, called the *centroid* or *centre of gravity* of the data.

If the variable X is taken as dependent instead of independent variable, we write (*18*) as $X = b_0 + b_1 Y$. Then the above results hold if X and Y are interchanged and a_0 and a_1 are replaced by b_0 and b_1 respectively. The resulting least square line, however, is in general not the same as that obtained above (see Problems 13.11 and 13.15(*d*)).

NON-LINEAR RELATIONSHIPS

Non-linear relationships can sometimes be reduced to linear relationships by appropriate transformation of variables. See Prob. 13.21.

THE LEAST SQUARE PARABOLA

The least square parabola approximating the set of points $(X_1, Y_1), (X_2, Y_2), \ldots, (X_N, Y_N)$ has the equation

$$Y = a_0 + a_1X + a_2X^2 \tag{23}$$

where the constants a_0, a_1 and a_2 are determined by solving simultaneously the equations

$$\begin{cases} \Sigma Y = a_0N + a_1 \Sigma X + a_2 \Sigma X^2 \\ \Sigma XY = a_0 \Sigma X + a_1 \Sigma X^2 + a_2 \Sigma X^3 \\ \Sigma X^2Y = a_0 \Sigma X^2 + a_1 \Sigma X^3 + a_2 \Sigma X^4 \end{cases} \tag{24}$$

called the *normal equations for the least square parabola* (23).

The equations (24) are easily remembered by observing that they can be obtained formally by multiplying equation (23) by 1, X, and X^2 respectively and summing on both sides of the resulting equations. This technique can be extended to obtain normal equations for least square cubic curves, least square quartic curves and in general any of the least square curves corresponding to equation (5).

As in the case of the least square line, simplifications of (24) occur if X is chosen so that $\Sigma X = 0$. Simplification also occurs by choosing the new variables $x = X - \bar{X}$, $y = Y - \bar{Y}$.

REGRESSION

Often, on the basis of sample data, we wish to estimate the value of a variable Y corresponding to a given value of a variable X. This can be accomplished by estimating the value of Y from a least square curve which fits the sample data. The resulting curve is called a *regression curve of Y on X*, since Y is estimated from X.

If we desired to estimate the value of X from a given value of Y we would use a *regression curve of X on Y*, which amounts to interchanging the variables in the scatter diagram so that X is the dependent variable and Y is the independent variable. This is equivalent to replacing vertical deviations in the definition of least square curve on Page 219 by horizontal deviations.

In general the regression line or curve of Y on X is not the same as the regression line or curve of X on Y.

APPLICATIONS TO TIME SERIES

If the independent variable X is time, the data show the values of Y at various times. Data arranged according to time are called *time series*. The regression line or curve of Y on X in this case is often called a *trend line* or *trend curve* and is often used for purposes of *estimation*, *prediction* or *forecasting*.

PROBLEMS INVOLVING MORE THAN TWO VARIABLES

Problems involving more than two variables can be treated in a manner analogous to that for two variables. For example, there may be a relationship between the three variables X, Y and Z which can be described by the equation

$$Z = a_0 + a_1X + a_2Y \tag{25}$$

which is called a *linear equation in the variables X, Y and Z*.

In a three dimensional rectangular co-ordinate system this equation represents a plane and the actual sample points $(X_1, Y_1, Z_1), (X_2, Y_2, Z_2), \ldots, (X_N, Y_N, Z_N)$ may "scatter" not too far from this plane which we can call an *approximating plane*.

By extension of the method of least squares, we can speak of a *least square plane* approximating the data. If we are estimating Z from given values of X and Y, this would be called a *regression plane of Z on X and Y*. The normal equations corresponding to the least square plane (25) are given by

$$\begin{cases} \Sigma Z = a_0 N + a_1 \Sigma X + a_2 \Sigma Y \\ \Sigma XZ = a_0 \Sigma X + a_1 \Sigma X^2 + a_2 \Sigma XY \\ \Sigma YZ = a_0 \Sigma Y + a_1 \Sigma XY + a_2 \Sigma Y^2 \end{cases} \qquad (26)$$

and can be remembered as obtained from (25) by multiplying by 1, X and Y successively and then summing.

More complicated equations than (25) can also be considered. These represent *regression surfaces*. If the number of variables exceeds three, geometric intuition is lost since we then require four, five, . . . dimensional spaces.

Problems involving estimation of a variable from two or more variables are called problems of *multiple regression* and will be considered in more detail in Chap. 15.

Solved Problems

STRAIGHT LINES

13.1. (a) Construct a straight line which approximates the data of Table 13.1. (b) Find an equation for this line.

Table 13.1

X	2	3	5	7	9	10
Y	1	3	7	11	15	17

Solution:

(a) Plot the points (2,1), (3,3), (5, 7), (7, 11), (9, 15) and (10, 17) on a rectangular co-ordinate system as shown in Fig. 13-4.

It is clear from this figure that all the points lie on a straight line (shown dashed). Thus a straight line fits the data *exactly*.

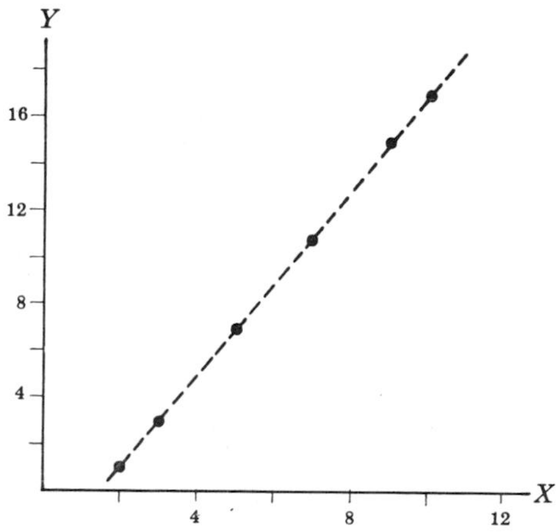

(b) To determine the equation of the line given by

$$(1) \qquad Y = a_0 + a_1 X$$

only two points are necessary. Choose the points (2, 1) and (3, 3), for example.

For the point (2, 1), $X = 2$ and $Y = 1$. Substituting these values into (1) yields

$$(2) \qquad 1 = a_0 + 2a_1$$

Similarly for the point (3, 3), $X = 3$ and $Y = 3$; and substitution into (1) yields

$$(3) \qquad 3 = a_0 + 3a_1$$

Fig. 13-4

Solving (2) and (3) simultaneously, $a_0 = -3$ and $a_1 = 2$; and the required equation is

$$Y = -3 + 2X \quad \text{or} \quad Y = 2X - 3$$

As a check, we can show that the points (5, 7), (7, 11), (9, 15) and (10, 17) also lie on the line.

13.2. In Problem 13.1, find (a) Y when $X = 4$, (b) Y when $X = 15$, (c) Y when $X = 0$, (d) X when $Y = 7{\cdot}5$, (e) X when $Y = 0$, (f) the increase in Y corresponding to a unit increase in X.

Solution:

We assume that the same law of relationship $Y = 2X - 3$ holds for values of X and Y other than those specified in Table 13.1 of Problem 13.1.

(a) If $X = 4$, $Y = 2(4) - 3 = 8 - 3 = 5$. Since we are finding the value of Y corresponding to a value of X included *between* two given values of X, this process is often called *linear interpolation*.

(b) If $X = 15$, $Y = 2(15) - 3 = 30 - 3 = 27$. Since we are finding the value of Y corresponding to a value of X *outside* or *exterior* to the given values of X, this process is often called *linear extrapolation*.

(c) If $X = 0$, $Y = 2(0) - 3 = 0 - 3 = -3$. The value of Y when $X = 0$ is called the Y *intercept*. It is the value of Y at the point where the line (extended if necessary) intersects the Y axis.

(d) If $Y = 7{\cdot}5$, $7{\cdot}5 = 2X - 3$; then $2X = 7{\cdot}5 + 3 = 10{\cdot}5$ and $X = 10{\cdot}5/2 = 5{\cdot}25$.

(e) If $Y = 0$, $0 = 2X - 3$; then $2X = 3$ and $X = 1{\cdot}5$. The value of X when $Y = 0$ is called the X *intercept*. It is the value of X at the point where the line (extended if necessary) intersects the X axis.

(f) If X increases 1 unit from 2 to 3, Y increases from 1 to 3, a change of 2 units.

If X increases from 2 to 10, i.e. $(10 - 2) = 8$ units, Y increases from 1 to 17 or $(17 - 1) = 16$ units. Then Y increases 16 units corresponding to an 8 unit increase in X, or 2 units for a unit increase in X.

In general if ΔY denotes the change in Y due to a change in X of ΔX, then the change in Y per unit change in X is given by $\Delta Y/\Delta X = 2$. This is called the *slope* of the line and is always equal to a_1 in the equation $Y = a_0 + a_1 X$. The constant a_0 is the Y *intercept* of the line (see part (c)).

The above questions can also be answered by direct reference to the graph, Fig. 13-4.

13.3. (a) Show that the equation of a straight line which passes through the points (X_1, Y_1) and (X_2, Y_2) is given by

$$Y - Y_1 = \frac{Y_2 - Y_1}{X_2 - X_1}(X - X_1)$$

(b) Find the equation of a straight line which passes through the points $(2, -3)$ and $(4, 5)$.

Solution:

(a) The equation of a straight line is (1) $Y = a_0 + a_1 X$.
Since (X_1, Y_1) lies on the line,   (2) $Y_1 = a_0 + a_1 X_1$.
Since (X_2, Y_2) lies on the line,   (3) $Y_2 = a_0 + a_1 X_2$.
Subtracting equation (2) from (1), (4) $Y - Y_1 = a_1(X - X_1)$.

Subtracting equation (2) from (3), $Y_2 - Y_1 = a_1(X_2 - X_1)$ or $a_1 = \dfrac{Y_2 - Y_1}{X_2 - X_1}$.

Substituting this value of a_1 into (4), we obtain $Y - Y_1 = \dfrac{Y_2 - Y_1}{X_2 - X_1}(X - X_1)$ as required.

The quantity $\dfrac{Y_2 - Y_1}{X_2 - X_1}$, often abbreviated by the letter m, represents the change in Y divided by the corresponding change in X and is the *slope* of the line. The required equation can be written $Y - Y_1 = m(X - X_1)$.

(b) **Method 1,** using the result of part (a).
Corresponding to the first point $(2, -3)$ we have $X_1 = 2$, $Y_1 = -3$.
Corresponding to the second point $(4, 5)$ we have $X_2 = 4$, $Y_2 = 5$.

Then the slope $m = \dfrac{Y_2 - Y_1}{X_2 - X_1} = \dfrac{5 - (-3)}{4 - 2} = \dfrac{8}{2} = 4$, and the required equation is

$$Y - Y_1 = m(X - X_1) \text{ or } Y - (-3) = 4(X - 2)$$

which can be written $Y + 3 = 4(X - 2)$ or $Y = 4X - 11$.

Method 2, using the method of Problem 13.1(b).

The equation of a straight line is $\quad Y = a_0 + a_1 X$.

Since the point $(2, -3)$ is on the line, $\quad$ (1) $\quad -3 = a_0 + 2a_1$.

Since the point $(4, 5)$ is on the line, $\quad$ (2) $\quad 5 = a_0 + 4a_1$.

Solving (1) and (2) simultaneously, we obtain $a_1 = 4$, $a_0 = -11$.

Hence the required equation is $Y = -11 + 4X$ or $Y = 4X - 11$.

13.4. Give a graphical interpretation of the derivation in Problem 13.3(a).

Solution:

In Fig. 13-5 we have indicated the line passing through points P and Q which have co-ordinates (X_1, Y_1) and (X_2, Y_2) respectively. The point R which co-ordinates (X, Y) represents any other point on this line.

From the similar triangles PRT and PQS,

(1) $\qquad \dfrac{RT}{TP} = \dfrac{QS}{SP} \quad$ or $\quad \dfrac{Y - Y_1}{X - X_1} = \dfrac{Y_2 - Y_1}{X_2 - X_1}$

Then multiplying both sides by $X - X_1$,

$$Y - Y_1 = \frac{Y_2 - Y_1}{X_2 - X_1}(X - X_1)$$

which is the required equation of the line.

Note that each of the ratios in (1) is the slope m; this can be written

$$Y - Y_1 = m(X - X_1)$$

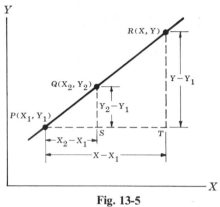

Fig. 13-5

13.5. Find (a) the slope, (b) the equation, (c) the Y intercept and (d) the X intercept of the line which passes through the points $(1, 5)$ and $(4, -1)$.

Solution:

(a) $(X_1 = 1, Y_1 = 5)$ and $(X_2 = 4, Y_2 = -1)$. Then

$$m = \text{slope} = \frac{Y_2 - Y_1}{X_2 - X_1} = \frac{-1 - 5}{4 - 1} = \frac{-6}{3} = -2$$

The negative sign of the slope indicates that as X increases, Y decreases, as shown in Fig. 13-6.

(b) The equation of the line is

$$Y - Y_1 = m(X - X_1) \quad \text{or} \quad Y - 5 = -2(X - 1)$$

i.e. $\qquad\qquad Y - 5 = -2X + 2 \quad$ or $\quad Y = 7 - 2X$

This can also be obtained by the second method of Prob. 13.3(b).

(c) The Y intercept, which is the value of Y when $X = 0$, is given by $Y = 7 - 2(0) = 7$. This can also be seen directly from the graph.

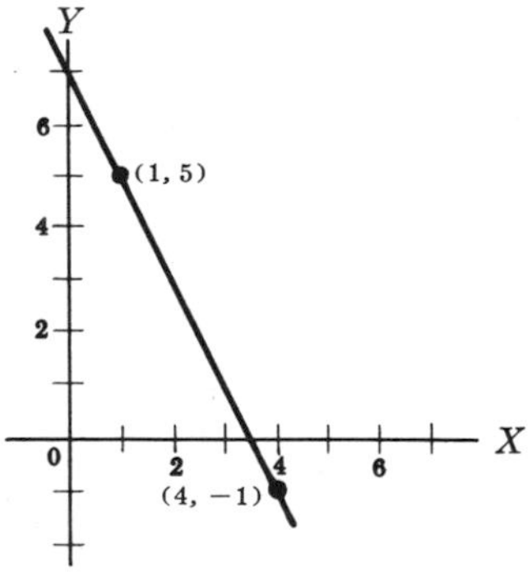

Fig. 13-6

(d) The X intercept is the value of X when $Y = 0$. Substituting $Y = 0$ in the equation $Y = 7 - 2X$, we have $0 = 7 - 2X$ or $2X = 7$, $X = 3.5$. This can also be seen directly from the graph.

13.6. Find the equation of a line passing through the point $(4, 2)$ which is parallel to the line $2X + 3Y = 6$.

Solution:

If two lines are parallel, their slopes are equal.

From $2X + 3Y = 6$ we have $3Y = 6 - 2X$ or $Y = 2 - \frac{2}{3}X$, so that the slope of the line is $m = -\frac{2}{3}$. Then the equation of the required line is

$$Y - Y_1 = m(X - X_1) \quad \text{or} \quad Y - 2 = -\tfrac{2}{3}(X - 4)$$

which can also be written $2X + 3Y = 14$.

Another method:

Any line parallel to $2X + 3Y = 6$ has the equation $2X \times 3Y = c$.

To find c, let $X = 4$, $Y = 2$. Then $2(4) + 3(2) = c$ or $c = 14$, and the required equation is $2X + 3Y = 14$.

13.7. Find the equation of a line whose slope is -4 and whose Y intercept is 16.

Solution:

In the equation $Y = a_0 + a_1 X$, $a_0 = 16$ is the Y intercept and $a_1 = -4$ is the slope. Hence $Y = 16 - 4X$ is the required equation.

13.8. (a) Construct a straight line which approximates the data of Table 13.2. (b) Find an equation for this line.

Table 13.2

X	1	3	4	6	8	9	11	14
Y	1	2	4	4	5	7	8	9

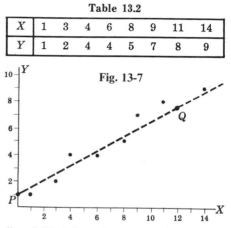

Fig. 13-7

Solution:

(a) Plot the points (1, 1), (3, 2), (4, 4), (6, 4), (8, 5), (9, 7), (11, 8) and (14, 9) on a rectangular co-ordinate system as shown in Fig. 13-7.

A straight line approximating the data is drawn *freehand* in the figure. For a method eliminating the need for individual judgment, see Prob. 13.11 which uses the method of least squares.

(b) To obtain the equation of the line constructed in (a), choose any two points on the line, such as P and Q, for example. The co-ordinates of these points as read from the graph are approximately (0, 1) and (12, 7·5).

The equation of the line is $Y = a_0 + a_1 X$. Using the points (0, 1) and (12, 7·5), we have respectively

$$(1)\ 1 = a_0 + a_1(0)\quad \text{and}\quad (2)\ 7\cdot5 = a_0 + 12a_1$$

From (1), $a_0 = 1$; then from (2), $a_1 = 6\cdot5/12 = 0\cdot542$. Thus the required equation is $Y = 1 + 0\cdot542X$.

Another method:

$$Y - Y_1 = \frac{Y_2 - Y_1}{X_2 - X_1}(X - X_1),\quad Y - 1 = \frac{7\cdot5 - 1}{12 - 0}(X - 0),\quad Y - 1 = 0\cdot542X\quad \text{or}\quad Y = 1 + 0\cdot542X.$$

13.9. (a) Compare the values of Y obtained from the approximating line with those given in the Table 13.2 of Problem 13.8. (b) What would be your estimated value of Y when $X = 10$?

Solution:

(a) For $X = 1$, $Y = 1 + 0\cdot542(1) = 1\cdot542$ or 1·5. For $X = 3$, $Y = 1 + 0\cdot542(3) = 2\cdot626$ or 2·6.

Similarly, the values of Y corresponding to other values of X can be obtained. We denote these values of Y as estimated from the equation $Y = 1 + 0\cdot542X$ by Y_{est}. These values are indicated in Table 13.3 together with the actual data in Table 13.2.

Table 13.3

X	1	3	4	6	8	9	11	14
Y	1	2	4	4	5	7	8	9
Y_{est}	1·5	2·6	3·2	4·3	5·3	5·9	7·0	8·6

(b) The estimated value of Y when $X = 10$ is $Y = 1 + 0\cdot542(10) = 6\cdot42$ or 6·4.

13.10. Table 13.4 shows the power to the nearest kilowatt and top speeds to the nearest km/h of a sample of 12 sports cars drawn at random from the stockists. (a) Obtain a scatter diagram of the data. (b) Construct a line which approximates the data. (c) Find the equation of the line constructed in (b). (d) Estimate the top speed of a car whose power is known to be 63 kW. (e) Estimate the power of a car whose top speed is known to be 168 km/h.

Table 13.4

Power X (kW)	70	63	72	60	66	70	74	65	62	67	65	68
Top Speed Y (km/h)	155	150	180	135	156	168	178	160	132	145	139	152

Solution:

(a) The scatter diagram, shown in Fig. 13-8, is obtained by plotting the points (70, 155), (63, 150), . . ., (68, 152).

(b) A straight line which approximates the data is shown dashed in the figure. This is but one of many possible lines which could have been constructed.

(c) Choose any two points on the line constructed in (b), such as P and Q, for example. The co-ordinates of these points as read from the graph are approximately (60, 130) and (72, 170). Then

$$Y - Y_1 = \frac{Y_2 - Y_1}{X_2 - X_1}(X - X_1)$$

$$Y - 130 = \frac{170 - 130}{72 - 60}(X - 60)$$

$$Y = \frac{10}{3}X - 70$$

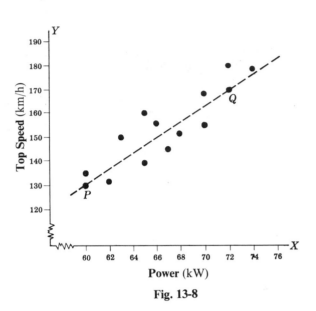

Fig. 13-8

(d) If $X = 63$, then $Y = \frac{10}{3}(63) - 70 = 140$ km/h.

(e) If $Y = 168$, then $168 = \frac{10}{3}X - 70$, $\frac{10}{3}X = 238$, and $X = 71\cdot4$ or 71 kW.

THE LEAST SQUARE LINE

13.11. Fit a least square line to the data of Problem 13.8 using (a) X as independent variable, (b) X as dependent variable.

Solution:

(a) The equation of the line is $Y = a_0 + a_1 X$. The normal equations are

$$\begin{cases} \Sigma Y = a_0 N + a_1 \Sigma X \\ \Sigma XY = a_0 \Sigma X + a_1 \Sigma X^2 \end{cases}$$

The work involved in computing the sums can be arranged as in Table 13.5. Although the last column is not needed for this part of the problem, it has been added to the table for use in part (b).

Table 13.5

X	Y	X^2	XY	Y^2
1	1	1	1	1
3	2	9	6	4
4	4	16	16	16
6	4	36	24	16
8	5	64	40	25
9	7	81	63	49
11	8	121	88	64
14	9	196	126	81
$\Sigma X = 56$	$\Sigma Y = 40$	$\Sigma X^2 = 524$	$\Sigma XY = 364$	$\Sigma Y^2 = 256$

Since there are 8 pairs of values of X and Y, $N = 8$ and the normal equations become

$$\begin{cases} 8a_0 + 56a_1 = 40 \\ 56a_0 + 524a_1 = 364 \end{cases}$$

Solving simultaneously, $a_0 = \frac{6}{11}$ or $0\cdot545$, $a_1 = \frac{7}{11}$ or $0\cdot636$; and the required least square line is $Y = \frac{6}{11} + \frac{7}{11}X$ or $Y = 0\cdot545 + 0\cdot636X$.

Another method:

$$a_0 = \frac{(\Sigma Y)(\Sigma X^2) - (\Sigma X)(\Sigma XY)}{N \Sigma X^2 - (\Sigma X)^2} = \frac{(40)(524) - (56)(364)}{(8)(524) - (56)^2} = \frac{6}{11} \text{ or } 0\cdot545$$

$$a_1 = \frac{N \Sigma XY - (\Sigma X)(\Sigma Y)}{N \Sigma X^2 - (\Sigma X)^2} = \frac{(8)(364) - (56)(40)}{(8)(524) - (56)^2} = \frac{7}{11} \text{ or } 0\cdot636$$

Then $Y = a_0 + a_1X$ or $Y = 0\cdot545 + 0\cdot636X$, as before.

(b) If X is considered as the dependent variable and Y as the independent variable, the equation of the least square line is $X = b_0 + b_1Y$ and the normal equations are

$$\begin{cases} \Sigma X = b_0 N + b_1 \Sigma Y \\ \Sigma XY = b_0 \Sigma Y + b_1 \Sigma Y^2 \end{cases}$$

Then using Table 13.5, the normal equations become

$$\begin{cases} 8b_0 + 40b_1 = 56 \\ 40b_0 + 256b_1 = 364 \end{cases}$$

from which $b_0 = -\frac{1}{2}$ or $-0\cdot50$, $b_1 = \frac{3}{2}$ or $1\cdot50$.

These values can also be obtained from

$$b_0 = \frac{(\Sigma X)(\Sigma Y^2) - (\Sigma Y)(\Sigma XY)}{N \Sigma Y^2 - (\Sigma Y)^2} = \frac{(56)(256) - (40)(364)}{(8)(256) - (40)^2} = 0\cdot50$$

$$b_1 = \frac{N \Sigma XY - (\Sigma X)(\Sigma Y)}{N \Sigma Y^2 - (\Sigma Y)^2} = \frac{(8)(364) - (56)(40)}{(8)(256) - (40)^2} = 1\cdot50$$

Thus the required equation of the least square line is $X = b_0 + b_1Y$ or $X = -0\cdot50 + 1\cdot50Y$.

Note that by solving this equation for Y we obtain $Y = \frac{1}{3} + \frac{2}{3}X$ or $Y = 0\cdot333 + 0\cdot667X$, which is not the same as the line obtained in part (a).

13.12. Graph the two lines obtained in the preceding problem.

Solution:

The graphs of the two lines, $Y = 0\cdot545 + 0\cdot636X$ and $X = -0\cdot500 + 1\cdot50Y$, are shown in Fig. 13-9. Note that the two lines in this case are practically coincident, which is an indication that the data are very well described by a linear relationship.

The line obtained in part (a) is often called the *regression line of Y on X* and is used for estimating Y for given values of X. The line obtained in part (b) is called the *regression line on X on Y* and is used for estimating X for given values of Y.

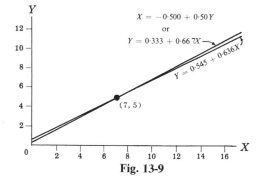

Fig. 13-9

13.13. (a) Show that the two least square lines obtained in Problem 13.11 intersect at point $(\bar{X}, \bar{Y})$.
(b) Estimate the value of Y when $X = 12$. (c) Estimate the value of X when $Y = 3$.

Solution:

$\bar{X} = \dfrac{\Sigma X}{N} = \dfrac{56}{8} = 7$, $\bar{Y} = \dfrac{\Sigma Y}{N} = \dfrac{40}{8} = 5$. Then point $(\bar{X}, \bar{Y})$, called the *centroid*, is $(7, 5)$.

(a) Point $(7, 5)$ lies on line $Y = 0\cdot545 + 0\cdot636X$ or, more exactly, $Y = \frac{6}{11} + \frac{7}{11}X$, since $5 = \frac{6}{11} + \frac{7}{11}(7)$.
Point $(7, 5)$ lies on line $X = \frac{1}{2} + \frac{3}{2}Y$, since $7 = -\frac{1}{2} + \frac{3}{2}(5)$.

Another method:
    The equations of the two lines are $Y = \frac{6}{11} + \frac{7}{11}X$ and $X = -\frac{1}{2} + \frac{3}{2}Y$.
    Solving simultaneously, we find $X = 7$, $Y = 5$. Thus the lines intersect in point $(7, 5)$.

(b) Putting $X = 12$ into the regression line of Y (Problem 13.11), $Y = 0{\cdot}545 + 0{\cdot}636(12) = 8{\cdot}2$.

(c) Putting $Y = 3$ into the regression line of X (Problem 13.11), $X = -0{\cdot}50 + 1{\cdot}50(3) = 4{\cdot}0$.

13.14. Prove that a least square line always passes through the point $(\bar{X}, \bar{Y})$.

Solution:
    Case 1: X is the independent variable.
        The equation of the least square is   (1) $Y = a_0 + a_1 X$
        A normal equation for the least square line is   (2) $\Sigma Y = a_0 N + a_1 \Sigma X$
        Dividing both sides of (2) by N gives   (3) $\bar{Y} = a_0 + a_1 \bar{X}$
        Subtracting (3) from (1), the least square line can be written

$$(4)\ \ Y - \bar{Y} = a_1(X - \bar{X})$$

    which shows that the line passes through the point $(\bar{X}, \bar{Y})$.

    Case 2: Y is the independent variable.
        Proceeding as in **Case 1** with X and Y interchanged and the constants a_0, a_1 replaced by b_0, b_1 respectively, we find that the least square line can be written

$$(5)\ X - \bar{X} = b_1(Y - \bar{Y})$$

    which indicates that the line passes through the point $(\bar{X}, \bar{Y})$.
        Note that lines (4) and (5) are not coincident, but they intersect in $(\bar{X}, \bar{Y})$.

13.15. (a) Considering X as the independent variable, show that the equation of the least square line can be written

$$y = \left(\frac{\Sigma xy}{\Sigma x^2}\right)x \quad \text{or} \quad y = \left(\frac{\Sigma xY}{\Sigma x^2}\right)x$$

where $x = X - \bar{X}$ and $y = Y - \bar{Y}$.

(b) If $\bar{X} = 0$, show that the least square line in (a) can be written

$$Y = \bar{Y} + \left(\frac{\Sigma XY}{\Sigma X^2}\right)X$$

(c) Write the equation of the least square line corresponding to that in part (a) if Y is the independent variable.

(d) Verify that the lines in (a) and (c) are not necessarily the same.

Solution:
(a) Equation (4) of Problem 13.14 can be written $y = a_1 x$, where $x = X - \bar{X}$ and $y = Y - \bar{Y}$. Also, from the simultaneous solution of the normal equations (see Page 220), we have

$$a_1 = \frac{N \Sigma XY - (\Sigma X)(\Sigma Y)}{N \Sigma X^2 - (\Sigma X)^2} = \frac{N \Sigma (x + \bar{X})(y + \bar{Y}) - \{\Sigma (x + \bar{X})\}\{\Sigma (y + \bar{Y})\}}{N \Sigma (x + \bar{X})^2 - \{\Sigma (x + \bar{X})\}^2}$$

$$= \frac{N \Sigma (xy + x\bar{Y} + \bar{X}y + \bar{X}\bar{Y}) - \{\Sigma x + N\bar{X}\}\{\Sigma y + N\bar{Y}\}}{N \Sigma (x^2 + 2x\bar{X} + \bar{X}^2) - \{\Sigma x + N\bar{X}\}^2}$$

$$= \frac{N \Sigma xy + N\bar{Y} \Sigma x + N\bar{X} \Sigma y + N^2 \bar{X}\bar{Y} - \{\Sigma x + N\bar{X}\}\{\Sigma y + N\bar{Y}\}}{N \Sigma x^2 + 2N\bar{X} \Sigma x + N^2 \bar{X}^2 - \{\Sigma x + N\bar{X}\}^2}$$

But $\Sigma x = \Sigma(X - \bar{X}) = 0$ and $\Sigma y = \Sigma(Y - \bar{Y}) = 0$; hence the above simplifies to

$$a_1 = \frac{N\,\Sigma\,xy + N^2\bar{X}\bar{Y} - N^2\bar{X}\bar{Y}}{N\,\Sigma\,x^2 + N^2\bar{X}^2 - N^2\bar{X}^2} = \frac{\Sigma xy}{\Sigma x^2}$$

This can also be written

$$a_1 = \frac{\Sigma xy}{\Sigma x^2} = \frac{\Sigma x(Y - \bar{Y})}{\Sigma x^2} = \frac{\Sigma xY - \bar{Y}\Sigma x}{\Sigma x^2} = \frac{\Sigma xY}{\Sigma x^2}.$$

Then the least square line is $y = a_1 x$, that is $y = \left(\dfrac{\Sigma xy}{\Sigma x^2}\right)x$ or $y = \left(\dfrac{\Sigma xY}{\Sigma x^2}\right)x.$

(b) If $\bar{X} = 0$, $x = X - \bar{X} = X$. Then from $y = \left(\dfrac{\Sigma xY}{\Sigma x^2}\right)$, $y = \left(\dfrac{\Sigma XY}{\Sigma X^2}\right)X$ or $Y = \bar{Y} + \left(\dfrac{\Sigma XY}{\Sigma X^2}\right)X.$

Another method:
The normal equations of the least square line $Y = a_0 + a_1 X$ are

$$\Sigma Y = a_0 N + a_1 \Sigma X \quad \text{and} \quad \Sigma XY = a_0 \Sigma X + a_1 \Sigma X^2$$

If $\bar{X} = (\Sigma X)/N = 0$, then $\Sigma X = 0$ and the normal equations become

$$\Sigma Y = a_0 N \quad \text{and} \quad \Sigma XY = a_1 \Sigma X^2$$

from which $a_0 = \dfrac{\Sigma Y}{N} = \bar{Y}$ and $a_1 = \dfrac{\Sigma XY}{\Sigma X^2}.$

Then the required equation of the least square line is $Y = a_0 + a_1 X$ or $Y = \bar{Y} + \left(\dfrac{\Sigma XY}{\Sigma X^2}\right)X.$

(c) By interchanging X and Y or x and y, we can show as in part (a) that $x = \left(\dfrac{\Sigma xy}{\Sigma y^2}\right)y.$

(d) From (a), the least square line is   (1) $y = \left(\dfrac{\Sigma xy}{\Sigma x^2}\right)x.$

From (c), the least square line is $x = \left(\dfrac{\Sigma xy}{\Sigma y^2}\right)y$ or   (2) $y = \left(\dfrac{\Sigma y^2}{\Sigma xy}\right)x.$

Since $\dfrac{\Sigma xy}{\Sigma x^2} \neq \dfrac{\Sigma y^2}{\Sigma xy}$, in general, the least square lines (1) and (2) are different in general. Note, however, that they intersect at $x = 0$, $y = 0$, i.e. at the point $(\bar{X}, \bar{Y})$.

13.16. If $X' = X + A$ and $Y' = Y + B$, where A and B are any constants, prove that

$$a_1 = \frac{N\,\Sigma\,XY - (\Sigma X)(\Sigma Y)}{N\,\Sigma\,X^2 - (\Sigma X)^2} = \frac{N\,\Sigma\,X'Y' - (\Sigma X')(\Sigma Y')}{N\,\Sigma\,X'^2 - (\Sigma X')^2} = a_1'$$

Solution:

$$x' = X' - \bar{X}' = (X + A) - (\bar{X} + A) = X - \bar{X} = x$$
$$y = Y' - \bar{Y}' = (X + B) - (\bar{Y} + B) = Y - \bar{Y} = y$$

Then $\dfrac{\Sigma xy}{\Sigma x^2} = \dfrac{\Sigma x'y'}{\Sigma x'^2}$ and the result follows from Prob. 13.15. A similar result holds for b_1.

This result is useful, since it enables us to simplify calculations in obtaining the regression line by subtracting suitable constants from the variables X and Y (see second method of Problem 13.17).
Note: The result does not hold if $X' = c_1 X + A$, $Y' = c_2 Y + B$ unless $c_1 = c_2$.

13.17. Fit a least square line to the data of Problem 13.10 using (a) X as the independent variable, (b) X as the dependent variable.

Solution:

(a) From Problem 13.15(a) the required line is $y = \left(\dfrac{\Sigma xy}{\Sigma x^2}\right)x$, where $x = X - \bar{X}$ and $y = Y - \bar{Y}.$

The work involved in computing the sums can be arranged as in Table 13.6. From the first two columns we find $\bar{X} = 802/12 = 66\cdot8$ and $\bar{Y} = 1850/12 = 154\cdot2$. The last column has been added for use in part (b).

Table 13.6

Power X	Top Speed Y	$x = X - \bar{X}$	$y = Y - \bar{Y}$	xy	x^2	y^2
70	155	3·2	0·8	2·56	10·24	0·64
63	150	−3·8	−4·2	15·96	14·44	17·64
72	180	5·2	25·8	134·16	27·04	665·64
60	135	−6·8	−19·2	130·56	46·24	368·64
66	156	−0·8	1·8	−1·44	0·64	3·24
70	168	3·2	13·8	44·16	10·24	190·44
74	178	7·2	23·8	171·36	51·84	566·44
65	160	−1·8	5·8	−10·44	3·24	33·64
62	132	−4·8	−22·2	106·56	23·04	492·84
67	145	0·2	−9·2	−1·84	0·04	84·64
65	139	−1·8	−15·2	27·36	3·24	231·04
68	152	1·2	−2·2	−2·64	1·44	4·84
$\Sigma X = 802$ $\bar{X} = 66·8$	$\Sigma Y = 1850$ $\bar{Y} = 154·2$			Σxy $= 616·32$	Σx^2 $= 191·68$	Σy^2 $= 2659·68$

The required least square line is

$$y = \left(\frac{\Sigma xy}{\Sigma x^2}\right) x = \frac{616·32}{191·68} x = 3·22x$$

or $Y - 154·2 = 3·22(X - 66·8)$, which can be written $Y = 3·22X - 60·9$. This equation is called the *regression line of Y on X* and is used for estimating Y from given values of X.

(b) If X is the dependent variable, the required line is

$$x = \left(\frac{\Sigma xy}{\Sigma y^2}\right) y = \frac{616·32}{2659·68} y = 0·232y$$

which can be written $X - 66·8 = 0·232(Y - 154·2)$ or $X = 31·0 + 0·232Y$. This equation is called the *regression line of X on Y* and is used for estimating X from given values of Y.

Note that the method of Problem 13.11 can also be used if desired.

Another method:

Using the result of Problem 13.16, we may subtract suitable constants from X and Y. We choose to subtract 65 from X and 150 from Y. Then the results can be arranged as in Table 13.7.

Table 13.7

X'	Y'	X'^2	$X'Y'$	Y'^2
5	5	25	25	25
−2	0	4	0	0
7	30	49	210	900
−5	−15	25	75	225
1	6	1	6	36
5	18	25	90	324
9	28	81	252	784
0	10	0	0	100
−3	−18	9	54	324
2	−5	4	−10	25
0	−11	0	0	121
3	2	9	6	4
$\Sigma X' = 22$	$\Sigma Y' = 50$	$\Sigma X'^2 = 232$	$\Sigma X'Y' = 708$	$\Sigma Y'^2 = 2868$

$$a_1 = \frac{N \Sigma X'Y' - (\Sigma X')(\Sigma Y')}{N \Sigma X'^2 - (\Sigma X')^2} = \frac{(12)(708) - (22)(50)}{(12)(232) - (22)^2} = 3·22$$

$$b_1 = \frac{N \Sigma X'Y' - (\Sigma Y')(\Sigma X')}{N \Sigma Y'^2 - (\Sigma Y')^2} = \frac{(12)(708) - (50)(22)}{(12)(2868) - (50)^2} = 0·232$$

Since $\bar{X} = 65 + 22/12 = 66·8$ and $\bar{Y} = 150 + 50/12 = 154·2$, the regression equations are
$$Y - 154·2 = 3·22 (X - 66·8) \quad \text{and} \quad X - 66·8 = 0·0232(Y - 154·2),$$
i.e. $Y = 3·22X - 60·9$ and $X = 0·232Y + 31·0$ in agreement with the first method.

13.18. (*a*) On the same set of axes draw the graphs of the two lines in Problem 13.17. (*b*) Estimate the top speed of a car whose power is known to be 63 kW. (*c*) Estimate the power of a car whose top speed is known to be 168 km/h.

Solution:

(*a*) The two lines are shown in Fig. 13-10 together with the original data points. Note that they intersect at $(\bar{X}, \bar{Y})$ or (66·8, 154·2).

(*b*) To estimate Y from X use the regression line of Y on X, given from Problem 13.17 by $Y = 3·22X - 60·9$. Then if $X = 63$, $Y = 3·22(63) - 60·9 = 142$ km/h.

(*c*) To estimate X from Y use the regression line of X on Y, given from Problem 13.17 by $X = 31·0 + 0·232Y$. Then if $Y = 168($ $X = 31·0 + 0·232(168) = 70·0$ kW.

The results in (*b*) and (*c*) should be compared with those in Problem 13.10(*d*) and 13.10(*e*).

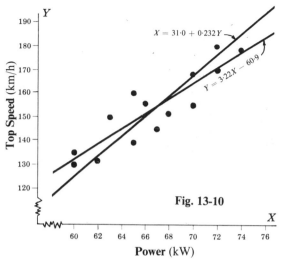

Fig. 13-10

APPLICATIONS TO TIME SERIES

13.19. The production of steel in millions of kilonewtons for a certain country during the years 1946–1956 is given in Table 13.8.

(*a*) Graph the data.

(*b*) Find the equation of a least square line fitting the data.

(*c*) Estimate the production of steel during the years 1957 and 1958 and compare with the true values of 112·7 and 85·3 millions of kilonewtons respectively.

(*d*) Estimate the production of steel during the years 1945 and 1944 and compare with the actual values of 79·7 and 89·6 millions of kilonewtons respectively.

Table 13.8

Year	Production of steel (millions of kilonewtons)
1946	66·6
1947	84·9
1948	88·6
1949	78·0
1950	96·8
1951	105·2
1952	93·2
1953	111·6
1954	88·3
1955	117·0
1956	115·2

Solution:

(*a*)

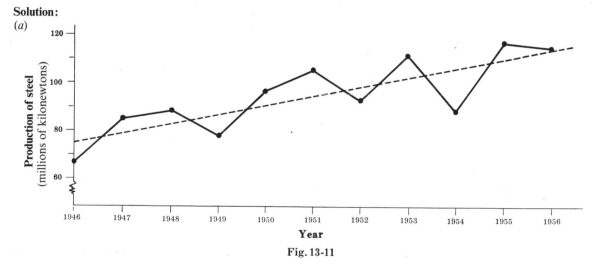

Fig. 13-11

(*b*)   **First method:**

Use the equation $y = \left(\dfrac{\Sigma xy}{\Sigma x^2}\right)x$, where $x = X - \bar{X}$ and $y = Y - \bar{Y}$. The work can be arranged as in Table 13.9.

Table 13.9

Year	X	Y	$x = X - \bar{X}$	$y = Y - \bar{Y}$	x^2	xy
1946	0	66·6	−5	−28·4	25	142·0
1947	1	84·9	−4	−10·1	16	40·4
1948	2	88·6	−3	−6·4	9	19·2
1949	3	78·0	−2	−17·0	4	34·0
1950	4	96·8	−1	1·8	1	−1·8
1951	5	105·2	0	10·2	0	0
1952	6	93·2	1	−1·8	1	−1·8
1953	7	111·6	2	16·6	4	33·2
1954	8	88·3	3	−6·7	9	−20·1
1955	9	117·0	4	22·0	16	88·0
1956	10	115·2	5	20·2	25	101·0
	$\Sigma X = 55$ $\bar{X} = 5$	$\Sigma Y = 1045\cdot4$ $\bar{Y} = 95\cdot0$			$\Sigma x^2 = 110$	$\Sigma xy = 434\cdot1$

The required equation $y = \left(\dfrac{\Sigma xy}{\Sigma x^2}\right)x$ becomes $y = \left(\dfrac{434\cdot1}{110}\right)x$ or $y = 3\cdot95x$ which can be written

$$Y - 95\cdot0 = 3\cdot95(X - 5) \quad \text{or} \quad Y = 75\cdot2 + 3\cdot95X$$

where the origin $X = 0$ is the year 1946 and the units of X are 1 year.

The graph of this line, sometimes called a *trend line*, is shown dashed in Fig. 13-11. The equation is often called the *trend equation* and the values of Y computed for various values of X are called the *trend values*.

Second method:

If we assign values of X to the years 1946–1956 so that $\Sigma X = 0$, the equation of the least square line can be written

$$Y = \bar{Y} + \left(\frac{\Sigma XY}{\Sigma X^2}\right)X$$

Since there are an odd number of years in the data, we can assign $X = 0$ to the middle year 1951, $X = 1, 2, 3, 4, 5$ to the successive years and $X = -1, -2, -3, -4, -5$ to the years which precede this middle year. The result is indicated in column 2 of Table 13.10 and is equivalent to use of column 4 in the table for the first method.

The middle year 1951 is called the *origin*. Unless otherwise specified we shall assume that the values of Y refer to mid-year values, i.e. as of July 1. Thus $X = 0$ corresponds to July 1, 1951, $X = -1$ corresponds to July 1, 1950, etc. The computations involved can be organized as in Table 13.10.

Table 13.10

Year	X	Y	X^2	XY
1946	−5	66·6	25	−333·0
1947	−4	84·9	16	−339·6
1948	−3	88·6	9	−265·8
1949	−2	78·0	4	−156·0
1950	−1	96·8	1	−96·8
1951	0	105·2	0	0
1952	1	93·2	1	93·2
1953	2	111·6	4	223·2
1954	3	88·3	9	264·9
1955	4	117·0	16	468·0
1956	5	115·2	25	576·0
	$\bar{X} = 0$	$\Sigma Y = 1045\cdot4$	$\Sigma X^2 = 110$	$\Sigma XY = 434\cdot1$

Then $\bar{Y} = (\Sigma Y/N) = 1\cdot45\cdot4/11 = 95\cdot0$ and the required equation is

$$Y = 95\cdot0 + (434\cdot1/110)X \ \text{ or } \ Y = 95\cdot0 + 3\cdot95X$$

where the origin $X = 0$ is the year 1951 and the unit of X is 1 year.

To shift the origin to 1946, five years earlier, we must replace X by $X - 5$, thus obtaining the equation $Y = 95\cdot0 + 3\cdot95(X - 5)$ or $Y = 75\cdot2 + 3\cdot95X$ as in the first method.

The second method is superior to the first method since the labour of computation is reduced. However, it must be modified in case the number of years in the data is even. For this modification see the method of Problem 13.20(*b*). The first method can be applied to all cases.

(*c*) Use the trend equation $Y = 95\cdot0 + 3\cdot95X$, where $X = 0$ corresponds to 1951. Then the years 1957 and 1958 correspond to $X = 6$ and $X = 7$ respectively.

For $X = 6$, $Y = 95\cdot0 + 0\cdot95(6) = 118\cdot7$, which compares well with the actual value $112\cdot7$.
For $X = 7$, $Y = 95\cdot0 + 3\cdot95(7) = 122\cdot6$ which does not compare well with the actual value $85\cdot3$ and illustrates the risk involved in the process of extrapolation.

The same results can be obtained by using the trend equation $Y = 75\cdot2 + 3\cdot95X$ which has as origin the year 1946, by placing $X = 11$ and $X = 12$ respectively.

(*d*) Using the trend line $Y = 75\cdot2 + 3\cdot95X$ with $X = -1$ and $X = -2$, we find the values

$$Y = 75\cdot2 + 3\cdot95(-1) = 71\cdot2 \ \text{ and } \ Y = 75\cdot2 + 3\cdot95(-2) = 67\cdot3$$

13.20. The U.S. production of small cigars during the years 1945–1954 is given in Table 13.11.

(*a*) Graph the data.

(*b*) Find the equation of a least square line fitting the data.

(*c*) Estimate the production of small cigars during the year 1955.

Table 13.11

Year	1945	1946	1947	1948	1949	1950	1951	1952	1953	1954
Number of small cigars (millions)	98·2	92·3	80·0	89·1	83·5	68·9	69·2	67·1	58·3	61·2

Solution:

(*a*)

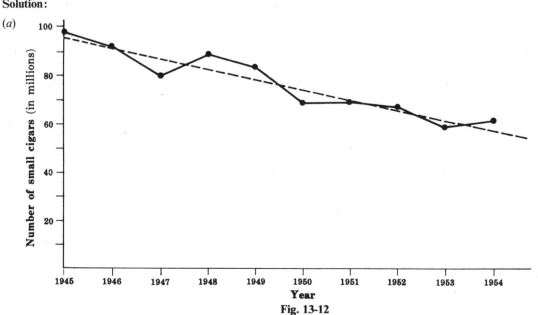

Fig. 13-12

(b) **First method:**

Table 13.12

Year	X	Y	$x = X - \bar{X}$	$y = Y - \bar{Y}$	x^2	xy
1945	0	98·2	−4·5	21·4	20·25	−96·30
1946	1	92·3	−3·5	15·5	12·25	−54·25
1947	2	80·0	−2·5	3·2	6·25	−8·00
1948	3	89·1	−1·5	12·3	2·25	−18·45
1949	4	83·5	−0·5	6·7	0·25	−3·35
1950	5	68·9	0·5	−7·9	0·25	−3·95
1951	6	69·2	1·5	−7·6	2·25	−11·40
1952	7	67·1	2·5	−9·7	6·25	−24·25
1953	8	58·3	3·5	−18·5	12·25	−64·75
1954	9	61·2	4·5	−15·6	20·25	−70·20
	$\Sigma X = 45$ $\bar{X} = 4\cdot5$	$\Sigma Y = 767\cdot8$ $\bar{Y} = 76\cdot8$			$\Sigma x^2 = 82\cdot5$	$\Sigma xy = -354\cdot9$

The required equation $y = \left(\dfrac{\Sigma xy}{\Sigma x^2}\right) x$ becomes $y = \dfrac{-354\cdot9}{82\cdot5}x$ or $y = -4\cdot30x$ which can be written

$$Y - 76\cdot8 = -4\cdot30(X - 4\cdot5) \text{ or } Y = 96\cdot2 - 4\cdot30X$$

where the origin $X = 0$ is the year 1945 and the units of X are 1 year.

The graph of this line, sometimes called the *trend line*, is shown dashed in Fig. 13-12.

Second method:

Table 13.13

Year	X	Y	X^2	XY
1945	−9	98·2	81	−883·8
1946	−7	92·3	49	−646·1
1947	−5	80·0	25	−400·0
1948	−3	89·1	9	−267·3
1949	−1	83·5	1	−83·5
1950	1	68·9	1	68·9
1951	3	69·2	9	207·6
1952	5	67·1	25	335·5
1953	7	58·3	49	408·1
1954	9	61·2	81	550·8
	$\Sigma X = 0$ $\bar{X} = 0$	$\Sigma Y = 767\cdot8$ $\bar{Y} = 76\cdot8$	$\Sigma X^2 = 330$	$\Sigma XY = -709\cdot8$

In this method we wish to assign values of X to the years so that $\Sigma X = 0$. Since an even number of years is present, there is no middle year and the second method of Problem 13.19 cannot be used. However, we can associate the numbers −0·5 and 0·5 to the two middle years 1949 and 1950, so that 1951, 1952, . . . are represented by 1·5, 2·5, . . . and 1948, 1947, . . . by −1·5, −2·5, ... This essentially amounts to the fourth column in Table 13.12 of the first method.

Also, to avoid fractions we double these values so as to obtain the second column in Table 13.13. Note that with these values of X the origin $X = 0$ is midway between July 1, 1949 and July 1, 1950, which is Jan. 1, 1950 or Dec. 31, 1949. Also, the units of X are in half years.

Since $X = 0$, the required equation has the form $Y = \bar{Y} + \left(\dfrac{\Sigma XY}{\Sigma X^2}\right) X$ which gives (see Table 13.13)

$$Y = 76\cdot8 + (-709\cdot8/330)X \text{ or } Y = 76\cdot8 - 2\cdot15X$$

where the origin $X = 0$ corresponds to Jan. 1, 1 950, and X is measured in half years.

If we wish to measure X in whole years instead of half years, we must replace X by $2X$, so that the equation is

$$Y = 76\cdot8 - 4\cdot30X$$

where the origin is Jan. 1, 1950 and X is measured in years.

If we now want the origin to be at July 1, 1945, we must replace X by $X - 4.5$ (since it is 4.5 years from July 1, 1945 to Jan. 1, 1950). The result is

$$Y = 76.8 - 4.30(X - 4.5) = 96.2 - 4.30X$$

where the origin is July 1, 1945 and X is measured in years. This agrees with the first method.

(c) Use the equation $Y = 96.2 - 4.30X$ with $X = 10$ corresponding to 1955. Then $Y = 53.2$, so that we can expect a production of 53.2 million small cigars if the trend continues.

NON-LINEAR EQUATIONS REDUCIBLE TO LINEAR FORM

13.21. Table 13.14 gives experimental values of the pressure P of a given mass of gas corresponding to various values of the volume V. According to thermodynamic principles a relationship having the form $PV^{\gamma} = C$, where γ and C are constants, should exist between the variables. (a) Find the values of γ and C. (b) Write the equation connecting P and V. (c) Estimate P when $V = 100.0$.

Table 13.14

Volume V	54.3	61.8	72.4	88.7	118.6	194.0
Pressure P	61.2	49.5	37.6	28.4	19.2	10.1

Solution:

Since $PV^{\gamma} = C$, we have

$$\log P + \gamma \log V = \log C \quad \text{or} \quad \log P = \log C - \gamma \log V$$

Calling $\log V = X$ and $\log P = Y$, the last equation can be written

(1) $$Y = a_0 + a_1 X$$

where $a_0 = \log C$ and $a_1 = -\gamma$.

Table 13.15 below gives $X = \log V$ and $Y = \log P$ corresponding to the values of V and P in Table 13.14 and also indicates the calculations involved in computing the least square line (1).

The normal equations corresponding to the least square line (1) are

$$\Sigma Y = a_0 N + a_1 \Sigma X \quad \text{and} \quad \Sigma XY = a_0 \Sigma X + a_1 \Sigma X^2$$

from which $a_0 = \dfrac{(\Sigma Y)(\Sigma X^2) - (\Sigma X)(\Sigma XY)}{N \Sigma X^2 - (\Sigma X)^2} = 4.20$, $a_1 = \dfrac{N \Sigma XY - (\Sigma X)(\Sigma Y)}{N \Sigma X^2 - (\Sigma X)^2} = -1.40$.

Then $Y = 4.20 - 1.40X$.

Table 13.15

$X = \log V$	$Y = \log P$	X^2	XY
1.7348	1.7868	3.0095	3.0997
1.7910	1.6946	3.2077	3.0350
1.8597	1.5752	3.4585	2.9294
1.9479	1.4533	3.7943	2.8309
2.0741	1.2833	4.3019	2.6617
2.2878	1.0043	5.2340	2.2976
$\Sigma X = 11.6953$	$\Sigma Y = 8.7975$	$\Sigma X^2 = 23.0059$	$\Sigma XY = 16.8543$

(a) Since $a_0 = 4.20 = \log C$ and $a_1 = -1.40 = -\gamma$, $C = 1.60 \times 10^4$ and $\gamma = 1.40$.

(b) The required equation in terms of P and V can be written $PV^{1.40} = 16\,000$.

(c) When $V = 100$, $X = \log V = 2$ and $Y = \log P = 4.20 - 1.40(2) = 1.40$. Then $P = $ antilog $1.40 = 25.1$.

13.22. Solve Problem 13.21 by plotting the data on log-log graph paper.

Solution:

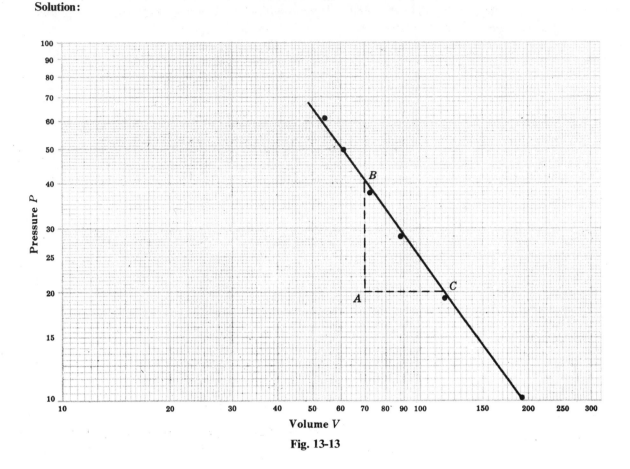

Fig. 13-13

For each pair of values of the pressure P and volume V in Table 13.14 of Problem 13.21, we obtain a point which is plotted on the specially constructed *log-log graph paper* shown in Fig. 13-13 above.

A line (drawn freehand) approximating these points is also indicated. The resulting graph shows that there is a linear relationship between log P and log V which can be represented by the equation

$$\log P = a_0 + a_1 \log V \quad \text{or} \quad Y = a_0 + a_1 X$$

The slope a_1, which is negative in this case, is given numerically by the ratio of the lengths of AB to AC (using an appropriate unit of length). Measurement in this case yields $a_1 = -1\cdot4$.

To obtain a_0, one point on the line is needed. For example when $V = 100$, $P = 25$ from the graph. Then

$$a_0 = \log P - a_1 \log V = \log 25 + 1\cdot4 \log 100 = 1\cdot4 + (1\cdot4)(2) = 4\cdot2$$

so that

$$\log P + 1\cdot4 \log V = 4\cdot2, \quad \log PV^{1\cdot4} = 4\cdot2, \quad \text{and} \quad PV^{1\cdot4} = 16\,000$$

THE LEAST SQUARE PARABOLA

13.23. Table 13.16 shows the population of the United States during the years 1850–1950 in ten year intervals.
(*a*) Find the equation of a least square parabola fitting the data. (*b*) Compute the trend values for the

years given in the table and compare with the actual values. (*c*) Estimate the population in 1945. (*d*) Estimate the population in 1960 and compare with the actual value. (*e*) Estimate the population in 1840 and compare with the actual value. See Problem 1.23, Chapter 1.

Table 13.16

Year	1850	1860	1870	1880	1890	1900	1910	1920	1930	1940	1950
U.S. Population (millions)	23·2	31·4	39·8	50·2	62·9	76·0	92·0	105·7	122·8	131·7	151·1

Source: Bureau of the Census

Solution:

(*a*) Let the variables X and Y denote respectively the year and the population during that year. The equation of a least square parabola fitting the data is

(*1*) $$Y = a_0 + a_1 X + a_2 X^2$$

where a_0, a_1 and a_2 are found from the normal equations

(*2*)
$$\begin{cases} \Sigma Y = a_0 N + a_1 \Sigma X + a_2 \Sigma X^2 \\ \Sigma XY = a_0 \Sigma X + a_1 \Sigma X^2 + a_2 \Sigma X^3 \\ \Sigma X^2 Y = a_0 \Sigma X^2 + a_1 \Sigma X^3 + a_2 \Sigma X^4 \end{cases}$$

It is convenient to choose X so that the middle year 1900 corresponds to $X = 0$, and the years 1910, 1920, 1930, 1940, 1950 and 1890, 1880, 1870, 1860, 1850 correspond to 1, 2, 3, 4, 5 and $-1, -2, -3, -4, -5$ respectively. With this choice ΣX and ΣX^3 are zero and equations (*2*) are simplified.

The work involved in computation can be arranged as in Table 13.17 below.

Using this table the normal equations (*2*) become

(*3*)
$$\begin{cases} 11a_0 + 110a_2 = 886·8 \\ 110a_1 = 1429·8 \\ 110a_0 + 1958a_2 = 9209·0 \end{cases}$$

From the second equation in (*3*), $a_1 = 13·00$; from the first and third equations, $a_0 = 76·64$, $a_2 = 0·3974$. Then the required equation is

(*4*) $$Y = 76·64 + 13·00X + 0·3974X^2$$

where the origin $X = 0$ is July 1, 1900 and the unit of X is ten years.

Table 13.17

Year	X	Y	X^2	X^3	X^4	XY	$X^2 Y$
1850	−5	23·2	25	−125	625	−116·0	580·0
1860	−4	31·4	16	−64	256	−125·6	502·4
1870	−3	39·8	9	−27	81	−119·4	358·2
1880	−2	50·2	4	−8	16	−100·4	200·8
1890	−1	62·9	1	−1	1	−62·9	62·9
1900	0	76·0	0	0	0	0	0
1910	1	92·0	1	1	1	92·0	92·0
1920	2	105·7	4	8	16	211·4	422·8
1930	3	122·8	9	27	81	368·4	1105·2
1940	4	131·7	16	64	256	526·8	2107·2
1950	5	151·1	25	125	625	755·5	3777·5
	$\Sigma X = 0$	$\Sigma Y = 886·8$	$\Sigma X^2 = 110$	$\Sigma X^3 = 0$	$\Sigma X^4 = 1958$	$\Sigma XY = 1429·8$	$\Sigma X^2 Y = 9209·0$

(*b*) The trend values, obtained by placing $X = -5, -4, -3, -2, -1, 0, 1, 2, 3, 4, 5$ in equation (*4*), are shown in Table 13.18 together with the actual values. It is seen that the agreement is good.

Table 13.18

Year	$X = -5$ 1850	$X = -4$ 1860	$X = -3$ 1870	$X = -2$ 1880	$X = -1$ 1890	$X = 0$ 1900	$X = 1$ 1910	$X = 2$ 1920	$X = 3$ 1930	$X = 4$ 1940	$X = 5$ 1950
Trend Value	21·6	31·0	41·2	52·2	64·0	76·6	90·0	104·2	119·2	135·0	151·6
Actual Value	23·2	31·4	39·8	50·2	62·9	76·0	92·0	105·7	122·8	131·7	151·1

(c) 1945 corresponds to $X = 4\cdot5$, for which $Y = 76\cdot64 + 13\cdot00(4\cdot5) + 0\cdot3974(4\cdot5)^2 = 143\cdot2$.

(d) 1960 corresponds to $X = 6$, for which $Y = 76\cdot64 + 13\cdot00(6) + 0\cdot3974(6)^2 = 168\cdot9$.

This does not agree too well with the actual value 179·3.

(e) 1840 corresponds to $X = -6$, for which $Y = 76\cdot64 + 13\cdot00(-6) + 0\cdot3974(-6)^2 = 12\cdot9$.

This does not agree with the actual value 17·1.

This example illustrates the fact that a relationship which is found to be satisfactory for a range of values need not necessarily be satisfactory for an extended range of values.

Supplementary Problems

STRAIGHT LINES

13.24. If $3X + 2Y = 18$, find (a) X when $Y = 3$, (b) Y when $X = 2$, (c) X when $Y = -5$, (d) Y when $X = -1$, (e) the X intercept, (f) the Y intercept.   *Ans.* (a) 4, (b) 6, (c) 28/3, (d) 10·5, (e) 6, (f) 9

13.25. Construct a graph of the equations (a) $Y = 3X - 5$ and (b) $X + 2Y = 4$ on the same set of axes. In what point do the graphs intersect?   *Ans.* (2, 1)

13.26. (a) Find an equation for the straight line passing through the points $(3, -2)$ and $(-1, 6)$. (b) Determine the X and Y intercepts of the line in (a). (c) Find the value of Y corresponding to $X = 3$ and to $X = 5$. (d) Verify your answers to (a), (b) and (c) directly from a graph.
Ans. (a) $2X + Y = 4$. (b) X intercept = 2, Y intercept = 4. (c) $-2, -6$

13.27. Find an equation for the straight line whose slope is $\frac{2}{3}$ and whose intercept is -3.
Ans. $Y = \frac{2}{3}X - 3$ or $2X - 3Y = 9$

13.28. (a) Find the slope and Y intercept of the line whose equation is $3X - 5Y = 20$. (b) What is the equation of a line which is parallel to the line in (a) and which passes through the point $(2, -1)$?
Ans. (a) slope = $\frac{3}{5}$, Y intercept = -4. (b) $3X - 5Y = 11$

13.29. Find (a) the slope, (b) the Y intercept and (c) the equation of the line passing through the points (5, 4) and (2, 8).
Ans. (a) $-4/3$, (b) 32/3, (c) $4X + 3Y = 32$

13.30. Find the equation of a straight line whose X and Y intercepts are 3 and -5 respectively.
Ans. $X/3 + Y/(-5) = 1$ or $5X - 3Y = 15$

13.31. A temperature of 100 degrees Celsius corresponds to 212 degrees Fahrenheit, while a temperature of 0 degrees Celsius corresponds to 32 degrees Fahrenheit. Assuming a linear relationship to exist between Celsius and Fahrenheit temperatures (denoted by C and F respectively), find (a) the equation connecting C and F, (b) the Fahrenheit temperature corresponding to 80 degrees Celsius, (c) the Celsius temperature corresponding to 68 degrees Fahrenheit.
Ans. (a) $F = \frac{9}{5}C + 32$, (b) 176°F, (c) 20°C

THE LEAST SQUARE LINE

13.32. Fit a least square line to the data in the following table using (a) X as the independent variable, (b) X as the dependent variable. Graph the data and the least square lines using the same set of co-ordinate axes.
Ans. (a) $Y = -\frac{1}{3} + \frac{5}{7}X$ or $Y = -0\cdot333 + 0\cdot714X$, (b) $X = 1 + \frac{9}{7}Y$ or $X = 1\cdot00 + 1\cdot29Y$

X	3	5	6	8	9	11
Y	2	3	4	6	5	8

13.33. For the data of the preceding problem find (a) the values of Y when $X = 5$ and $X = 12$, (b) the value of X when $Y = 7$. *Ans.* (a) 3·24, 8·24, (b) 10·00

13.34. (a) Use the freehand method to obtain an equation for a line fitting the data of Problem 13.32. (b) Answer Problem 13.33 by using the result of part (a).

13.35. The following table shows the final marks in Algebra and Physics obtained by 10 students selected at random from a large group of students. (a) Graph the data. (b) Find the least square line fitting the data, using X as the independent variable. (c) Find the least square line fitting the data, using Y as independent variable. (d) If a student receives a mark of 75 in Algebra, what is his expected mark in Physics? (e) If a student receives a mark of 95 in Physics, what is his expected mark in Algebra?

Algebra (X)	75	80	93	65	87	71	98	68	84	77
Physics (Y)	82	78	86	72	91	80	95	72	89	74

Ans. (b) $Y = 29 \cdot 13 + 0 \cdot 661X$, (c) $X = -14 \cdot 39 + 1 \cdot 15Y$, (d) 79, (e) 95

13.36. The following table shows the number of farm workers in the United States (in millions) during the years 1949–1957. (a) Graph the data. (b) Find a least square line fitting this time series and construct its graph. (c) Compute the trend values and compare with the actual values. (d) Estimate the number of farm workers in the year 1948 and compare with the actual value (10·36 million). (e) Predict the number of farm workers in 1958 (true value is 7·53 million). Discuss the possible sources of error in such prediction.

Year	1949	1950	1951	1952	1953	1954	1955	1956	1957
Number of Farm Workers (millions)	9·96	9·93	9·55	9·15	8·86	8·64	8·36	7·82	7·58

Source: Department of Agriculture

Ans. (b) $Y = 8 \cdot 872 - 0 \cdot 312X$, where Y is the number of farm workers in millions, is expressed in years and the origin is at July 1, 1953. (d) 10·43 million. (e) 7·31 million.

13.37. The consumer price index for medical care in the United States is given in the following table for the years 1950–1957. (The *reference period* or *base period* 1947–1949 is assigned the value 100 which actually means 100%. The index for 1952, for example, is 117·2 and shows that during 1952 the average price of medical care was 117·2% of what it was in the base period, i.e. it increased by 17·2%.)

(a) Graph the data.

(b) Find a least square fitting the data and construct its graph.

(c) Compute the trend values and compare with the actual values.

(d) Predict the price index for medical care during 1958 and compare with the true value (144·4).

(e) In what year can we expect the index of medical costs to be double that of 1947–1949 assuming present trends continue?

Year	1950	1951	1952	1953	1954	1955	1956	1957
Consumer Price Index for medical care (1947–1949 = 100)	106·0	111·1	117·2	121·3	125·2	128·0	132·6	138·0

Source: Bureau of Labor Statistics

Ans. (b) $Y = 122 \cdot 42 + 21 \cdot 19X$ if X units are $\frac{1}{2}$ year and origin is at Jan. 1, 1954, or $Y = 107 \cdot 09 + 4 \cdot 38X$ if X units are 1 year and origin is at July 1, 1950. (d) 142·1. (e) 1971.

LEAST SQUARE CURVES

13.38. Fit a least square parabola, $Y = a_0 + a_1X + a_2X^2$, to the data in the adjoining table.
Ans. $Y = 5 \cdot 51 + 3 \cdot 20(X - 3)0 \cdot 733(X - 3)^2$ or $Y = 2 \cdot 51 - 1 \cdot 20X + 0 \cdot 733X^2$

X	0	1	2	3	4	5	6
Y	2·4	2·1	3·2	5·6	9·3	14·6	21·9

13.39. The total time required to bring a vehicle to a stop after perceiving danger is composed of the *reaction time* (time between recognition of danger and application of brakes) plus braking time (time for stopping after application of brakes). The following table gives the stopping distance d (metres) of a car travelling at speeds v (metres per second) at the instant danger is sighted. (*a*) Graph d against v. (*b*) Fit a least square parabola of the form $d = a_0 + a_1 v + a_2 v^2$ to the data. (*c*) Estimate d when $v = 45$ m/s and 80 m/s.

Speed v (m/s)	20	30	40	50	60	70
Stopping distance d (m)	54	90	138	206	292	396

Ans. (*b*) $d = 41 \cdot 77 - 1 \cdot 096v + 0 \cdot 087\,86v^2$. (*c*) 170 m, 516 m

13.40. The following table shows the birth rate per 1000 population in the United States during the years 1915-1955 in 5 year intervals. (*a*) Graph the data. (*b*) Find a least square parabola fitting the data. (*c*) Compute the trend values and compare with the actual values. (*d*) Explain why the equation obtained in (*b*) is not useful for extrapolation purposes.

Year	1915	1920	1925	1930	1935	1940	1945	1950	1955
Birth rate per 1000 population	25·0	23·7	21·3	18·9	16·9	17·9	19·5	23·6	24·6

Source: Department of Health, Education and Welfare

Ans. $Y = 18 \cdot 16 - 0 \cdot 1083X + 0 \cdot 4653X^2$, where Y is the birth rate per 1000 population and X units are 5 years with origin at July 1, 1935.

13.41. The number Y of bacteria per unit volume present in a culture after X hours is given in the following table. (*a*) Graph the data on semi-logarithmic graph paper where the logarithmic scale is used for Y and the arithmetic scale is used for X. (*b*) Fit a least square curve having the form $Y = ab^X$ to the data and explain why this particular equation should yield good results. (*c*) Compare the values of Y obtained from this equation with the actual values. (*d*) Estimate the value of Y when $X = 7$.

Number of hours (X)	0	1	2	3	4	5	6
Number of bacteria per unit volume (Y)	32	47	65	92	132	190	275

Ans. (*b*) $Y = 32 \cdot 14(1 \cdot 427)^X$ or $Y = 32 \cdot 14(10)^{0 \cdot 1544X}$ or $Y = 32 \cdot 14e^{0 \cdot 3556X}$ where $e = 2 \cdot 718 \ldots$ is the natural logarithmic base. (*d*) 387

13.42. In the preceding problem show how a graph on semi-logarithmic graph paper can be used to obtain the required equation without employing the method of least squares.

CHAPTER 14

Correlation Theory

CORRELATION AND REGRESSION

In the last chapter we considered the problem of *regression* or *estimation* of one variable (the dependent variable) from one or more related variables (the independent variables). In this chapter we consider the closely related problem of *correlation*, or the degree of relationship between variables, which seeks to determine *how well* a linear or other equation describes or explains the relationship between variables.

If all the values of the variables satisfy an equation exactly we say that the variables are *perfectly correlated* or that there is *perfect correlation* between them. Thus the circumferences C and radii r of all circles are perfectly correlated since $C = 2\pi r$. If two dice are tossed simultaneously 100 times there is no relationship between corresponding points on each die (unless the dice are loaded), i.e. they are *uncorrelated*. The variable height and weight of individuals would show *some* correlation.

When only two variables are involved we speak of *simple correlation* and *simple regression*. When more than two variables are involved we speak of *multiple correlation* and *multiple regression*. In this chapter, only simple correlation is considered. Multiple correlation and regression are considered in Chapter 15.

LINEAR CORRELATION

If X and Y denote the two variables under consideration, a *scatter diagram* shows the location points (X, Y) on a rectangular co-ordinate system. If all points in this scatter diagram seem to lie near a line, as in (a) and (b) of Fig. 14-1, the correlation is called *linear*. In such cases, as we have seen in Chapter 13, a linear equation is appropriate for purposes of regression or estimation.

If Y tends to increase as X increases, as in (a), the correlation is called *positive* or *direct correlation*. If Y tends to decrease as X increases, as in (b), the correlation is called *negative* or *inverse correlation*.

If all points seem to lie near some curve, the correlation is called *non-linear* and a non-linear equation is appropriate for regression or estimation, as we have seen in Chapter 13. It is clear than non-linear correlation can be sometimes positive and sometimes negative.

If there is no relationship indicated between the variables, as in Fig. 14-1(c), we say that there is *no correlation* between them, i.e. they are *uncorrelated*.

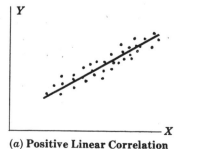
(a) Positive Linear Correlation

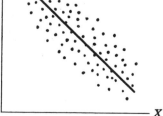
(b) Negative Linear Correlation

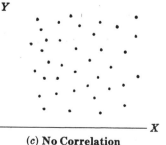
(c) No Correlation

Fig. 14-1

MEASURES OF CORRELATION

We can determine in a *qualitative* manner how well a given line or curve describes the relationship between variables by direct observation of the scatter diagram itself. For example, it is seen that a straight line is far more helpful in describing the relation between X and Y for the data of Fig. 14-1(a) than for the data of Fig. 14-1(b) due to the fact that there is less scattering about the line of Fig. 14-1(a).

If we are to deal with the problem of scattering of sample data about lines or curves in a *quantitative* manner it will be necessary for us to devise *measures of correlation*.

THE LEAST SQUARE REGRESSION LINES

We first consider the problem of how well a straight line explains the relationship between two variables. To do this we shall need the equations for the least square regression lines obtained in Chapter 13. As we have seen, the least square regression line of Y on X is

$$Y = a_0 + a_1 X \tag{1}$$

where a_0 and a_1 are obtained from the normal equations

$$\begin{cases} \Sigma Y = a_0 N + a_1 \Sigma X \\ \Sigma XY = a_0 \Sigma X + a_1 \Sigma X^2 \end{cases} \tag{2}$$

which yield

$$\begin{cases} a_0 = \dfrac{(\Sigma Y)(\Sigma X^2) - (\Sigma X)(\Sigma XY)}{N \Sigma X^2 - (\Sigma X)^2} \\[3mm] a_1 = \dfrac{N \Sigma XY - (\Sigma X)(\Sigma Y)}{N \Sigma X^2 - (\Sigma X)^2} \end{cases} \tag{3}$$

Similarly, the regression line of X on Y is given by

$$X = b_0 + b_1 Y \tag{4}$$

where b_0 and b_1 are obtained from the normal equations

$$\begin{cases} \Sigma X = b_0 N + b_1 \Sigma Y \\ \Sigma XY = b_0 \Sigma X + b_1 \Sigma Y^2 \end{cases} \tag{5}$$

which yield

$$\begin{cases} b_0 = \dfrac{(\Sigma X)(\Sigma Y^2) - (\Sigma Y)(\Sigma XY)}{N \Sigma Y^2 - (\Sigma Y)^2} \\[3mm] b_1 = \dfrac{N \Sigma XY - (\Sigma X)(\Sigma Y)}{N \Sigma Y^2 - (\Sigma Y)^2} \end{cases} \tag{6}$$

The equations (1) and (4) can also be written respectively as

$$y = \left(\frac{\Sigma xy}{\Sigma x^2}\right)x \quad \text{and} \quad x = \left(\frac{\Sigma xy}{\Sigma y^2}\right)y \tag{7}$$

where $x = X - \bar{X}$ and $y = Y - \bar{Y}$.

The regression equations are identical if and only if all points of the scatter diagram lie on a line. In such case there is *perfect linear correlation* between X and Y.

STANDARD ERROR OF ESTIMATE

If we let $Y_{est.}$ represent the value of Y for given values of X as estimated from (1), a measure of the scatter about the regression line of Y on X is supplied by the quantity

$$s_{Y.X} = \sqrt{\frac{\Sigma (Y - Y_{est.})^2}{N}} \tag{8}$$

which is called the *standard error of estimate of Y on X*.

If the regression line (4) is used, an analogous standard error of estimate of X on Y is defined by

$$s_{X.Y} = \sqrt{\frac{\Sigma (X - X_{est.})^2}{N}} \tag{9}$$

In general, $s_{Y.X} \neq s_{X.Y}$

Equation (8) can be written

$$s_{Y.X}^2 = \frac{\Sigma Y^2 - a_0 \Sigma Y - a_1 \Sigma XY}{N} \tag{10}$$

which may be more suitable for computation (see Problem 14.3). A similar expression exists for (9).

The standard error of estimate has properties analogous to those of the standard deviation. For example, if we construct lines parallel to the regression line of Y on X at respective vertical distances $s_{Y.X}$, $2s_{Y.X}$ and $3s_{Y.X}$ from it, we should find, if N is large enough, that there would be included between these lines about 68%, 95% and 99·7% of the sample points.

Just as a modified standard deviation given by $\hat{s} = \sqrt{\dfrac{N}{N-1}} s$ was found useful for small samples, so a

modified standard error of estimate given by $\hat{s}_{Y.X} = \sqrt{\dfrac{N}{N-2}} s_{Y.X}$ is useful. For this reason some statisticians prefer to define (8) or (9) with $N - 2$ replacing N in the denominator.

EXPLAINED AND UNEXPLAINED VARIATION

The *total variation* of Y is defined as $\Sigma (Y - \bar{Y})^2$, i.e. the sum of the squares of the deviations of the values of Y from the mean $\bar{Y}$. As shown in Problem 14.7 this can be written

$$\Sigma (Y - \bar{Y})^2 = \Sigma (Y - Y_{est.})^2 + \Sigma (Y_{est.} - \bar{Y})^2 \tag{11}$$

The first term on the right of (11) is called the *unexplained variation* while the second term is called the *explained variation*, so called because the deviations $Y_{est.} - \bar{Y}$ have a definite pattern while the deviations $Y - Y_{est.}$ behave in a random or unpredictable manner. Similar results hold for the variable X.

COEFFICIENT OF CORRELATION

The ratio of the explained variation to the total variation is called the *coefficient of determination*. If there is zero explained variation, i.e. the total variation is all unexplained, this ratio is zero. If there is zero unexplained variation, i.e. the total variation is all explained, the ratio is one. In other cases the ratio lies between zero and one. Since the ratio is always non-negative, we denote it by r^2. The quantity r, called the *coefficient of correlation*, is given by

$$r = \pm \sqrt{\frac{\text{explained variation}}{\text{total variation}}} = \pm \sqrt{\frac{\Sigma (Y_{est.} - \bar{Y})^2}{\Sigma (Y - \bar{Y})^2}} \tag{12}$$

and varies between -1 and $+1$. The signs $\pm$ are used for positive correlation and negative linear correlation respectively. Note that r is a dimensionless quantity, i.e. it does not depend on the units employed.

By using (8) and (11) and the fact that the standard deviation of Y is

$$s_Y = \sqrt{\frac{\Sigma(Y - \bar{Y})^2}{N}} \tag{13}$$

we find that (12) can be written, disregarding the sign, as

$$r = \sqrt{1 - \frac{s_{Y.X}^2}{s_Y^2}} \quad \text{or} \quad s_{Y.X} = s_Y\sqrt{1 - r^2} \tag{14}$$

Similar equations exist when X and Y are interchanged.

For the case of linear correlation the quantity r is the same regardless of whether X or Y is considered the independent variable. Thus r is a very good measure of the linear correlation between two variables.

REMARKS CONCERNING THE CORRELATION COEFFICIENT

The definitions (12) or (14) for the correlation coefficient are quite general and can be used for non-linear relationship as well as linear, the only differences being that $Y_{est.}$ is computed from a non-linear regression equation in place of a linear regression equation and the signs $\pm$ are omitted. In such case equation (8) defining the standard error of estimate is perfectly general. Equation (10), however, which applies to linear regression only, must be modified. If, for example, the estimating equation is

$$Y = a_0 + a_1 X + a_2 X^2 + \cdots + a_{n-1} X^{n-1} \tag{15}$$

equation (10) is replaced by

$$s_{Y.X}^2 = \frac{\Sigma Y^2 - a_0 \Sigma Y - a_1 \Sigma XY - \cdots - a_{n-1} \Sigma X^{n-1} Y}{N} \tag{16}$$

In such case the *modified standard error of estimate* (see discussion on Page 243) is $\hat{s}_{Y.X} = \sqrt{\frac{N}{N-n}}\, s_{Y.X}$, where the quantity $N - n$ is called the number of *degrees of freedom*.

It must be emphasized that the value of r computed in any case measures the degree of the relationship relative to the type of equation which is actually assumed. Thus if a linear equation is assumed (12) or (14) yields a value of r near zero, it means that there is almost no *linear correlation* between the variables. However, it does not mean that there is no correlation at all, since there may actually be a high *non-linear correlation* between the variables. In other words the correlation coefficient measures the goodness of fit of the equation actually assumed to the data. Unless otherwise specified, the term correlation coefficient is used to mean *linear* correlation coefficient.

It should also be pointed out that a high correlation coefficient (i.e. near 1 or -1) does not necessarily indicate a direct dependence of the variables. Thus there may be a high correlation between the number of books published each year and the number of baseball games played each year. Such examples are sometimes referred to as *nonsense* or *spurious correlations*.

PRODUCT-MOMENT FORMULA FOR THE LINEAR CORRELATION COEFFICIENT

If a linear relationship between two variables is assumed, equation (12) becomes

$$r = \frac{\Sigma xy}{\sqrt{(\Sigma x^2)(\Sigma y^2)}} \tag{17}$$

where $x = X - \bar{X}$ and $y = Y - \bar{Y}$ (see Problem 14.10). This formula, which automatically gives the proper sign of r, is called the *product-moment formula* and clearly shows the symmetry between X and Y.

If we write

$$s_{XY} = \frac{\Sigma xy}{N}, \qquad s_X = \sqrt{\frac{\Sigma x^2}{N}}, \qquad s_Y = \sqrt{\frac{\Sigma y^2}{N}} \qquad (18)$$

then s_X and s_Y will be recognized as the standard deviations of the variables X and Y repectively, while s_X^2 and s_Y^2 are their variances. The new quantity s_{XY} is called the *covariance* of X and Y. In terms of the symbols of (18), (17) can be written

$$r = \frac{s_{XY}}{s_X s_Y} \qquad (19)$$

Note that r is not only independent of the choice of units of X and Y but is also independent of the choice of origin.

SHORT COMPUTATIONAL FORMULAE

Formula (17) can be written in the equivalent form

$$r = \frac{N \Sigma XY - (\Sigma X)(\Sigma Y)}{\sqrt{[N \Sigma X^2 - (\Sigma X)^2][N \Sigma Y^2 - (\Sigma Y)^2]}} \qquad (20)$$

which is often used in computing r (see Problems 14.15 and 14.16).

For data grouped as in a *bivariate table* or *bivariate frequency distribution* (see Problem 14.17), it is convenient to use a *coding method* as in previous chapters. In such case (20) can be written

$$r = \frac{N \Sigma f u_X u_Y - (\Sigma f_X u_X)(\Sigma f_Y u_Y)}{\sqrt{[N \Sigma f_X u_X^2 - (\Sigma f_X u_X)^2][N \Sigma f_Y u_Y^2 - (\Sigma f_Y u_Y)^2]}} \qquad (21)$$

See Problem 14.18. For convenience in calculations using this formula, a *correlation table* is used (see Problem 14.19).

For grouped data, formulae (18) can be written

$$s_{XY} = c_X c_Y \left[\frac{\Sigma f u_X u_Y}{N} - \left(\frac{\Sigma f_X u_X}{N}\right)\left(\frac{\Sigma f_Y u_Y}{N}\right) \right] \qquad (22)$$

$$s_X = c_X \sqrt{\frac{\Sigma f_X u_X^2}{N} - \left(\frac{\Sigma f_X u_X}{N}\right)^2} \qquad (23)$$

$$s_Y = c_Y \sqrt{\frac{\Sigma f_Y u_Y^2}{N} - \left(\frac{\Sigma f_Y u_Y}{N}\right)^2} \qquad (24)$$

where c_X and c_Y are the class interval widths (assumed constant) corresponding to the variables X and Y respectively. Note that (23) and (24) are equivalent to formula (11) of Chapter 4, Page 71.

Formula (19) is seen to be equivalent to (21) if the results (22)–(24) are used.

REGRESSION LINES AND THE LINEAR CORRELATION COEFFICIENT

The equation of the least square line $Y = a_0 + a_1 X$, or regression line of Y on X, can be written as

$$Y - \bar{Y} = \frac{r s_Y}{s_X}(X - \bar{X}) \quad \text{or} \quad y = \frac{r s_Y}{s_X} x \qquad (25)$$

Similarly the regression line of X on Y, $X = b_0 + b_1 Y$, can be written

$$X - \bar{X} = \frac{r s_X}{s_Y}(Y - \bar{Y}) \quad \text{or} \quad x = \frac{r s_X}{s_Y} y \qquad (26)$$

The slopes of the lines (25) and (26) are equal if and only if $r = \pm 1$. In such case the two lines are identical and there is perfect linear correlation between the variables X and Y. If $r = 0$ the lines are at right angles and there is no linear correlation between X and Y. Thus the linear correlation coefficient measures the departure of the two regression lines.

Note that if the equations (25) and (26) are written $Y = a_0 + a_1 X$ and $X = b_0 + b_1 Y$ respectively, then $a_1 b_1 = r^2$ (see Problem 14.22).

RANK CORRELATION

Instead of using precise values of the variables, or when such precision is unavailable, the data may be ranked in order of size, importance, etc., using the numbers $1, 2, \ldots, N$. If two variables X and Y are ranked in such manner the *coefficient of rank correlation* is given by

$$r_{\text{rank}} = 1 - \frac{6\,\Sigma D^2}{N(N^2 - 1)} \tag{27}$$

where $D =$ differences between ranks of corresponding values of X and Y

$N =$ number of pairs of values (X, Y) in the data.

The formula (27) is called *Spearman's formula for rank correlation*.

CORRELATION OF TIME SERIES

If each of the variables X and Y depends on time, it is possible that a relationship may exist between X and Y even though such relationship is not necessarily one of direct dependence and may produce "nonsense correlation". The correlation coefficient is obtained by simply considering the pairs of values (X, Y) corresponding to the various times and proceeding as usual, making use of the above formulae. See Problem 14.18.

It is possible to attempt to correlate values of a variable X at certain times with corresponding values of X at earlier times. Such correlation is often called *autocorrelation*.

CORRELATION OF ATTRIBUTES

The methods described in this chapter do not enable us to consider the correlation of variables which are non-numerical by nature, such as the *attributes* of individuals (e.g. hair colour, eye colour, etc.). For a discussion of correlation of attributes, see Chap. 12.

SAMPLING THEORY OF CORRELATION

The N pairs of values (X, Y) of two variables can be thought of as a sample from a population of all possible such pairs. Since two variables are involved this is called a *bivariate population*, which we assume is a *bivariate normal distribution*.

We can think of a theoretical population coefficient of correlation denoted by ρ, which is estimated by the sample correlation coefficient r. Tests of significance or hypotheses concerning various values of ρ require knowledge of the sampling distribution of r. For $\rho = 0$ this distribution is symmetric and a statistic involving Student's distribution can be used. For $\rho \neq 0$ the distribution is skewed. In such case a transformation due to *Fisher* produces a statistic which is approximately normally distributed. The following tests summarize the procedures involved.

1. Test of Hypothesis $\rho = 0$.

Here we use the fact that the statistic

$$t = \frac{r\sqrt{N-2}}{\sqrt{1-r^2}} \tag{28}$$

has Student's distribution with $v = N - 2$ degrees of freedom. See Problems 14.33 and 14.34.

2. Test of Hypothesis $\rho = \rho_0 \neq 0$.

Here we use the fact that the statistic

$$Z = \tfrac{1}{2} \log_e \left(\frac{1+r}{1-r}\right) = 1.1513 \log_{10} \left(\frac{1+r}{1-r}\right) \tag{29}$$

where $e = 2.71828\ldots$, is approximately normally distributed with mean and standard deviation given by

$$\mu_Z = \tfrac{1}{2} \log_e \left(\frac{1+\rho_0}{1-\rho_0}\right) = 1.1513 \log_{10} \left(\frac{1+\rho_0}{1-\rho_0}\right), \qquad \sigma_Z = \frac{1}{\sqrt{N-3}} \tag{30}$$

These facts can also be used to find confidence limits for correlation coefficients (see Problems 14.35, 14.36). The transformation (29) is called *Fisher's Z transformation*.

3. Significance of a Difference between Correlation Coefficients.

To determine whether two correlation coefficients r_1 and r_2, drawn from samples of sizes N_1 and N_2 respectively, differ significantly from each other, we compute Z_1 and Z_2 corresponding to r_1 and r_2 using (29). We then use the fact that the test statistic

$$z = \frac{Z_1 - Z_2 - \mu_{Z_1 - Z_2}}{\sigma_{Z_1 - Z_2}} \tag{31}$$

where $\mu_{Z_1 - Z_2} = \mu_{Z_1} - \mu_{Z_2}$ and $\sigma_{Z_1 - Z_2} = \sqrt{\sigma_{Z_1}^2 + \sigma_{Z_2}^2} = \sqrt{\frac{1}{N_1 - 3} + \frac{1}{N_2 - 3}}$

is normally distributed (see Problem 13.37).

SAMPLING THEORY OF REGRESSION

The regression equation $Y = a_0 + a_1 X$ is obtained on the basis of sample data. We are often interested in the corresponding regression equation for the population from which the sample was drawn. The following are two tests concerning such a population.

1. Test of Hypothesis $a_1 = A_1$.

To test the hypothesis that the regression coefficient a_1 is equal to some specified value A_1, we use the fact that the statistic

$$t = \frac{a_1 - A_1}{s_{Y.X}/s_X} \sqrt{N-2} = \frac{a_1 - A_1}{\sqrt{1 - r^2}} \sqrt{N-2} \tag{32}$$

has Student's distribution with $N - 2$ degrees of freedom. This can also be used to find confidence intervals for population regression coefficients from sample values. See Problems 14.38 and 14.39.

2. Test of Hypothesis for Predicted Values.

Let Y_0 denote the predicted value of Y corresponding to $X = X_0$ as estimated from the sample regression equation, i.e. $Y_0 = a_0 + a_1 X_0$. Let Y_p denote the predicted value of Y corresponding to $X = X_0$ for the population. Then the statistic

$$t = \frac{Y_0 - Y_p}{s_{Y.X}\sqrt{N + 1 + (X_0 - \bar{X})^2/s_X^2}} \sqrt{N-2} = \frac{Y_0 - Y_p}{\hat{s}_{Y.X}\sqrt{1 + 1/N + (X_0 - \bar{X})^2/(Ns_X^2)}} \tag{33}$$

has Student's distribution with $N - 2$ degrees of freedom. From this, confidence limits for predicted population values can be found. See Problem 14.40.

3. Test of Hypothesis for Predicted Mean Values.

Let Y_0 denote the predicted value of Y corresponding to $X = X_0$ as estimated from the sample regression equation, i.e. $Y_0 = a_0 + a_1 X_0$. Let $\bar{Y}_p$ denote the predicted *mean value* of Y corresponding to $X = X_0$ for the population. Then the statistic

$$ t = \frac{Y_0 - \bar{Y}_p}{s_{Y.X}\sqrt{1 + (X_0 - \bar{X})^2/s_X^2}}\sqrt{N-2} = \frac{Y_0 - \bar{Y}_p}{\hat{s}_{Y.X}\sqrt{1/N + (X_0 - \bar{X})^2/(Ns_X^2)}} \tag{34} $$

has Student's distribution with $N - 2$ degrees of freedom. From this, confidence limits for predicted mean population values can be found. See Problem 14.41.

Solved Problems

SCATTER DIAGRAMS AND REGRESSION LINES

14.1. Table 14.1 shows the respective masses X and Y of a sample of 12 fathers and their oldest sons.

(a) Construct a scatter diagram.

(b) Find the least square regression line of Y on X.

(c) Find the least square regression line of X on Y.

Table 14.1

Mass X of Father (kg)	65	63	67	64	68	62	70	66	68	67	69	71
Mass Y of Son (kg)	68	66	68	65	69	66	68	65	71	67	68	70

Solution:

(a) The scatter diagram is obtained by plotting the points (X, Y) on a rectangular co-ordinate system as shown in Fig. 14-2.

(b) The regression line of Y on X is given by $Y = a_0 + a_1 X$, where a_0 and a_1 are obtained by solving the normal equations

$$ \begin{cases} \Sigma Y = a_0 N + a_1 \Sigma X \\ \Sigma XY = a_0 \Sigma X + a_1 \Sigma X^2 \end{cases} $$

The sums are shown in Table 14.2, and so the normal equations become

$$ \begin{cases} 12a_0 + 800a_1 = 811 \\ 800a_0 + 53\,418a_1 = 54\,107 \end{cases} $$

from which we find $a_0 = 35 \cdot 82$ and $a_1 = 0 \cdot 476$, so that $Y = 35 \cdot 82 + 0 \cdot 476X$. The graph of this equation is shown in Fig. 14-2.

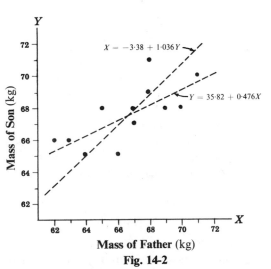

$X = -3 \cdot 38 + 1 \cdot 036Y$

$Y = 35 \cdot 82 + 0 \cdot 476X$

Mass of Father (kg)

Fig. 14-2

Table 14.2

X	Y	X^2	XY	Y^2
65	68	4225	4420	4624
63	66	3969	4158	4356
67	68	4489	4556	4624
64	65	4096	4160	4225
68	69	4624	4692	4761
62	66	3844	4092	4356
70	68	4900	4760	4624
66	65	4356	4290	4225
68	71	4624	4828	5041
67	67	4489	4489	4489
69	68	4761	4692	4624
71	70	5041	4970	4900
$\Sigma X = 800$	$\Sigma Y = 811$	$\Sigma X^2 = 53\,418$	$\Sigma XY = 54\,107$	$\Sigma Y^2 = 54\,849$

Another method:

$$a_0 = \frac{(\Sigma Y)(\Sigma X^2) - (\Sigma X)(\Sigma XY)}{N\,\Sigma\,Y^2 - (\Sigma Y)^2} = 35{\cdot}82, \qquad a_1 = \frac{N\Sigma\,XY - (\Sigma Y)(\Sigma Y)}{N\,\Sigma\,X^2 - (\Sigma X)^2}$$

(c) The regression line of X on Y is given by $X = b_0 + b_1 Y$, where b_0 and b_1 are obtained by solving the normal equations

$$\begin{cases} \Sigma X = b_0 N + b_1 \Sigma\,Y \\ \Sigma XY = b_0\,\Sigma\,Y + b_1\,\Sigma\,Y^2 \end{cases}$$

Using the sums in Table 14.2, these become

$$\begin{cases} 12b_0 + 811b_1 = 800 \\ 811b_0 + 54\,849b_1 = 54\,107 \end{cases}$$

from which we find $b_0 = -3{\cdot}38$ and $b_1 + 1{\cdot}036$, so that $X = -3{\cdot}38 + 1{\cdot}036\,Y$. The graph of this equation is shown in Fig. 14-2.

Another method:

$$b_0 = \frac{(\Sigma X)(\Sigma Y^2) - (\Sigma Y)(\Sigma XY)}{N\,\Sigma\,Y^2 - (\Sigma\,Y)^2} = -3{\cdot}38, \qquad b_1 = \frac{N\,\Sigma\,XY - (\Sigma Y)\Sigma X)}{N\,\Sigma\,Y^2 - (\Sigma Y)^2} = 1{\cdot}036$$

14.2. Work Prob. 14.1(b) and (c) by using the regression lines $y = \left(\dfrac{\Sigma xy}{\Sigma x^2}\right) x$ and $x = \left(\dfrac{\Sigma xy}{\Sigma y^2}\right) y$ where $x = X - \bar{X},\, y = Y - \bar{Y}$.

Solution:

First method: The work may be arranged as in Table 14.3.

Table 14.3

X	Y	$x = X - \bar{X}$	$y = Y - \bar{Y}$	x^2	xy	y^2
65	68	$-1{\cdot}7$	$0{\cdot}4$	$2{\cdot}89$	$-0{\cdot}68$	$0{\cdot}16$
63	66	$-3{\cdot}7$	$-1{\cdot}6$	$13{\cdot}69$	$5{\cdot}92$	$2{\cdot}56$
67	68	$0{\cdot}3$	$0{\cdot}4$	$0{\cdot}09$	$0{\cdot}12$	$0{\cdot}16$
64	65	$-2{\cdot}7$	$-2{\cdot}6$	$7{\cdot}29$	$7{\cdot}02$	$6{\cdot}76$
68	69	$1{\cdot}3$	$1{\cdot}4$	$1{\cdot}69$	$1{\cdot}82$	$1{\cdot}96$
62	66	$-4{\cdot}7$	$-1{\cdot}6$	$22{\cdot}09$	$7{\cdot}52$	$2{\cdot}56$
70	68	$3{\cdot}3$	$0{\cdot}4$	$10{\cdot}89$	$1{\cdot}32$	$0{\cdot}16$
66	65	$-0{\cdot}7$	$-2{\cdot}6$	$0{\cdot}49$	$1{\cdot}82$	$6{\cdot}76$
68	71	$1{\cdot}3$	$3{\cdot}4$	$1{\cdot}69$	$4{\cdot}42$	$11{\cdot}56$
67	67	$0{\cdot}3$	$-0{\cdot}6$	$0{\cdot}09$	$-0{\cdot}18$	$0{\cdot}36$
69	68	$2{\cdot}3$	$0{\cdot}4$	$5{\cdot}29$	$0{\cdot}92$	$0{\cdot}16$
71	70	$4{\cdot}3$	$2{\cdot}4$	$18{\cdot}49$	$10{\cdot}32$	$5{\cdot}76$
$\Sigma X = 800$ $\bar{X} = 800/12$ $= 66{\cdot}7$	$\Sigma Y = 811$ $\bar{Y} = 811/12$ $= 67{\cdot}6$			$\Sigma x^2 = 84{\cdot}68$	$\Sigma xy = 40{\cdot}34$	$\Sigma y^2 = 38{\cdot}92$

Regression line of Y on X is $y = \left(\dfrac{\Sigma xy}{\Sigma x^2}\right) x = \left(\dfrac{40\cdot34}{84\cdot68}\right) x = 0\cdot476x$ or $Y - 67\cdot6 = 0\cdot476(X - 66\cdot7)$.

Regression line of X on Y is $x = \left(\dfrac{\Sigma xy}{\Sigma y^2}\right) y = \left(\dfrac{40\cdot34}{38\cdot92}\right) y = 1\cdot036y$ or $X - 66\cdot7 = 1\cdot036(Y - 67\cdot6)$.

These agree with the results of Problem 14.1.

Second method:

Subtract a suitable constant, say 60, from each of the values of X and Y and proceed as in the second method of Prob. 13.17, Chap. 13.

Table 14.4

X'	Y'	X'^2	$X'Y'$	Y'^2
5	8	25	40	64
3	6	9	18	36
7	8	49	56	64
4	5	16	20	25
8	9	64	72	81
2	6	4	12	36
10	8	100	80	64
6	5	36	30	25
8	11	64	88	121
7	7	49	49	49
9	8	81	72	64
11	10	121	110	100
$\Sigma X' = 80$	$\Sigma Y' = 91$	$\Sigma X'^2 = 618$	$\Sigma X'Y' = 647$	$\Sigma Y'^2 = 729$

Then $\quad a^1 = \dfrac{N \Sigma X'Y' - (\Sigma Y')(\Sigma Y')}{N \Sigma X'^2 - (\Sigma X')^2} \qquad\qquad b^1 = \dfrac{N \Sigma X'Y' - (\Sigma Y')(\Sigma X')}{N \Sigma Y'^2 - (\Sigma Y')^2} = 1\cdot036$

Since $\bar X = 60 + 80/12 = 66\cdot7$ and $\bar Y = 60 + 91/12 = 67\cdot6$, the required regression equations are as before.

Note that if a_0 and b_0 were computed by this method, we would *not* get the same results as before since they depend on the choice of origin. Thus this method is used *only* to obtain a_1 and b_1, which are independent of choice of origin.

STANDARD ERROR OF ESTIMATE

14.3.   If the regression line of Y on X is given by $Y = a_0 + a_0X$, prove that the standard error of estimate $s_{Y.X}$ is given by

$$s^2_{Y.X} = \frac{\Sigma Y^2 - a_0 \Sigma Y - a_1 \Sigma XY}{N}$$

Solution:

The values of Y as estimated from the regression line are given by $Y_{est.} = a_0 + a_1X$. Then

$$s^2_{Y.X} = \frac{\Sigma (Y - Y_{est.})^2}{N} = \frac{\Sigma (Y - a_0 - a_1X)^2}{N}$$

$$= \frac{\Sigma Y(Y - a_0 - a_1X) - a_0 \Sigma (Y - a_0 - a_1X) - a_1 \Sigma X(Y - a_0 - a_1X)}{N}$$

But $\qquad\qquad\qquad \Sigma (Y - a_0 - a_1X) = \Sigma Y - a_0N - a_1 \Sigma X = 0$

and $\qquad\qquad\qquad \Sigma X(Y - a_0 - a_1X) = \Sigma XY - a_0 \Sigma X - a_1 \Sigma X^2 = 0$

since from the normal equations

$$\begin{cases} \Sigma Y = a_0N + a_1 \Sigma X \\ \Sigma XY = a_0 \Sigma X + a_1 \Sigma X^2. \end{cases}$$

Then $\qquad\qquad s^2_{Y.X} = \dfrac{\Sigma Y(Y - a_0 - a_1X)}{N} = \dfrac{\Sigma Y^2 - a_0 \Sigma Y - a_1 \Sigma XY}{N}$

This result can be extended to non-linear regression equations.

14.4. If $x = X - \bar{X}$ and $y = Y - \bar{Y}$, show that the result of Problem 14.3 can be written

$$s_{Y.X}^2 = \frac{\Sigma y^2 - a_1 \Sigma xy}{N}$$

Solution:

From Problem 14.3, with $X = x + \bar{X}$ and $Y = y + \bar{Y}$, we have

$$N s_{Y.X}^2 = \Sigma Y^2 - a_0 \Sigma Y - a_1 \Sigma XY = \Sigma (y + \bar{Y})^2 - a_0 \Sigma (y + \bar{Y}) - a_1 \Sigma (x + \bar{X})(y + \bar{Y})$$

$$= \Sigma (y^2 + 2y\bar{Y} + \bar{Y}^2) - a_0(\Sigma y + N\bar{Y}) - a_1 \Sigma (xy + \bar{X}y + x\bar{Y} + \bar{X}\bar{Y})$$

$$= \Sigma y^2 + 2\bar{Y} \Sigma y + N\bar{Y}^2 - a_0 N\bar{Y} - a_1 \Sigma xy - a_1 \bar{X} \Sigma y - a_1 \bar{Y} \Sigma x - a_1 N\bar{X}\bar{Y}$$

$$= \Sigma y^2 + N\bar{Y}^2 - a_0 N\bar{Y} - a_1 \Sigma xy - a_1 N\bar{X}\bar{Y} = \Sigma y^2 - a_1 \Sigma xy + N\bar{Y}(\bar{Y} - a_0 - a_1 \bar{X})$$

$$= \Sigma y^2 - a_1 \Sigma xy$$

where we have used the results $\Sigma x = 0$, $\Sigma y = 0$ and $\bar{Y} = a_0 + a_1 \bar{X}$ (which follows on division of both sides of the normal equation $\Sigma Y = a_0 N + a_1 \Sigma X$ by N).

14.5. Compute the standard error of estimate, $s_{Y.X}$, for the data of Problem 14.1 by using (*a*) the definition, (*b*) the result of the Problem 14.4.

Solution:

(*a*) From Problem 14.1(*b*) the regression line of Y on X is $Y = 35 \cdot 82 + 0 \cdot 476X$. In Table 14.5 are listed the actual values of Y (from the table of Problem 14.1) and the estimated values of Y, denoted by $Y_{est.}$ as obtained from the regression line. For example, corresponding to $X = 65$ we have $Y_{est.} = 35 \cdot 82 + 0 \cdot 476(65) = 66 \cdot 76$.

Also listed are the values $Y - Y_{est.}$ which are needed in computing $s_{Y.X}$

Table 14.5

X	65	63	67	64	68	62	70	66	68	67	69	71
Y	68	66	68	65	69	66	68	65	71	67	68	70
$Y_{est.}$	66·76	65·81	67·71	66·28	68·19	65·33	69·14	67·24	68·19	67·71	68·66	69·62
$Y - Y_{est.}$	1·24	0·19	0·29	−1·28	0·81	0·67	−1·14	−2·24	2·81	−0·71	−0·66	0·38

$$s_{Y.X}^2 = \frac{\Sigma(Y - Y_{est.})^2}{N} = \frac{(1 \cdot 24)^2 + (0 \cdot 19)^2 + \cdots + (0 \cdot 38)^2}{12} = 1 \cdot 642$$

and $s_{Y.X} = \sqrt{1 \cdot 642} = 1 \cdot 28$ kg.

(*b*) From Problems 1, 2 and 4,

$$s_{Y.X}^2 = \frac{\Sigma y^2 - a_1 \Sigma xy}{N} = \frac{38 \cdot 92 - 0 \cdot 476(40 \cdot 34)}{12} = 1 \cdot 643$$

and $s_{Y.X} = \sqrt{1 \cdot 643} = 1 \cdot 28$ kg.

14.6. (*a*) Construct two lines parallel to the regression line of Problem 14.1 and having vertical distance $s_{Y.X}$ from it. (*b*) Determine the percentage of data points falling between these two lines.

Solution:

(a) The regression line $Y = 35 \cdot 82 + 0 \cdot 476X$ as obtained in Problem 14.1 is shown heavy in Fig. 14-3. The two parallel lines, each having vertical distance $s_{Y.X} = 1 \cdot 28$ (see Problem 14.5) from it, are shown dashed in Fig. 14-3.

(b) From the figure it is seen that of the 12 data points 7 fall between the lines while 3 appear to lie on the lines. Further examination using the last line in Table 14.5 of Problem 14.5, for example, reveals that 2 of these 3 points lie between the lines. Then the required percentage $= 9/12 = 75\%$.

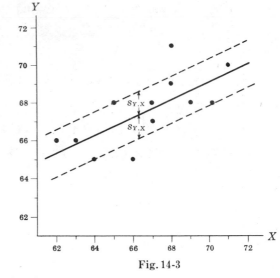

Fig. 14-3

Another method:

From the last line in Table 14.5 of Problem 14.5, $Y - Y_{est.}$ lies between $-1 \cdot 28$ and $1 \cdot 28$ (i.e. $\pm s_{Y.X}$) for 9 points (X, Y). Then the required percentage is $9/12 = 75\%$.

If the points are normally distributed about the regression line, theory predicts that about 68% of the points lie between the lines. This would have been more nearly the case if the sample size were large.

Note: A better estimate of the standard error of estimate of the population from which the sample heights were taken is given by $\hat{s}_{Y.X} = \sqrt{N/(N-2)}\, s_{Y.X} = \sqrt{12/10}\,(1 \cdot 28) = 1 \cdot 40$ kg.

EXPLAINED AND UNEXPLAINED VARIATION

14.7. Prove that $\Sigma(Y - \bar{Y})^2 = \Sigma(Y - Y_{est.})^2 + \Sigma(Y_{est.} - \bar{Y})^2$.

Solution:

Squaring both sides of $Y - \bar{Y} = (Y - Y_{est.}) + (Y_{est.} - \bar{Y})$ and then summing, we have

$$\Sigma(Y - \bar{Y})^2 = \Sigma(Y - Y_{est.})^2 + \Sigma(Y_{est.} - \bar{Y})^2 + 2\,\Sigma(Y - Y_{est.})(Y_{est.} - \bar{Y})$$

The required result follows at once if we can show that the last sum is zero. In the case of linear regression this is so since

$$\Sigma(Y - Y_{est.})(Y_{est.} - \bar{Y}) = \Sigma(Y - a_0 - a_1X)(a_0 + a_1X - \bar{Y})$$
$$= a_0\,\Sigma(Y - a_0 - a_1X) + a_1\,\Sigma X(Y - a_0 - a_1X) - \bar{Y}\,\Sigma(Y - a_0 - a_1X) = 0$$

because of the normal equations $\Sigma(Y - a_0 - a_1X) = 0$ and $\Sigma X(Y - a_0 - a_1X) = 0$.

The result can similarly be shown valid for non-linear regression using a least square curve given by $Y_{est.} = a_0 + a_1X + a_2X^2 + \ldots + a_nX^n$.

14.8. Compute (a) the total variation, (b) the unexplained variation and (c) the explained variation for the data of Problem 14.1.

Solution:

(a) Total variation $= \Sigma(Y - \bar{Y})^2 = \Sigma y^2 = 38 \cdot 92$ from Problem 14.2.

(b) Unexplained variation $= \Sigma(Y - Y_{est.})^2 = Ns_{Y.X}^2 = 19 \cdot 70$ from Problem 14.5.

(c) Explained variation $= \Sigma(Y_{est.} - \bar{Y})^2 = 38 \cdot 92 - 19 \cdot 70 = 19 \cdot 22$ using Problem 14.7.

Another method:

Since $\bar{Y} = 811/12 = 67 \cdot 58$, we can construct the following table using the values $Y_{est.}$ as obtained in Table 14.5 of Problem 14.5.

$Y_{est.} - \bar{Y}$ or $Y_{est.} - 67.58$	-0.82 -1.77 0.13 -1.30 0.61 -2.25 1.56 -0.34 0.61 0.13 1.08 2.04

Then $\Sigma(Y_{est.} - \bar{Y})^2 = (-0.82)^2 + (-1.77)^2 + \ldots + (2.04)^2 = 19.21$.
The results of (a) and (b) can also be obtained directly.

COEFFICIENT OF CORRELATION

14.9. Find (a) the coefficient of determination and (b) the coefficient of correlation for the data of Problem 14.1. Use the results of Problem 14.8.

Solution:

(a) Coefficient of determination $= r^2 = \dfrac{\text{explained variation}}{\text{total variation}} = \dfrac{19.22}{38.92} = 0.4938$.

(b) Coefficient of correlation $= r = \pm\sqrt{0.4938} = \pm 0.7027$.

Since the variable $Y_{est.}$ increases as X increases, the correlation is positive and we therefore write $r = 0.7027$, or 0.70 to two significant figures.

14.10. Prove that for linear regression the coefficient of correlation between the variables X and Y can be written

$$r = \frac{\Sigma xy}{\sqrt{(\Sigma x^2)(\Sigma y^2)}}$$

where $x = X - \bar{X}$ and $y = Y - \bar{Y}$.

Solution:

The least square regression line of Y on X can be written $Y_{est.} = a_0 + a_1 X$ or $y_{est.} = a_1 x$, where $a_1 = \dfrac{\Sigma xy}{\Sigma x^2}$

and $y_{est.} = Y_{est.} - \bar{Y}$ (see Problem 15(a) of Chapter 13). Then

$$r^2 = \frac{\text{explained variation}}{\text{total variation}} = \frac{\Sigma(Y_{est.} - \bar{Y})^2}{\Sigma(Y - \bar{Y})^2} = \frac{\Sigma y_{est.}^2}{\Sigma y^2}$$

$$= \frac{\Sigma a_1^2 x^2}{\Sigma y^2} = \frac{a_1^2 \Sigma x^2}{\Sigma y^2} = \left(\frac{\Sigma xy}{\Sigma x^2}\right)^2 \frac{\Sigma x^2}{\Sigma y^2} = \frac{(\Sigma xy)^2}{(\Sigma x^2)(\Sigma y^2)}$$

and $r = \pm\dfrac{\Sigma xy}{\sqrt{(\Sigma x^2)(\Sigma y^2)}}$. However, since the quantity $\dfrac{\Sigma xy}{\sqrt{(\Sigma x^2)(\Sigma y^2)}}$ is positive when $y_{est.}$ increases as

x increases (i.e. positive linear correlation) and negative when $y_{est.}$ decreases as x increases (i.e. negative linear correlation), it *automatically* has the correct sign associated with it. Hence we define the coefficient of linear correlation to be

$$r = \frac{\Sigma xy}{\sqrt{(\Sigma x^2)(\Sigma y^2)}}$$

This is often called the *product-moment formula* for the linear correlation coefficient.

PRODUCT-MOMENT FORMULA FOR LINEAR CORRELATION COEFFICIENT

14.11. Find the coefficient of linear correlation between the variables X and Y presented in Table 14.6.

Table 14.6

X	1	3	4	6	8	9	11	14
Y	1	2	4	4	5	7	8	9

Solution:

The work involved in the computation can be organized as in Table 14.7.

Table 14.7

X	Y	$x = X - \bar{X}$	$y = Y - \bar{Y}$	x^2	xy	y^2
1	1	-6	-4	36	24	16
3	2	-4	-3	16	12	9
4	4	-3	-1	9	3	1
6	4	-1	-1	1	1	1
8	5	1	0	1	0	0
9	7	2	2	4	4	4
11	8	4	3	16	12	9
14	9	7	4	49	28	16
$\Sigma X = 56$ $\bar{X} = 56/8 = 7$	$\Sigma Y = 40$ $\bar{Y} = 40/8 = 5$			$\Sigma x^2 = 132$	$\Sigma xy = 84$	$\Sigma y^2 = 56$

$$r = \frac{\Sigma xy}{\sqrt{(\Sigma x^2)(\Sigma y^2)}} = \frac{84}{\sqrt{(132)(56)}} = 0{\cdot}977$$

This shows that there is a very high linear correlation between the variables, as we have already observed in Problems 13.8 and 13.12 of Chapter 13.

14.12. Find (*a*) the standard deviation of X, (*b*) the standard deviation of Y, (*c*) the variance of X, (*d*) the variance of Y and (*e*) the covariance of X and Y for the data of Prob. 14.11.

Solution:
(*a*)   Standard deviation of $X = s_X = \sqrt{\dfrac{\Sigma (X - \bar{X})^2}{N}} = \sqrt{\dfrac{\Sigma x^2}{N}} = \sqrt{\dfrac{132}{8}} = 4{\cdot}06$

(*b*)   Standard deviation of $Y = s_Y = \sqrt{\dfrac{\Sigma (Y - \bar{Y})^2}{N}} = \sqrt{\dfrac{\Sigma y^2}{N}} = \sqrt{\dfrac{56}{8}} = 2{\cdot}65$

(*c*)   Variance of $X = s_X^2 = 16{\cdot}50$ (*d*)   Varaince of $Y = s_Y^2 = 7{\cdot}00$

(*e*)   Covariance of X and $Y = s_{XY} = \dfrac{\Sigma xy}{N} = \dfrac{84}{8} = 10{\cdot}50$

14.13. Verify the formula $r = \dfrac{s_{XY}}{s_X s_Y}$ for the data of Problem 14.11.

Solution:
From Problem 14.12, $r = \dfrac{s_{XY}}{s_X s_Y} = \dfrac{10{\cdot}50}{(4{\cdot}06)(2{\cdot}65)} = 0{\cdot}976$, in agreement (except for rounding errors) with the result of Problem 14.11.

14.14. By using the product-moment formula, obtain the linear correlation coefficient for the data of Problem 14.1.

Solution:
The work involved in the computation can be organized as in Table 14.3 of Problem 14.2. Then

$$r = \frac{\Sigma xy}{\sqrt{(\Sigma x^2)(\Sigma y^2)}} = \frac{40{\cdot}34}{\sqrt{(84{\cdot}68)(38{\cdot}92)}} = 0{\cdot}7027$$

agreeing with the longer method of Problem 14.9.

14.15. Show that the linear correlation coefficient is given by

$$r = \frac{N \Sigma XY - (\Sigma X)(\Sigma Y)}{\sqrt{[N \Sigma X^2 - (\Sigma X)^2][N \Sigma Y^2 - (\Sigma Y)^2]}}$$

Solution:

Writing $x = X - \bar{X}$, $y = Y - \bar{Y}$ in the result of Problem 10, we have

(1)
$$r = \frac{\Sigma xy}{\sqrt{(\Sigma x^2)(\Sigma y^2)}} = \frac{\Sigma (X - \bar{X})(Y - \bar{Y})}{\sqrt{[\Sigma (X - \bar{X})^2][\Sigma (Y - \bar{Y})^2]}}$$

But
$$\Sigma (X - \bar{X})(Y - \bar{Y}) = \Sigma (XY - \bar{X}Y - X\bar{Y} + \bar{X}\bar{Y}) = \Sigma XY - \bar{X} \Sigma Y - \bar{Y} \Sigma X + N\bar{X}\bar{Y}$$
$$= \Sigma XY - N\bar{X}\bar{Y} - N\bar{Y}\bar{X} + N\bar{X}\bar{Y} = \Sigma XY - N\bar{X}\bar{Y}$$
$$= \Sigma XY - \frac{(\Sigma X)(\Sigma Y)}{N}$$

since $\bar{X} = (\Sigma X)/N$ and $\bar{Y} = (\Sigma Y)/N$.

Similarly,
$$\Sigma (X - \bar{X})^2 = \Sigma (X^2 - 2X\bar{X} + \bar{X}^2) = \Sigma X^2 - 2\bar{X} \Sigma X + N\bar{X}^2$$
$$= \Sigma X^2 - \frac{2(\Sigma X)^2}{N} + \frac{(\Sigma X)^2}{N} = \Sigma X^2 - \frac{(\Sigma X)^2}{N}$$

and
$$\Sigma (Y - \bar{Y})^2 = \Sigma Y^2 - \frac{(\Sigma Y)^2}{N}. \quad \text{Then (1) becomes}$$

$$r = \frac{\Sigma XY - (\Sigma X)(\Sigma Y)/N}{\sqrt{[\Sigma X^2 - (\Sigma X)^2/N][\Sigma Y^2 - (\Sigma Y)^2/N]}} = \frac{N \Sigma XY - (\Sigma X)(\Sigma Y)}{\sqrt{[N \Sigma X^2 - (\Sigma X)^2][N \Sigma Y^2 - (\Sigma Y)^2]}}$$

14.16. Use the formula of Prob. 14.15 to obtain the linear correlation coefficient for the data of Prob. 14.1.

Solution:

From Table 14.2 of Problem 14.1, we have

$$r = \frac{N \Sigma XY - (\Sigma X)(\Sigma Y)}{\sqrt{[N \Sigma X^2 - (\Sigma X)^2][N \Sigma Y^2 - (\Sigma Y)^2]}}$$

$$= \frac{(12)(54\ 107) - (800)(811)}{\sqrt{[(12)(53\ 418) - (800)^2][(12)(54\ 849) - (811)^2]}} = 0.7027$$

as in Problems 14.9 and 14.14.

Another method:

The value of r is independent of the choice of origin of X and Y. We can thus use the results of the second method of Prob. 14.2 to obtain

$$r = \frac{N \Sigma X'Y' - (\Sigma X')(\Sigma Y')}{\sqrt{[N \Sigma X'^2 - (\Sigma X')^2][N \Sigma Y'^2 - (\Sigma Y')^2]}} = \frac{12(647) - (80)(91)}{\sqrt{[(12)(618) - (80)^2][(12)(729) - (91)^2]}} = 0.7027$$

CORRELATION COEFFICIENT FOR GROUPED DATA

14.17. Table 14.8 shows the frequency distributions of the final marks of 100 students in mathematics and physics. With reference to this table determine

(a) the number of students who received marks 70–79 in mathematics and 80–89 in physics,

(b) the percentage of students with mathematics marks below 70,

(c) the number of students who received a mark of 70 or more in physics and less than 80 in mathematics,

(d) the percentage of students who passed both physics and mathematics assuming 60 to be the minimum pass mark.

Table 14.8
Mathematics Marks

Physics Marks	40—49	50—59	60—69	70—79	80—89	90—99	Totals
90—99				2	4	4	10
80—89			1	4	6	5	16
70—79			5	10	8	1	24
60—69	1	4	9	5	2		21
50—59	3	6	6	2			17
40—49	3	5	4				12
Totals	7	15	25	23	20	10	100

Solution:

(*a*) Proceed down the column headed 70–79 (mathematics mark) to the row marked 80–89 (physics mark). The entry 4 gives the required number of students.

(*b*) Total number of students with mathematics marks below 70
= number with marks 40–49 + number with marks 50–59 + number with marks 60–69
= 7 + 15 + 25 = 47.
Percentage of students with mathematics marks below 70 = 47/100 = 47%.

(*c*) The required number of students is the total of the entries in Table 14.9, which represents part of Table 14.8.
Required number of students = 1 + 5 + 2 + 4 + 10 = 22.

Table 14.9
Mathematics Marks

Physics Marks	60—69	70—79
90—99		2
80—89	1	4
70—79	5	10

Table 14.10
Mathematics Marks

Physics Marks	40—49	50—59
50—59	3	6
40—49	3	5

(*d*) Referring to Table 14.10 which is taken from Table 14.8, it is seen that the number of students with marks below 60 in both mathematics and physics is 3 + 3 + 6 + 5 = 17. Then the number of students with marks 60 or over in both physics and mathematics = 100 − 17 = 83, and the required percentage is 83/100 = 83%.

Table 14.8 is sometimes called a *bivariate frequency table* or *bivariate frequency distribution.* Each square in the table is called a *cell* and corresponds to a pair of classes or class intervals. The number indicated in the cell is called the *cell frequency*. For example, in part (*a*) the number 4 is the frequency of the cell corresponding to the pair of class intervals 70–79 in mathematics and 80–89 in physics.

The totals indicated in the last row and last column are called *marginal totals* or *marginal frequencies.* They correspond respectively to the class frequencies of the separate frequency distributions of mathematics and physics grades.

14.18. Show how to modify the formula of Problem 14.15 for the case of data grouped as in the bivariate frequency table (Table 14.8) of Problem 14.17.

Solution:

For grouped data, we can consider the various values of the variables X and Y as coinciding with the class marks, while f_X and f_Y are the corresponding *class* frequencies or marginal frequencies indicated in the last row and column of the bivariate frequency table. If we let f represent the various cell frequencies corresponding to the pairs of class marks (X, Y), then we can replace the formula of Problem 14.15 by

$$(1) \qquad r = \frac{N \Sigma fXY \quad (\Sigma f_X X)(\Sigma f_Y Y)}{\sqrt{[N \Sigma f_X X^2 - (\Sigma f_X X)^2][N \Sigma f_Y Y^2 - (\Sigma f_Y Y)^2]}}$$

If we let $X = A + c_X u_X$ and $Y = B + c_Y u_Y$, where c_X and c_Y are the class interval widths (assumed constant) and A and B are arbitrary class marks corresponding to the variables, the above formula becomes

$$(2) \qquad r = \frac{N \Sigma f u_X u_Y - (\Sigma f_X u_X)(\Sigma f_Y u_Y)}{\sqrt{[N \Sigma f_X u_X^2 - (\Sigma f_X u_X)^2][N \Sigma f_Y u_Y^2 - (\Sigma f_Y u_Y)^2]}}$$

This is the *coding method* used in previous chapters as a short method for computing means, standard deviations and higher moments.

14.19. Find the coefficient of linear correlation of the mathematics and physics marks of Problem 14.17.

Solution:

We use formula (2) of Problem 14.18. The work can be arranged as in Table 14.11 which is called a *correlation table*. The sums Σf_X, $\Sigma f_X u_X$, $\Sigma f_X u_X^2$, Σf_Y, $\Sigma f_Y u_Y$ and $\Sigma f_Y u_Y^2$ are obtained by using the coding method as in earlier chapters.

Table 14.11

		Mathematics Marks X									
	X	44·5	54·5	64·5	74·5	84·5	94·5	f_Y	$f_Y u_Y$	$f_Y u_Y^2$	Sum of corner numbers in each row
Y	u_X \ u_Y	−2	−1	0	1	2	3				
94·5	2				2 [4]	4 [16]	4 [24]	10	20	40	44
84·5	1			1 [0]	4 [4]	6 [12]	5 [15]	16	16	16	31
74·5	0			5 [0]	10 [0]	8 [0]	1 [0]	24	0	0	0
64·5	−1	1 [2]	4 [4]	9 [0]	5 [−5]	2 [−4]		21	−21	21	−3
54·5	−2	3 [12]	6 [12]	6 [0]	2 [−4]			17	−34	68	20
44·5	−3	3 [18]	5 [15]	4 [0]				12	−36	108	33
f_X		7	15	25	23	20	10	$\Sigma f_X = \Sigma f_Y$ $= N = 100$	$\Sigma f_Y u_Y$ $= -55$	$\Sigma f_Y u_Y^2$ $= 253$	$\Sigma f u_X u_Y$ $= 125$
$f_X u_X$		−14	−15	0	23	40	30	$\Sigma f_X u_X$ $= 64$			
$f_X u_X^2$		28	15	0	23	80	90	$\Sigma f_X u_X^2$ $= 236$			
Sum of corner numbers in each column		32	31	0	−1	24	39	$\Sigma f u_X u_Y$ $= 125$			

Check

The number in the corner of each cell represents the product $f u_X u_Y$, where f is the cell frequency. The sum of these corner numbers in each row is indicated in the corresponding row of the last column. The sum of these corner numbers in each column is indicated in the corresponding column of the last row. The final totals of the last row and last column are equal and represent $\Sigma f u_X u_Y$.

From Table 14.11 we have

$$r = \frac{N \Sigma f u_X u_Y - (\Sigma f_X u_X)(\Sigma f_Y u_Y)}{\sqrt{[N \Sigma f_X u_X^2 - (\Sigma f_X u_X)^2][N \Sigma f_Y u_Y^2 - (\Sigma f_Y u_Y)^2]}}$$

$$= \frac{(100)(125) - (64)(-55)}{\sqrt{[(100)(236) - (64)^2][(100)(253) - (-55)^2]}} = \frac{16\,020}{\sqrt{(19\,504)(22\,275)}} = 0.7686$$

14.20. Use the correlation table of Problem 14.19 to compute (a) s_X, (b) s_Y and (c) s_{XY} and verify the formula $r = s_{XY}/(s_X s_Y)$.

Solution:

(a) $s_X = c_X \sqrt{\dfrac{\Sigma f_X u_X^2}{N} - \left(\dfrac{\Sigma f_X u_X}{N}\right)^2} = 10 \sqrt{\dfrac{236}{100} - \left(\dfrac{64}{100}\right)^2} = 13.966$

(b) $s_Y = c_Y \sqrt{\dfrac{\Sigma f_Y u_Y^2}{N} - \left(\dfrac{\Sigma f_Y u_Y}{N}\right)^2} = 10 \sqrt{\dfrac{253}{100} - \left(\dfrac{-55}{100}\right)^2} = 14.925$

(c) $s_{XY} = c_X c_Y \left[\dfrac{\Sigma f u_X u_Y}{N} - \left(\dfrac{\Sigma f_X u_X}{N}\right)\left(\dfrac{\Sigma f_Y u_Y}{N}\right)\right] = (10)(10)\left[\dfrac{125}{100} - \left(\dfrac{64}{100}\right)\left(\dfrac{-55}{100}\right)\right] = 160.20$

Thus the standard deviation of mathematics marks and physics marks are 14·0 and 14·9 respectively, while their covariance is 160·2.

The correlation coefficient $r = \dfrac{s_{XY}}{s_X s_Y} = \dfrac{160.20}{(13.966)(14.925)} = 0.7686$, agreeing with Problem 14.19.

REGRESSION LINES AND THE CORRELATION COEFFICIENT

14.21. Prove that the regression lines of Y on X and X on Y have equations given respectively by

$$(a) \quad Y - \bar{Y} = \frac{r s_Y}{s_X}(X - \bar{X}) \quad \text{and} \quad (b) \quad X - \bar{X} = \frac{r s_X}{s_Y}(Y - \bar{Y})$$

Solution:
(a) By Problem 15(a) of Chapter 13, the regression line of Y on X has the equation

$$y = \left(\frac{\Sigma xy}{\Sigma x^2}\right)x \quad \text{or} \quad Y - \bar{Y} = \left(\frac{\Sigma xy}{\Sigma x^2}\right)(X - \bar{X})$$

Then since $r = \dfrac{\Sigma xy}{\sqrt{(\Sigma x^2)(\Sigma y^2)}}$ (see Prob. 14.10), $\dfrac{\Sigma xy}{\Sigma x^2} = \dfrac{r\sqrt{(\Sigma x^2)(\Sigma y^2)}}{\Sigma x^2} = \dfrac{r\sqrt{\Sigma y^2}}{\sqrt{\Sigma x^2}} = \dfrac{r s_Y}{s_X}$

and the required result follows.

(b) This follows by interchanging X and Y in part (a).

14.22. If the regression lines of Y and X are given respectively by $Y = a_0 + a_1 X$ and $X = b_0 + b_1 Y$, prove that $a_1 b_1 = r^2$.

Solution:
By Probs. 21(a) and 21(b), $a_1 = \dfrac{r s_Y}{s_X}$ and $b_1 = \dfrac{r s_X}{s_Y}$. Then $a_1 b_1 = \left(\dfrac{r s_Y}{s_X}\right)\left(\dfrac{r s_X}{s_Y}\right) = r^2$.

This result can be taken as starting point for a definition of the linear correlation coefficient.

14.23. Use the result of Problem 14.22 to find the linear correlation coefficient for the data of Problem 14.1.

Solution:

From Prob. 14.1(b) and 14.1(c) respectively, $a_1 = 484/1016 = 0.476$ and $b_1 = 484/467 = 1.036$.
Then $r^2 = a_1 b_1 = (484/1016)(484/467)$ and $r = 0.7027$, agreeing with Problems 14.9, 14.14 and 14.16.

14.24. Write the equations of the regression lines of (a) Y on X and (b) X on Y for the data of Problem 14.19.

Solution:

From the correlation table of Problem 14.19, we have

$$\bar{X} = A + c_X \frac{\Sigma f_X u_X}{N} = 64.5 + \frac{(10)(64)}{100} = 70.9$$

$$\bar{Y} = B + c_Y \frac{\Sigma f_Y u_Y}{N} = 75.4 + \frac{(10)(-55)}{100} = 69.0$$

From the results of Problem 14.20, $s_X = 13.966$, $s_Y = 14.925$ and $r = 0.7686$.

We now use Problems 14.21(a) and 14.21(b) to obtain the equations of the regression lines.

(a) $Y - \bar{Y} = \frac{rs_Y}{s_X}(X - \bar{X})$, $Y - 69.0 = \frac{(0.7686)(14.925)}{13.966}(X - 70.9)$, or $Y - 69.0 = 0.821(X - 70.9)$

(b) $X - \bar{X} = \frac{rs_X}{s_Y}(Y - \bar{Y})$, $X - 70.9 = \frac{(0.7686)(13.966)}{14.925}(Y - 69.0)$, or $X - 70.9 = 0.719(Y - 69.0)$

14.25. Compute the standard errors of estimate (a) $s_{Y.X}$ and (b) $s_{X.Y}$ for the data of Prob. 14.19. Use the results of Prob. 14.20.

Solution:

(a) $s_{Y.X} = s_Y\sqrt{1 - r^2} = 14.925\sqrt{1 - (0.7686)^2} = 9.548$

(b) $s_{X.Y} = s_X\sqrt{1 - r^2} = 13.966\sqrt{1 - (0.7686)^2} = 8.934$

RANK CORRELATION

14.26. The following table shows how 10 students, arranged in alphabetical order, were ranked according to their achievements in both laboratory and lecture portions of a biology course. Find the coefficient of rank correlation.

Laboratory	8	3	9	2	7	10	4	6	1	5
Lecture	9	5	10	1	8	7	3	4	2	6

Solution:

The difference of ranks D in laboratory and lecture for each student is given in the following table. Also given in the table are D^2 and ΣD^2.

Difference of ranks, D	−1	−2	−1	1	−1	3	1	2	−1	−1	
D^2	1	4	1	1	1	9	1	4	1	1	$\Sigma D^2 = 24$

Then

$$r_{rank} = 1 - \frac{6\Sigma D^2}{N(N^2 - 1)} = 1 - \frac{6(24)}{10(10^2 - 1)} = 0.8545$$

indicating that there is a marked relationship between achievements in laboratory and lecture.

14.27. Calculate the coefficient of rank correlation for the data of Problem 14.1 and compare your result with the correlation coefficient obtained by other methods.

> **Solution:**
> Arranged in ascending order of magnitude, the fathers' masses are
>
> (1)   62, 63, 64, 65, 66, 67, 67, 68, 68, 69, 70, 71
>
> Since the 6th and 7th places in this array represent the same mass (67 kg), we assign a *mean rank* 6·5 to these places. Similarly the 8th and 9th places are assigned the rank 8·5. Thus the fathers' masses are assigned the ranks
>
> (2)   1, 2, 3, 4, 5, 6·5, 6·5, 8·5, 8·5, 10, 11, 12
>
> Similarly the sons' masses arranged in ascending order of magnitude are
>
> (3)   65, 65, 66, 66, 67, 68, 68, 68, 68, 69, 70, 71
>
> and since the 6th, 7th, 8th and 9th places represent the same mass (68 kg), we assign the *mean rank* 7·5 [(6 + 7 + 8 + 9)/4] to these places. Thus the sons' masses are assigned the ranks
>
> (4)   1·5, 1·5, 3·5, 3·5, 5, 7·5, 7·5, 7·5, 7·5, 10, 11, 12
>
> Using the correspondences (1) and (2), (3) and (4), Table 14.1 of Problem 14.1 becomes

Rank of Father	4	2	6·5	3	8·5	1	11	5	8·5	6·5	10	12
Rank of Son	7·5	3·5	7·5	1·5	10	3·5	7·5	1·5	12	5	7·5	11

> The difference in ranks D, and the computations of D^2 and ΣD^2 are shown in the table below.

D	−3·5	−1·5	−1·0	1·5	−1·5	−2·5	3·5	3·5	−3·5	1·5	2·5	1·0	
D^2	12·25	2·25	1·00	2·25	2·25	6·25	12·25	12·25	12·25	2·25	6·25	1·00	ΣD^2 = 72·50

> Then
>
> $$r_{\text{rank}} = 1 - \frac{6 \Sigma D^2}{N(N^2 - 1)} = 1 - \frac{6(72 \cdot 50)}{12(12^2 - 1)} = 0 \cdot 7465$$
>
> which agrees well with the value $r = 0 \cdot 7027$ obtained in Problems 14.9, 14.14, 14.16 or 14.23 of Chapter 14.

CORRELATION OF TIME SERIES

14.28. The average price of stocks and bonds listed on the New York Stock Exchange during the years 1950–1959 are given in Table 14.12. (*a*) Find the correlation coefficient and (*b*) interpret the results.

Table 14.12

Year	1950	1951	1952	1953	1954	1955	1956	1957	1958	1959
Average price of stocks (dollars)	35·22	39·87	41·85	43·23	40·06	53·29	54·14	49·12	40·71	55·15
Average price of bonds (dollars)	102·43	100·93	97·43	97·81	98·32	100·07	97·08	91·59	94·85	94·65

Source: New York Stock Exchange

> **Solution:**
> (*a*) Denoting by X and Y the average prices of stocks and bonds, the calculation of the correlation coefficient can be organized as in Table 14.13. Note that the year is used only to specify the corresponding values of X and Y.

Table 14.13

X	Y	$x = X - \bar{X}$	$y = Y - \bar{Y}$	x^2	xy	y^2
35·22	102·43	−10·04	4·91	100·80	−49·30	24·11
39·87	100·93	−5·39	3·41	29·05	−18·38	11·63
41·85	97·43	−3·41	−0·09	11·63	0·31	0·01
43·23	97·81	−2·03	0·29	4·12	−0·59	0·08
40·06	98·32	−5·20	0·80	27·04	−4·16	0·64
53·29	100·07	8·03	2·55	64·48	20·48	6·50
54·14	97·08	8·88	−0·44	78·85	−3·91	0·19
49·12	91·59	3·86	−5·93	14·90	−22·89	35·16
40·71	94·85	−4·55	−2·67	20·70	12·15	7·13
55·15	94·65	9·89	−2·87	97·81	−28·38	8·24
ΣX = 452·64 X = 45·26	ΣY = 975·16 Y = 97·52			Σx^2 = 449·38	Σxy − 94·67	Σy^2 = 93·69

Then by the product-moment formula, $r = \dfrac{\Sigma xy}{\sqrt{(\Sigma x^2)(\Sigma y^2)}} = \dfrac{-94\cdot67}{\sqrt{(449\cdot38)(93\cdot69)}} = -0\cdot4614.$

(b) We conclude that there is some negative correlation between stock and bond prices (i.e. a tendency for stock prices to go down when bond prices go up, and vice versa) although this relationship is not marked.

Another method: using rank correlation (as in Problems 14.26 and 14.27).

Table 14.14 shows the ranks of the average prices of stocks and bonds for the years 1950–1959 in order of increasing prices. Also shown in the table are the differences in rank D and ΣD^2.

Table 14.14

Year	1950	1951	1952	1953	1954	1955	1956	1957	1958	1959	
Stock prices in order of rank	1	2	5	6	3	8	9	7	4	10	
Bond prices in order of rank	10	9	5	6	7	8	4	1	3	2	
Differences in rank D	−9	−7	0	0	−4	0	5	6	1	8	
D^2	81	49	0	0	16	0	25	36	1	64	$\Sigma D^2 =$ 272

Then
$$r_{\text{rank}} = 1 - \frac{6\,\Sigma\,D^2}{N(N^2 - 1)} = 1 - \frac{6(272)}{10(10^2 - 1)} = -0\cdot6485$$

This compares favourably with the result of the first method.

We can also proceed by subtracting suitable constants from the variables and then use the second method of Problem 14.16.

NON-LINEAR CORRELATION

14.29. Fit a least square parabola having the form $Y = a_0 + a_1X + a_2X^2$ to the following set of data.

Table 14.15

X	1·2	1·8	3·1	4·9	5·7	7·1	8·6	9·8
Y	4·5	5·9	7·0	7·8	7·2	6·8	4·5	2·7

Solution:

The normal equations are (see Chapter 13, Page 221),

$$(1) \quad \begin{cases} \Sigma Y = a_0 N + a_1 \Sigma X + a_2 \Sigma X^2 \\ \Sigma XY = a_0 \Sigma X + a_1 \Sigma X^2 + a_2 \Sigma X^3 \\ \Sigma X^2 Y = a_0 \Sigma X^2 + a_1 \Sigma X^3 + a_2 \Sigma X^4 \end{cases}$$

The work involved in computing the sums can be arranged as in Table 14.16.

Table 14.16

X	Y	X^2	X^3	X^4	XY	X^2Y
1·2	4·5	1·44	1·73	2·08	5·40	6·48
1·8	5·9	3·24	5·83	10·49	10·62	19·12
3·1	7·0	9·61	29·79	92·35	21·70	67·27
4·9	7·8	24·01	117·65	576·48	38·22	187·28
5·7	7·2	32·49	185·19	1055·58	41·04	233·93
7·1	6·8	50·41	357·91	2541·16	48·28	342·79
8·6	4·5	73·96	636·06	5470·12	38·70	332·82
9·8	2·7	96·04	941·19	9223·66	26·46	259·31
$\Sigma X = 42\cdot2$	$\Sigma Y = 46\cdot4$	ΣX^2 $= 291\cdot20$	ΣX^3 $= 2275\cdot35$	ΣX^4 $= 18\,971\cdot92$	ΣXY $= 230\cdot42$	ΣX^2Y $= 1449\cdot00$

Then the normal equations (1) become, since $N = 8$,

$$(2) \quad \begin{cases} 8a_0 + 42\cdot2a_1 + 291\cdot20a_2 = 46\cdot4 \\ 42\cdot2a_0 + 291\cdot20a_1 + 2275\cdot35a_2 = 230\cdot42 \\ 291\cdot20a_0 + 2275\cdot35a_1 + 18\,971\cdot92a_2 = 1449\cdot00 \end{cases}$$

Solving, $a_0 = 2\cdot588$, $a_1 = 2\cdot065$, $a_2 = -0\cdot2110$; hence the required least square parabola has the equation

$$Y = 2\cdot588 + 2\cdot065X - 0\cdot2110X^2$$

14.30. Use the least square parabola of Problem 14.29 to estimate the values of Y from the given values of X.

Solution:

For $X = 1\cdot2$, $Y_{\text{est.}} = 2\cdot588 + 2\cdot065(1\cdot2) - 0\cdot2110(1\cdot2)^2 = 4\cdot762$. Similarly other estimated values are obtained. The results are shown in Table 14.17 together with the actual values of Y.

Table 14.17

$Y_{\text{est.}}$	4·762	5·621	6·962	7·640	7·503	6·613	4·741	2·561
Y	4·5	5·9	7·0	7·8	7·2	6·8	4·5	2·7

14.31. (a) Find the linear correlation coefficient between the variables X and Y of Problem 14.29.

(b) Find the non-linear correlation coefficient between these variables, assuming the parabolic relationship obtained in Problem 14.29.

(c) Explain the difference between the correlation coefficients obtained in (a) and (b).

(d) What percentage of the total variation remains unexplained by the assumption of parabolic relationship between X and Y?

Solution:

(a) Using the calculations already obtained in Table 14.16 of Problem 14.29 an- the added fact that $\Sigma Y^2 = 290\cdot52$, we find

$$r = \frac{N\,\Sigma\,XY - (\Sigma X)(\Sigma Y)}{\sqrt{[N\,\Sigma\,X^2 - (\Sigma X)^2][N\,\Sigma\,Y^2 - (\Sigma Y)^2]}} = \frac{(8)(230\cdot42) - (42\cdot2)(46\cdot4)}{\sqrt{[(8)(291\cdot20) - (42\cdot2)^2][(8)(290\cdot52) - (46\cdot4)^2]}} = -0\cdot3743$$

(b) From Table 14.16 of Problem 14.29, $\bar{Y} = (\Sigma Y)/N = (46\cdot4)/8 = 5\cdot80$. Then, total variation $= \Sigma(Y - \bar{Y})^2 = 21\cdot40$. From Table 14.17 of Problem 14.30, explained variation $= \Sigma(Y_{est.} - \bar{Y})^2 = 21\cdot02$.

Thus $r^2 = \dfrac{\text{explained variation}}{\text{total variation}} = \dfrac{21\cdot02}{21\cdot40} = 0\cdot9822$, and $r = 0\cdot9911$ or $0\cdot99$

(c) The fact that part (a) shows a linear correlation coefficient of only $-0\cdot3743$ indicates practically no *linear relationship* between X and Y. However, there is a very good *non-linear relationship* supplied by the parabola of Problem 14.29 as is indicated by the fact that the correlation coefficient in (b) is $0\cdot99$.

(d) $\dfrac{\text{unexplained variation}}{\text{total variation}} = 1 - r^2 = 1 - 0\cdot9822 = 0\cdot0178$

Thus $1\cdot78\%$ of the total variation remains unexplained. This could be due to random fluctuations or to an additional variable which has not been considered.

14.32. Find (a) s_Y and (b) $s_{Y.X}$ for the data of Problem 14.29.

Solution:

(a) From Problem 14.31(a), $\Sigma(Y - \bar{Y})^2 = 21\cdot40$. Then the standard deviation of Y is

$$s_Y = \sqrt{\frac{\Sigma(Y - \bar{Y})^2}{N}} = \sqrt{\frac{21\cdot40}{8}} = 1\cdot636 \text{ or } 1\cdot64$$

(b) **First method:**

Using (a) and Problem 14.31(b), the standard error of estimate of Y on X is

$$s_{Y.X} = s_Y\sqrt{1 - r^2} = 1\cdot636\sqrt{1 - (0\cdot991\,1)^2} = 0\cdot218 \text{ or } 0\cdot22$$

Second method:

Using Problem 14.31,

$$s_{Y.X} = \sqrt{\frac{\Sigma(Y - Y_{est.})^2}{N}} = \sqrt{\frac{\text{unexplained variation}}{N}} = \sqrt{\frac{21\cdot40 - 21\cdot02}{8}} = 0\cdot218 \text{ or } 0\cdot22$$

Third method:

Using Problem 14.29 and the additional calculation $\Sigma Y^2 = 290\cdot52$, we have

$$s_{Y.X} = \sqrt{\frac{\Sigma Y^2 - a_0 \Sigma Y - a_1 \Sigma XY - a_2 \Sigma X^2 Y}{N}} = 0\cdot218 \text{ or } 0\cdot22.$$

SAMPLING THEORY OF CORRELATION

14.33. A correlation coefficient based on a sample of size 18 was computed to be $0\cdot32$. Can we conclude at a significance level of (a) $0\cdot05$ and (b) $0\cdot01$ that the corresponding population correlation coefficient differs from zero?

Solution:

We wish to decide between the hypotheses $H_0: \rho = 0$ and $H_1: \rho > 0$.

$$t = \frac{r\sqrt{N - 2}}{\sqrt{1 - r^2}} = \frac{0\cdot32\sqrt{18 - 2}}{\sqrt{1 - (0\cdot32)^2}} = 1\cdot35$$

(a) On the basis of a one-tailed test of Student's distribution at a $0\cdot05$ level, we would reject H_0 if $t > t_{0.95} = 1\cdot75$ for $(18 - 2) = 16$ degrees of freedom. Thus we cannot reject H_0 at a $0\cdot05$ level.

(b) Since we cannot reject H_0 at a $0\cdot05$ level, we certainly cannot reject it at a $0\cdot01$ level.

14.34. What is the minimum sample size necessary in order that we may conclude that a correlation coefficient of $0\cdot32$ differs significantly from zero at a $0\cdot05$ level?

Solution:

At a 0·05 level using a one-tailed test of Student's distribution, the minimum value of N must be such that

$$\frac{0\cdot32\sqrt{N-2}}{\sqrt{1-(0\cdot32)^2}} = t_{0\cdot95} \text{ for } N-2 \text{ degrees of freedom}$$

For an infinite number of degrees of freedom, $t_{0\cdot95} = 1\cdot64$ and hence $N = 25\cdot6$.

For $N = 26$, $v = 24$, $t_{0\cdot95} = 1\cdot71$, $t = 0\cdot32\sqrt{24}/\sqrt{1-(0\cdot32)^2} = 1\cdot65$.
For $N = 27$, $v = 25$, $t_{0\cdot95} = 1\cdot71$, $t = 0\cdot32\sqrt{25}/\sqrt{1-(0\cdot32)^2} = 1\cdot69$.
For $N = 28$, $v = 26$, $t_{0\cdot95} = 1\cdot71$, $t = 0\cdot32\sqrt{26}/\sqrt{1-(0\cdot32)^2} = 1\cdot72$.

Then the minimum sample size is $N = 28$.

14.35. A correlation coefficient based on a sample of size 24 was computed to be $r = 0\cdot75$. Can we reject the hypothesis that the population correlation coefficient is as small as (a) $\rho = 0\cdot60$, (b) $\rho = 0\cdot50$, at a 0·05 significance level?

Solution:

(a) $Z = 1\cdot1513 \log \left(\dfrac{1+0\cdot75}{1-0\cdot75}\right)$ $\mu_Z = 1\cdot1513 \log \left(\dfrac{1+0\cdot60}{1-0\cdot60}\right)$ $\sigma_Z = \dfrac{1}{\sqrt{N-3}} = \dfrac{1}{\sqrt{21}}$

    $= 0\cdot9730,$ $= 0\cdot6932,$ $= 0\cdot2182$

Then $z = (Z - \mu_Z)/\sigma_Z = (0\cdot9730 - 0\cdot6932)/0\cdot2182 = 1\cdot28$

At a 0·05 level of significance using a one-tailed test of the normal distribution, we would reject the hypothesis only if z were greater than 1·64. Thus we cannot reject the hypothesis that the population correlation coefficient is as small as 0·60.

(b) If $\rho = 0\cdot50$, $\mu_Z = 1\cdot1513 \log 3 = 0\cdot5493$ and $z = (0\cdot9730 - 0\cdot5493)/0\cdot2182 = 1\cdot94$. Thus we can reject the hypothesis that the population correlation coefficient is as small as $\rho = 0\cdot50$ at a 0·05 level of significance.

14.36. The correlation coefficient between physics and mathematics final marks for a group of 21 students was computed to be 0·80. Find 95% confidence limits for this coefficient.

Solution:

Since $r = 0\cdot80$ and $N = 21$, 95% confidence limits for μ_Z are given by

$$Z \pm 1\cdot96\sigma_Z = 1\cdot1513 \log \left(\frac{1+r}{1-r}\right) \pm 1\cdot96 \left(\frac{1}{\sqrt{N-3}}\right) = 1\cdot0986 \pm 0\cdot4620$$

Then μ_Z has the 95% confidence interval 0·5366 to 1·5606.

If $\mu_Z = 1\cdot1513 \log \left(\dfrac{1+\rho}{1-\rho}\right) = 0\cdot5366$, $\rho = 0\cdot4904$. If $\mu_Z = 1\cdot1513 \log \left(\dfrac{1+\rho}{1-\rho}\right) = 1\cdot5606$, $\rho = 0\cdot9155$.

Thus the 95% confidence limits for ρ are 0·49 and 0·92.

14.37. Two correlation coefficients obtained from samples of size $N_1 = 28$ and $N_2 = 35$ were computed to be $r_1 = 0\cdot50$ and $r_2 = 0\cdot30$ respectively. Is there a significant difference between the two coefficients at a 0·05 level?

Solution:

$$Z_1 = 1\cdot1513 \log \left(\frac{1+r_1}{1-r_1}\right) = 0\cdot5493, \quad Z_2 = 1\cdot1513 \log \left(\frac{1+r_2}{1-r_2}\right) = 0\cdot3095$$

and

$$\sigma_{Z_1 - Z_2} = \sqrt{\frac{1}{N_1 - 3} + \frac{1}{N_2 - 3}} = 0\cdot2669$$

We wish to decide between the hypotheses $H_0: \mu_{Z_1} = \mu_{Z_2}$ and $H_1: \mu_{Z_1} \neq \mu_{Z_2}$.

Under hypothesis H_0, $z = \dfrac{Z_1 - Z_2 - (\mu_{Z_1} - \mu_{Z_2})}{\sigma_{Z_1 - Z_2}} = \dfrac{0\cdot5493 - 0\cdot3095 - 0}{0\cdot2669} = 0\cdot8985$.

Using a two-tailed test of the normal distribution, we would reject H_0 only if $z > 1\cdot96$ or $z < -1\cdot96$. Thus we cannot reject H_0 and we conclude that the results are not significantly different at a 0·05 level.

SAMPLING THEORY OF REGRESSION

14.38. In Problem 14.1 we found the regression equation of Y on X to be $Y = 35 \cdot 82 + 0 \cdot 476X$. Test the hypothesis at a $0 \cdot 05$ significance level that the regression coefficient of the population regression equation is as low as $0 \cdot 180$.

Solution:

$$t = \frac{a_1 - A_1}{s_{Y.X}/s_X} \sqrt{N-2} = \frac{0 \cdot 476 - 0 \cdot 180}{1 \cdot 28/2 \cdot 66} \sqrt{12-2} = 1 \cdot 95$$

since $s_{Y.X} = 1 \cdot 28$ (computed in Problem 14.5) and $s_X = \sqrt{(\Sigma x^2)/N} = \sqrt{84 \cdot 68/12} = 2 \cdot 66$ (from Problem 14.2).

On the basis of a one-tailed test of Student's distribution at a $0 \cdot 05$ level we would reject the hypothesis that the regression coefficient is as low as $0 \cdot 180$ if $t > t_{0.95} = 1 \cdot 81$ for $(12 - 2) = 10$ degrees of freedom. Thus we cannot reject the hypothesis.

14.39. Find 95% confidence limits for the regression coefficient of the preceding problem.

Solution:

$A_1 = a_1 - \dfrac{t}{\sqrt{N-2}} \dfrac{s_{Y.X}}{s_X}$. Then 95% confidence limits for A_1 (obtained by putting $t = \pm t_{0.975} = \pm 2 \cdot 23$ for $12 - 2 = 10$ degrees of freedom) are given by

$$a_1 \pm \frac{2 \cdot 23}{\sqrt{12-2}} \left(\frac{s_{Y.X}}{s_X} \right) = 0 \cdot 476 \pm \frac{2 \cdot 23}{\sqrt{10}} \left(\frac{1 \cdot 28}{2 \cdot 66} \right) = 0 \cdot 476 \pm 0 \cdot 340$$

i.e. we are 95% confident that A_1 lies between $0 \cdot 136$ and $0 \cdot 816$.

14.40. In Problem 14.1, find 95% confidence limits for the masses of sons whose fathers' masses are (a) $65 \cdot 0$ and (b) $70 \cdot 0$ kg.

Solution:

Since $t_{0.975} = 2 \cdot 23$ for $(12 - 2) = 10$ degrees of freedom, the 95% confidence limits for Y_p (see Page 247) are given by

$$Y_0 \pm \frac{2 \cdot 23}{\sqrt{N-2}} s_{Y.X} \sqrt{N + 1 + \frac{(X_0 - \bar{X})^2}{s_X^2}}.$$

where $Y_0 = 35 \cdot 82 + 0 \cdot 476X_0$ (Prob. 14.1), $s_{Y.X} = 1 \cdot 28$, $s_X = 2 \cdot 26$ (Prob. 14.38), and $N = 12$.

(a) If $X_0 = 65 \cdot 0$, $Y_0 = 66 \cdot 76$ kg. Also, $(X_0 - \bar{X})^2 = (65 \cdot 0 - 800/12)^2 = 2 \cdot 78$. Then 95% confidence limits are

$$66 \cdot 76 \pm \frac{2 \cdot 23}{\sqrt{10}} (1 \cdot 28) \sqrt{12 + 1 + \frac{2 \cdot 78}{(2 \cdot 66)^2}} = 66 \cdot 76 \pm 3 \cdot 31 \text{ kg}$$

i.e. we can be about 95% confident that the sons' masses are between $63 \cdot 4$ and $70 \cdot 1$ kg.

(b) If $X_0 = 70 \cdot 0$, $Y_0 = 69 \cdot 14$ kg. Also, $(X_0 - \bar{X})^2 = (70 \cdot 0 - 800/12)^2 = 11 \cdot 11$. Then the 95% confidence limits are computed to be $69 \cdot 14 \pm 3 \cdot 45$ kg, i.e. we can be about 95% confident that the sons' masses are between $65 \cdot 7$ and $72 \cdot 6$ kg.

Note that for large values of N, 95% confidence limits are given approximately by $Y_0 \pm 1 \cdot 96 s_{Y.X}$ or $Y_0 \pm 2 s_{Y.X}$ provided $(X_0 - \bar{X})$ is not too large. This agrees with the approximate results mentioned on Page 157. The methods of this problem hold regardless of the size of N or $(X_0 - \bar{X})$, i.e. the sampling methods are exact.

14.41. In Problem 14.1, find 95% confidence limits for the mean masses of sons whose fathers' masses are (a) $65 \cdot 0$ and (b) $70 \cdot 0$ kg.

Solution:

Since $t_{0.975} = 2 \cdot 23$ for 10 degrees of freedom, the 95% confidence limits for $\bar{Y}_p$ (see Page 247) are given by

$$Y_0 \pm \frac{2 \cdot 23}{\sqrt{10}} s_{Y.X} \sqrt{1 + \frac{(X_0 - \bar{X})^2}{s_X^2}}$$

where $Y_0 = 35 \cdot 82 + 0 \cdot 476X_0$ (Problem 14.1), $s_{X.Y} = 1 \cdot 28$, $s_X = 2 \cdot 66$ (Problem 14.38).

(a) If $X_0 = 65 \cdot 0$, we find [compare Prob. 14.40(a)] the 95% confidence limits $(66 \cdot 76 \pm 1 \cdot 07)$ kg, i.e. we can be about 95% confident that the *mean mass* of all sons whose fathers' masses are $65 \cdot 0$ kg will lie between $65 \cdot 7$ and $67 \cdot 8$ kg.

(b) If $X_0 = 70 \cdot 0$, we find [compare Prob. 14.40(b)] the 95% confidence limits $(69 \cdot 14 \pm 1 \cdot 45)$ kg, i.e. we can be about 95% confident that the mean mass of all sons whose fathers' masses are $70 \cdot 0$ kg will lie between $67 \cdot 7$ and $70 \cdot 6$ kg.

Supplementary Problems

LINEAR REGRESSION AND CORRELATION

14.42. The following table shows the first two marks, denoted by X and Y respectively, of 10 students on two short quizzes in biology.

    (a) Construct a scatter diagram.

    (b) Find the least square regression line of Y on X.   *Ans.* $Y = 4.000 + 0.500X$

    (c) Find the least square regression line of X on Y.   *Ans.* $X = 2.408 + 0.612Y$

    (d) Graph the two regression lines of (b) and (c) on the scatter diagram of (a).

Mark on first quiz (X)	6	5	8	8	7	6	10	4	9	7
Mark on second quiz (Y)	8	7	7	10	5	8	10	6	8	6

14.43. Find (a) $s_{Y.X}$ and (b) $s_{X.Y}$ for the data in the preceding problem.   *Ans.* (a) 1.304, (b) 1.443

14.44. Compute the (a) total variation in Y, (b) unexplained variation in Y, and (c) explained variation in Y for the data of Prob. 14.42.   *Ans.* (a) 24.50, (b) 17.00, (c) 7.50

14.45. Use the results of Prob. 14.44 to find the correlation coefficient between the two sets of quiz marks of Problem 14.42. *Ans.* 0.5533

14.46. (a) Find the correlation coefficient between the two sets of quiz marks in Problem 14.42 by using the product-moment formula and compare with the result of Problem 14.45. (b) Obtain the correlation directly from the slopes of the regression lines of Problem 14.42(b) and (c).

14.47. Find the covariance for the data of Problem 14.42 (a) directly and (b) by using the formula $s_{X.Y} = rs_X s_Y$ and the result of Problems 14.45 or 14.46.   *Ans.* 1.5

14.48. The following table shows the ages X and systolic blood pressures Y of 12 women.

    (a) Find the correlation coefficient between X and Y.

    (b) Determine the least square regression equation of Y on X.

    (c) Estimate the blood pressure of a woman whose age is 45 years.

Age (X)	56	42	72	36	63	47	55	49	38	42	68	60
Blood pressure (Y)	147	125	160	118	149	128	150	145	115	140	152	155

    Ans. (a) 0.8961, (b) $Y = 80.78 + 1.138X$, (c) 132

14.49. Find the correlation coefficients for the data of (a) Prob. 13.32 of Chap. 13, (b) Prob. 13.35 of Chap 13.   *Ans.* (a) 0.958, (b) 0.872

14.50. The correlation coefficient between two variables X and Y is $r = 0.60$. If $s_X = 1.50$, $s_Y = 2.00$, $\bar{X} = 10$ and $\bar{Y} = 20$, find the equations of the regression lines of (a) Y on X, (b) X on Y.
    Ans. (a) $Y = 0.8X + 12$, (b) $X = 0.45Y + 1$

14.51. Compute (a) $s_{Y.X}$ and (b) $s_{X.Y}$ for the data of Problem 14.50.   *Ans.* (a) 1.60, (b) 1.20

14.52. If $s_{Y.X} = 3$ and $s_Y = 5$, find r.   *Ans.* ± 0.80

14.53. If the correlation coefficient between X and Y is 0.50, what percentage of the total variation remains unexplained by the regression equation?   *Ans.* 75%

14.54. Prove that the equation of the regression line of Y on X can be written $Y - \bar{Y} = \frac{s_{XY}}{s_X^2}(X - \bar{X})$. Write the analogous equation for the regression line of X on Y.

14.55. (a) Compute the correlation coefficient between the corresponding values of X and Y given in the adjoining table.

    (b) Multiply each X value in the table by 2 and add 6. Multiply each Y value in the table by 3 and subtract 15. Find the correlation coefficient between the two new sets of values, explaining why you do or do not obtain the same result as in part (a).   *Ans.* (a) -0.9203

X	2	4	5	6	8	11
Y	18	12	10	8	7	5

14.56. (a) Find the regression equations of Y on X for the data considered in parts (a) and (b) of the preceding problem. (b) Discuss the relationship between these regression equations.
Ans. (a) $Y = 18 \cdot 04 - 1 \cdot 34X$, $Y = 51 \cdot 18 - 2 \cdot 01X$

14.57. Prove that the correlation coefficient between X and Y can be written

$$r = \frac{\overline{XY} - \bar{X}\bar{Y}}{\sqrt{[\overline{X^2} - \bar{X}^2][\overline{Y^2} - \bar{Y}^2]}}$$

15.58. Prove that a correlation coefficient is independent of the choice of origin of the variables or the units in which they are expressed. (Hint: Assume $X' = c_1X + A$, $Y' = c_2Y + B$, where c_1, c_2, A, B are any constants, and prove that the correlation between X' and Y' is the same as that between X and Y.)

14.59. Prove that for linear regression $\dfrac{s_{Y.X}^2}{s_Y^2} = \dfrac{s_{X.Y}^2}{s_X^2}$. Does the result hold for non-linear regression?

CORRELATION COEFFICIENT FOR GROUPED DATA

14.60. Find the correlation coefficient between the variables X and Y having the values as given in the following frequency table.

X

	59 — 62	63 — 66	67 — 70	71 — 74	75 — 78
90 — 109	2	1			
110 — 129	7	8	4	2	
130 — 149	5	15	22	7	1
150 — 169	2	12	63	19	5
170 — 189		7	28	32	12
190 — 209		2	10	20	7
210 — 229			1	4	2

Y

Ans. 0·5402

14.61. (a) Find the least square regression equation of Y on X for the data of the preceding problem. (b) Estimate Y when $X = 64$ and 72.
Ans. (a) $Y = 3 \cdot 33X - 66 \cdot 4$, (b) 146·7 and 173·4

14.62. Find (a) $s_{Y.X}$ and (b) $s_{X.Y}$ for the data of Problem 14.60.   *Ans.* (a) 20·36, (b) 3·30

14.63. Establish formula (21), Page 245, for the correlation coefficient of grouped data.

CORRELATION OF TIME SERIES

14.64. Find the correlation coefficient between U.S. consumer price indexes and wholesale price indexes for all commodities given for the years 1949–1958 in the following table. The base period 1947–1949 = 100. (See Prob. 13.37, Chap. 13.)

Year	1949	1950	1951	1952	1953	1954	1955	1956	1957	1958
Consumer Price Index	101·8	102·8	111·0	113·5	114·4	114·8	114·5	116·2	120·2	123·5
Wholesale Price Index	99·2	103·1	114·8	111·6	110·1	110·3	110·7	114·3	117·6	119·2

Source: Bureau of Labor Statistics

Ans. 0·9254

14.65. Find the correlation for the data in Prob. 1.66, Chap. 1.   *Ans.* 0·1608

RANK CORRELATION

14.66. Two judges in a contest, who were asked to rank 8 candidates A, B, C, D, E, F, G and H in order of their preference, submitted the choices shown in the following table. Find the coefficient of rank correlation and decide how well the judges agreed in their choices.

	A	B	C	D	E	F	G	H
First Judge	5	2	8	1	4	6	3	7
Second Judge	4	5	7	3	2	8	1	6

Ans. $r_{\text{rank}} = \frac{2}{3}$

14.67. Find the coefficient of rank correlation for the data of (a) Prob. 14.42, (b) Prob. 14.48. *Ans.* (a) 0·5606, (b) 0·9318

14.68. (a) Find the coefficient of rank correlation for the data of Prob. 14.55. (b) From your observations in (a), discuss a possible disadvantage of the method of rank correlation. *Ans.* (a) $-1\cdot0000$

14.69. (a) Find the coefficient of rank correlation for the data of Prob. 14.64. (b) Compare with the correlation coefficient obtained in that problem. *Ans.* (a) 0·7333

SAMPLING THEORY OF CORRELATION

14.70. A correlation coefficient based on a sample of size 27 was computed to be 0·40. Can we conclude at a significance level of (a) 0·05, (b) 0·01, that the corresponding population correlation coefficient differs from zero? *Ans.* (a) Yes, (b) No

14.71. A correlation coefficient based on a sample of size 35 was computed to be 0·50. Can we reject the hypothesis that the population correlation coefficient is (a) as small as $\rho = 0\cdot30$, (b) as large as $\rho = 0\cdot70$, using a 0·05 significance level? *Ans.* (a) No, (b) Yes

14.72. Find (a) 95%, (b) 99%, confidence limits for a correlation coefficient which is computed to be 0·60 from a sample of size 28. *Ans.* (a) 0·2923 and 0·7951, (b) 0·1763 and 0·8361

14.73. Work Problem 14.72 if the sample size is 52. *Ans.* (a) 0·3912 and 0·7500, (b) 0·3146 and 0·7861

14.74. Find 95% confidence limits for the correlation coefficient computed in (a) Problem 14.48, (b) Problem 14.60. *Ans.* (a) 0·7096 and 0·9653, (b) 0·4547 and 0·6158

14.75. Two correlation coefficients obtained from samples of size 23 and 28 were computed to be 0·80 and 0·95 respectively. Can we conclude at a level of (a) 0·05, (b) 0·01, that there is a significant difference between the two coefficients? *Ans.* (a) Yes, (b) No

SAMPLING THEORY OF REGRESSION

14.76. On the basis of a sample of size 27 a regression equation of Y on X was found to be $Y = 25\cdot0 + 2\cdot00X$. If $s_{Y.X} = 1\cdot50$, $s_X = 3\cdot00$ and $\bar{X} = 7\cdot50$, find (a) 95% (b) 99%, confidence limits for the regression coefficient. *Ans.* (a) $2\cdot00 \pm 0\cdot21$, (b) $2\cdot00 \pm 0\cdot28$

14.77. In Prob. 14.76 test the hypothesis that the population regression coefficient is (a) as low as 1·70, (b) as high as 2·20, at a 0·01 level of significance.
Ans. (a) Using a one-tailed test we can reject the hypothesis
(b) Using a one-tailed test we cannot reject the hypothesis

14.78. In Prob. 14.76 find (a) 95%, (b) 99%, confidence limits for Y when $X = 6\cdot00$. *Ans.* $37\cdot0 \pm 3\cdot28$, (b) $37\cdot0 \pm 4\cdot45$

14.79. In Prob. 14.76 find (a) 95%, (b) 99%, confidence limits for the mean of all values of Y corresponding to $X = 6\cdot00$. *Ans.* (a) $37\cdot0 \pm 0\cdot69$, (b) $37\cdot0 \pm 0\cdot94$

14.80. Referring to Prob. 14.48, find 95% confidence limits for (a) the regression coefficient of Y on X, (b) the blood pressures of all women who are 45 years old, (c) the mean of the blood pressures of all women who are 45 years old. *Ans.* $1\cdot138 \pm 0\cdot398$, (b) $132\cdot0 \pm 16\cdot6$, (c) $132\cdot0 \pm 5\cdot4$

CHAPTER 15

Multiple and Partial Correlation

MULTIPLE CORRELATION

The degree of relationship existing between three or more variables is called *multiple correlation*. The fundamental principles involved in problems of multiple correlation are analogous to those of simple correlation as treated in Chapter 14.

SUBSCRIPT NOTATION

To allow for generalizations to large numbers of variables, it is convenient to adopt a notation involving subscripts.

We shall let $X_1, X_2, X_3, \ldots$ denote the variables under consideration. Then we can let $X_{11}, X_{12}, X_{13}, \ldots$ denote the values assumed by the variable X_1, and $X_{21}, X_{22}, X_{23}, \ldots$ denote the values assumed by the variable X_2, and so on. With this notation a sum such as $X_{21} + X_{22} + X_{23} + \ldots + X_{2N}$, for example, could be written $\sum_{j=1}^{N} X_{2j}$, $\sum_{j} X_{2j}$ or simply ΣX_2. When no ambiguity can result we use the last notation. In such case the mean of X_2 is written $\bar{X}_2 = \dfrac{\Sigma X_2}{N}$.

REGRESSION EQUATION. REGRESSION PLANE

A *regression equation* is an equation for estimating a dependent variable, say X_1, from the independent variables $X_2, X_3, \ldots$ and is called a *regression equation of X_1 on $X_2, X_3, \ldots$*. In functional notation this is sometimes written briefly as $X_1 = F(X_2, X_3, \ldots)$ read "X_1 is a function of X_2, X_3, and so on".

For the case of three variables, the simplest regression equation of X_1 on X_2 and X_3 has the form

$$X_1 = b_{1.23} + b_{12.3}X_2 + b_{13.2}X_3 \tag{1}$$

where $b_{1.23}$, $b_{12.3}$ and $b_{13.2}$ are constants.

If we keep X_3 constant in equation (1), the graph of X_1 vs. X_2 is a straight line with slope $b_{12.3}$. If we keep X_2 constant the graph of X_1 vs. X_3 is a straight line with slope $b_{13.2}$. It is clear that the subscripts after the dot indicate the variables held constant in each case.

Due to the fact that X_1 varies partially because of variation in X_2 and partially because of variation in X_3, we call $b_{12.3}$ and $b_{13.2}$ the *partial regression coefficients* of X_1 on X_2 keeping X_3 constant and of X_1 on X_3 keeping X_2 constant, respectively.

Equation (1) is called a *linear regression equation* of X_1 on X_2 and X_3. In a three dimensional rectangular co-ordinate system it represents a plane called a *regression plane* and is a generalization of the regression line for two variables as considered in Chapter 13.

NORMAL EQUATIONS FOR THE LEAST SQUARE REGRESSION PLANE

Just as there exist least square regression lines approximating a set of N data points (X, Y) in a two dimensional scatter diagram, so also there exist *least square regression planes* fitting a set of N data points (X_1, X_2, X_3) in a three dimensional scatter diagram.

The least square regression plane of X_1 on X_2 and X_3 has the equation (*1*) where $b_{1.23}$, $b_{12.3}$ and $b_{13.2}$ are determined by solving simultaneously the *normal equations*

$$\begin{cases} \Sigma X_1 &= b_{1.23}N + b_{12.3}\Sigma X_2 + b_{13.2}\Sigma X_3 \\ \Sigma X_1 X_2 &= b_{1.23}\Sigma X_2 + b_{12.3}\Sigma X_2^2 + b_{13.2}\Sigma X_2 X_3 \\ \Sigma X_1 X_3 &= b_{1.23}\Sigma X_3 + b_{12.3}\Sigma X_2 X_3 + b_{13.2}\Sigma X_3^2 \end{cases} \tag{2}$$

These can be obtained formally by multiplying both sides of equation (*1*) by 1, X_2 and X_3 successively and summing on both sides.

Unless otherwise specified, whenever we refer to a regression equation it will be assumed that the least square regression equation is meant.

If $x_1 = X_1 - \bar{X}_1$, $x_2 = X_2 - \bar{X}_2$ and $x_3 = X_3 - \bar{X}_3$, the regression equation of X_1 on X_2 and X_3 can be written more simply as

$$x_1 = b_{12.3}x_2 + b_{13.2}x_3 \tag{3}$$

where $b_{12.3}$ and $b_{13.2}$ are obtained by solving simultaneously the equations

$$\begin{cases} \Sigma x_1 x_2 &= b_{12.3}\Sigma x_2^2 + b_{13.2}\Sigma x_2 x_3 \\ \Sigma x_1 x_3 &= b_{12.3}\Sigma x_2 x_3 + b_{13.2}\Sigma x_3^2 \end{cases} \tag{4}$$

These equations, which are equivalent to the normal equations (*2*) can be obtained formally by multiplying both sides of (*3*) by x_2 and x_3 successively and summing both sides. See Problem 15.8.

REGRESSION PLANES AND CORRELATION COEFFICIENTS

If the linear correlation coefficients between variables X_1 and X_2, X_1 and X_3, X_2 and X_3, as computed in Chapter 14, are denoted respectively by r_{12}, r_{13}, r_{23} (sometimes called *zero order correlation coefficients*), the least square regression plane has the equation

$$\frac{x_1}{s_1} = \left(\frac{r_{12} - r_{13}r_{23}}{1 - r_{23}^2}\right)\frac{x_2}{s_2} + \left(\frac{r_{13} - r_{12}r_{23}}{1 - r_{23}^2}\right)\frac{x_3}{s_3} \tag{5}$$

where $x_1 = X_1 - \bar{X}_1$, $x_2 = X_2 - \bar{X}_2$, $x_3 = X_3 - \bar{X}_3$; and s_1, s_2 and s_3 are the standard deviations of X_1, X_2 and X_3 respectively (see Problem 15.9).

Note that if the variable X_3 is non-existent and $X_1 = Y$ and $X_2 = X$, equation (5) reduces to equation (*25*) on Page 245 of Chapter 14.

STANDARD ERROR OF ESTIMATE

By an obvious generalization of equation (*8*), Page 243 of Chapter 14, we can define the *standard error of estimate* of X_1 on X_2 and X_3 by

$$s_{1.23} = \sqrt{\frac{\Sigma(X_1 - X_{1\,\text{est.}})^2}{N}} \tag{6}$$

where $X_{1\,\text{est.}}$ indicates the estimated values of X_1 as calculated from the regression equations (*1*) or (*5*).

In terms of the correlation coefficients r_{12}, r_{13} and r_{23}, the standard error of estimate can also be computed from the result

$$s_{1.23} = s_1\sqrt{\frac{1 - r_{12}^2 - r_{13}^2 - r_{23}^2 + 2r_{12}r_{13}r_{23}}{1 - r_{23}^2}} \tag{7}$$

The sampling interpretations of the standard error of estimate for two variables as given on Page 243 for the case where N is large can be extended to three dimensions by replacing the lines parallel to the regression line by planes parallel to the regression plane. A better estimate of the population standard error of estimate is given by $\hat{s}_{1.23} = \sqrt{N/(N-3)}\, s_{1.23}$.

THE COEFFICIENT OF MULTIPLE CORRELATION

The coefficient of multiple correlation is defined by an extension of equations (*12*) or (*14*) on Pages 243 and 244 of Chapter 14. In the case of two independent variables, for example, the *coefficient of multiple correlation* is given by

$$R_{1.23} = \sqrt{1 - \frac{s_{1.23}^2}{s_1^2}} \tag{8}$$

where s_1 is the standard deviation of the variable X_1 and $s_{1.23}$ is given by (*6*) or (*7*). The quantity $R_{1.23}^2$ is called the *coefficient of multiple determination*.

When a linear regression equation is used, the coefficient of multiple correlation is called the *coefficient of linear multiple correlation*. Unless otherwise specified, whenever we refer to multiple correlation we shall imply linear multiple correlation.

In terms of r_{12}, r_{13} and r_{23}, (*9*) can also be written

$$R_{1.23} = \sqrt{\frac{r_{12}^2 + r_{13}^2 - 2r_{12}r_{13}r_{23}}{1 - r_{23}^2}} \tag{9}$$

A coefficient of multiple correlation, such as $R_{1.23}$, lies between 0 and 1. The closer it is to 1 the better is the linear relationship between the variables. The closer it is to 0 the worse is the linear relationship. If the coefficient of multiple correlation is 1, the correlation is called *perfect*. Although a correlation coefficient of 0 indicates no linear relationship between the variables, it is possible that a *non-linear relationship* may exist.

CHANGE OF DEPENDENT VARIABLE

The above results hold when X_1 is considered as the dependent variable. However, if we want to consider X_3, for example, as dependent variable instead of X_1, we would only have to replace the subscripts 1 by 3, and 3 by 1, in the formulae already obtained.

For example, the regression equation of X_3 on X_1 and X_2 would be

$$\frac{x_3}{s_3} = \left(\frac{r_{23} - r_{13}r_{12}}{1 - r_{12}^2}\right)\frac{x_2}{s_2} + \left(\frac{r_{13} - r_{23}r_{12}}{1 - r_{12}^2}\right)\frac{x_1}{s_1} \tag{10}$$

as obtained from equation (*5*), using the results $r_{32} = r_{23}$, $r_{31} = r_{13}$, $r_{21} = r_{12}$.

GENERALIZATIONS TO MORE THAN THREE VARIABLES

These are obtained by analogy with above results. For example, the linear regression equation of X_1 on X_2, X_3 and X_4 can be written

$$X_1 = b_{1.234} + b_{12.34}X_2 + b_{13.24}X_3 + b_{14.23}X_4 \tag{11}$$

and represents a *hyperplane in four dimensional space*. On formal multiplication of both sides of equation (*11*) by 1, X_2, X_3 and X_4 successively and then summing on both sides we obtain the normal equations for determination of $b_{1.234}$, $b_{12.34}$, $b_{13.24}$ and $b_{14.23}$ which when substituted in (*11*) gives the *least square regression equation of X_1 on X_2, X_3 and X_4*. This can be written in a form similar to that of equation (*5*). (See Problem 14.41.)

PARTIAL CORRELATION

It is often important to measure the correlation between a dependent· variable and one particular independent variable when all other variables involved are kept constant, i.e. when the effects of all other variables are removed (often indicated by the phrase "other things being equal"). This can be obtained by defining a *coefficient of partial correlation* as in equation (*12*) on Page 243 of Chapter 14, except that we must consider the explained and unexplained variations which arise both with and without the particular independent variable.

If we denote by $r_{12.3}$ the coefficient of partial correlation between X_1 and X_2 keeping X_3 constant, we find that

$$r_{12.3} \; = \; \frac{r_{12} - r_{13}r_{23}}{\sqrt{(1 - r_{13}^2)(1 - r_{23}^2)}} \tag{12}$$

Similarly if $r_{12.34}$ is the coefficient of partial correlation between X_1 and X_2 keeping X_1 and X_4 constant, then

$$r_{12.34} \; = \; \frac{r_{12.4} - r_{13.4}r_{23.4}}{\sqrt{(1 - r_{13.4}^2)(1 - r_{23.4}^2)}} \; = \; \frac{r_{12.3} - r_{14.3}r_{24.3}}{\sqrt{(1 - r_{14.3}^2)(1 - r_{24.3}^2)}} \tag{13}$$

These results are useful since by means of them any partial correlation coefficient can ultimately be made to depend on the correlation coefficients r_{12}, r_{23}, etc. (i.e. the *zero order correlation coefficients*).

In the case of two variables X and Y, if the two regression lines have equations $Y = a_0 + a_1 X$ and $X = b_0 + b_1 Y$, we have seen that $r^2 = a_1 b_1$ (see Problem 14.22, Chapter 14). This result can be generalized. For example, if

$$X_1 = b_{1.234} + b_{12.34}X_2 + b_{13.24}X_3 + b_{14.23}X_4 \tag{14}$$

and

$$X_4 = b_{4.123} + b_{41.23}X_1 + b_{42.13}X_2 + b_{43.12}X_3 \tag{15}$$

are linear regression equations of X_1 on X_2, X_3, X_4 and X_4 on X_1, X_2, X_3 respectively, then

$$r_{14.23}^2 = b_{14.23}b_{41.23} \tag{16}$$

(See Problem 15.18.) This can be taken as the starting point for a definition of linear partial correlation coefficients.

RELATIONSHIPS BETWEEN MULTIPLE AND PARTIAL CORRELATION COEFFICIENTS

Interesting results connecting the multiple correlation coefficients and the various partial correlation coefficients can be found. For example, we find

$$1 - R_{1.23}^2 = (1 - r_{12}^2)(1 - r_{13.2}^2) \tag{17}$$

$$1 - R_{1.234}^2 = (1 - r_{12}^2)(1 - r_{13.2}^2)(1 - r_{14.23}^2) \tag{18}$$

Generalizations of these results are easily made.

NON-LINEAR MULTIPLE REGRESSION

The above results for linear multiple regression can be extended to non-linear multiple regression. Coefficients of multiple and partial correlation can then be defined by methods similar to those given above.

Solved Problems

REGRESSION EQUATIONS INVOLVING THREE VARIABLES

15.1. Using an appropriate subscript notation, write the regression equations of (a) X_2 on X_1 and X_3, (b) X_3 on X_1, X_2 and X_4, (c) X_5 on X_1, X_2, X_3 and X_4.

Solution:

(a) $X_2 = b_{2.13} + b_{21.3}X_1 + b_{23.1}X_3$

(b) $X_3 = b_{3.124} + b_{31.24}X_1 + b_{32.14}X_2 + b_{34.12}X_4$

(c) $X_5 = b_{5.1234} + b_{51.234}X_1 + b_{52.134}X_2 + b_{53.124}X_3 + b_{54.123}X_4$

15.2. Write the normal equations corresponding to the regression equations

(a) $X_3 = b_{3.12} + b_{31.2}X_1 + b_{32.1}X_2$

(b) $X_1 = b_{1.234} + b_{12.34}X_2 + b_{13.24}X_3 + b_{14.23}X_4$.

Solution:

(a) Multiply the equation successively by 1, X_1 and X_2, and sum on both sides. The normal equations are

$$\begin{cases} \Sigma X_3 = b_{3.12}N + b_{31.2}\Sigma X_1 + b_{32.1}\Sigma X_2 \\ \Sigma X_1 X_3 = b_{3.12}\Sigma X_1 + b_{31.2}\Sigma X_1^2 + b_{32.1}\Sigma X_1 X_2 \\ \Sigma X_2 X_3 = b_{3.12}\Sigma X_2 + b_{31.2}\Sigma X_1 X_2 + b_{32.1}\Sigma X_2^2 \end{cases}$$

(b) Multiply the equation successively by 1, X_2, X_3 and X_4, and sum on both sides. The normal equations are

$$\begin{cases} \Sigma X_1 = b_{1.234}N + b_{12.34}\Sigma X_2 + b_{13.24}\Sigma X_3 + b_{14.23}\Sigma X_4 \\ \Sigma X_1 X_2 = b_{1.234}\Sigma X_2 + b_{12.34}\Sigma X_2^2 + b_{13.24}\Sigma X_2 X_3 + b_{14.23}\Sigma X_2 X_4 \\ \Sigma X_1 X_3 = b_{1.234}\Sigma X_3 + b_{12.34}\Sigma X_2 X_3 + b_{13.24}\Sigma X_3^2 + b_{14.23}\Sigma X_3 X_4 \\ \Sigma X_1 X_4 = b_{1.234}\Sigma X_4 + b_{12.34}\Sigma X_2 X_4 + b_{13.24}\Sigma X_3 X_4 + b_{14.23}\Sigma X_4^2 \end{cases}$$

Note that these are not derivations of the normal equations but only formal means for remembering them. Derivations are obtained most simply by using calculus as in Appendix VIII, Page 350.

The number of normal equations is equal to the number of unknown constants.

15.3. The variable X_1 is thought to be a linear function of X_2 and X_3. A sample of 12 pairs of readings (X_2, X_3) produced the values of X_1 shown in Table 15.1.

(a) Find the least square regression equation of X_1 on X_2 and X_3.

(b) Determine the estimated values of X_1 from the given values of X_2 and X_3.

(c) Estimate X_1 when $X_2 = 54$ and $X_3 = 9$.

Table 15.1

X_1	64	71	53	67	55	58	77	57	56	51	76	68
X_2	57	59	49	62	51	50	55	48	52	42	61	57
X_3	8	10	6	11	8	7	10	9	10	6	12	9

Solution:

(a) The linear regression equation of X_1 on X_2 and X_3 can be written

$$X_1 = b_{1.23} + b_{12.3}X_2 + b_{13.2}X_3$$

The normal equations of the least square regression equation are

(1)
$$\begin{cases} \Sigma X_1 = b_{1.23}N + b_{12.3}\Sigma X_2 + b_{13.2}\Sigma X_3 \\ \Sigma X_1 X_2 = b_{1.23}\Sigma X_2 + b_{12.3}\Sigma X_2^2 + b_{13.2}\Sigma X_2 X_3 \\ \Sigma X_1 X_3 = b_{1.23}\Sigma X_3 + b_{12.3}\Sigma X_2 X_3 + b_{13.2}\Sigma X_3^2 \end{cases}$$

The work involved in computing the sums can be arranged as in Table 15.2. Although the column headed X_1^2 is not needed at present, we have added it for future reference.

Table 15.2

X_1	X_2	X_3	X_1^2	X_2^2	X_3^2	X_1X_2	X_1X_3	X_2X_3
64	57	8	4096	3249	64	3648	512	456
71	59	10	5041	3481	100	4189	710	590
53	49	6	2809	2401	36	2597	318	294
67	62	11	4489	3844	121	4154	737	682
55	51	8	3025	2601	64	2805	440	408
58	50	7	3364	2500	49	2900	406	350
77	55	10	5929	3025	100	4235	770	550
57	48	9	3249	2304	81	2736	513	432
56	52	10	3136	2704	100	2912	560	520
51	42	6	2601	1764	36	2142	306	252
76	61	12	5776	3721	144	4636	912	732
68	57	9	4624	3249	81	3876	612	513
$\Sigma X_1 =$ 753	$\Sigma X_2 =$ 643	$\Sigma X_3 =$ 106	$\Sigma X_1^2 =$ 48 139	$\Sigma X_2^2 =$ 34 843	$\Sigma X_3^2 =$ 976	$\Sigma X_1X_2 =$ 40 830	$\Sigma X_1X_3 =$ 6796	$\Sigma X_2X_3 =$ 5779

Using Table 15.2, the normal equations (*1*) become

$$(2) \quad \begin{cases} 12b_{1.23} + 643b_{12.3} + 106b_{13.2} = 753 \\ 643b_{1.23} + 34\,843b_{12.3} + 5\,779b_{13.2} = 40\,830 \\ 106b_{1.23} + 5\,779b_{12.3} + 976b_{13.2} = 6\,796 \end{cases}$$

Solving, $b_{1.23} = 3.6512$, $b_{12.3} = 0.8546$, $b_{13.2} = 1.5063$, and the required regression equation is

$$(3) \quad X_1 = 3.6512 + 0.8546X_2 + 1.5063X_3 \text{ or } X_1 = 3.65 + 0.855X_2 + 1.506X_3$$

For another method which avoids solution of simultaneous equations, see Problem 15.6.

(b) Using the regression equation (*3*) we obtain the estimated values of X_1, denoted by $X_{1\text{est.}}$, by substituting the corresponding values of X_2 and X_3. For example, substituting $X_2 = 57$ and $X_3 = 8$ in (*3*) we find $X_{1\text{est.}} = 64.414$.

　　Similarly the other estimated values of X_1 are obtained and are given in the Table 15.3 together with the sample values of X_1.

Table 15.3

$X_{1\text{est.}}$	64.414	69.136	54.564	73.206	59.286	56.925	65.717	58.229	63.153	48.582	73.857	65.920
X_1	64	71	53	67	55	58	77	57	56	51	76	68

(c) Putting $X_2 = 54$ and $X_3 = 9$ in (*3*), the estimate is $X_{1\text{est.}} = 63.356$, or about 63.

15.4. Calculate the standard deviations (*a*) s_1, (*b*) s_2 and (*c*) s_3 for the data of Problem 15.3.

Solution:

(*a*) The quantity s_1 is the standard deviation of the variable X_1. Then using Table 15.2 of Problem 15.3(*a*) we find, by the methods of Chapter 4,

$$s_1 = \sqrt{\frac{\Sigma X_1^2}{N} - \left(\frac{\Sigma X_1}{N}\right)^2} = \sqrt{\frac{48\,139}{12} - \left(\frac{753}{12}\right)^2} = 8.6035 \text{ or } 8.6$$

(*b*)
$$s_2 = \sqrt{\frac{\Sigma X_2^2}{N} - \left(\frac{\Sigma X_2}{N}\right)^2} = \sqrt{\frac{34\,843}{12} - \left(\frac{643}{12}\right)^2} = 5.6930 \text{ or } 5.7$$

(*c*)
$$s_3 = \sqrt{\frac{\Sigma X_3^2}{N} - \left(\frac{\Sigma X_3}{N}\right)^2} = \sqrt{\frac{976}{12} - \left(\frac{106}{12}\right)^2} = 1.8181 \text{ or } 1.8$$

15.5. Compute (a) r_{12}, (b) r_{13} and (c) r_{23} for the data of Problem 15.3.

Solution:

(a) The quantity r_{12} is the linear correlation coefficient between the variables X_1 and X_2, ignoring the variable X_3. Then by the methods of Chapter 14, we have

$$r_{12} = \frac{N \Sigma X_1 X_2 - (\Sigma X_1)(\Sigma X_2)}{\sqrt{[N \Sigma X_1^2 - (\Sigma X_1)^2][N \Sigma X_2^2 - (\Sigma X_2)^2]}}$$

$$= \frac{(12)(40\,830) - (753)(643)}{\sqrt{[(12)(48\,139) - (753)^2][(12)(34\,843) - (643)^2]}} = 0.8196 \text{ or } 0.82$$

(b), (c). Using corresponding formulae, we obtain $r_{13} = 0.7698$ or 0.77, and $r_{23} = 0.7984$ or 0.80.

15.6. Work Problem 15.3(a) by using equation (5) on Page 270 and the results of Problems 15.4 and 15.5.

Solution:

The regression equation of X_1 on X_2 and X_3 is, on multiplying both sides of equation (5), Page 270, by s_1,

(1)
$$x_1 = \left(\frac{r_{12} - r_{13}r_{23}}{1 - r_{23}^2}\right)\left(\frac{s_1}{s_2}\right)x_2 + \left(\frac{r_{13} - r_{12}r_{23}}{1 - r_{23}^2}\right)\left(\frac{s_1}{s_3}\right)x_3$$

where $x_1 = X_1 - \bar{X}_1$, $x_2 = X_2 - \bar{X}_2$, $x_3 = X_3 - \bar{X}_3$. Using the results of Problems 15.4 and 15.5, equation (1) becomes

$$x_1 = 0.8546x_2 + 1.5063x_3$$

Since $\bar{X}_1 = \frac{\Sigma X_1}{N} = \frac{753}{12} = 62.750$, $\bar{X}_2 = \frac{\Sigma X_2}{N} = 53.583$, $\bar{X}_3 = 8.833$ (from Table 15.2 of Problem 15.3), the required equation can be written

$$X_1 - 62.750 = 0.8546(X_2 - 53.583) + 1.506(X_3 - 8.833)$$

agreeing with the result of Problem 15.3(a).

15.7. For the data of Problem 15.3 determine (a) the average increase in X_1 per unit increase in X_2 for constant X_3 (b) the average increase in X_1 per unit increase in X_3 for constant X_2.

Solution:

From the regression equation obtained in Problems 15.3(a) or 15.6 we see that the answer to (a) is 0.8546 or about 0.9, and the answer to (b) is 1.5063 or about 1.5.

15.8. Show that equations (3) and (4), Page 270, follow from (1) and (2), Pages 269 and 270.

Solution:

From the first of equations (2), Page 270, we have on dividing both sides by N,

(1)
$$\bar{X}_1 = b_{1.23} + b_{12.3}\bar{X}_2 + b_{13.2}\bar{X}_3$$

Subtracting this equation from equation (1), Page 269, gives

(2)
$$X_1 - \bar{X}_1 = b_{12.3}(X_2 - \bar{X}_2) + b_{13.2}(X_3 - \bar{X}_3) \quad \text{or}$$
$$x_1 = b_{12.3}x_2 + b_{13.2}x_3$$

which is equation (3), Page 270.

Let $X_1 = x_1 + \bar{X}_1$, $X_2 = x_2 + \bar{X}_2$, $X_3 = x_3 + \bar{X}_3$ in the second and third of equations (2), Page 270. Then after some algebraic simplifications, using the results $\Sigma x_1 = \Sigma x_2 = \Sigma x_3 = 0$, they become

(3)
$$\Sigma x_1 x_2 = b_{12.3} \Sigma x_2^2 + b_{13.2} \Sigma x_2 x_3 + N\bar{X}_2[b_{1.23} + b_{12.3}\bar{X}_2 + b_{13.2}\bar{X}_3 - \bar{X}_1]$$

(4)
$$\Sigma x_1 x_3 = b_{12.3} \Sigma x_2 x_3 + b_{13.2} \Sigma x_3^2 + N\bar{X}_3[b_{1.23} + b_{12.3}\bar{X}_2 + b_{13.2}\bar{X}_3 - \bar{X}_1]$$

which reduce to equations (4), Page 270, since the quantities in brackets on the right hand sides of (3) and (4) are zero because of equation (1).

Another method: See Problem 15.30.

15.9. Establish equation (5), Page 270: $\dfrac{x_1}{s_1} = \Big(\dfrac{r_{12} - r_{13}r_{23}}{1 - r_{23}^2}\Big)\dfrac{x_2}{s_2} + \Big(\dfrac{r_{13} - r_{12}r_{23}}{1 - r_{23}^2}\Big)\dfrac{x_3}{s_3}.$

Solution:

From equations (3) and (4) of Problem 15.8,

$$(1) \qquad \begin{cases} b_{12.3}\,\Sigma x_2^2 + b_{13.2}\,\Sigma x_2 x_3 = \Sigma x_1 x_2 \\ b_{12.3}\,\Sigma x_2 x_3 + b_{13.2}\,\Sigma x_3^2 = \Sigma x_1 x_3 \end{cases}$$

Since $s_2^2 = \dfrac{\Sigma x_2^2}{N}$ and $s_3^2 = \dfrac{\Sigma x_3^2}{N}$, $\Sigma x_2^2 = N s_2^2$ and $\Sigma x_3^2 = N s_3^2$.

Since $r_{23} = \dfrac{\Sigma x_2 x_3}{\sqrt{(\Sigma x_2^2)(\Sigma x_3^2)}} = \dfrac{\Sigma x_2 x_3}{N s_2 s_3}$, $\Sigma x_2 x_3 = N s_2 s_3 r_{23}$.

Similarly, $\Sigma x_1 x_2 = N s_1 s_2 r_{12}$ and $\Sigma x_1 x_3 = N s_1 s_3 r_{13}$.

Substituting in (1) and simplifying, we find

$$(2) \qquad \begin{cases} b_{12.3}\,s_2 + b_{13.2}\,s_3 r_{23} = s_1 r_{12} \\ b_{12.3}\,s_2 r_{23} + b_{13.2}\,s_3 = s_1 r_{13} \end{cases}$$

Solving equations (2) simultaneously, $b_{12.3} = \Big(\dfrac{r_{12} - r_{13}r_{23}}{1 - r_{23}^2}\Big)\Big(\dfrac{s_1}{s_2}\Big)$ and $b_{13.2} = \Big(\dfrac{r_{13} - r_{12}r_{23}}{1 - r_{23}^2}\Big)\Big(\dfrac{s_1}{s_3}\Big).$

Substituting these in the equation $x_1 = b_{12.3}x_2 + b_{13.2}x_3$ (equation (2), Problem 15.8) and dividing by s_1 yields the required result.

STANDARD ERROR OF ESTIMATE

15.10. Compute the standard error of estimate of X_1 on X_2 and X_3 for the data of Problem 15.3.

Solution:

From Table 15.3 of Problem 15.3(b), we have

$$s_{1.23} = \sqrt{\frac{\Sigma(X_1 - X_{1\,\text{est.}})^2}{N}}$$

$$= \sqrt{\frac{(64 - 64.414)^2 + (71 - 69.136)^2 + \cdots + (68 - 65.920)^2}{12}} = 4.6447 \text{ or } 4.6$$

The population standard error of estimate is estimated by $\hat{s}_{1.23} = \sqrt{N/(N - 3)}\, s_{1.23} = 5.3$ in this case.

15.11. Use $s_{1.23} = s_1 \sqrt{\dfrac{1 - r_{12}^2 - r_{13}^2 - r_{23}^2 + 2 r_{12} r_{13} r_{23}}{1 - r_{23}^2}}$ to obtain the result of Problem 15.10.

Solution:

From Problems 15.4(a) and 15.5, we have

$$s_{1.23} = 8.6035 \sqrt{\frac{1 - (0.8196)^2 - (0.7698)^2 - (0.7984)^2 + 2(0.8196)(0.7698)(0.7984)}{1 - (0.7984)^2}} = 4.6$$

Note that by the method of this problem the standard error of estimate can be found without use of the regression equation.

COEFFICIENT OF MULTIPLE CORRELATION

15.12. Compute the coefficient of linear multiple correlation of X_1 on X_2 and X_3 for the data of Problem 15.3.

Solution:

First method: From the results of Problems 15.4(a) and 15.10, we have

$$R_{1.23} = \sqrt{1 - \frac{s_{1.23}^2}{s_1^2}} = \sqrt{1 - \frac{(4.6447)^2}{(8.6035)^2}} = 0.8418$$

Second method: From the results of Problem 15.5,

$$R_{1.23} = \sqrt{\frac{r_{12}^2 + r_{13}^2 - 2r_{12}r_{13}r_{23}}{1 - r_{23}^2}} = \sqrt{\frac{(0 \cdot 8196)^2 + (0 \cdot 7698)^2 - 2(0 \cdot 8196)(0 \cdot 7698)(0 \cdot 7984)}{1 - (0 \cdot 7984)^2}} = 0 \cdot 8418$$

Note that the coefficient of multiple correlation $R_{1.23}$ is larger than either of the coefficients r_{12} or r_{13} (see Problem 15.5). This is always true and is in fact to be expected, since by taking into account additional relevant independent variables we should arrive at a better relationship between the variables.

15.13. Compute the coefficient of multiple determination of X_1 on X_2 and X_3 for the data of Problem 15.3.

Solution:

Coefficient of multiple determination of X_1 on X_2 and X_3 is

$$R_{1.23}^2 = (0 \cdot 8418)^2 = 0 \cdot 7086$$

using Problem 15.12. Thus about 71 % of the total variation in X_1 is explained by use of the regression equation.

15.14. Calculate (a) $R_{2.13}$ and (b) $R_{3.12}$ for the data of Problem 15.3 and compare with the value of $R_{1.23}$.

Solution:

(a) $R_{2.13} = \sqrt{\dfrac{r_{12}^2 + r_{23}^2 - 2r_{12}r_{13}r_{23}}{1 - r_{13}^2}} = \sqrt{\dfrac{(0 \cdot 8196)^2 + (0 \cdot 7984)^2 - 2(0 \cdot 8196)(0 \cdot 7698)(0 \cdot 7984)}{1 - (0 \cdot 7698)^2}} = 0 \cdot 8606$

(b) $R_{3.12} = \sqrt{\dfrac{r_{13}^2 + r_{23}^2 - 2r_{12}r_{13}r_{23}}{1 - r_{12}^2}} = \sqrt{\dfrac{(0 \cdot 7698)^2 + (0 \cdot 7984)^2 - 2(0 \cdot 8196)(0 \cdot 7698)(0 \cdot 7984)}{1 - (0 \cdot 8196)^2}} = 0 \cdot 8234$

This problem illustrates the fact that, in general, $R_{2.13}$, $R_{3.12}$ and $R_{1.23}$ are not necessarily equal, as seen by comparison with Problem 15.12.

15.15. If $R_{1.23} = 1$, prove that (a) $R_{2.13} = 1$, (b) $R_{3.12} = 1$.

Solution:

$$(1) \quad R_{1.23} = \sqrt{\frac{r_{12}^2 + r_{13}^2 - 2r_{12}r_{13}r_{23}}{1 - r_{23}^2}} \quad \text{and} \quad (2) \quad R_{2.13} = \sqrt{\frac{r_{12}^2 + r_{23}^2 - 2r_{12}r_{13}r_{23}}{1 - r_{13}^2}}$$

(a) In (1) setting $R_{1.23} = 1$ and squaring both sides, $r_{12}^2 + r_{13}^2 - 2r_{12}r_{13}r_{23} = 1 - r_{23}^2$. Then

$$r_{12}^2 + r_{23}^2 - 2r_{12}r_{13}r_{23} = 1 - r_{13}^2 \quad \text{or} \quad \frac{r_{12}^2 + r_{23}^2 - 2r_{12}r_{13}r_{23}}{1 - r_{13}^2} = 1$$

i.e. $R_{2.13}^2 = 1$ or $R_{2.13} = 1$, since the coefficient of multiple correlation is considered non-negative.

(b) $R_{3.12} = 1$ follows from part (a) by interchanging subscripts 2 and 3 in the result $R_{2.13} = 1$.

15.16. If $R_{1.23} = 0$, does it necessarily follow that $R_{2.13} = 0$?

Solution:

From equation (1) of Problem 15.15, $R_{1.23} = 0$ if and only if

$$r_{12}^2 + r_{13}^2 - 2r_{12}r_{13}r_{23} = 0 \quad \text{or} \quad 2r_{12}r_{13}r_{23} = r_{12}^2 + r_{13}^2$$

Then from the equation (2) of Problem 15.15, we have

$$R_{2.13} = \sqrt{\frac{r_{12}^2 + r_{23}^2 - (r_{12}^2 + r_{13}^2)}{1 - r_{13}^2}} = \sqrt{\frac{r_{23}^2 - r_{13}^2}{1 - r_{13}^2}}$$

which is not necessarily zero.

PARTIAL CORRELATION

15.17. Compute the coefficients of linear partial correlation (a) $r_{12.3}$, (b) $r_{13.2}$ and (c) $r_{23.1}$ for the data of Problem 15.3.

Solution:

$$r_{12.3} = \frac{r_{12} - r_{13}r_{23}}{\sqrt{(1 - r_{13}^2)(1 - r_{23}^2)}}, \quad r_{13.2} = \frac{r_{13} - r_{12}r_{23}}{\sqrt{(1 - r_{12}^2)(1 - r_{23}^2)}}, \quad r_{23.1} = \frac{r_{23} - r_{12}r_{13}}{\sqrt{(1 - r_{12}^2)(1 - r_{13}^2)}}$$

Using the results of Problem 15.5, we find $r_{12.3} = 0.5334$, $r_{13.2} = 0.3346$, $r_{23.1} = 0.4580$.

It follows that for constant X_3 the correlation coefficient between X_1 and X_2 is 0.53. For constant X_2 the correlation coefficient between X_1 and X_3 is only 0.33. Since these results are based on a small sample of only 12 sets of values, they are of course not as reliable as those which would be obtained from a larger sample.

15.18. If $X_1 = b_{1.23} + b_{12.3}X_2 + b_{13.2}X_3$ and $X_3 = b_{3.12} + b_{32.1}X_2 + b_{31.2}X_1$ are the regression equations of X_1 on X_2 and X_3, and X_3 on X_2 and X_1, respectively, prove that $r_{13.2}^2 = b_{13.2}b_{31.2}$.

Solution:

The regression equation of X_1 on X_2 and X_3 can be written (see equation (5), Page 270)

$$(1) \qquad X_1 - \bar{X}_1 = \left(\frac{r_{12} - r_{13}r_{23}}{1 - r_{23}^2}\right)\left(\frac{s_1}{s_2}\right)(X_2 - \bar{X}_2) + \left(\frac{r_{13} - r_{12}r_{23}}{1 - r_{23}^2}\right)\left(\frac{s_1}{s_3}\right)(X_3 - \bar{X}_3)$$

The regression equation of X_3 on X_2 and X_1 can be written (see equation (10), Page 270)

$$(2) \qquad X_3 - \bar{X}_3 = \left(\frac{r_{23} - r_{13}r_{12}}{1 - r_{12}^2}\right)\left(\frac{s_3}{s_2}\right)(X_2 - \bar{X}_2) + \left(\frac{r_{13} - r_{23}r_{12}}{1 - r_{12}^2}\right)\left(\frac{s_3}{s_1}\right)(X_1 - \bar{X}_1)$$

From (1) and (2), the coefficients of X_3 and X_1 are respectively

$$b_{13.2} = \left(\frac{r_{13} - r_{12}r_{23}}{1 - r_{23}^2}\right)\left(\frac{s_1}{s_3}\right) \quad \text{and} \quad b_{31.2} = \left(\frac{r_{13} - r_{23}r_{12}}{1 - r_{12}^2}\right)\left(\frac{s_3}{s_1}\right)$$

Then

$$b_{13.2}b_{31.2} = \frac{(r_{13} - r_{12}r_{23})^2}{(1 - r_{23}^2)(1 - r_{12}^2)} = r_{13.2}^2$$

15.19. If $r_{12.3} = 0$, prove that (a) $r_{13.2} = r_{13}\sqrt{\dfrac{1 - r_{23}^2}{1 - r_{12}^2}}$,   (b) $r_{23.1} = r_{23}\sqrt{\dfrac{1 - r_{13}^2}{1 - r_{12}^2}}$.

Solution:

If $r_{12.3} = \dfrac{r_{12} - r_{13}r_{23}}{\sqrt{(1 - r_{13}^2)(1 - r_{23}^2)}} = 0$, we have $r_{12} = r_{13}r_{23}$.

(a) $r_{13.2} = \dfrac{r_{13} - r_{12}r_{23}}{\sqrt{(1 - r_{12}^2)(1 - r_{23}^2)}} = \dfrac{r_{13} - (r_{13}r_{23})r_{23}}{\sqrt{(1 - r_{12}^2)(1 - r_{23}^2)}} = \dfrac{r_{13}(1 - r_{23}^2)}{\sqrt{(1 - r_{12}^2)(1 - r_{23}^2)}} = r_{13}\sqrt{\dfrac{1 - r_{23}^2}{1 - r_{12}^2}}$

(b) Interchange the subscripts 1 and 2 in the result of part (a).

MULTIPLE AND PARTIAL CORRELATION WITH FOUR OR MORE VARIABLES

15.20. A college entrance examination consisted of three tests, in mathematics, English and general knowledge. To test the ability of the examination to predict performance in a statistics course, data concerning a sample of 200 students were gathered and analysed. Letting

X_1 = mark in statistics course X_3 = score on English test

X_2 = score on mathematics test X_4 = score on general knowledge test

the following calculations were obtained:

$$\bar{X}_1 = 75, \, s_1 = 10, \, \bar{X}_2 = 24, \, s_2 = 5, \, \bar{X}_3 = 15, \, s_3 = 3, \, \bar{X}_4 = 36, \, s_4 = 6$$
$$r_{12} = 0.90, \, r_{13} = 0.75, \, r_{14} = 0.80, \, r_{23} = 0.70, \, r_{24} = 0.70, \, r_{34} = 0.85$$

Find the least square regression equation of X_1 on X_2, X_3 and X_4.

Solution:

Generalizing the result of Problem 15.8, we can write the least square regression equation of X_1 on X_2, X_3 and X_4 in the form

(1)
$$x_1 = b_{12.34}x_2 + b_{13.24}x_3 + b_{14.23}x_4$$

where $b_{12.34}$, $b_{13.24}$, $b_{14.23}$ can be obtained from the normal equations

(2)
$$\begin{cases} \Sigma x_1 x_2 = b_{12.34}\,\Sigma x_2^2 + b_{13.24}\,\Sigma x_2 x_3 + b_{14.23}\,\Sigma x_2 x_4 \\ \Sigma x_1 x_3 = b_{12.34}\,\Sigma x_2 x_3 + b_{13.24}\,\Sigma x_3^2 + b_{14.23}\,\Sigma x_3 x_4 \\ \Sigma x_1 x_4 = b_{12.34}\,\Sigma x_2 x_4 + b_{13.24}\,\Sigma x_3 x_4 + b_{14.23}\,\Sigma x_4^2 \end{cases}$$

and where $x_1 = X_1 - \bar{X}_1$, $x_2 = X_2 - \bar{X}_2$, $x_3 = X_3 - \bar{X}_3$, $x_4 = X_4 - \bar{X}_4$.

From the given data, we find

$$\Sigma x_2^2 = Ns_2^2 = 5000 \qquad \Sigma x_1 x_2 = Ns_1 s_2 r_{12} = 9000 \qquad \Sigma x_2 x_3 = Ns_2 s_3 r_{23} = 2100$$
$$\Sigma x_3^2 = Ns_3^2 = 1800 \qquad \Sigma x_1 x_3 = Ns_1 s_3 r_{13} = 4500 \qquad \Sigma x_2 x_4 = Ns_2 s_4 r_{24} = 4200$$
$$\Sigma x_4^2 = Ns_4^2 = 7200 \qquad \Sigma x_1 x_4 = Ns_1 s_4 r_{14} = 9600 \qquad \Sigma x_3 x_4 = Ns_3 s_4 r_{34} = 3060$$

Putting these results into equations (2) and solving, we obtain

(3)
$$b_{12.34} = 1.3333, \quad b_{13.24} = 0.0000, \quad b_{14.23} = 0.5556$$

which when substituted in (1) yields the required regression equation

(4)
$$x_1 = 1.3333x_2 + 0.0000x_3 + 0.5556x_4$$
or
$$X_1 - 75 = 1.3333(X_2 - 24) + 0.5556(X_4 - 27)$$
or
$$X_1 = 22.9999 + 1.3333X_2 + 0.5556X_4$$

An exact solution of (2) yields $b_{12.34} = \frac{4}{3}$, $b_{13.24} = 0$ and $b_{14.23} = \frac{5}{9}$, so that the regression equation can also be written

(5)
$$X_1 = 23 + \tfrac{4}{3}X_2 + \tfrac{5}{9}X_4$$

It is interesting to note that the regression equation does not involve the score in English, namely X_3. This does not mean that one's knowledge of English has no bearing on proficiency in statistics. Instead it means that the need for English, insofar as prediction of the statistics grade is concerned, is amply evidenced by the scores achieved on the other tests.

15.21. Two students taking the college entrance examination of Problem 15.20 receive respective scores of: (a) 30 in mathematics, 18 in English and 32 in general knowledge; (b) 18 in mathematics, 20 in English and 36 in general knowledge. What would be their predicted marks in statistics?

Solution:

(a) Substituting $X_2 = 30$, $X_3 = 18$ and $X_4 = 32$ in equation (5) of Problem 15.19, the predicted mark in statistics is $X_1 = 81$.

(b) Proceeding as in part (a) with $X_2 = 18$, $X_3 = 20$ and $X_4 = 36$, we find $X_1 = 67$.

15.22. Find the partial correlation coefficients (a) $r_{12.34}$, (b) $r_{13.24}$ and (c) $r_{14.23}$ for the data of Problem 15.20.

Solution:

(a), (b)
$$r_{12.4} = \frac{r_{12} - r_{14}r_{24}}{\sqrt{(1 - r_{14}^2)(1 - r_{24}^2)}}, \qquad r_{13.4} = \frac{r_{13} - r_{14}r_{34}}{\sqrt{(1 - r_{14}^2)(1 - r_{34}^2)}}, \qquad r_{23.4} = \frac{r_{23} - r_{24}r_{34}}{\sqrt{(1 - r_{21}^2)(1 - r_{34}^2)}}$$

Substituting values from Prob. 15.20, we obtain $r_{12.4} = 0.7935$, $r_{13.4} = 0.2215$, $r_{23.4} = 0.2791$. Then

$$r_{12.34} = \frac{r_{12.4} - r_{13.4}r_{23.4}}{\sqrt{(1 - r_{13.4}^2)(1 - r_{23.4}^2)}} = 0.7814 \quad \text{and} \quad r_{13.24} = \frac{r_{13.4} - r_{12.4}r_{23.4}}{\sqrt{(1 - r_{12.4}^2)(1 - r_{23.4}^2)}} = 0.0000$$

(c)
$$r_{14.3} = \frac{r_{14} - r_{13}r_{34}}{\sqrt{(1 - r_{13}^2)(1 - r_{34}^2)}}, \qquad r_{12.3} = \frac{r_{12} - r_{13}r_{23}}{\sqrt{(1 - r_{13}^2)(1 - r_{23}^2)}}, \qquad r_{24.3} = \frac{r_{24} - r_{23}r_{34}}{\sqrt{(1 - r_{23}^2)(1 - r_{34}^2)}}$$

Substituting values from Prob. 15.20, we obtain $r_{14.3} = 0.4664$, $r_{12.3} = 0.7939$, $r_{24.3} = 0.2791$. Then

$$r_{14.23} = \frac{r_{14.3} - r_{12.3}r_{24.3}}{\sqrt{(1 - r_{12.3}^2)(1 - r_{24.3}^2)}} = 0.4193$$

15.23. Interpret the partial correlation coefficients (a) $r_{12.4}$, (b) $r_{13.4}$, (c) $r_{12.34}$, (d) $r_{14.3}$ and (e) $r_{14.23}$ obtained in Problem 15.22.

Solution:

(a) $r_{12.4} = 0.7935$ represents the (linear) correlation coefficient between statistics marks and mathematics scores for students having the same general knowledge scores. In obtaining this coefficient, scores in English (as well as other factors which have not been taken into account) are not considered, as is evidenced by the fact that the subscript 3 is omitted.

(b) $r_{13.4} = 0.2215$ represents the correlation coefficient between statistics marks and English scores for students having the same general knowledge scores. Here, scores in mathematics have not been considered.

(c) $r_{12.34} = 0.7814$ represents the correlation coefficient between statistics marks and mathematics scores for students having both the same English scores and general knowledge scores.

(d) $r_{14.3} = 0.4664$ represents the correlation coefficient between statistics marks and general knowledge scores for students having the same English scores.

(e) $r_{14.23} = 0.4193$ represents the correlation coefficient between statistics marks and general knowledge scores for students having both the same mathematics scores and English scores.

15.24. (a) For the data of Problem 15.20, show that

(1)
$$\frac{r_{12.4} - r_{13.4}r_{23.4}}{\sqrt{(1 - r_{13.4}^2)(1 - r_{23.4}^2)}} = \frac{r_{12.3} - r_{14.3}r_{24.3}}{\sqrt{(1 - r_{14.3}^2)(1 - r_{24.3}^2)}}$$

(b) Explain the significance of the equality in part (a).

Solution:

(a) The left side of (1) is evaluated in Prob. 15.22(a), yielding the result 0.7814. To evaluate the right side of (1), use the results of Prob. 15.22(c) and obtain again 0.7814. Thus the equality holds in this special case. It can be shown by direct algebraic processes that the equality holds in general.

(b) The left side of (1) is $r_{12.34}$, and the right side is $r_{12.43}$. Since $r_{12.34}$ is the correlation between variables X_1 and X_2 keeping X_3 and X_4 constant, while $r_{12.43}$ is the correlation between X_1 and X_2 keeping X_4 and X_3 constant, it is at once evident why the equality should hold.

15.25. Find (a) the multiple correlation coefficient $R_{1.234}$ and (b) the standard error of estimate $s_{1.234}$ for the data of Problem 15.20.

Solution:

(a)
$$1 - R_{1.234}^2 = (1 - r_{12}^2)(1 - r_{13.2}^2)(1 - r_{14.23}^2) \quad \text{or} \quad R_{1.234} = 0.9310$$

since $r_{12} = 0.90$ from Prob. 15.20, $r_{14.23} = 0.4193$ from Prob. 15.22(c), and

$$r_{13.2} = \frac{r_{13} - r_{12}r_{23}}{\sqrt{(1 - r_{12}^2)(1 - r_{23}^2)}} = \frac{0.75 - (0.90)(0.70)}{\sqrt{[1 - (0.90)^2][1 - (0.70)^2]}} = 0.3855$$

Another method:

Interchanging subscripts 2 and 4 in the first equation yields
$$1 - R_{1.234}^2 = (1 - r_{14}^2)(1 - r_{13.4}^2)(1 - r_{12.34}^2) \quad \text{or} \quad R_{1.234} = 0.9310$$
where the results of Prob. 15.22(a) are used directly.

(b) $R_{1.234} = \sqrt{1 - s_{1.234}^2/s_1^2}$ or $s_{1.234} = s_1\sqrt{1 - R_{1.234}^2} = 10\sqrt{1 - (0.9310)^2} = 3.650$

Compare with equation (8), Page 271.

Supplementary Problems

REGRESSION EQUATIONS INVOLVING THREE VARIABLES

15.26. Using an appropriate subscript notation, write the regression equations of (a) X_3 on X_1 and X_2, (b) X_4 on X_1, X_2, X_3 and X_5.
Ans. (a) $X_3 = b_{3.12} + b_{31.2}X_1 + b_{32.1}X_2$, (b) $X_4 = b_{4.1235} + b_{41.235}X_1 + b_{42.135}X_2 + b_{43.125}X_3$

15.27. Write the normal equations corresponding to the regression equations of (a) X_2 on X_1 and X_3, (b) X_5 on X_1, X_2, X_3 and X_4.

15.28. The table shows the corresponding values of three variables X_1, X_2 and X_3. (a) Find the least square regression equation of X_3 on X_1 and X_2. (b) Estimate X_3 when $X_1 = 10$ and $X_2 = 6$.
Ans. (a) $X_3 = 61\cdot40 - 3\cdot65X_1 + 2\cdot54X_2$, (b) 40

X_1	3	5	6	8	12	14
X_2	16	10	7	4	3	2
X_3	90	72	54	42	30	12

15.29. An instructor of mathematics wished to determine the relationship of marks on a final examination to marks on two quizzes given during the term. Calling X_1, X_2 and X_3 the marks of a student on the first quiz, second quiz and final examination respectively, he made the following computations for a total of 120 students.

$$\bar{X}_1 = 6\cdot8 \qquad \bar{X}_2 = 7\cdot0 \qquad \bar{X}_3 = 74$$
$$s_1 = 1\cdot0 \qquad s_2 = 0\cdot80 \qquad s_3 = 9\cdot0$$
$$r_{12} = 0\cdot60 \qquad r_{13} = 0\cdot70 \qquad r_{23} = 0\cdot65$$

(a) Find the least square regression equation of X_3 on X_1 and X_2.

(b) Estimate the final marks of two students who scored respectively 9 and 7, 4 and 8, on the two quizzes.

Ans. (a) $X_3 - 74 = 4\cdot36(X_1 - 6\cdot8) + 4\cdot04(X_2 - 7\cdot0)$ or $X_3 = 16\cdot07 + 4\cdot36X_1 + 4\cdot04X_2$. (b) 84 and 66

15.30. Work Prob. 15.8 by choosing the variables X_2 and X_3 so that $\Sigma X_2 = \Sigma X_3 = 0$.

STANDARD ERROR OF ESTIMATE

15.31. Find the standard error of estimate of X_3 on X_1 and X_2 for the data of Problem 15.28.   *Ans.* $3\cdot12$

15.32. Find the standard error of estimate of (a) X_3 on X_1 and X_2, (b) X_1 on X_2 and X_3, for the data of Problem 15.29.
Ans. (a) $5\cdot883$, (b) $0\cdot6882$

COEFFICIENT OF MULTIPLE CORRELATION

15.33. Compute the coefficient of linear multiple correlation of X_3 on X_1 and X_2 for the data of Prob. 15.28.   *Ans.* $0\cdot9927$

15.34. Compute (a) $R_{3.12}$, (b) $R_{1.23}$ and (c) $R_{2.13}$ for the data of Prob. 15.29.
Ans. (a) $0\cdot7567$, (b) $0\cdot7255$, (c) $0\cdot6810$

15.35. If $r_{12} = r_{13} = r_{23} = r \neq 1$, show that $R_{1.23} = R_{2.31} = R_{3.12} = \dfrac{r\sqrt{2}}{\sqrt{1+r}}$. Discuss the case $r = 1$.

15.36. If $R_{1.23} = 0$, prove that $|r_{23}| \geq |r_{12}|$ and $|r_{23}| \geq |r_{13}|$ and interpret.

PARTIAL CORRELATION

15.37. Compute the coefficients of linear partial correlation (a) $r_{12.3}$, (b) $r_{13.2}$ and (c) $r_{23.1}$ for the data of Prob. 15.28 and interpret your answers.   *Ans.* (a) $0\cdot5950$, (b) $-0\cdot8995$, (c) $0\cdot8727$

15.38. Work Prob. 15.37 for the data of Prob. 15.29.   *Ans.* (a) $0\cdot2672$, (b) $0\cdot5099$, (c) $0\cdot4026$

15.39. If $r_{12} = r_{13} = r_{23} = r \neq 1$, show that $r_{12.3} = r_{13.2} = r_{23.1} = r/(1+r)$. Discuss the case $r = 1$.

15.40. If $r_{12.3} = 1$, show that (a) $|r_{13.2}| = 1$, (b) $|r_{23.1}| = 1$, (c) $R_{1.23} = 1$, (d) $s_{1.23} = 0$.

MULTIPLE AND PARTIAL CORRELATION INVOLVING FOUR OR MORE VARIABLES

15.41. Show that the regression equation of X_4 on X_1, X_2 and X_3 can be written

$$\frac{x_4}{s_4} = a_1\left(\frac{x_1}{s_1}\right) + a_2\left(\frac{x_2}{s_2}\right) + a_3\left(\frac{x_3}{s_3}\right)$$

where a_1, a_2 and a_3 are determined by solving simultaneously the equations

$$\begin{cases} a_1 r_{11} + a_2 r_{12} + a_3 r_{13} = r_{14} \\ a_1 r_{21} + a_2 r_{22} + a_3 r_{23} = r_{24} \\ a_1 r_{31} + a_2 r_{32} + a_3 r_{33} = r_{34} \end{cases}$$

and where $x_j = X_j - \bar{X}_j$, $r_{jj} = 1$, $j = 1, 2, 3, 4$. Generalize to the case of more than 4 variables.

15.42. Given $\bar{X}_1 = 20$, $\bar{X}_2 = 36$, $\bar{X}_3 = 12$, $\bar{X}_4 = 80$, $s_1 = 1\cdot0$, $s_2 = 2\cdot0$, $s_3 = 1\cdot5$, $s_4 = 6\cdot0$, $r_{12} = -0\cdot20$, $r_{13} = 0\cdot40$, $r_{23} = 0\cdot50$, $r_{14} = 0\cdot40$, $r_{24} = 0\cdot30$, $r_{34} = -0\cdot10$. (a) Find the regression equation of X_4 on X_1, X_2 and X_3. (b) Estimate X_4 when $X_1 = 15$, $X_2 = 40$ and $X_3 = 14$.   *Ans.* (a) $X_4 = 6X_1 + 3X_2 - 4X_3 - 100$, (b) 54

15.43. Find (a) $r_{41.23}$, (b) $r_{42.13}$ and (c) $r_{43.12}$ for the data of Prob. 15.42 and interpret your results. *Ans.* (a) $0\cdot8710$, (b) $0\cdot8587$, (c) $-0\cdot8426$

15.44. Find (a) $R_{4.123}$ and (b) $s_{4.123}$ for the data of Prob. 15.42.   *Ans.* (a) $0\cdot8947$, (b) $2\cdot680$

15.45. A scientist collected data concerning four variables T, U, V and W. He believed that an equation of the form $W = aT^bU^cV^d$, where a, b, c and d are unknown constants, could be found from which he could determine W by knowing T, U and V. Outline clearly a procedure by means of which this aim may be accomplished. [Hint: Take logarithms of both sides of the equation.]

CHAPTER 16

Analysis of Time Series

TIME SERIES

A time series is a set of observations taken at specified times, usually at equal intervals.

Examples of time series are the total annual production of steel in the United States over a number of years, the daily closing price of a share on the Stock Exchange, the hourly temperatures announced by the weather bureau of a city, and the total of monthly sales receipts in a department store.

Mathematically a time series is defined by the values Y_1, Y_2, . . . of a variable Y (temperature, closing price of a share, etc.) at times t_1, t_2, ... Thus Y is a function of t, symbolized by $Y = F(t)$.

GRAPHS OF TIME SERIES

A time series involving a variable Y is represented pictorially by constructing a graph of Y vs. t, as done many times in previous chapters. The graph of a time series showing the cattle population of the United States during the years 1870–1960 is given in Fig. 16-1.

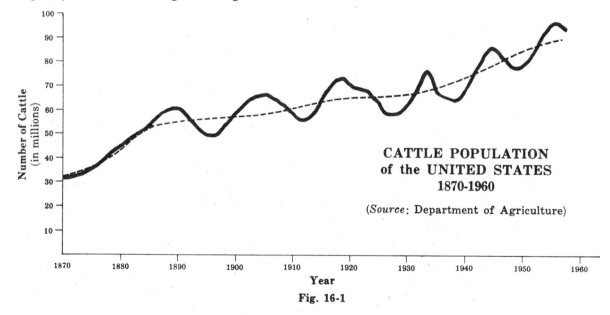

Fig. 16-1

CHARACTERISTIC MOVEMENTS OF TIME SERIES

It is interesting to think of the graph of a time series, such as shown in Fig. 16-1, as described by a point moving with the passage of time, in many ways analogous to that of a physical particle moving under the influence of physical forces. However, in place of physical forces the motion may be due to a combination of economic, sociological, psychological and other forces.

Experience with many examples of time series has revealed certain *characteristic movements* or *varia-*

283

tions, some or all of which are present to varying degrees. Analysis of such movements is of great value in many connections, one of which is the problem of *forecasting* future movements. It should thus come as no surprise that many industries and governmental agencies are vitally concerned with this important subject.

CLASSIFICATION OF TIME SERIES MOVEMENTS

Characteristic movements of time series may be classified into four main types, often called *components* of a time series:

1. **Long-term or secular movements** refer to the general direction in which the graph of a time series appears to be going over a long interval of time. In the above graph this secular movement, or as it is sometimes called *secular variation* or *secular trend*, is indicated by a *trend curve*, shown dashed. For some time series a *trend line* may be appropriate. Determination of such trend lines and curves by the method of least squares has been considered in Chap. 13. Other methods are discussed below.

2. **Cyclical movements** or *cyclical variations* refer to the long-term oscillations or swings about a trend line or curve. These *cycles*, as they are sometimes called, may or may not be *periodic*, i.e. they may or may not follow exactly similar patterns after equal intervals of time. In business and economic activity, movements are considered cyclical only if they recur after time intervals of more than a year.

     An important example of cyclical movements are the so-called *business cycles* representing intervals of prosperity, recession, depression and recovery.

     In Fig. 16-1 the cyclical movements about the trend curve are quite apparent.

3. **Seasonal movements** or *seasonal variations* refer to the identical, or almost identical, patterns which a time series appears to follow during corresponding months of successive years. Such movements are due to recurring events which take place annually, as for instance the sudden increase of department store sales before Christmas.

     In Fig. 16-1 no seasonal movements are present, since in obtaining the graph only annual figures were used.

     Although seasonal movements in general refer to *annual* periodicity in business or economic theory, the ideas involved can be extended to include periodicity over any interval of time such as daily, hourly, weekly, etc., depending on the type of data available.

4. **Irregular or random movements** refer to the sporadic motions of time series due to chance events such as floods, strikes, elections, etc. Although it is ordinarily assumed that such events produce variations lasting only a short time, it is conceivable that they may be so intense as to result in new cyclical or other movements.

THE ANALYSIS OF TIME SERIES

The analysis of time series consists of a description (generally mathematical) of component movements present. To motivate procedures involved in such description, consider Fig. 16-2 which refers to an *ideal* time series.

Fig. (*a*) shows a graph of a long-term or secular trend line (we could have used a trend curve as well). Fig. (*b*) shows this long-term trend line with a superimposed cyclical movement (assumed periodic). Fig. (*c*) shows a superposition of a seasonal movement on the graph of Fig. (*b*). If we were to superimpose on the graph (*c*) some random or irregular movements, the result would look more like a time series occurring in practice.

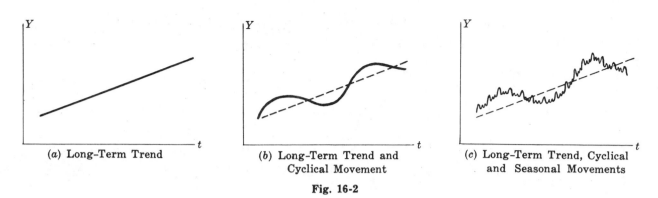

(a) Long-Term Trend   (b) Long-Term Trend and   (c) Long-Term Trend, Cyclical
                           Cyclical Movement   and Seasonal Movements

Fig. 16-2

The ideas presented above provide us with a possible technique for analysing time series. We assume that the time series variable Y is a product of the variables, T, C, S and I which produce respectively the trend, cyclical, seasonal and irregular movements. In symbols,

$$Y = T \times C \times S \times I = TCSI \tag{1}$$

The analysis of time series consists in an investigation of the factors T, C, S and I and is often referred to as a *decomposition* of a time series into its basic component movements.

It should be mentioned that some statisticians prefer to consider Y as a sum $T + C + S + I$ of the basic variables involved. Although we shall assume the decomposition (*1*) in methods of this chapter, analogous procedures are available when a sum is assumed. In practice the decision as to which method of decomposition should be assumed depends on the degree of success achieved in applying the assumption.

MOVING AVERAGES. SMOOTHING OF TIME SERIES

Given a set of numbers

$$Y_1, Y_2, Y_3, \ldots \tag{2}$$

we define a *moving average of order N* to be given by the sequence of arithmetic means,

$$\frac{Y_1 + Y_2 + \ldots + Y_N}{N}, \quad \frac{Y_2 + Y_3 + \ldots + Y_{N+1}}{N}, \quad \frac{Y_3 + Y_4 + \ldots + Y_{N+2}}{N}, \ldots \tag{3}$$

The sums in the numerators of (*3*) are called *moving totals of order N*.

Example 1. Given the numbers 2, 6, 1, 5, 3, 7, 2 a moving average of order 3 is given by the sequence

$$\frac{2+6+1}{3}, \frac{6+1+5}{3}, \frac{1+5+3}{3}, \frac{5+3+7}{3}, \frac{3+7+2}{3} \text{ i.e. } 3, 4, 3, 5, 4$$

It is customary to locate each number in the moving average at its appropriate position relative to the original data. In this example we would write

Original data   2, 6, 1, 5, 3, 7, 2
Moving average of order 3   3, 4, 3, 5, 4

each number in the moving average being the mean of the 3 numbers immediately above it.

If data are given annually or monthly, a moving average of order N is called respectively an *N year moving average* or *N month moving average*. Thus we speak of 5 year moving averages, 12 month moving averages, etc. Clearly any other unit of time can also be used.

Moving averages have the property that they tend to reduce the amount of variation present in a set of data. In the case of time series this property is often used to eliminate unwanted fluctuations and the process is called *smoothing of time series*.

If in (3), weighted arithmetic means are used, the weights being specified in advance, the resulting sequence is called a *weighted moving average of order N*.

Example 2. If the weights 1, 4, 1 are used in Example 1, a weighted moving average of order 3 is given by the sequence

$$\frac{1(2) + 4(6) + 1(1)}{1 + 4 + 1}, \quad \frac{1(6) + 4(1) + 1(5)}{1 + 4 + 1}, \quad \frac{1(1) + 4(5) + 1(3)}{1 + 4 + 1},$$

$$\frac{1(5) + 4(3) + 1(7)}{1 + 4 + 1}, \quad \frac{1(3) + 4(7) + 1(2)}{1 + 4 + 1}$$

or 4·5, 2·5, 4·0, 4·0, 5·5.

ESTIMATION OF TREND

Estimation of trend can be achieved in several possible ways.

1. **The Method of Least Squares** of Chap. 13 can be used to find the equation of an appropriate trend line or trend curve. From this equation we can compute the trend values T.

2. **The Freehand Method,** which consists of fitting a trend line or curve simply by looking at the graph, can be used to estimate T. However, this has the obvious disadvantage of depending too much on individual judgment.

3. **The Moving Average Method.** By using moving averages of appropriate orders, cyclical, seasonal and irregular patterns may be eliminated, thus leaving only the trend movement.

    One disadvantage of this method is that data at the beginning and end of a series are lost. Thus in Example 1 above, we started with 7 numbers, and with a moving average of order 3 we arrived at 5 numbers. Another disadvantage is that moving averages may generate cycles or other movements which were not present in the original data. A third disadvantage is that moving averages are strongly affected by extreme values. To overcome this somewhat, a weighted moving average with appropriate weights is sometimes used. In such case the central item (or items) is given the largest weight and extreme values are given small weights.

4. **The Method of Semi-Averages** consists of separating the data into two parts (preferably equal) and averaging the data in each part, thus obtaining two points on the graph of the time series. A trend line is then drawn between these two points and trend values can be determined. The trend values can also be determined directly without a graph (see Prob. 16.5).

    Although this method is simple to apply, it may lead to poor results when used indiscriminately. Also it is applicable only where the trend is linear or approximately linear, although it can be extended to cases where the data can be broken up into several parts in each of which the trend is linear.

ESTIMATION OF SEASONAL VARIATIONS. SEASONAL INDEX

To determine the seasonal factor S in equation (1), we must estimate how the data in the time series vary from month to month throughout a typical year. A set of numbers showing relative values of a variable during the months of the year is called a *seasonal index* for the variable. If for example we know that sales during January, February, March, etc., are 50, 120, 90, . . . per cent of the average monthly sales for the whole year, the numbers 50, 120, 90, . . . provide the seasonal index for the year and are sometimes referred to as *seasonal index numbers*. The average (mean) seasonal index for the whole year should be 100%, i.e. the sum of the index numbers should be 1200%.

Various methods are available for computing a seasonal index.

1. **The Average Percentage Method.** In this method the data for each month are expressed as percentages of the average for the year. The percentages for corresponding months of different years are then averaged using either a mean or median. If the mean is used it is best to avoid extreme values which may occur.

   The resulting 12 percentages give the seasonal index. If their mean is not 100% (i.e. if the sum is not 1200%) these should be adjusted by multiplying by a suitable factor.

2. **The Percentage Trend or Ratio to Trend Method.** In this method the data for each month are expressed as percentages of monthly trend values. An appropriate average of the percentages for corresponding months then gives the required index. As in method 1, we adjust these if they do not average to 100%.

   Note that division of each monthly value Y by the corresponding trend value T yields $Y/T = CSI$, from equation (*1*). Subsequent averaging of Y/T produces seasonal indexes which may include cyclical and irregular variations especially when they are large. This may be an important disadvantage of the method.

3. **Percentage Moving Average or Ratio to Moving Average Method.** In this method a 12 month moving average is computed. Since results thus obtained fall between successive months instead of in the middle of the month as for the original data, we compute a 2 month moving average of this 12 month moving average. The result is often called a *centred 12 month moving average*.

   After this has been done, the original date for each month is expressed as a percentage of the centred 12 month moving average corresponding to it. Percentages for corresponding months are then averaged, giving the required index. As before, we adjust these if they do not average to 100%.

   Note that the logical reasoning behind this method follows from equation (*1*). A centred 12 month moving average of Y serves to eliminate seasonal and irregular movements S and I, and is thus equivalent to values given by TC. Then division of the original data by TC yields SI. The subsequent averages over corresponding months serve to eliminate the irregularity I and thus result in a suitable index S.

4. **The Link Relative Method.** In this method, data for each month are expressed as percentages of data for the previous month. These percentage are called *link relatives*, since they *link* each month to the preceding one. An appropriate average of the link relatives for corresponding months is then taken.

   From these 12 average link relatives we can obtain the relative percentages of each month with respect to January which is considered as 100%. After this has been done it will usually be found that the next January will have an associated percentage which is either higher or lower than 100% depending on whether there has been an increase or decrease in trend. Using this, the various percentages obtained can then be adjusted for this trend. These final percentages, adjusted so that they average to 100%, provide the required seasonal index.

DESEASONALIZATION OF DATA

If the original monthly data are divided by the corresponding seasonal index numbers, the resulting data are said to be *deseasonalized* or *adjusted for seasonal variation*. Such data still include trend, cyclical and irregular movements.

ESTIMATION OF CYCLICAL VARIATIONS

After data have been deseasonalized, they can also be adjusted for trend by simply dividing the data by corresponding trend values. According to equation (*1*) the process of adjusting for seasonal variation and trend corresponds to division of Y by ST, which gives CI, i.e. cyclical and irregular variations. An appropriate moving average of a few months duration (3, 5 or 7 months, for example, so that subsequent centring is not necessary) then serves to smooth out the irregular variations I and leaves only the cyclical variations. Once these have been isolated they may be studied in detail. If periodicity (or approximate periodicity) of cycles occurs, *cyclical indexes* can be constructed in much the same manner as seasonal indexes.

ESTIMATION OF IRREGULAR OR RANDOM VARIATIONS

Estimation of irregular or random variations can be achieved by adjusting data for trend, seasonal and cyclical variations. This amounts to division of original data Y by T, S and C which, by equation (*1*), yields I. In practice it is found that irregular movements tend to be of small magnitude and that they often tend to follow the pattern of a normal distribution, i.e. small deviations occur with large frequency, large deviations occur with small frequency.

COMPARABILITY OF DATA

One must always be careful when comparing data that such comparison is justified.

For example, in comparing data for March with that of February, we must realize that March has 31 days while February has 28 or 29 days. Similarly, in comparing months of February for different years, we must remember that during leap year February has 29 rather than 28 days. The number of working days during various months of the same or different years may also differ due to holidays, strikes, layoffs, etc.

In practice, no definite rule is followed for making adjustments due to such variations. The need for such adjustment is left to the discretion of the investigator.

FORECASTING

The ideas presented above can be used to aid in the important problem of *forecasting* time series. However, it must be realized that mathematical treatment of data does not in itself solve all problems. Coupled with the common sense, experience, ingenuity and good judgment of an investigator, such mathematical analysis can nevertheless be of value both for *long range* and *short range forecasting*.

SUMMARY OF FUNDAMENTAL STEPS IN TIME SERIES ANALYSIS

1. Collect data for the time series, making every effort to insure that data are reliable. In collection of data one should always keep in mind the eventual purpose of the time series analysis. For example, if one wishes to *forecast* a given time series, it may be helpful to obtain related time series as well as other information. If necessary, adjust the data for comparability, i.e. adjust for leap years, etc.

2. Graph the time series, noting qualitatively the presence of long-term trend, cyclical variations and seasonal variations.

3. Construct the long-term trend curve or line and obtain appropriate trend values by use of either the method of least squares, the freehand method, the method of moving averages, or the method of semi-averages.

4. If seasonal variations are present, obtain a seasonal index and adjust the data for these seasonal variations, i.e. deseasonalize the data.

5. Adjust the deseasonalized data for trend. The resulting data contains (theoretically) only the cyclical and irregular variations. A moving average of 3, 5 or 7 months serves to remove irregular variations and reveal the cyclical variations.

6. Graph the cyclical variations obtained in step 5, noting any periodicities (or approximate periodicities) which may occur.

7. By combining the results of steps 1–6 and using any other available information, make a forecast (if desired) and if possible discuss sources of error and their magnitude.

Solved Problems

CHARACTERISTIC MOVEMENTS OF TIME SERIES

16.1. With which characteristic movement of a time series would you mainly associate each of the following:

(a) a fire in a factory delaying production for 3 weeks.   *Ans.* irregular

(b) an era of prosperity.   *Ans.* cyclical

(c) an after Easter sale in a department store.   *Ans.* seasonal

(d) a need for increased wheat production due to a constant
      increase in population.   *Ans.* long-term

(e) the monthly number of millimetres of rainfall in a city over
      a 5 year period.   *Ans.* seasonal

MOVING AVERAGES

16.2. Table 16.1 shows the average monthly production, for a certain country, of bituminous coal in millions of kilogrammes for the years 1948–1958. Construct a (a) 5 year moving average, (b) 4 year moving average, (c) 4 year centred moving average.

Table 16.1

Year	1948	1949	1950	1951	1952	1953	1954	1955	1956	1957	1958
Average monthly production of bituminous coal (millions of kilogrammes)	50·0	36·5	43·0	44·5	38·9	38·1	32·6	38·7	41·7	41·1	33·8

Solution:

(a) Refer to Table 16.2.

The first moving total 212·9 of column 3 is the sum of the 1st through 5th entries of column 2. The second moving total 201·0 is the sum of the 2nd through 6th entries in column 2, etc.

In practice, after obtaining the first moving total 212·9, the second moving total is easily obtained by subtracting 50·0 (1st entry of column 2) and adding 38·1 (6th entry of column 2), the result being 201·0. Succeeding moving totals are obtained similarly.

Dividing each moving total by 5 yields the required moving average.

Table 16.2

Year	Data	5 year moving total	5 year moving average
1948	50·0		
1949	36·5		
1950	43·0	212·9	42·6
1951	44·5	201·0	40·2
1952	38·9	197·1	39·4
1953	38·1	192·8	39·6
1954	32·6	190·0	38·0
1955	38·7	192·2	38·4
1956	41·7	187·9	37·6
1957	41·1		
1958	33·8		

(b) Refer to Table 16.3.

The 4 year moving totals are obtained as in part (a), except that 4 entries of column 2 are added instead of 5. Note that the moving totals are centred *between* successive years, unlike part (a). This is always the case when an *even* number of years is taken in the moving average. If we consider that 1949, for example, stands for July 1, 1949, the first 4 year moving total is centred at Jan. 1, 1950 or Dec. 31, 1949.

The 4 year moving averages are obtained by dividing the 4 year moving totals by 4.

Table 16.3

Year	Data	4 year moving total	4 year moving average
1948	50·0		
1949	36·5		
1950	43·0	174·0	43·5
1951	44·5	162·9	40·7
1952	38·9	164·5	41·1
1953	38·1	154·1	38·5
1954	32·6	148·3	37·1
1955	38·7	151·1	37·8
1956	41·7	154·1	38·5
1957	41·1	155·3	38·8
1958	33·8		

(c) **First method:** See Table 16.4.

First we compute a 4 year moving average as in part (b). These values are centred *between* successive years as shown.

If we now compute a 2 year moving total of these 4 year moving averages, the results are centred at the required years.

Dividing the results in column 4 by 2 yields the required 4 year *centred* moving averages.

Table 16.4

Year	Data	4 year moving average	2 year moving total of col. 3	4 year centred moving average (col. 4 ÷ 2)
1948	50·0			
1949	36·5			
		43·5		
1950	43·0		84·2	42·1
		40·7		
1951	44·5		81·8	40·9
		41·1		
1952	38·9		79·6	39·8
		38·5		
1953	38·1		75·6	37·8
		37·1		
1954	32·6		74·9	37·5
		37·8		
1955	38·7		76·3	38·2
		38·5		
1956	41·7		77·3	38·7
		38·8		
1957	41·1			
1958	33·8			

Second method: See Table 16.5.

First we compute a 4 year moving total as in part (b). These values are centred between successive years as shown.

If we now compute a 2 year moving total of these 4 year moving totals, the results become centred at the required years.

Dividing the results in column 4 by 8 (2 × 4) yields the required moving average.

Table 16.5

Year	Data	4 year moving total	2 year moving total of col. 3	4 year centred moving average (col. 4 ÷ 8)
1948	50·0			
1949	36·5	174·0		
1950	43·0		336·9	42·1
1951	44·5	162·9	327·4	40·9
1952	38·9	164·5	318·6	39·8
1953	38·1	154·1	302·4	37·8
1954	32·6	148·3	299·4	37·4
1955	38·7	151·1	305·2	38·2
1956	41·7	154·1	309·4	38·7
1957	41·1	155·3		
1958	33·8			

16.3. Show that the 4 year centred moving average of Prob. 16.2(c) is equivalent to a 5 year weighted moving average with weights 1, 2, 2, 2, 1 respectively.

Solution:

Let $Y_1, Y_2, \ldots, Y_{11}$ denote the values corresponding to the years 1948, 1949, ..., 1958 respectively. Then proceeding as in the second method of Prob. 16.2(c), we obtain Table 16.6.

Table 16.6

Year	Y	4 year moving total	2 year moving total of col. 3	4 year centred moving average (col. 4 ÷ 8)
1948	Y_1			
1949	Y_2	$Y_1 + Y_2 + Y_3 + Y_4$		
1950	Y_3	$Y_2 + Y_3 + Y_4 + Y_5$	$Y_1 + 2Y_2 + 2Y_3 + 2Y_4 + Y_5$	$\frac{1}{8}(Y_1 + 2Y_2 + 2Y_3 + 2Y_4 + Y_5)$
1951	Y_4	$Y_3 + Y_4 + Y_5 + Y_6$	$Y_2 + 2Y_3 + 2Y_4 + 2Y_5 + Y_6$	$\frac{1}{8}(Y_2 + 2Y_3 + 2Y_4 + 2Y_5 + Y_6)$
1952	Y_5	$Y_4 + Y_5 + Y_6 + Y_7$	$Y_3 + 2Y_4 + 2Y_5 + 2Y_6 + Y_7$	$\frac{1}{8}(Y_3 + 2Y_4 + 2Y_5 + 2Y_6 + Y_7)$
1953	Y_6			
.	.	.	.	.
.	.	.	.	.
.	.	.	.	.
1958	Y_{11}			

From the last column it follows that the 4 year centred moving average is a 5 year weighted moving average with respective weights 1, 2, 2, 2, 1. Note that 8 is the sum of these weights, i.e. $1 + 2 + 2 + 2 + 1 = 8$.

This method can be used to obtain the results of Prob. 16.2(c). For example, the first entry (corresponding to 1950)

$$\frac{(1)(50 \cdot 0) + (2)(36 \cdot 5) + (2)(43 \cdot 0) + (2)(44 \cdot 5) + (1)(38 \cdot 9)}{8} = 42 \cdot 1$$

16.4. Graph the moving average of Problem 16.2(*a*) together with the original data.

Solution:

The graph of the original data is shown in Fig. 16-3 by the solid line graph. The graph of the moving average is shown dashed.

Note how the moving average has smoothed the graph of the original data, showing clearly the trend line.

A disadvantage of the moving average is that data is lost at the end and the beginning of the time series. This can be serious when the amount of data is not very large.

Fig. 16-3

ESTIMATION OF TREND

16.5. Obtain the trend values for the data of Prob. 16.2 by using the method of semi-averages, where the average is taken as (*a*) the mean, (*b*) the median.

Table 16.7

1948	50·2	1954	32·6
1949	36·5	1955	38·7
1950	43·0	1956	41·7
1951	44·5	1957	41·1
1952	38·9	1958	33·8
	Total 212·9		Total 187·9

Solution:

(*a*) Divide the data into two equal parts (omitting the middle year 1953) as shown. Compute the mean of the data in each part.

From the results obtained it follows that in 6 years (1950 to 1956) there has been a *decrease* of 5·0 (42·6 − 37·6) million kilogrammes, or a decrease of 5·0/6 = 0·83 million kilogrammes per year.

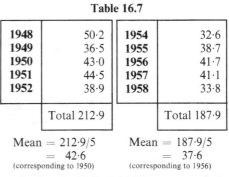

Mean = 212·9/5
= 42·6
(corresponding to 1950)

Mean = 187·9/5
= 37·6
(corresponding to 1956)

By knowing this, the trend values can be computed. Thus the trend values for 1951 and 1952 are respectively 42·6 − 0·83 = 41·8 and 42·6 − 2(0·83) = 40·9, etc., as shown in Table 16.8.

Table 16.8

Year	1948	1949	1950	1951	1952	1953	1954	1955	1956	1957	1958
Trend value	44·3	43·4	42·6	41·8	40·9	40·1	39·3	38·5	37·6	36·8	36·0

The results can also be obtained by drawing the graph of a line connecting the points (1950, 42·6) and (1956, 37·6) and reading the trend values from the graph.

(*b*) The medians for each of the two parts in (*a*) are 43·0 and 38·7 respectively. Thus there is a decrease of (43·0 − 38·7)/6 = 0·72 per year, and the trend values are as indicated in Table 16.9.

Table 16.9

Year	1948	1949	1950	1951	1952	1953	1954	1955	1956	1957	1958
Trend value	44·4	43·7	43·0	42·3	41·6	40·8	40·1	39·4	38·7	38·0	37·2

When medians are used the method is sometimes called the method of *semi-medians*. If the type of average is not specified the mean is implied.

16.6. Describe how you would use (*a*) the freehand method and (*b*) the method of moving averages to compute the trend values for the data of Prob. 16.2.

Solution:

(*a*) In this method we would simply construct a line or curve which closely approximated the data given in the graph of Prob. 16.4. From this graph we could then read the trend values.

(*b*) By use of a 5 year moving average, we have seen (Prob. 16.4) that the time series data have been considerably smoothed. We can use the averages obtained as the trend values for the years 1950-1956. Thus from Prob. 16.2(*a*) we see that the trend values corresponding to 1950, 1951, 1952, etc., are 42·6, 40·2, 39·4, etc. By this method, however, the trend values for the years 1948, 1949, 1957 and 1958 are unavailable. If desired these could be obtained by extrapolation from the graph of Prob. 16.4.

16.7. (a) Use the method of least squares to fit a line to the data of Prob. 16.2.

(b) From the result in (a) find the trend values.

Solution:

(a) We use the second method of Prob. 13.19(b), Chapter 13, since an odd number of years is present.

Table 16.10

Year	X	Y	X²	XY
1948	−5	50·0	25	−250·0
1949	−4	36·5	16	−146·0
1950	−3	43·0	9	−129·0
1951	−2	44·5	4	−89·0
1952	−1	38·9	1	−38·9
1953	0	38·1	0	0
1954	1	32·6	1	32·6
1955	2	38·7	4	77·4
1956	3	41·7	9	125·1
1957	4	41·1	16	164·4
1958	5	33·8	25	169·0
		$\Sigma Y = 438\cdot9$	$\Sigma X^2 = 110$	$\Sigma XY = -84\cdot4$

Then the required least square line is

$$Y = \bar{Y} + \left(\frac{\Sigma XY}{\Sigma X^2}\right)X = \frac{438\cdot9}{11} + \left(\frac{-84\cdot4}{110}\right)X \text{ or } Y = 39\cdot9 - 0\cdot767X$$

where the origin $X = 0$ is the year 1953 and the unit of X is 1 year.

(b) Placing $X = -5, -4, -3, \ldots, 5$ in the least square equation found in part (a), the trend values given in Table 16.11 are obtained.

Table 16.11

Year	1948	1949	1950	1951	1952	1953	1954	1955	1956	1957	1958
Trend value	43·7	43·0	42·2	41·4	40·7	39·9	39·1	38·4	37·6	36·8	36·1

The results compare favourably with those of Prob. 16.5.

ESTIMATION OF SEASONAL VARIATIONS SEASONAL INDEX

16.8. Table 16.12 shows the monthly electric power in millions of kilowatt hours used for street and highway lighting in the United States during the years 1952-1958. (a) Construct a graph of the data. (b) Obtain a seasonal index using the average percentage method.

Table 16.12

	Jan	Feb	Mar	Apr	May	June	July	Aug	Sept	Oct	Nov	Dec
1951	318	281	278	250	231	216	223	245	269	302	325	347
1952	342	309	299	268	249	236	242	262	288	321	342	364
1953	367	328	320	287	269	251	259	284	309	345	367	394
1954	392	349	342	311	290	273	282	305	328	364	389	417
1955	420	378	370	334	314	296	305	330	356	396	422	452
1956	453	412	398	362	341	322	335	359	392	427	454	483
1957	487	440	429	393	370	347	357	388	415	457	491	516
1958	529	477	463	423	398	380	389	419	448	493	526	560

Source: Survey of Current Business

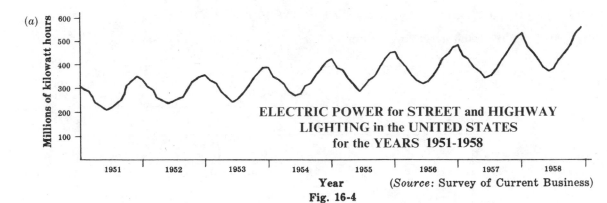

Fig. 16-4   (*Source*: Survey of Current Business)

(*b*) The totals and monthly averages (means) for the years 1951–1958 are as follows.

Table 16.13	1951	1952	1953	1954	1955	1956	1957	1958
Totals	3285	3522	3780	4042	4373	4738	5090	5505
Monthly averages	273·7	293·5	315·0	336·8	364·4	394·8	424·2	458·7

Dividing the given monthly data by the corresponding monthly averages for each year and expressing the result as a per cent yields the entries in Table 16.14. For example, the first entry in the table is given by $318/273·7 = 116·2\%$.

Table 16.14

	Jan	Feb	Mar	Apr	May	June	July	Aug	Sept	Oct	Nov	Dec
1951	116·2	102·7	101·6	91·3	84·4	78·9	81·5	89·5	98·3	110·3	118·7	126·8
1952	116·5	105·3	101·9	91·3	84·8	80·4	82·5	89·3	98·1	109·4	116·5	124·0
1953	116·5	104·1	101·6	91·1	85·4	79·7	82·2	90·2	98·1	109·5	116·5	125·1
1954	116·4	103·6	101·5	92·3	86·1	81·1	83·7	90·6	97·4	108·1	115·5	123·8
1955	115·3	103·7	101·5	91·7	86·2	81·2	83·7	90·6	97·7	108·7	115·8	124·0
1956	114·7	104·4	100·8	91·7	86·4	81·6	84·9	90·9	99·3	108·2	115·0	122·3
1957	114·8	103·7	101·1	92·6	87·2	81·8	84·2	91·5	97·8	107·7	115·7	121·6
1958	115·3	104·0	100·9	92·2	86·8	82·8	84·8	91·3	97·7	107·5	114·7	122·1
Total	925·7	831·5	810·9	734·2	687·3	647·5	667·5	723·9	784·4	869·4	928·4	989·7
Mean	115·7	103·9	101·4	91·8	85·9	80·9	83·4	90·5	98·1	108·7	116·1	123·7

The mean percentage for each month is given in the last line of Table 16.14. The total of these percentages is $1200·1\%$ which is so close to the required 1200% that no adjustment is necessary. Then the figures in the last line represent the required seasonal index.

16.9. Obtain a seasonal index for the data of Prob. 16.8 by using the percentage trend or ratio to trend method. In applying this method use the method of least squares to obtain the monthly trend values.

Solution:

From a graph of the actual data, Prob. 16.8(*a*), it appears that the long-term trend can be suitably approximated by a straight line. Instead of obtaining this line from the given monthly data we shall obtain it from the monthly averages for the years 1951–1958 given in Table 16.15, reproduced from Table 16.13 of Prob. 16.8(*b*).

Table 16.15

Year	1951	1952	1953	1954	1955	1956	1957	1958
Monthly average	273·7	293·5	315·0	336·8	364·4	394·8	424·2	458·7

Assuming that the given monthly figures correspond to the middle of the month, the averages in this table correspond to June 30 or July 1 of the corresponding year.

We use the second method of Prob. 13.20(b), Chap. 13, since there is an even number of years.

Table 16.16

Year	X	Y	X^2	XY
1951	−7	273·7	49	−1915·9
1952	−5	293·5	25	−1467·5
1953	−3	315·0	9	−945·0
1954	−1	336·8	1	−336·8
1955	1	364·4	1	364·4
1956	3	394·8	9	1184·4
1957	5	424·2	25	2121·0
1958	7	458·7	49	3210·9
		$\Sigma Y = 2861\cdot1$	$\Sigma X^2 = 168$	$\Sigma XY = 2215\cdot5$

The required least square line is

$$Y = \bar{Y} + \left(\frac{\Sigma XY}{\Sigma X^2}\right)X = \frac{2861\cdot1}{8} + \left(\frac{2215\cdot5}{168}\right)X = 357\cdot6 + 13\cdot188X$$

where X is measured in half years and the origin is Dec. 31, 1954 or Jan. 1, 1955.

From this equation it follows that the values of Y increase by 13·188 after every half year or $13\cdot188/6 = 2\cdot20$ every month. Thus when $X = 0$ (Jan. 1, 1955), $Y = 357\cdot6$. A half month later (Jan. 15, 1955), the value of Y is $357\cdot6 + \frac{1}{2}(2\cdot20) = 358\cdot7$, which is the trend value corresponding to Jan. 1955. By successively adding 2·20 to 358·7 the trend values for Feb. 1955, Mar. 1955, etc., are found to be $358\cdot7 + 2\cdot20 = 360\cdot9$, $360\cdot9 + 2\cdot20 = 363\cdot1$, etc. Similarly, by successively subtracting 2·20 from 358·7 the trend values for Dec. 1954, Nov. 1954, etc., are found to be $358\cdot7 - 2\cdot20 = 356\cdot5$, $356\cdot5 - 2\cdot20 = 354\cdot3$, etc. In this manner the monthly trend values in Table 16.17 are obtained.

Table 16.17

	Jan	Feb	Mar	Apr	May	June	July	Aug	Sept	Oct	Nov	Dec
1951	253·1	255·3	257·5	259·7	261·9	264·1	266·3	268·5	270·7	272·9	275·1	277·3
1952	279·5	281·7	283·9	286·1	288·3	290·5	292·7	294·9	297·1	299·3	301·5	303·7
1953	305·9	308·1	310·3	312·5	314·7	316·9	319·1	321·3	323·5	325·7	327·9	330·1
1954	332·3	334·7	336·7	338·9	341·1	343·3	345·5	347·7	349·9	352·1	354·3	356·5
1955	358·7	360·9	363·1	365·3	367·5	369·7	371·9	374·1	376·3	378·5	380·7	382·9
1956	385·1	387·3	389·5	391·7	393·9	396·1	398·3	400·5	402·7	404·9	407·1	409·3
1957	411·5	413·7	415·9	418·1	420·3	422·5	424·7	426·9	429·1	431·3	433·5	435·7
1958	437·9	440·1	442·3	444·5	446·7	448·9	451·1	453·3	455·5	457·5	459·9	462·1

We now divide each of the given monthly values in Table 16.12 of Prob. 16.8 by the corresponding trend values in Table 16.17. The results expressed as a percentage are given in Table 16.18. For example, the first entry of the table is given by $318/253\cdot1 = 125\cdot6\%$.

Table 16.18

	Jan	Feb	Mar	Apr	May	June	July	Aug	Sept	Oct	Nov	Dec
1951	125·6	110·1	108·0	96·3	88·2	81·8	83·7	91·2	99·4	110·7	118·1	125·1
1952	122·4	110·0	105·3	93·7	86·4	81·2	82·7	88·8	96·9	107·3	113·4	119·9
1953	120·0	106·5	103·1	91·8	85·5	79·2	81·2	88·4	95·5	105·9	111·9	119·4
1954	118·0	104·3	101·6	91·8	85·0	79·5	81·6	87·7	93·7	103·4	109·8	117·0
1955	117·1	104·7	101·9	91·4	85·4	80·1	82·0	88·2	94·6	104·6	110·8	118·0
1956	117·6	106·4	102·2	92·4	86·6	81·3	84·1	89·6	97·3	105·5	111·5	118·0
1957	118·3	106·4	103·1	94·0	88·0	82·1	84·1	90·9	96·7	106·0	113·3	118·4
1958	120·8	108·4	104·7	95·2	89·1	84·7	86·2	92·4	98·4	107·7	114·4	121·2
Median	119·2	106·4	103·1	93·0	86·5	81·2	83·2	89·2	96·8	106·0	112·6	118·9

To obtain the average percentage for each month of the various years, the median has been used, as indicated in the last row of the table, because of the presence of extreme values. Since the sum of these medians is 1196·1, we adjust them by multiplying by 1200/1196·1 so that their sum is 1200. In this manner we obtain the required seasonal index shown in Table 16.19.

Table 16.19

	Jan	Feb	Mar	Apr	May	June	July	Aug	Sept	Oct	Nov	Dec
Seasonal index	119·6	106·7	103·4	93·3	86·8	81·5	83·5	89·5	97·1	106·3	113·0	119·3

It is interesting to note that for the first seven months the above seasonal index numbers are consistently higher than those obtained in Prob. 16.8, while for the last five months they are consistently lower.

The seasonal index can also be obtained by using the mean instead of the median in the last row of Table 16.18. In such case extreme values in any column should be eliminated before the mean is computed.

16.10. Obtain a seasonal index for the data of Prob. 16.8 by using the percentage moving average or ratio to moving average method.

Solution:

We first obtain a centre 12 month moving average by using the second method of Prob. 16.2(c) as shown in Table 16.20.

Table 16.20

Year and Month	Data	12 month moving total	2 month moving total of col. 3	12 month centred moving average (col. 4 ÷ 24)	Year and Month	Data	12 month moving total	2 month moving total of col. 3	12 month centred moving average (col. 4 ÷ 24)
1951					**1953**				
Jan	318				Jan	367	3641	7299	304·1
Feb	281				Feb	328	3658	7338	305·7
Mar	278				Mar	320	3680	7381	307·5
Apr	250				Apr	287	3701	7426	309·4
May	231				May	269	3725	7475	311·5
June	216				June	251	3750	7530	313·7
July	223	3285	6594	274·7	July	259	3780	7585	316·0
Aug	245	3309	6646	276·9	Aug	284	3805	7631	318·0
Sept	269	3337	6695	279·0	Sept	309	3826	7674	319·7
Oct	302	3358	6734	280·6	Oct	345	3848	7720	321·7
Nov	325	3376	6770	282·1	Nov	367	3872	7765	323·5
Dec	347	3394	6808	283·7	Dec	394	3893	7808	325·3
		3414					3915		
1952					**1954**				
Jan	342	3433	6847	285·3	Jan	392	3938	7853	327·2
Feb	309	3450	6883	286·8	Feb	349	3959	7897	329·0
Mar	299	3469	6919	288·3	Mar	342	3978	7937	330·7
Apr	268	3488	6957	289·9	Apr	311	3997	7975	332·3
May	249	3505	6993	291·4	May	290	4019	8016	334·0
June	236	3522	7027	292·9	June	273	4042	8061	335·9
July	242	3547	7069	294·5	July	282	4070	8112	338·0
Aug	262	3566	7113	296·4	Aug	305	4099	8169	340·4
Sept	288	3587	7153	298·0	Sept	328	4127	8226	342·7
Oct	321	3606	7193	299·7	Oct	364	4150	8277	344·9
Nov	342	3626	7232	301·3	Nov	389	4174	8324	346·8
Dec	364		7267	302·8	Dec	417		8371	348·8

Table 16.20 (cont.)

Year and Month	Data	12 month moving total	2 month moving total of col. 3	12 month centred moving average (col. 4 ÷ 24)	Year and Month	Data	12 month moving total	2 month moving total of col. 3	12 month centred moving average (col. 4 ÷ 24)
1955		4197			**1957**		4916		
Jan	420	4220	8417	350·7	Jan	487	4938	9854	410·6
Feb	378	4245	8465	352·7	Feb	440	4967	9905	412·7
Mar	370	4273	8518	354·9	Mar	429	4990	9957	414·9
Apr	334	4305	8578	375·4	Apr	393	5020	10 010	417·1
May	314	4338	8643	360·1	May	370	5057	10 077	419·9
June	296	4373	8711	363·0	June	347	5090	10 147	422·8
July	305	4406	8779	365·8	July	357	5132	10 222	425·9
Aug	330	4440	8846	368·6	Aug	388	5169	10 301	429·2
Sept	356	4468	8908	371·2	Sept	415	5203	10 372	432·2
Oct	396	4496	8964	373·5	Oct	457	5233	10 436	434·8
Nov	422	4523	9019	375·8	Nov	491	5261	10 494	437·2
Dec	452	4549	9072	378·0	Dec	516	5294	10 555	439·8
1956					**1958**				
Jan	453	4579	9128	380·3	Jan	529	5326	10 620	442·5
Feb	412	4608	9187	382·8	Feb	477	5357	10 683	445·1
Mar	398	4644	9252	385·5	Mar	463	5390	10 747	447·8
Apr	362	4675	9319	388·3	Apr	423	5426	10 816	450·7
May	341	4707	9382	390·9	May	398	5461	10 887	453·6
June	322	4738	9445	393·5	June	380	5505	10 966	456·9
July	335	4772	9510	396·2	July	389			
Aug	359	4800	9572	398·8	Aug	419			
Sept	392	4831	9631	401·3	Sept	448			
Oct	427	4862	9693	403·9	Oct	493			
Nov	454	4891	9753	406·4	Nov	526			
Dec	483		9807	408·6	Dec	560			

We now divide each of the actual monthly values by the corresponding 12 month centred moving average and express each result as a percentage. For example, corresponding to July 1951 we obtain 223/274·7 = 81·2(%). The results are given in Table 16.21. Note that the entries for the first six months of 1951 and the last six months of 1958 are not available by this method.

Table 16.21

	Jan	Feb	Mar	Apr	May	June	July	Aug	Sept	Oct	Nov	Dec
1951							81·2	88·5	96·4	107·6	115·2	122·3
1952	119·9	107·7	103·7	92·4	85·4	80·6	82·2	88·4	96·6	107·1	113·5	120·2
1953	120·7	107·3	104·1	92·8	86·4	80·0	82·0	89·3	96·7	107·2	113·4	121·1
1954	119·8	106·1	103·4	93·6	86·8	81·3	83·4	89·6	95·7	105·5	112·2	119·6
1955	119·8	107·2	104·3	93·5	87·2	81·5	83·4	89·5	95·9	106·0	112·3	119·6
1956	119·1	107·6	103·2	93·2	87·2	81·8	84·6	90·0	97·7	105·7	111·7	118·2
1957	118·6	106·6	103·4	94·2	88·1	82·1	83·8	90·4	96·0	105·1	112·3	117·3
1958	119·5	107·2	103·4	93·9	87·7	83·2						
Median	119·8	107·2	103·4	93·5	87·2	81·5	83·4	89·5	96·4	106·0	112·3	119·6

To obtain the average percentage for each month of the various years, the median has been used, as indicated in Table 16.21, because of the presence of extreme values in some cases (e.g. Nov., Dec.). The mean could also have been used; in such extreme values in any column should be eliminated.

The sum of the medians, 1199·8, is so close to the required 1200 that no adjustment is necessary. The required seasonal index is therefore given by the last row in Table 16.21.

The results agree very well with those of Prob. 16.9.

16.11. Obtain a seasonal index for the data of Prob. 16.8 by using the link relative method.

Solution:

We first express the data for each month as a percentage of the data for the previous month, as shown in Table 16.22. Each of these percentages is called a *link relative*. For example, to obtain the entries for February and March 1951, we have from the data of Prob. 16.6,

$$\text{Link relative for Feb. 1951} = \frac{\text{value for Feb. 1951}}{\text{value for Jan. 1951}} = \frac{281}{318} = 88\cdot4\%$$

$$\text{Link relative for Mar. 1951} = \frac{\text{value for Mar. 1951}}{\text{value for Feb. 1951}} = \frac{278}{281} = 98\cdot9\%$$

Table 16.22

	Jan	Feb	Mar	Apr	May	June	July	Aug	Sept	Oct	Nov	Dec
1951		88·4	98·9	89·9	92·4	93·5	103·2	109·9	109·8	112·3	107·6	106·8
1952	98·6	90·4	96·8	89·6	92·9	94·8	102·5	108·3	109·9	111·5	106·5	106·4
1953	100·8	89·4	97·6	89·7	93·7	93·3	103·2	109·7	108·8	111·7	106·4	107·4
1954	99·5	89·0	98·0	90·9	93·2	94·1	103·3	108·2	107·5	111·0	106·9	107·2
1955	100·7	90·0	97·9	90·3	94·0	94·3	103·0	108·2	107·9	111·2	106·6	107·1
1956	100·2	90·9	96·6	91·0	94·2	94·4	104·0	107·2	109·2	108·9	106·3	106·4
1957	100·8	90·3	97·5	91·6	94·1	93·8	102·9	108·7	107·0	110·1	107·4	105·1
1958	102·5	90·2	97·1	91·4	94·1	95·5	102·4	107·7	106·9	110·0	106·7	106·5
Median	100·7	90·1	97·6	90·6	93·8	94·2	103·1	108·2	108·4	111·1	106·6	106·6

The average of the link relatives for the various months (in this case the median) is shown in the last row of Table 16.22. We can also use the mean (see Prob. 16.12).

Consider January as having the value 100% (see Table 16.23). Since the average link relative for February is 90·1 (from Table 16.22), the data for February are, on the average, 90·1% of the data for January, i.e. 90·1% of 100 = 90·1. Similarly the average link relative for March is 97·6% of that of February, i.e. 97·6% of 90·1 = 87·9, etc. Thus we obtain Table 16.23 whose entries are sometimes called *chain relatives*.

Table 16.23

Jan	Feb	Mar	Apr	May	June	July	Aug	Sept	Oct	Nov	Dec	Jan
100·0	90·1	87·9	79·6	74·7	70·4	72·6	78·6	85·2	94·7	101·0	107·7	108·5

In Table 16.23 the results for the second January (last column) is 108·5, an increase of 8·5 over that of the first January. This increase is due to the long-term trend in the data. To adjust for this trend we must subtract (12/12)(8·5) = 8·5 from the last column entry (thus making the last January value 100), (11/12)(8·5) = 7·8 from the December value, (10/12)(8·5) = 7·1 from the November value, etc. The values adjusted for trend are given in Table 16.24. [Strictly speaking we should *multiply* the entries from right to left by $(100\cdot0/108\cdot5)^{12/12}$, $(100\cdot0/108\cdot5)^{11/12}$, $(100\cdot0/108\cdot5)^{10/12}$, etc. This, however, yields practically the same results as in Table 16.24.]

Table 16.24

Jan	Feb	Mar	Apr	May	June	July	Aug	Sept	Oct	Nov	Dec
100·0	89·4	86·5	77·5	71·9	66·9	68·4	73·6	79·5	88·3	93·9	99·9

Since these percentages total 995·8, we adjust them by multiplying by 1200/995·8 to obtain the seasonal index in Table 16.25.

Table 16.25

	Jan	Feb	Mar	Apr	May	June	July	Aug	Sept	Oct	Nov	Dec
Seasonal index	120·5	107·7	104·2	93·4	86·6	80·6	82·4	88·7	95·8	106·4	113·2	120·4

16.12. Work Problem 16.11 if the mean of the link relatives is used instead of the median.

Solution:

The mean link relatives are as shown in Table 16.26.

Table 16.26

	Jan	Feb	Mar	Apr	May	June	July	Aug	Sept	Oct	Nov	Dec
Mean	100·4	89·8	97·6	90·5	93·6	94·2	103·1	108·5	108·4	110·8	106·8	106·6

If we consider January as having the value 100(%), the value for February is 89·8% of 100 = 89·8, the value for March is 97·6% of 89·8 = 87·6, etc., as shown in Table 16.27.

Table 16.27

Jan	Feb	Mar	Apr	May	June	July	Aug	Sept	Oct	Nov	Dec	Jan
100·0	89·8	87·6	79·3	74·2	69·9	72·1	78·2	84·8	94·0	100·4	107·0	107·4

Here the result for the last January is 107·4, an increase due to trend of 7.4 over that of the first January. To adjust for this we subtract (12/12)(7·4) = 7·4 from the last column entry, (11/12)(7·4) = 6·8 from the December entry, (10/12)(7·4) = 6·2 from the November entry, etc., so that the values are as given in Table 16.28.

Table 16.28

Jan	Feb	Mar	Apr	May	June	July	Aug	Sept	Oct	Nov	Dec
100·0	89·2	86·4	77·5	71·7	66·8	68·4	73·9	79·9	88·4	94·2	100·0

Since the sum of the entries in the last row of Table 16.28 is 996·6, we adjust them by multiplying by 1200/996·6 and obtain the seasonal index given in Table 16.29.

Table 16.29

	Jan	Feb	Mar	Apr	May	June	July	Aug	Sept	Oct	Nov	Dec
Seasonal index	120·4	107·4	104·0	93·3	86·3	80·4	82·4	89·0	96·2	106·4	113·4	120·7

DESEASONALIZATION OF DATA

16.13. Adjust the data of Prob. 16.8 for seasonal variation, i.e. deseasonalize the data.

Solution:

To adjust the data for seasonal variation, we must divide every entry in the original data of Prob. 16.8 by the seasonal index of the corresponding month as found by any of the above methods.

If for example we use the seasonal index of Prob. 16.10, we divide all January values by 119·8% (i.e. 1·198), all February values by 107·2% (i.e. 1·072), etc. Then the deseasonalized data are as given in Table 16.30.

Table 16.30

	Jan	Feb	Mar	Apr	May	June	July	Aug	Sept	Oct	Nov	Dec
1951	265	262	269	267	265	265	267	274	279	285	289	290
1952	285	288	289	287	286	290	290	293	299	303	305	304
1953	306	306	309	307	308	308	311	317	321	325	327	329
1954	327	326	331	333	333	335	338	341	340	343	346	349
1955	351	353	358	357	360	363	366	369	369	374	376	378
1956	378	384	385	387	391	395	402	401	407	403	404	404
1957	407	410	415	420	424	426	428	434	430	431	437	431
1958	442	445	448	452	456	466	466	468	465	465	468	468

16.14. (*a*) Graph the deseasonalized data obtained in the preceding problem.

    (*b*) Compare the graph with that of Prob. 16.8(*a*).

Solution:

(*a*)

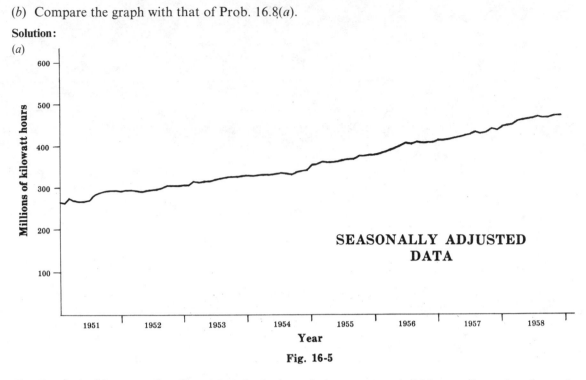

Fig. 16-5

(*b*) The graph of the seasonally adjusted data clearly shows the long-term trend which, apart from minor fluctuations, closely approximate a straight line although there is a slight upward tendency.

    If we denote the data of Prob. 16.8 by $Y = TCSI$, the graph in (*a*) is that of the variable $Y/S = TCI$ plotted against time t and thus contains the long-term trend, cyclical and irregular movements. Since the graph indicates the long-term trend so well, it appears that the product CI of cyclical and irregular factors must be practically 100%. This fact is confirmed in Prob. 16.16.

ESTIMATION OF CYCLICAL AND IRREGULAR VARIATIONS

16.15. Adjust the data of Prob. 16.13 for trend.

Solution:

    To remove the trend from the data of Prob. 16.13 we divide each entry by the corresponding monthly trend value, computed by any of the methods considered. Here we shall use the monthly trend values, obtained in Prob. 16.10 by the method of moving averages. The results are shown in Table 16.31. To obtain the entry for July 1951, for example, we divide the corresponding entry 267 in Table 16.30 of Prob. 16.13 by the value of 274·7 (see Prob. 16.10, first entry in column 5 of Table 16.20), which gives $267/274·7 = 97·2(\%)$. Other entries are obtained in a similar manner. A disadvantage of this method, as with all methods involving moving averages, is that data at both ends of the time series are lost.

Table 16.31

	Jan	Feb	Mar	Apr	May	June	July	Aug	Sept	Oct	Nov	Dec
1951							97·2	99·0	100·0	101·6	102·4	102·2
1952	99·9	100·4·	100·2	99·0	98·1	99·0	98·5	97·8	100·3	101·1	101·2	100·4
1953	100·6	100·1	100·5	99·2	98·9	98·2	98·4	99·7	100·4	101·0	101·1	101·1
1954	99·9	99·1	100·1	100·2	99·7	99·7	100·0	100·2	99·2	99·4	99·8	100·1
1955	100·1	100·1	100·9	100·0	100·0	100·0	100·1	100·2	99·4	100·1	100·1	100·0
1956	99·4	100·3	99·9	99·7	100·0	100·4	101·5	100·6	101·4	99·8	99·4	98·9
1957	99·1	99·3	100·0	100·7	101·0	100·8	100·5	101·1	99·5	99·1	100·0	98·0
1958	99·9	100·0	100·0	100·3	100·5	102·0						

16.16. (*a*) Graph the data obtained in Prob. 16.15.

(*b*) Explain the significance of the graph.

Solution:

(*a*) It is convenient to subtract 100(%) from the data of the previous problem and graph the resulting deviations. The resulting graph, using a greatly magnified vertical scale, is shown in Fig. 16-6.

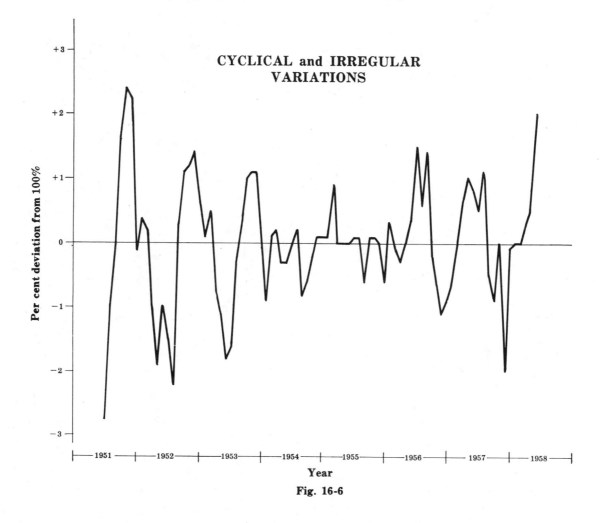

Fig. 16-6

(*b*) The original data is represented by $Y = TCSI$. Adjusting for seasonal variation as in Prob. 16.13 amounted to dividing both sides by the seasonal index S to obtain $Y/S = TCI$. Subsequent adjustment for trend amounted to division by T to obtain $Y/ST = CI$. Subtraction of 100(%) gives $Y/ST - 100 = CI - 100$. Thus the dependent variable in the above graph is $Y/SY - 100$, and the independent variable is time t.

The graph is theoretically composed of only the cyclical and irregular movements, represented by the corresponding factors C and I respectively. Note that the product CI varies between 97% and 103% confirming the statement made at the end of Prob. 16.14.

16.17. (*a*) Obtain a 3 month and 7 month average of the data in Prob. 16.15.

(*b*) Construct the graphs of the moving averages of part (*a*).

(*c*) Interpret the graphs.

Solution:

(*a*) The required moving averages are shown in Table 16.32.

Table 16.32

Year and Month	Data	3 month moving total	3 month moving average	7 month moving total	7 month moving average
1951					
July	97·2				
Aug	99·0	296·2	98·7		
Sept	100·0	300·6	100·2		
Oct	101·6	304·0	101·3	702·3	100·3
Nov	102·4	306·2	102·1	705·5	100·8
Dec	102·2	304·5	101·5	706·7	101·0
1952					
Jan	99·9	302·5	100·8	705·7	100·8
Feb	100·4	300·5	100·2	702·2	100·3
Mar	100·2	299·6	99·9	698·8	99·5
Apr	99·0	297·3	99·1	695·1	99·3
May	98·1	296·1	98·7	693·0	99·0
June	99·0	295·6	98·5	692·9	99·0
July	98·5	295·3	98·4	693·8	99·1
Aug	97·8	296·6	98·9	696·0	99·4
Sept	100·3	299·2	99·7	698·3	99·8
Oct	101·1	302·6	100·9	699·9	100·0
Nov	101·2	302·7	100·9	701·5	100·2
Dec	100·4	302·2	100·7	704·2	100·6
1953					
Jan	100·6	301·1	100·4	703·1	100·4
Feb	100·1	301·2	100·4	700·9	100·1
Mar	100·5	299·8	99·9	697·9	99·7
Apr	99·2	298·6	99·5	695·9	99·4
May	98·9	296·3	98·8	695·0	99·3
June	98·2	295·5	98·5	695·3	99·3
July	98·4	296·3	98·8	695·8	99·4
Aug	99·7	298·5	99·5	697·7	99·7
Sept	100·4	301·1	100·4	699·9	100·0
Oct	101·0	302·5	100·8	701·6	100·2
Nov	101·1	303·2	101·1	702·3	100·3
Dec	101·1	302·1	100·7	702·7	100·4
1954					
Jan	99·9	300·9	100·3	702·5	100·4
Feb	99·1	299·1	99·7	701·2	100·2
Mar	100·1	299·4	99·8	699·8	100·0
Apr	100·2	300·0	100·0	698·7	99·8
May	99·7	299·6	99·9	699·0	99·9
June	99·7	299·4	99·8	699·1	99·9
July	100·0	299·9	100·0	698·4	99·8
Aug	100·2	299·4	99·8	698·0	99·7
Sept	99·2	298·8	99·6	698·4	99·8
Oct	99·4	298·4	99·5	698·8	99·8
Nov	99·8	299·3	99·8	698·9	99·8
Dec	100·1	300·0	100·0	699·6	100·0

Table 16.32 (cont.)

Year and Month	Data	3 month moving total	3 month moving average	7 month moving total	7 month moving average
1955					
Jan	100·1	300·3	100·1	700·4	100·1
Feb	100·1	301·1	100·4	701·0	100·1
Mar	100·9	301·0	100·3	701·2	100·2
Apr	100·0	300·9	100·3	701·2	100·2
May	100·0	300·0	100·0	701·2	100·2
June	100·0	300·1	100·0	700·5	100·1
July	100·1	300·2	100·1	699·7	100·0
Aug	100·1	299·6	99·9	699·8	100·0
Sept	99·4	299·6	99·9	699·8	100·0
Oct	100·1	299·6	99·9	699·2	100·0
Nov	100·1	300·2	100·1	699·4	100·0
Dec	100·0	299·5	99·8	699·2	100·0
1956					
Jan	99·4	299·7	99·9	699·5	100·0
Feb	100·3	299·6	99·9	699·4	100·0
Mar	99·9	299·9	100·0	699·7	100·0
Apr	99·7	299·6	99·9	701·2	100·2
May	100·0	300·1	100·0	702·4	100·3
June	100·4	301·9	100·6	703·5	100·5
July	101·5	302·5	100·8	703·4	100·5
Aug	100·6	302·5	100·8	703·1	100·4
Sept	101·4	301·8	100·6	702·0	100·3
Oct	99·8	300·6	100·2	700·7	100·1
Nov	99·4	298·1	99·4	698·5	99·5
Dec	98·9	297·4	99·1	697·9	99·0
1957					
Jan	99·1	297·3	99·1	697·2	99·6
Feb	99·3	298·4	99·5	698·4	99·8
Mar	100·0	300·0	100·0	699·8	100·0
Apr	100·7	301·7	100·6	701·4	100·2
May	101·0	302·5	100·8	703·4	100·5
June	100·8	302·3	100·8	703·6	100·5
July	100·5	302·4	100·8	702·7	100·4
Aug	101·1	301·1	100·4	702·0	100·3
Sept	99·5	299·7	99·9	699·0	99·9
Oct	99·1	298·6	99·5	698·1	99·7
Nov	100·0	297·1	99·9	697·6	99·7
Dec	98·0	297·9	99·3	696·5	99·5
1958					
Jan	99·9	297·9	99·3	697·3	99·6
Feb	100·0	299·8	100·0	698·7	99·8
Mar	100·0	300·3	100·1	700·7	100·1
Apr	100·3	300·8	100·2		
May	100·5	302·8	100·9		
June	102·0				

(*b*) As in Prob. 16.16 it is convenient to subtract 100(%) from the moving averages and graph the resulting deviations as shown below.

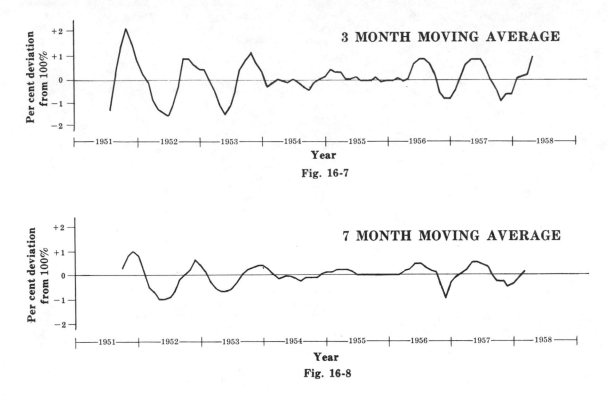

Fig. 16-7

Fig. 16-8

(*c*) As is to be expected, the moving averages serve to smooth out irregularities in the data of Prob. 16.15, as is evidenced by comparison of the graphs in (*b*) with the graph of Prob. 16.16. It is also clear from the graphs that the 7 month moving average provides better smoothing of data than the 3 month average in this case.

It is of interesting to note that the three maxima at the left and the two minima at the right in the graphs of (*b*) all occur near December. Also, the two minima at the left and the two maxima at the right occur near June. These observations seem to indicate a small residual seasonal variation at the beginning and end of the 8 year period, which act in *opposite* directions, thus indicating a possible changing seasonal pattern. Over the entire 8 year period these would naturally cancel out. The small residual seasonality present is further evidenced if a centred 12 month moving average is taken.

Ordinarily the method of this problem is used for investigating cyclical patterns. We would expect that this is the case since if the original data, given $Y = TCSI$, is adjusted for trend and seasonal variation we obtain new data $Y/ST = CI$, which (theoretically) contain only cyclical and irregular movements. A suitable moving average then serves to eliminate irregularities and reveals the cyclical pattern, if one exists. For this purpose a centred 12 month moving average is perhaps best since it eliminates residual seasonal variations as well as irregularities.

In the present problem no cyclical effect appears to be present, or if it is present it is negligible. In economic theory we often require data of as much as 20 years duration before cycles make their appearance (see Fig. 16-1).

COMPARABILITY OF DATA

16.18. How should the data of Prob. 16.8 be modified so as to make allowances for the leap years 1952 and 1956?

Solution:

In a leap year, February has 29 days instead of the usual 28 days. To achieve comparability of data we would multiply data for a leap year February by 28/29. Then in Table 16.12 of Prob. 16.8,

Value for Feb. 1952 is replaced by (28/29)(309) = 298
Value for Feb. 1956 is replaced by (28/29)(412) = 398

These adjustments have not been made in obtaining the seasonal index (see Problems 16.8–16.12). Their effect on the results, however, is negligible (see Prob. 16.52).

FORECASTING

16.19. (*a*) Using the data in Table 16.12 of Prob. 16.8, predict the monthly electric power used for street and highway lighting in the United States during the year 1959. (*b*) Compare the predicted values with the actual values.

Solution:

(*a*) The future monthly values are given by $Y = TCSI$, where we must estimate T, C, S and I.

To estimate the trend T, many possible methods can be used. From the graph of Prob. 16.14 (see Fig. 16-5) it appears that we should be able to get fairly accurate estimates of future trend values by fitting a line to the trend values for the last two years, for example. This can be done by using the method of least squares or any of the other methods which have been discussed.

We shall obtain the trend values by the relatively simple method of semi-averages applied to the results obtained in Prob. 16.10. In the adjacent table we have divided into two equal parts the 12 month centred moving averages for the months from July 1956 to June 1958.

From the means of the data in each part, it follows that there has been an increase of $441\cdot3 - 409\cdot4 = 31\cdot9$ in 12 months or $31\cdot9/12 = 2\cdot66$ per month.

By successively adding $2\cdot66$ to $456\cdot9$, the last available figure corresponding to June 1958, we can obtain the trend values for 1959 as shown in the third row of Table 16.34(*a*) below.

To estimate the seasonal factor S we use the seasonal index obtained in Prob. 16.10, although the seasonal index found by the other methods can be used as well. This seasonal index has been repeated in the fourth row of Table 16.34(*a*).

From Fig. 16-6 of Problem 16.16 it is seen that the estimate of the cylical and irregular factors CI differs from 100% by less than $2\cdot5\%$. Thus if we assume that $CI = 100\% = 1$, i.e. $Y = T \times C \times S \times I = (T \times S)(C \times I) = T \times S$, we should not be off by more than $2\cdot5\%$ in Y.

Table 16.33

July 1956	396·2	July 1957	425·9
Aug 1956	398·8	Aug 1957	429·2
Sept 1956	401·3	Sept 1957	432·2
Oct 1956	403·9	Oct 1957	434·8
Nov 1956	406·4	Nov 1957	437·2
Dec 1956	408·6	Dec 1957	439·8
Jan 1957	410·6	Jan 1958	442·5
Feb 1957	412·7	Feb 1958	445·1
Mar 1957	414·9	Mar 1958	447·8
Apr 1957	417·1	Apr 1958	450·7
May 1957	419·9	May 1958	453·6
June 1957	422·8	June 1958	456·9
Total	4913·2	Total	5295·7
Mean	409·4	Mean	441·3

Table 16.34(*a*)	Jan	Feb	Mar	Apr	May	June	July	Aug	Sept	Oct	Nov	Dec
·1958 Trend values (*T*)						456·9	459·6	462·2	464·9	467·5	470·2	472·9
1959 Trend values (*T*)	475·5	478·2	480·8	483·5	486·2	488·8	491·5	494·1	496·8	499·5	502·1	504·8
Seasonal index (*S*%)	119·8	107·2	103·4	93·5	87·2	81·5	83·4	89·5	96·4	106·0	112·3	119·6
1959 Predicted power (*T* × *S*) (millions kW h)	570	513	497	452	424	398	410	442	479	529	564	604

Multiplying the values of T for 1959 by the corresponding values of S (remembering that S is in per cent), we obtain the predicted monthly values or *projections* for 1959 given in the last line of Table 16.34(*a*) above. For example, the predicted value for Jan. 1959 is $(475\cdot5)(1\cdot198) = 570$, etc.

(*b*) The actual monthly values for the year 1959, shown in the following Table 16.34(*b*) are seen to agree quite well with the predicted values.

Table 16.34(*b*)	Jan	Feb	Mar	Apr	May	June	July	Aug	Sept	Oct	Nov	Dec
1959 Actual power (millions kW h)	563	509	497	454	424	404	415	446	478	524	561	594

Source: Survey of Current Business

Supplementary Problems

CHARACTERISTIC MOVEMENTS OF TIME SERIES

16.20. With which characteristic movement of a time series would you mainly associate each of the following: (*a*) a recession, (*b*) an increase in employment during summer months, (*c*) the decline in the death rate due to advances in science, (*d*) a steel strike, (*e*) a continually increasing demand for smaller automobiles.
Ans. (*a*) cyclical, (*b*) seasonal, (*c*) long-term trend, (*d*) irregular, (*e*) long-term trend

MOVING AVERAGES

16.21. Given the numbers $1, 0, -1, 0, 1, 0, -1, 0, 1$, determine a moving average of order (*a*) two, (*b*) three, (*c*) four, (*d*) five.
Ans. (*a*) $0.5, -0.5, -0.5, 0.5, 0.5, -0.5, -0.5, 0.5$; (*b*) $0, -\frac{1}{3}, 0, \frac{1}{3}, 0, -\frac{1}{3}, 0$; (*c*) $0, 0, 0, 0, 0, 0$; (*d*) $\frac{1}{5}, 0, -\frac{1}{5}, 0, \frac{1}{5}$

16.22. Prove that if a sequence of numbers has period N (i.e. the sequence repeats itself after N terms) then every moving average of order less than N has period N. Illustrate with reference to Prob. 16.21.

16.23. (*a*) In Problem 16.22 what happens in the case of a moving average of order N? (*b*) What happens if the order is greater than N? Illustrate with reference to Prob. 16.21.

16.24. Prove that if every number in a sequence is increased (or decreased) by a constant, the moving average is also increased (or decreased) by this constant.

16.25. Prove that if every number in a sequence is multiplied (or divided) by a non-zero constant, the moving average is also multiplied (or divided) by this constant.

16.26. Find the weighted moving average of the numbers in Prob. 16.21(*b*), (*c*) and (*d*) if the weights are respectively: (*b*) 1, 2, 1; (*c*) 1, 2, 2, 1; (*d*) 1, 2, 2, 2, 1. Compare with the results of Prob. 16.21.
Ans. (*b*) $0, -0.5, 0, 0.5, 0, -0.5, 0$; (*c*) $-\frac{1}{6}, -\frac{1}{6}, \frac{1}{6}, \frac{1}{6}, -\frac{1}{6}, -\frac{1}{6}$; (*d*) $0, 0, 0, 0, 0,$

16.27. (*a*) Prove the properties in Problems 16.24 and 16.25 for weighted moving averages. (*b*) Does the result of Prob. 16.22 hold for weighted moving averages?

16.28. A sequence has (*a*) 24, (*b*) 25 and (*c*) 200 numbers. How many numbers will there be in a moving average of order 5?
Ans. (*a*) 20, (*b*) 21, (*c*) 196

16.29. A sequence has M numbers. (*a*) Prove that in a moving average of order N there will be $M - N + 1$ numbers. Illustrate by several examples using different values of M and N. (*b*) Discuss the case where $M = N$.

16.30. Table 16.35 gives the average monthly consumption, in thousands of bales, of domestic and foreign cotton in the United States for the years 1949-1958. Construct a (*a*) 2 year moving average, (*b*) 2 year centred moving average, (*c*) 3 year moving average, (*d*) 4 year centred moving average, (*e*) 6 year centred moving average.

Table 16.35

Year	1949	1950	1951	1952	1953	1954	1955	1956	1957	1958
U.S. consumption of cotton (thousands of bales)	656	804	836	765	777	711	755	747	696	677

Source: Survey of Current Business

 Ans. (*a*) 730, 820, 800, 771, 744, 733, 751, 722, 686. (*b*) 775, 810, 786, 758, 738, 742, 736, 704. (*c*) 765, 802, 793, 751, 748, 738, 733, 707. (*d*) 780, 784, 762, 750, 737, 723. (*e*) 766, 770, 753, 734.

16.31. Graph the moving averages of Prob. 16.30 together with the original data and discuss the results obtained.

16.32. (*a*) Show that the 2 year centred moving average of Prob. 16.30(*b*) is equivalent to a 3 year weighted moving average with weights 1, 2, 1 respectively. Illustrate by direct numerical calculation. (*b*) Show that the 6 year centred moving average of Prob. 16.30(*e*) is equivalent to a suitable weighted moving average.

16.33. (*a*) For the data of Prob. 16.30 determine a weighted moving average of order 3 if the weights 1, 4, 1 are used. (*b*) Graph this moving average and compare with Prob. 16.30(*c*). *Ans.* (*a*) 785, 819, 779, 764, 729, 746, 740, 701

16.34. Table 16.36 shows the total monthly factory sales in thousands of passenger cars in the United States during the years 1953–1958. Construct a (*a*) 12 month moving average, (*b*) 12 month centred moving average, (*c*) 6 month centred moving average. In parts (*b*) and (*c*) graph the moving average together with the original data and compare the results.

Table 16.36

	Jan	Feb	Mar	Apr	May	June	July	Aug	Sept	Oct	Nov	Dec
1953	452·6	485·3	566·1	595·8	548·3	585·7	596·9	512·7	476·2	528·8	378·9	389·6
1954	454·6	446·7	531·5	534·7	497·1	507·1	451·7	445·3	301·0	221·2	498·2	669·9
1955	635·5	677·7	791·3	753·4	721·1	647·7	658·7	620·6	467·8	505·2	746·0	695·1
1956	591·0	560·9	583·2	552·9	474·0	445·8	441·0	417·0	203·9	352·1	576·7	617·6
1957	628·0	570·0	585·7	541·7	537·1	496·3	484·7	521·3	318·3	291·1	583·8	555·2
1958	478·4	396·2	359·5	322·5	352·1	342·2	316·4	195·0	102·7	272·2	511·9	608·7

Source: Survey of Current Business

ESTIMATION OF TREND

16.35. Obtain the trend values for the data of Prob. 16.30 by using the method of semi-averages where the average is taken as (*a*) the mean, (*b*) the median. Construct a graph illustrating the results obtained.
Ans. (*a*) 788, 778, 768, 758, 747, 737, 727, 717, 707, 697
(*b*) 803, 790, 777, 764, 751, 737, 724, 711, 698, 685

16.36. Work Prob. 16.30 by using (*a*) the freehand method, (*b*) a moving average of suitable order. Compare with the results of Prob. 16.35.

16.37. (*a*) Use the method of least squares to fit a line to the data of Prob. 16.30.
(*b*) From the result in (*a*) find the trend values. Compare with the results in Problems 16.35 and 16.36.
Ans. (*a*) $Y = 742·4 - 3·358X$, where X is in half years and the origin is Jan. 1, 1954.
(*b*) 758·0, 754·7, 751·3, 747·9, 744·6, 741·2, 737·9, 734·5, 731·1, 727·8

16.38. (*a*) Fit a parabola $Y = a_0 + a_1X + a_2X^2$ to the data of Prob. 16.8 using the monthly averages in Table 16.13 of Prob. 16.9.
(*b*) Compare the result in (*a*) with the least square line of Prob. 16.9 and compute the trend values.
Ans. (*a*) $Y = 351·1 + 13·188X + 0·3110X^2$, where X is measured in units of a half year and the origin is at Jan. 1, 1955.

16.39. Obtain trend values for the data of Prob. 16.34 by using (*a*) the method of semi-averages, (*b*) the freehand method, (*c*) a centred 12 month moving average, (*d*) a suitable least square curve (to determine this use the graph of the original data constructed in Prob. 16.34). Discuss the advantages and disadvantages of each method.

ESTIMATE OF SEASONAL VARIATIONS. SEASONAL INDEX

16.40. Table 16.37 shows, for a certain country, the monthly production of creamery butter in millions of kilogrammes during the years 1951–1958. (*a*) Graph the data. (*b*) Construct a seasonal index using the average percentage method. Adjust the data for leap years before obtaining this index.

Table 16.37

	Jan	Feb	Mar	Apr	May	June	July	Aug	Sept	Oct	Nov	Dec
1951	85·6	80·9	92·2	101·8	132·6	141·2	130·5	119·0	93·6	86·6	68·4	70·4
1952	78·7	78·8	91·5	102·5	135·0	128·0	117·7	105·7	92·1	87·7	75·9	94·6
1953	103·9	101·9	121·4	133·5	156·0	154·0	135·6	118·7	95·0	91·6	91·3	109·0
1954	118·7	116·6	143·3	142·0	164·5	160·9	129·7	109·4	92·6	87·8	86·8	97·0
1955	108·1	104·3	121·1	129·4	157·9	151·9	123·0	102·1	91·9	94·7	92·7	105·8
1956	114·6	114·1	129·6	135·4	151·9	149·0	127·6	109·8	92·4	93·1	92·3	103·4
1957	115·3	110·3	124·6	132·3	159·3	148·1	125·8	106·9	90·1	100·3	94·1	105·7
1958	118·6	113·4	129·5	130·3	150·6	144·7	126·9	97·7	86·7	91·9	90·9	107·2

16.41. Obtain a seasonal index for the data of Prob. 16.40 by using the percentage trend or ration to trend method. To obtain trend values, fit a suitable least square curve to the monthly averages of the given years.

16.42. Obtain a seasonal index for the data of Prob. 16.40 by using the percentage moving average or ratio to moving average method.

16.43. Obtain a seasonal index for the data of Prob. 16.40 by using the link relative method.

16.44. Table 16.38 shows the estimated sales in billions of dollars of all retail stores in the United States during the years 1951–1958. (a) Graph the data. (b) Obtain a seasonal index using the average percentage method.

Table 16.38

	Jan	Feb	Mar	Apr	May	June	July	Aug	Sept	Oct	Nov	Dec
1951	12·63	11·72	13·43	12·53	13·29	13·27	12·36	13·27	13·10	13·86	13·39	15·38
1952	11·84	11·74	12·74	13·40	14·85	13·81	13·40	13·45	13·62	14·82	14·01	16·91
1953	13·05	12·33	13·96	14·17	14·66	14·58	14·38	14·18	14·08	14·95	13·96	16·44
1954	12·34	12·06	13·54	14·32	14·25	14·66	14·39	13·90	14·14	14·66	14·53	17·87
1955	13·15	12·64	14·57	15·49	15·33	15·60	15·26	15·48	15·76	15·68	15·75	19·12
1956	13·73	13·55	15·72	14·89	16·11	16·58	15·38	16·19	15·58	16·13	16·49	19·38
1957	14·74	14·06	15·79	16·44	17·20	17·11	16·86	17·49	16·37	16·95	17·13	19·84
1958	15·29	13·78	15·55	16·27	17·36	16·60	16·60	17·00	16·33	17·36	17·04	21·17

Source: Survey of Current Business

16.45. Obtain a seasonal index for the data of Prob. 16.44 by using the percentage trend or ratio to trend method.

16.46. Obtain a seasonal index for the data of Prob. 16.44 by using the ratio to moving average method.

16.47. Obtain a seasonal index for the data of Prob. 16.44 by using the link relative method.

16.48. Table 16.39 shows the United States freight carloadings in thousands of railroad cars during the years 1951–1958. (a) Construct a graph of the data. (b) Obtain a seasonal index by using the average percentage method.

Table 16.39

	Jan	Feb	Mar	Apr	May	June	July	Aug	Sept	Oct	Nov	Dec
1951	3661	2834	2999	3152	3977	3295	3807	3307	3312	4317	3139	2700
1952	3562	2911	2868	2912	3678	2606	2969	3149	3364	4156	3139	2672
1953	3351	2730	2801	2957	3883	3204	3758	3229	3153	4024	2797	2413
1954	2967	2462	2412	2445	3345	2730	3251	2708	2711	3629	2685	2518
1955	2505	2556	3256	2757	3754	3052	3015	3883	3148	3282	3758	2669
1956	2713	2751	3517	2971	3835	3143	2397	3700	3155	3284	3740	2641
1957	2565	2616	3446	2696	3558	2959	2708	3737	2849	2920	3223	2221
1948	2164	2108	2702	2105	2729	2489	2138	3146	2570	2733	2462	2188

Source: Survey of Current Business

16.49. Work Prob. 16.48 by the ratio to trend method.

16.50. Work Prob. 16.48 by the ratio to moving average method.

16.51. Work Prob. 16.48 by the link relative method.

16.52. Rework (a) Prob. 16.8, (b) Prob. 16.9, (c) Prob. 16.10 and (d) Prob. 16.11 using data adjusted for leap year and determine whether the adjustment has any significant bearing on the final seasonal index obtained.

16.53. (a) Calculate a seasonal index for the last four years and the first four years for the data of Prob. 16.8 using any method. (b) Compare the two indexes obtained and explain the difference if any.

DESEASONALIZATION OF DATA

16.54. (a) Deseasonalize the data of Prob. 16.40, using any seasonal index obtained in Problems 16.40–16.43. (b) Graph the deseasonalized data and explain the results.

16.55. (a) Adjust the data of Prob. 16.44 for seasonal variation using any of the results of Problems 16.44–16.47. (b) Graph the seasonally adjusted data and interpret the results obtained.

16.56. (a) Deseasonalize the data of Prob. 16.48 using any of the seasonal indexes of Problems 16.48–16.51. (b) Graph the seasonally adjusted data and explain the results obtained.

ESTIMATION OF CYCLICAL AND IRREGULAR VARIATIONS

16.57. (*a*) Adjust the data of Prob. 16.54 for trend, using any method. (*b*) Graph the data obtained. (*c*) Take a 3 or 5 month moving average of the data in (*a*). (*d*) Graph the results of (*c*) explaining the variations observed and in particular pointing out the existence of any cyclical movements which may be present.

16.58. Table 16.40 shows, for the country in Prob. 16.40, the average monthly production of butter in millions of kilogrammes during the years 1930–1958. (*a*) Graph the data and discuss the possible existence of cycles. (*b*) Compare the results arrived at in (*a*) with the conclusions reached in Prob. 16.57(*d*) and explain any discrepancies.

Table 16.40

Year	1930	1931	1932	1933	1934	1935	1936	1937	1938	1939	1940	1941	1942	1943	1944
Monthly average	133·1	139·0	141·1	146·9	141·2	136·0	135·8	135·3	148·8	148·5	153·1	156·0	147·0	139·5	124·0

Year	1945	1946	1947	1948	1949	1950	1951	1952	1953	1954	1955	1956	1957	1958
Monthly average	113·6	97·6	110·8	100·9	117·7	115·5	100·2	99·0	117·7	120·7	115·2	117·8	117·7	115·5

16.59. Work Problems 16.57 and 16.58 for the deseasonalized data of Prob. 16.56. The monthly averages of freight carloadings in thousands of railroad cars are given for the years 1930–1958 in the following table.

Table 16.41

Year	1930	1931	1932	1933	1934	1935	1936	1937	1938	1939	1940	1941	1942	1943	1944
Monthly average	3823	3096	2348	2435	2570	2625	3009	3139	2538	2826	3030	3529	3564	3537	3617

Year	1945	1946	1947	1948	1949	1950	1951	1952	1953	1954	1955	1956	1957	1958
Monthly average	3493	3445	3709	3560	2993	3242	3375	3165	3185	2826	3136	3154	2958	2517

Source: Survey of Current Business

16.60. In adjusting data for trend and seasonal variation, does it make any difference which is done first? Illustrate your answer (*a*) theoretically, (*b*) by employing one of the time series in Problems 16.40, 16.44 or 16.48.

16.61. (*a*) Work Prob. 16.17 using a centred 12 month average and construct the graph. (*b*) What conclusions do you draw from the results in (*a*)?

16.62. (*a*) Obtain a frequency distribution for the magnitudes of the irregular variations found in Problems 16.15 and 16.16. (*b*) Does the frequency distribution found in (*a*) approximate a normal distribution? If so give a possible reason for this.

FORECASTING

16.63. (*a*) Use any results of Problems 16.40–16.43, 16.57 and 16.58 to forecast the production of creamery butter during the year 1959. (*b*) Discuss possible sources of error. (*c*) Compare your predictions with the actual values for 1959 given in Table 16.42.

Table 16.42

Jan	Feb	Mar	Apr	May	June	July	Aug	Sept	Oct	Nov	Dec
116·3	108·2	121·4	126·8	143·4	135·6	112·5	90·9	82·6	92·1	91·2	108·0

16.64. (*a*) Use any results of Problems 16.44–16.47 to predict the sales of all retail stores in the United States during the year 1959. (*b*) Discuss possible sources of error. (*c*) Compare your predictions with the actual values for 1959 given in Table 16.43.

Table 16.43

Jan	Feb	Mar	Apr	May	June	July	Aug	Sept	Oct	Nov	Dec
16·23	14·96	17·19	17·59	18·60	18·71	18·33	18·05	17·57	19·10	17·64	21·45

Source: Survey of Current Business

16.65. (*a*) Use any results of Problems 16.48–16.51 and 16.59 to forecast the freight carloadings in the United States during the year 1959. (*b*) Discuss possible sources of error. (*c*) Compare your predictions with the actual values for 1959 given in Table 16.44.

Table 16.44

Jan	Feb	Mar	Apr	May	June	July	Aug	Sept	Oct	Nov	Dec
2742	2291	2398	2489	3419	2813	2249	2712	2190	2908	2403	2376

Source: Survey of Current Business

MISCELLANEOUS PROBLEMS

16.66–
16.70   Analyse each of the time series given in Table 16.45 which refers to data for a country. If desired use only the data for the years through 1958 and forecast results for 1959, comparing with the actual data for that year. Note that data for the years 1929–1958 in the first part of the table give *monthly averages* for each year, while data at the end of the table give *monthly values* for each year.

Table 16.45

Year	Permanent non-farm dwelling units under construction (thousands)	Total newspaper advertising in 52 cities (millions of lines)	Production of lumber (thousands of cubic metres)	Production of primary aluminium (thousands of tonnes)	Value of new public construction activity (millions of dollars)
1929	42·4	158·1	3074	9·50	207
1930	27·5	137·9	2171	9·54	238
1931	21·2	122·2	1377	7·40	222
1932	11·2	97·1	902	4·37	155
1933	7·8	88·8	1225	3·55	137
1934	10·5	98·2	1291	3·09	184
1935	18·4	103·9	1628	4·97	186
1936	26·6	115·0	2030	9·37	293
1937	28·0	117·5	2166	12·20	258
1938	33·8	102·1	1804	11·95	285
1939	42·9	103·6	2096	13·63	317
1940	50·2	105·7	2411	17·19	302
1941	58·8	109·4	2801	25·76	479
1942	29·7	103·5	3028	43·43	888
1943	15·9	116·4	2857	76·68	527
1944	11·8	113·4	2745	64·70	256
1945	17·4	116·0	2344	41·26	200
1946	55·9	144·1	2843	34·14	186
1947	70·8	167·4	2950	47·65	277
1948	77·6	188·6	3064	51·96	392
1949	85·4	191·8	2742	50·29	522
1950	116·3	203·3	3242	59·89	572
1951	90·9	206·5	3126	69·74	771
1952	93·9	208·8	3122	78·11	898
1953	92·0	217·6	3062	104·33	936
1954	101·7	215·1	3030	121·71	973
1955	110·7	237·0	3154	130·48	977
1956	93·2	242·6	3219	139·91	1059
1957	86·8	235·8	2851	137·31	1168
1958	100·8	223·8	2798	130·46	1284

Above data refers to *monthly averages*.

Table 16.45 (cont.)

Year and Month	Permanent non-farm dwelling units under construction (thousands)	Total newspaper advertising in 52 cities (millions of lines)	Production of lumber (thousands of cubic metres)	Production of primary aluminium (thousands of tonnes)	Value of new public construction activity (millions of dollars)
1952					
Jan	64·9	178·1	2694	76·93	671
Feb	77·7	184·6	2766	72·37	636
Mar	103·9	213·2	2872	77·07	722
Apr	106·2	218·4	3123	76·88	829
May	109·6	225·6	3049	80·80	924
June	103·5	209·3	3214	77·48	1002
July	102·6	175·4	3213	78·37	1037
Aug	99·1	186·6	3489	85·18	1089
Sept	100·8	214·5	3569	76·88	1109
Oct	101·1	245·0	3596	77·31	1071
Nov	86·1	234·9	3052	74·64	922
Dec	71·5	219·8	2825	83·42	769
1953					
Jan	72·1	182·7	2769	89·90	732
Feb	79·2	186·1	2754	92·65	719
Mar	105·8	231·7	3091	104·46	798
Apr	111·4	232·6	3280	102·07	880
May	108·3	244·4	3071	105·46	953
June	104·6	216·0	3219	104·15	1034
July	96·7	188·0	3141	109·29	1089
Aug	93·2	198·6	3237	110·55	1097
Sept	95·1	219·6	3266	109·33	1143
Oct	90·1	244·4	3326	108·22	1084
Nov	81·5	241·3	2893	105·64	933
Dec	65·8	224·3	2695	110·29	774
1954					
Jan	66·4	182·9	2746	116·25	745
Feb	75·2	180·7	2906	110·48	730
Mar	95·2	216·2	3361	122·34	792
Apr	107·7	233·3	3307	120·43	888
May	108·5	234·6	3324	125·14	998
June	116·5	216·6	3124	120·76	1088
July	116·0	185·8	2724	126·16	1159
Aug	114·3	199·4	2956	125·30	1202
Sept	115·7	218·9	3279	120·33	1205
Oct	110·7	244·9	3363	125·09	1103
Nov	103·6	238·5	3154	121·25	964
Dec	90·6	229·5	3085	127·04	804
1955					
Jan	87·6	196·2	2707	128·20	742
Feb	89·9	194·4	2845	116·24	697
Mar	113·8	242·5	3268	130·27	776
Apr	132·0	243·8	3147	126·39	898
May	137·6	260·4	3327	131·13	1030
June	134·5	243·7	3491	127·63	1107
July	122·7	212·3	2946	132·67	1165
Aug	124·7	219·8	3554	133·55	1216
Sept	114·9	246·2	3442	130·61	1208
Oct	105·8	273·1	3334	134·66	1131
Nov	89·2	268·5	3009	133·69	971
Dec	76·2	242·5	2788	140·75	783

Above data refers to *monthly values.*

Table 16.45 (cont.)

Year and Month	Permanent non-farm dwelling units under construction (thousands)	Total newspaper advertising in 52 cities (millions of lines)	Production of lumber (thousands of cubic metres)	Production of primary aluminium (thousands of tonnes)	Value of new public construction activity (millions of dollars)
1956					
Jan	75·1	212·2	2991	140·39	741
Feb	78·4	218·3	2993	132·76	700
Mar	98·6	251·3	3182	145·90	774
Apr	111·4	261·0	3245	144·73	932
May	113·7	268·5	3545	150·80	1099
June	107·4	239·3	3437	145·73	1223
July	101·1	214·0	3175	151·62	1290
Aug	103·9	227·3	3669	92·41	1349
Sept	93·9	244·1	3263	132·32	1341
Oct	93·6	269·9	3496	149·13	1296
Nov	77·4	262·0	3036	145·08	1066
Dec	63·6	243·1	2597	148·39	901
1957					
Jan	64·2	210·5	2693	147·03	896
Feb	65·8	207·1	2687	119·06	793
Mar	87·0	249·5	2914	135·71	885
Apr	93·7	245·4	3003	139·15	1055
May	103·0	265·6	3113	145·17	1204
June	99·9	240·6	2952	138·01	1326
July	97·8	204·0	2793	142·04	1303
Aug	100·0	216·4	3194	143·45	1436
Sept	91·9	241·3	2970	129·28	1473
Oct	97·0	259·0	3097	133·76	1453
Nov	78·2	250·0	2559	135·02	1170
Dec	63·4	239·5	2239	140·04	1023
1958					
Jan	67·9	197·1	2526	139·91	951
Feb	66·1	188·3	2388	121·98	861
Mar	81·4	227·8	2548	134·02	938
Apr	99·1	228·0	2676	125·00	1109
May	108·5	240·9	2824	126·33	1274
June	113·0	226·2	2889	115·33	1422
July	112·8	198·0	2810	118·54	1486
Aug	124·0	211·6	3056	125·42	1555
Sept	121·0	224·6	3143	125·94	1604
Oct	115·0	259·2	3272	139·84	1600
Nov	109·4	252·9	2731	140·96	1403
Dec	91·2	231·0	2716	152·30	1209
1959					
Jan	87·0	193·5	2650	156·7	1130
Feb	94·5	196·1	2642	142·1	1032
Mar	121·0	236·5	2964	157·2	1126
Apr	142·2	255·0	3121	155·2	1285
May	137·0	263·8	3163	163·9	1468
June	136·7	237·0	3216	167·3	1637
July	128·8	220·4	3136	179·2	1611
Aug	129·3	234·4	3171	172·8	1608
Sept	120·3	246·9	3324	168·2	1528
Oct	105·5	271·3	3304	173·7	1420
Nov	92·5	259·5	2892	153·7	1119
Dec	83·7	250·9	2947	163·0	1013

Above data refers to *monthly values.*

CHAPTER 17

Index Numbers

INDEX NUMBER

An index number is a statistical measure designed to show changes in a variable or group of related variables with respect to time, geographic location or other characteristic such as income, profession, etc. A collection of index numbers for different years, locations, etc., is sometimes called an *index series*.

APPLICATIONS OF INDEX NUMBERS

By using index numbers we can, for example, compare food or other living costs in a city during one year with those of a previous year, or we can compare steel production during a given year in one part of a country with that in another part. Although mainly used in business and economics, index numbers can be applied in many other fields. In education for example, we can use index numbers to compare intelligence of students in different locations or for different years.

Many governmental and private agencies are engaged in computation of index numbers or *indexes*, as they are often called, for purpose of forecasting business and economic conditions, providing general information, etc. Thus we have wage indexes, production indexes, unemployment indexes, and many others. Perhaps the most well-known is the *cost of living index* or *consumer price index* prepared by the *Bureau of Labor Statistics*. In many labour contracts there appear certain *escalator clauses* which provide automatic increases in wages corresponding to increases in the cost of living index.

In this chapter we shall be mainly interested in index numbers showing changes with respect to time, although methods described can be applied to other cases.

PRICE RELATIVES .

One of the simplest examples of an index number is a *price relative*, which is the ratio of the price of a single commodity in a *given period* to its price in another period called the *base period* or *reference period*. For simplicity we assume prices to be constant for any one period. If they are not, an appropriate average for the period can be taken to make this assumption valid.

If p_o and p_n denote the commodity prices during the base period and given period respectively, then by definition

$$\text{Price relative} = \frac{p_n}{p_o} \qquad (1)$$

and is generally expressed as a percentage by multiplying by 100.

More generally if p_a and p_b are prices of a commodity during periods a and b respectively, the price relative in period b *with respect to period a* is defined as p_b/p_a and is denoted by $p_{a|b}$, a notation which will be found useful. With this notation the price relative in equation (1) can be denoted by $p_{o|n}$.

Example 1. Suppose the consumer prices of a certain item in the years 1955 and 1960 were 25 and 30 new pence respectively. Taking 1955 as the *base year* and 1960 as the *given year*, we have

$$\text{Price relative} = p_{1955|1950} = \frac{\text{price in 1960}}{\text{price in 1955}} = \frac{30\text{ p}}{25\text{ p}} = 1\cdot 2 = 120\%$$

or briefly 120, omitting the % sign as is often done in statistical literature. This result simply means that in 1960 the price of the item was 120% of that in 1955, i.e. it *increased* by 20%.

Example 2. Taking 1960 as the base year and 1955 as the given year in Example 1, we have

$$\text{Price relative} = p_{1960|1955} = \frac{\text{price in 1955}}{\text{price in 1960}} = \frac{25\text{ p}}{30\text{ p}} = \frac{5}{6} = 83\tfrac{1}{3}\%$$

or briefly $83\tfrac{1}{3}$. This means that in 1955 the price of the item was $83\tfrac{1}{3}$% of that in 1960, i.e. it *decreased* by $16\tfrac{2}{3}$%.

Note that the price relative for a given period with respect to the *same period* is always 100% or 100. In particular the price relative corresponding to a base period is always 100. This accounts for the notation often used in statistical literature of writing, for example, 1955 = 100 to indicate that the year 1955 is taken as base period.

PROPERTIES OF PRICE RELATIVES

If $p_a, p_b, p_c, \ldots$ denote prices in periods $a, b, c, \ldots$ respectively, the following properties exist for the associated price relatives. Proofs follow directly from definitions.

1. **Identity Property** $\qquad\qquad\qquad\qquad p_{a|a} = 1$
   This simply states that the price relative for a given period with respect to the same period is 1 or 100%.

2. **Time Reversal Property** $\qquad\qquad p_{a|b}\, p_{b|a} = 1 \quad \text{or} \quad p_{a|b} = \dfrac{1}{p_{b|a}}$

   This states that if two periods are interchanged, the corresponding price relatives are reciprocals of each other. See Examples 1 and 2 above.

3. **Cyclical or Circular Property** $\qquad p_{a|b}\, p_{b|c}\, p_{c|a} = 1$
   $$p_{a|b}\, p_{b|c}\, p_{c|d}\, p_{d|a} = 1, \text{ etc.}$$

4. **Modified Cyclical or Circular Property** $\qquad p_{a|b}\, p_{b|c} = p_{a|c}$
   $$p_{a|b}\, p_{b|c}\, p_{c|d} = p_{a|d}, \text{ etc.}$$

   These follow directly from properties 2 and 3.

QUANTITY OR VOLUME RELATIVES

Instead of comparing prices of a commodity, we may be interested in comparing *quantities* or *volumes* of the commodity, such as quantity or volume of production, consumption, exports, etc. In such cases we speak of *quantity relatives* or *volume relatives*. For simplicity, as in the case of prices, we assume that quantities are constant for any period. If they are not, an appropriate average for the period can be taken to make this assumption valid.

If q_o denotes the quantity or volume of a commodity produced, consumed, exported, etc., during a base period, while q_n denotes the corresponding quantity produced, consumed, etc., during a given period, we define

$$\text{Quantity or volume relative} = \frac{q_n}{q_o} \qquad\qquad\qquad (2)$$

which is generally expressed as a percentage.

As in the case of price relatives, we use the notation $q_{a|b} = q_b/q_a$ to denote the quantity relative in period b with respect to period a. The same remarks and properties pertaining to price relatives are applicable to quantity relatives.

VALUE RELATIVES

If p is the price of a commodity during a period and q is the quantity or volume produced, sold, etc., during the period, then pq is called the *total value*. Thus if 1000 items are sold at 30 new pence each the total value is $(£0{\cdot}30)(1000) = £300$.

If p_o and q_o denote the price and quantity of a commodity during a base period while p_n and q_n denote the corresponding price and quantity during a given period, the total values during these periods are given by v_o and v_n respectively and we define

$$\text{Value relative} = \frac{v_n}{v_o} = \frac{p_n q_n}{p_o q_o} = \left(\frac{p_n}{p_o}\right)\left(\frac{q_n}{q_o}\right) \tag{3}$$

$$= \text{price relative} \times \text{quantity relative}$$

The same remarks, notation and properties pertaining to price and quantity relatives can be applied to value relatives.

In particular if $p_{a|b}$, $q_{a|b}$ and $v_{a|b}$ denote the price, quantity and value relatives of period b with respect to period a then, as in equation (3)

$$v_{a|b} = p_{a|b} q_{a|b}$$

which is called the *factor reversal property*.

LINK AND CHAIN RELATIVES

Let $p_1, p_2, p_3, \ldots$ represent prices during successive intervals of time 1, 2, 3, $\ldots$. Then $p_{1|2}, p_{2|3}, p_{3|4}, \ldots$ represent price relatives of each time interval with respect to the preceding time interval and are called *link relatives*.

Example 1. If the prices of a commodity during 1953, 1954, 1955, 1956 are 8, 12, 15, 18 cents respectively, the link relatives are $p_{1953|1954} = 12/8 = 150(\%)$, $p_{1954|1955} = 15/12 = 125(\%)$, $p_{1955|1956} = 18/15 = 120(\%)$.

The price relative for a given period with respect to any other period taken as base can always be expressed in terms of link relatives. This is a consequence of the cyclical or circular property of relatives. Thus, for example, $p_{5|2} = p_{5|4} p_{4|3} p_{3|2}$.

Example 2. From Example 1, the price relative for 1956 with respect to the base year 1953 is

$$p_{1953|1956} = p_{1953|1954} \; p_{1954|1955} \; p_{1955|1956} = \frac{12}{8} \cdot \frac{15}{12} \cdot \frac{18}{15} = \frac{18}{8} = 225(\%)$$

The price relatives with respect to a *fixed* base period, which as we have seen can be obtained by using link relatives, are sometimes called the *chain relatives* with respect to this base, or the relatives *chained* to the fixed base.

Example 3. In Examples 1 and 2, the collection of chain relatives for the years 1954, 1955 and 1956 with respect to the base 1953 are given by

$$p_{1953|1954} = \frac{12}{8} = 150(\%)$$

$$p_{1953|1955} = p_{1953|1954} p_{1954|1955} = \frac{12}{8} \cdot \frac{15}{12} = 187{\cdot}5(\%)$$

$$p_{1953|1956} = p_{1953|1954} p_{1954|1955} \; p_{1955|1956} = \frac{12}{8} \cdot \frac{15}{12} \cdot \frac{18}{15} = 225(\%)$$

The above ideas are also applicable to quantity and value relatives.

PROBLEMS INVOLVED IN COMPUTATION OF INDEX NUMBERS

In actual practice we are not so much interested in comparisons of prices, quantities or values of single commodities as in comparisons of large groups of such commodities. For example, in computing a cost of living index we not only wish to compare prices of milk in one period with respect to another but we also want to compare prices of eggs, meat, bread, rent, clothes, etc., so as to obtain some general picture. Of course we could merely list the price relatives of *all* commodities. This, however, would not be satisfactory. What we would like is a *single price* index number which would compare prices in the two periods *on the average*.

It is not difficult to foresee that computations of index numbers involving groups of commodities involve many problems which must be resolved. Thus in computing a cost of living index, for example, we must decide what commodities should be included as well as how to weight their relative importance. We must collect data concerning prices and quantities of such commodities. We are faced with problems as to what to do insofar as different *qualities* of the same type of commodity are concerned, or what to do when certain appliances or materials are available during a given year which were not available during the base year. Finally we must decide how to put all of this information together and come up with a single cost of living index number which has practical significance.

THE USE OF AVERAGES

Since we must arrive at a single index number summarizing a large amount of information, it is easy to realize that averages, such as considered in Chapter 3, play an important role in computing index numbers.

Just as many methods exist for computing averages, so there are many methods for computing index numbers, each with its own advantages and disadvantages.

In the following we shall examine a few methods commonly employed in practice and using various types of averaging procedures. Although we restrict ourselves to price indexes at first, we shall see how suitable modifications are easily made for quantity or value indexes.

THEORETICAL TESTS FOR INDEX NUMBERS

It is desirable from a theoretical viewpoint that index numbers for groups of commodities have the properties satisfied by relatives (i.e. index numbers for single commodities). Any index number which has such a property is said to meet the *test* associated with the property. Thus, for example, index numbers having the time reversal property are said to satisfy the *time reversal test*, etc.

No index number yet discovered meets *all* tests, although in many instances the tests are satisfied approximately. Fisher's ideal index (Page 318), which in particular satisfies the *time reversal test* and *factor reversal test*, comes closer than any other useful index number in satisfying the properties considered important, hence the name *ideal*.

From a practical point of view, however, other index numbers serve as well and we shall examine some of these.

NOTATION

It is customary to let $p_n^{(1)}, p_n^{(2)}, p_n^{(3)}, \ldots$ denote the prices of a first, second, third, $\ldots$ commodity during a given period n. The corresponding prices during a base period are denoted by $p_o^{(1)}, p_o^{(2)}, p_o^{(3)}$, etc. The numbers

1, 2, 3, . . . are *superscripts* and should not be confused with exponents. With this notation the price of commodity j during period n can then be indicated by $p_n^{(j)}$.

As in previous chapters, we can use the summation notation by summing over the index j. For example, assuming that there are a total of N commodities, the sum of their prices during the period n could be written $\sum_{j=1}^{N} p_n^{(j)}$ or $\Sigma p_n^{(j)}$. It is simpler, however, to omit the superscript altogether and to write Σp_n, which we shall do when no confusion can result. One should, however, keep in mind the fact that the more complete symbolism is implied. With this notation, Σp_0 would denote the sum of prices of all commodities during the base period.

A similar notation is used for quantities and values.

SIMPLE AGGREGATE METHOD

In this method of computing a price index, we express the total of commodity prices in the given year as a percentage of total commodity prices in the base year. In symbols, we have

$$\text{Simple aggregate price index} = \frac{\Sigma p_n}{\Sigma p_o} \tag{4}$$

where Σp_o = sum of all commodity prices in the base year

Σp_n = sum of corresponding commodity prices in the given year

and where the result is expressed as a percentage, as are index numbers in general.

Although this method has the advantage of being easy to apply, it has two great disadvantages which make it unsatisfactory.

1. It does not take into account the relative importance of the various commodities. Thus, according to this method, equal weight or importance would be attached to milk and shaving cream in computing a cost of living index.

2. The particular units used in price quotations, such as litres, grammes, etc., affect the value of the index. See Problem 17.12.

SIMPLE AVERAGE OF RELATIVES METHOD

In this method several possibilities exist depending on the procedure used to average price relatives, such as arithmetic mean, geometric mean, harmonic mean, median, etc. Using the arithmetic mean, for example, we would have

$$\text{Simple arithmetic mean of relatives price index} = \frac{\Sigma p_n/p_o}{N} \tag{5}$$

where $\Sigma p_n/p_o$ = sum of all commodity price relatives
N = number of commodity price relatives used.

For indexes using other types of averages, see Problems 17.14 and 17.15.

Although this method does not have the second disadvantage of the simple aggregate method, it still has the first disadvantage.

WEIGHTED AGGREGATE METHOD

To overcome the disadvantages of the simple aggregate method, we *weight* the price of each commodity by a suitable factor often taken as the quantity or volume of the commodity sold during the base year, the given year or some typical year (which may involve an average over several years). Such weights indicate the importance of the particular commodity. Three possible formulae occur depending on whether base year, given year or typical year quantities, denoted by q_o, q_n and q_t respectively, are used.

1. **Laspeyres' index or base year method**
   Weighted aggregate price index with base year quantity weights $= \dfrac{\Sigma\, p_n q_o}{\Sigma\, p_o q_o}$ (6)

2. **Paasche's index, or given year method**
   Weighted aggregate price index with given year quantity weights $= \dfrac{\Sigma\, p_n q_n}{\Sigma\, p_o q_n}$ (7)

3. **Typical Year method**
   If we let q_t denote the quantity weight during some typical period t, then we define

   Weighted aggregate price index with typical year quantity weights $= \dfrac{\Sigma\, p_n q_t}{\Sigma\, p_o q_t}$ (8)

   For $t = o$ and $t = n$ this reduces to (6) and (7) respectively.

FISHER'S IDEAL INDEX

We define

$$\text{Fisher's ideal price index} = \sqrt{\left(\frac{\Sigma\, p_n q_o}{\Sigma\, p_o q_o}\right)\left(\frac{\Sigma\, p_n q_n}{\Sigma\, p_o q_n}\right)} \tag{9}$$

This index is the geometric mean of the Laspeyres and Paasche index numbers given in equations (6) and (7). As already remarked, Fisher's ideal index satisfies both the *time reversal* and *factor reversal tests*, which gives it certain theoretical advantages over other index numbers.

THE MARSHALL–EDGEWORTH INDEX

The Marshall Edgeworth index uses the weighted aggregate typical year method where the weights are taken as the arithmetic mean of base year and given year quantities, i.e. $q_t = \tfrac{1}{2}(q_o + q_n)$. Substituting this value of q_t into equation (8), we have

$$\text{Marshall–Edgeworth price index} = \frac{\Sigma\, p_n (q_o + q_n)}{\Sigma\, p_o (q_o + q_n)} \tag{10}$$

WEIGHTED AVERAGE OF RELATIVES METHOD

To overcome the disadvantages of the simple average of relatives method, we can use a *weighted average of relatives*. The weighted average most often used in this connection is the *weighted arithmetic mean*, although other weighted averages such as the weighted geometric mean (Chap. 3) can be used as well.

In this method we weight each price, relative by the total value of the commodity in terms of some monetary unit such as the dollar. Since the value of a commodity is obtained by multiplying the price p of the commodity by the quantity q, the weights are given by pq.

Three possible formulae occur depending on whether base year, given year or typical year values, denoted by $p_o q_o$, $p_n q_n$ and $p_t q_t$ respectively, are used.

Weighted arithmetic mean of price relatives using base year value weights

$$= \frac{\Sigma\,(p_n/p_o)(p_o q_o)}{\Sigma\, p_o q_o} = \frac{\Sigma\, p_n q_o}{\Sigma\, p_o q_o} \tag{11}$$

Weighted arithmetic mean of price relatives using given year value weights

$$= \frac{\Sigma (p_n/p_o)(p_n q_n)}{\Sigma p_n q_n} \tag{12}$$

Weighted arithmetic mean of price relatives using typical year value weights

$$= \frac{\Sigma (p_n/p_o)(p_t q_t)}{\Sigma p_t q_t} \tag{13}$$

Note that (11) yields the same result as Laspeyres' formula given in (6).

QUANTITY OR VOLUME INDEX NUMBERS

The formulae described above for obtaining price index numbers are easily modified to obtain quantity or volume index numbers by simply interchanging p and q. For example, a replacement of p by q in (5) gives

$$\text{Simple arithmetic mean of relatives volume index} \quad = \frac{\Sigma q_n/q_o}{N} \tag{14}$$

where $\Sigma q_n/q_o$ = sum of all commodity quantity relatives
N = number of commodity quantity relatives used.

Similarly, formulae (6) and (7) become

Weighted aggregate volume index with base year price weights $\quad = \dfrac{\Sigma q_n p_o}{\Sigma q_o p_o}$ $\quad$ (15)
This is sometimes called a *Laspeyres volume index*.

Weighted aggregate volume index with given year price weights $\quad = \dfrac{\Sigma q_n p_n}{\Sigma q_o p_n}$ $\quad$ (16)
This is sometimes called a *Paasche volume index*.

In these formulae the weights used are prices. However, any other suitable weights can be used instead. Formulae (8)–(13) could be modified in the same manner.

VALUE INDEX NUMBERS

Just as we have obtained formulae for price and volume indexes, so we can obtain formulae for *value indexes*. The simplest such index is

$$\text{Value index} \quad = \frac{\Sigma p_n q_n}{\Sigma p_o q_o} \tag{17}$$

where $\Sigma p_o q_o$ = total value of all commodities in the base period
$\Sigma p_n q_n$ = total value of all commodities in the given period.

This is a *simple aggregate index*, since the values have not been weighted. Other formulae in which we use weights to indicate the relative importance of the items can be formulated.

CHANGING THE BASE PERIOD OF INDEX NUMBERS

In practice it is desirable that the base period chosen for comparison purposes be a period of economic stability which is not too far distant in the past. From time to time it may therefore be necessary to change this base period.

One possibility is to recompute all index numbers using the new base period. A simpler approximate method is to divide all index numbers for the various years corresponding to the old base period by the index number corresponding to the new base period, expressing the results as percentages. These results represent the new index numbers, the index number for the new base period being 100(%), as it should be.

Mathematically speaking, this method is strictly applicable only if the index numbers satisfy the *circular test* (see Prob. 17.37). However, for many types of index numbers the method fortunately yields results which in practice are close enough to those which would be obtained theoretically.

DEFLATION OF TIME SERIES

Although incomes of individuals may theoretically be rising over a period of years, their *real incomes* may actually be declining due to increases in living costs and therefore decreases in *purchasing power*. These real incomes can be obtained by dividing the *apparent or physical incomes* for the various years by the *cost of living or consumer index numbers* for the years, using an appropriate base period.

For example, if an individual's income in 1960 is 150% of his income in 1950 (i.e. it has increased by 50%) while the cost of living index has doubled over the same period, the individual's real income in 1960 is only 150/2 = 75% of what it was in 1950.

We have described above the process of *deflating* a time series involving income. A similar process may be used to deflate other time series. In Chapter 16, for example, we used an analogous procedure to *deseasonalize* data by employing *seasonal index numbers*.

Mathematically speaking, this method of deflating time series is strictly applicable only if the index numbers satisfy the *factor reversal test*, and for this reason Fisher's ideal index is appropriate. However, other index numbers can be used since they yield results which are correct for most practical purposes.

Solved Problems

PRICE RELATIVES

17.1. The average retail prices in dollars per tonne of bituminous coal sold in a certain country during the years 1953–1958 are given in Table 17.1. (*a*) Using 1953 as a base, find the price relatives corresponding to the years 1956 and 1958. (*b*) Using 1956 as a base, find the price relatives corresponding to all the given years. (*c*) Using 1953–1955 as a base, find the price relatives corresponding to all the given years.

Table 17.1

Year	1953	1954	1955	1956	1957	1958
Average retail price of bituminous coal (dollars per tonne)	14·95	14·94	15·10	15·65	16·28	16·53

Solution:

(a) Price relative for 1956 using 1953 as base

$$= p_{1953|1956} = \frac{\text{price in 1956}}{\text{price in 1953}} = \frac{15\cdot65}{14\cdot95} = 1\cdot047 = 104\cdot7\%$$

Price relative for 1958 using 1953 as base

$$= p_{1953|1958} = \frac{\text{price in 1958}}{\text{price in 1953}} = \frac{16\cdot53}{14\cdot95} = 0\cdot106 = 110\cdot6\%$$

In statistical literature it is the custom to omit % signs when quoting index numbers, although these signs are understood. With this convention the above price relatives are quoted as 104·7 and 110·6 respectively.

(b) Divide each retail price in Table 17·1 by 15·65 (dollars), the price for the year 1956. Then the required price relatives expressed in percentage form are as shown in Table 17.2.

Table 17.2

Year	1953	1954	1955	1956	1957	1958
Price Relative (1956 = 100)	95·5	95·5	96·5	100·0	104·0	105·6

These represent *index numbers* of bituminous coal retail prices for the years 1953-1958, the entire collection being called an *index series*. Note that the price relative (or price index number) corresponding to the year 1956 is in percentage form equal to 100·0, as is always true for a base period. This is often written symbolically in statistical literature as 1956 = 100.

(c) Arithmetic mean of prices for years 1953–1955 $= \dfrac{\$14\cdot95 + \$14\cdot94 + \$15\cdot10}{3} = \$15\cdot00$

Divide each retail price in Table 17.1 by this average base period price of $15·00. Then the required price relatives expressed in percentage form are as shown in Table 17.3.

Table 17.3

Year	1953	1954	1955	1956	1957	1958
Price Relative (1953 = 100)	99·7	99·6	100·7	104·3	108·5	110·2

These represent index numbers of bituminous coal retail prices for the years 1953-1958 using 1953-1955 as base period. Note that the arithmetic mean of the index numbers corresponding to the base period 1953-1955 is (99·7 + 99·6 + 100·7)/3 = 100·0, as is always true for a base period. This is often written symbolically in statistical literature as 1953–1955 = 100.

17.2. Prove that (a) $p_{a|b}\,p_{b|c} = p_{a|c}$, (b) $p_{a|b}\,p_{b|a} = 1$.

Solution:

(a) By definition, $\quad p_{a|b}\,p_{b|c} = \dfrac{p_b}{p_a}\cdot\dfrac{p_c}{p_b} = \dfrac{p_c}{p_a} = p_{a|c}.$

(b) By definition, $\quad p_{a|b}\,p_{b|a} = \dfrac{p_b}{p_a}\cdot\dfrac{p_a}{p_b} = 1.$

17.3. Using Table 17.3 of Problem 17.1(c) with 1953–1955 as a base, obtain the price relatives with 1956 as a base.

Solution:

Divide each price relative in Table 17.3 by the price relative 104·3 corresponding to 1956. The resulting numbers expressed as a per cent are the required price relatives and are given, apart from rounding errors, in Table 17.2 of Problem 17.1(b).

This example shows that given an index series corresponding to one base period, we can obtain the index series corresponding to another base period without using the original price data. The process involved is known as *changing the base period* or *shifting the base*.

For a proof of the method used here see Problem 17.36.

17.4. In 1956 the average price of a commodity was 20% more than in 1955, 20% less than in 1954 and 50% more than in 1957. Reduce the data to price relatives using (*a*) 1955, (*b*) 1956 and (*c*) 1954-1955 as base.

Solution:

(*a*)   Taking 1955 as base, the price relative (or index number) corresponding to it is 100. (Symbolically, 1955 = 100 or 100%).

Since the price in 1956 is 20% more than in 1955, the price relative corresponding to 1956 is 100 + 20 = 120, i.e. the price in 1956 is 120% of the price in 1955.

Since the price in 1956 is 20% less than in 1954 it must be 100 − 20 = 80% of the 1954 price. Then the 1954 price is 1/0·80 = 5/4 = 125% of the 1956 price, i.e. the 1954 price relative = 125% of the 1956 price relative = 125% of 120 = 150.

Since the price in 1956 is 50% more than in 1957, it must be 100 + 50 = 150% of the 1957 price. Then the 1957 price is 1/1·50 = $\frac{2}{3}$ of the 1956 price, i.e. the 1957 price relative = $\frac{2}{3}$ of the 1956 price relative = $\frac{2}{3}$ of 120 = 80.

Thus the required price relatives are as in Table 17.4.

Table 17.4

Year	1954	1955	1956	1957
Price Relative (1955 = 100)	150	100	120	80

(*b*)   We use the method of changing the base period given in Prob. 17.3. Divide each price relative in Table 17.4 by 120 (the price relative corresponding to the new base year 1956) and express the result as a percentage. Then the required price relatives with base year 1956 are as given in Table 17.5.

Table 17.5

Year	1954	1955	1956	1957
Price Relative (1956 = 100)	125	83·3	100	66·7

We can also proceed directly by reasoning as in part (*a*), choosing 1956 = 100.

(*c*)   **First method,** using part (*a*).

From Table 17.4, the arithmetic mean of price relatives for 1954 and 1955 is $\frac{1}{2}$(1·50 + 100) = 125. Then dividing each price relative in Table 17.4 by 125, we obtain the required price relatives as given in Table 17.6.

Table 17.6

Year	1954	1955	1956	1957
Price Relative (1954-1955 = 100)	120	80	96	64

Second method, using part (*b*).

From Table 17.5, the arithmetic mean of price relatives for 1954 and 1955 is $\frac{1}{2}$(125 + 83·3) = 104·2. Then dividing each price relative in Table 17.5 by 104·2, we obtain the same results as in the first method.

QUANTITY OR VOLUME RELATIVES

17.5. In Table 17.7 are given data on the production of wheat, of a country, in millions of litres for the years 1950-1958. Reduce the data to quantity relatives using (a) 1955 and (b) 1950-1953 as a base.

Table 17.7

Year	1950	1951	1952	1953	1954	1955	1956	1957	1958
Production of Wheat (millions of litres)	1019	988	1306	1173	984	935	1004	951	1462

Solution:

(a) Dividing the production figure of each year by 935 (the production figure for the base year 1955), the required quantity relatives (or quantity index numbers) for the various years expressed in percentage form are as given in Table 17.8.

Table 17.8

Year	1950	1951	1952	1953	1954	1955	1956	1957	1958
Quantity Relative (1955 = 100)	109·0	105·7	139·7	125·5	105·2	100·0	107·4	101·7	156·4

(b) Arithmetic mean of production for the years 1950-1953 $= \frac{1}{4}(1019 + 988 + 1306 + 1173) = 1122$.

Dividing the production figure for each year by 1122, the required quantity relatives expressed in percentage form are as given in Table 17.9.

Table 17.9

Year	1950	1951	1952	1953	1954	1955	1956	1957	1958
Quantity Relative (1950-1953 = 100)	90·8	88·1	116·4	104·5	87·7	83·3	89·5	84·8	130·3

17.6. The quantity relative for the year 1958 with 1949 as a base is 105, while the quantity relative for 1958 with 1953 as a base is 140. Find the quantity relative for 1953 with 1949 as a base.

Solution:
First method:

From the properties of quantity relatives, we have

$$q_{a|b} \, q_{b|c} = q_{a|c}$$

Let $a = 1949$, $b = 1953$, $c = 1958$. Then

$$q_{1949|1953} = q_{1949|1958} \, q_{1958|1953} = (1·05)(1/1·40) = 0·75 = 75\%$$

and the required quantity relative is 75.

Second method:

Let q_{1949}, q_{1953} and q_{1958} denote the actual quantities in the years 1949, 1953 and 1958 respectively. Then

$$\text{Quantity relative for 1958 with base 1949} = \frac{q_{1958}}{q_{1949}} = 105\% = 1·05$$

$$\text{Quantity relative for 1958 with base 1953} = \frac{q_{1958}}{q_{1953}} = 140\% = 1·40$$

Then the quantity relative for 1953 with base 1949 is

$$\frac{q_{1953}}{q_{1949}} = \frac{q_{1953}/q_{1958}}{q_{1949}/q_{1958}} = \frac{1/1·40}{1/1·05} = \frac{1·05}{1·40} = 75\%$$

Third method:

Since $q_{1958} = 1·05 q_{1949} = 1·40 q_{1953}$, then $\dfrac{q_{1953}}{q_{1949}} = \dfrac{1·05}{1·40} = 75\%$. Thus the required quantity relative is 75.

VALUE RELATIVES

17.7. In January 1960 a factory paid out a total of $40 000 to 120 employees on the payroll. In July of the same year the factory had 30 more employees on the payroll and paid out $6000 more than in January. Using January 1960 as a base, find (*a*) the employment index number (quantity relative) for July, (*b*) the labour expense index number (value relative) for July. (*c*) Using the result Price Relative × Quantity Relative = Value Relative, what interpretation could be given to the price relative in this case?

Solution:

(*a*) Employment index number = quantity relative = $\dfrac{120 + 30}{120} = 1.25 = 125\%$ or 125

(*b*) Labour expense index number = value relative = $\dfrac{\$40\,000 + \$6000}{\$40\,000} = 1.15 = 115\%$ or 115

(*c*) Price relative = $\dfrac{\text{value relative}}{\text{quantity relative}} = \dfrac{115}{125} = 0.92 = 92\%$ or 92

We can interpret this as a *cost per employee index number*. The significance is that in July 1960 the cost per employee was 92% of the cost per employee in the base period January 1960. Sometimes this is called a *per capita* labour cost index number.

17.8. A company expects its sales of a commodity to increase by 50% in the coming year. By what percentage should it increase the selling price in order that the gross income be doubled?

Solution:

$$\text{Price relative} \times \text{quantity relative} = \text{value relative}$$
or
$$\text{Price relative} \times 150\% = 200\%$$

Then price relative = $200/150 = 4/3 = 133\frac{1}{3}\%$, so that it should increase its selling price by $133\frac{1}{3} - 100 = 33\frac{1}{3}\%$.

LINK AND CHAIN RELATIVES

17.9. The link relatives for prices in 1956-1960 are 125, 120, 135, 150 and 175 respectively. (*a*) Find the price relative for 1957 with 1955 as base. (*b*) Chain the link relatives to a 1956 base.

Solution:

$$p_{1955|1956} = 1.25, \quad p_{1956|1957} = 1.20, \quad p_{1957|1958} = 1.35, \quad p_{1958|1959} = 1.50, \quad p_{1959|1960} = 1.75$$

(*a*) $p_{1955|1957} = p_{1955|1956}\, p_{1956|1957} = (1.25)(1.20) = 1.50 = 150\%$

(*b*) $p_{1956|1955} = \dfrac{1}{p_{1955|1956}} = \dfrac{1}{1.25} = 80\%$

$p_{1956|1956} = 100\% \qquad p_{1956|1957} = 120\%$

$p_{1956|1958} = p_{1956|1957}\, p_{1957|1958} = (1.20)(1.35) = 1.62 = 162\%$

$p_{1956|1959} = p_{1956|1957}\, p_{1957|1958}\, p_{1958|1959} = (1.20)(1.35)(1.50) = 2.43 = 243\%$

$p_{1956|1960} = p_{1956|1957}\, p_{1957|1958}\, p_{1958|1959}\, p_{1959|1960} = (1.20)(1.35)(1.50)(1.75) = 425\%$

INDEX NUMBERS. SIMPLE AGGREGATE METHOD

17.10. Table 17.10 shows a country's average wholesale prices and production of fluid milk, butter and cheese for the years 1949, 1950 and 1958. Compute a simple aggregate wholesale price index of these dairy products for the year 1958 using (*a*) 1949 and (*b*) 1949-1950 as a base.

Table 17.10

	Prices (per kilogramme)			Quantities produced (millions of kilogrammes)		
	1949	**1950**	**1958**	**1949**	**1950**	**1958**
Milk	3·95	3·89	4·13	9675	9717	10 436
Butter	61·5	62·2	59·7	117·7	115·5	115·5
Cheese	34·8	35·4	38·9	77·93	74·39	82·79

Solution:

(a) Simple aggregate price index $= \dfrac{\Sigma p_n}{\Sigma p_o} = \dfrac{\text{sum of prices in given year (1958)}}{\text{sum of prices in base year (1949)}}$

$$= \frac{4·13 + 59·7 + 38·9}{3·95 + 61·5 + 34·8} = 102·5(\%)$$

i.e. average wholesale prices in 1958 are 102·5% of those in 1949 or (2·5% greater).

(b) Average (mean) price of milk in base period 1949–1950 $= \frac{1}{2}(3·95 + 3·89) = 3·92$
Average (mean) price of butter in base period 1949–1950 $= \frac{1}{2}(61·5 + 62·2) = 61·85$
Average (mean) price of cheese in base period 1949–1950 $= \frac{1}{2}(34·8 + 35·4) = 35·1$

Simple aggregate price index $= \dfrac{\Sigma p_n}{\Sigma p_o} = \dfrac{\text{sum of prices in given year (1958)}}{\text{sum of prices in base period (1949-1950)}}$

$$= \frac{4·13 + 59·7 + 38·9}{3·92 + 61·85 + 35·1} = 101·8(\%)$$

Note that this method does not make use of the *quantities* produced but only the *prices* of the commodities. For illustrative purposes, we have used only 3 commodities to compute an index number. In actual practice many more commodities would be included.

17.11. Explain why the index numbers obtained in Problem 17.10 may be inappropriate measures of price changes in the given commodities.

Solution:

The index computed in Problem 17.10 does not take into account the relative importance of the commodities as would be determined, for example, by *how much* is used by the consumer or *how much* is produced for consumption purposes. These considerations are taken up in later problems.

17.12. Table 17.11 shows the average retail prices and production of anthracite coal and petrol during the years 1949 and 1958. Explain why a simple aggregate price index for 1958 with base year 1949 is an inappropriate measure of price changes in the given commodities.

Table 17.11

	Prices		Quantities	
	1949	**1958**	**1949**	**1958**
Anthracite Coal	$20·13 per tonne	$28·20 per tonne	3·559 million tonnes	1·821 million tonnes
Petrol	20·3¢ per litre	21·4¢ per litre	80·2 million barrels*	118·6 million barrels*

*Each barrel contains 159 litres

Solution:

If a simple aggregate price index is used the result is

$$\frac{\Sigma p_n}{\Sigma p_o} = \frac{\text{sum of prices in given year (1958)}}{\text{sum of prices in base year (1949)}} = \frac{\$28\cdot20 + \$0\cdot214}{\$20\cdot13 + \$0\cdot203} = 139\cdot7(\%)$$

indicating that average retail prices of these commodities in 1958 were 39·7% greater than in 1949.

If we express the price of anthracite coal in cents per kg instead of dollars per tonne, the price in 1949 is \$20·13/(1000 kg) = 2·013¢/kg, while the price in 1958 is \$28·20/(1000 kg) = 2·820¢/kg. In this case the simple aggregate price index is

$$\frac{\Sigma p_n}{\Sigma p_o} = \frac{2\cdot820¢ + 21\cdot4¢}{2\cdot013¢ + 20\cdot3¢} = 108\cdot5(\%)$$

indicating that average retail prices of these commodities in 1958 were 8·5% greater than in 1949.

Since the simple aggregate index is so sensitive to the units used in price quotations, it is clearly an inappropriate measure for such cases. This together with the disadvantage expressed in Problem 17.11 provide good reasons for not using this index in practice.

The remark made at the end of Problem 17.10 applies in this problem as well.

SIMPLE AVERAGE OF RELATIVES METHOD

17.13. Use the simple average (mean) of relatives method to compute a wholesale price index of the dairy products in Problem 17.10 for the year 1958 with (a) 1949 and (b) 1949-1950 as base.

Solution:

(a) The price relatives for milk, butter and cheese in 1958 with 1949 as base are as follows:

$$\text{Price relative for milk} = \frac{\text{price of milk in 1958}}{\text{price of milk in 1949}} = \frac{4\cdot13}{3\cdot95} = 104\cdot6(\%)$$

$$\text{Price relative for butter} = \frac{\text{price of butter in 1958}}{\text{price of butter in 1949}} = \frac{59\cdot7}{61\cdot5} = 97\cdot1(\%)$$

$$\text{Price relative for cheese} = \frac{\text{price of cheese in 1958}}{\text{price of cheese in 1949}} = \frac{38\cdot9}{34\cdot8} = 111\cdot8(\%)$$

$$\text{Average (mean) of price relatives} = \frac{\Sigma p_n/p_o}{N} = \frac{104\cdot6 + 97\cdot1 + 111\cdot8}{3} = 104\cdot5(\%).$$

(b) Referring to Prob. 17.10(b), the price relatives in 1958 with 1949–1950 as base are:

$$\text{Price relative for milk} = \frac{\text{price of milk in 1958}}{\text{price of milk in 1949–50}} = \frac{4\cdot13}{3\cdot92} = 105\cdot4(\%)$$

$$\text{Price relative for butter} = \frac{\text{price of butter in 1958}}{\text{price of butter in 1949–50}} = \frac{59\cdot7}{61\cdot85} = 96\cdot5(\%)$$

$$\text{Price relative for cheese} = \frac{\text{price of cheese in 1958}}{\text{price of cheese in 1949–50}} = \frac{38\cdot9}{35\cdot1} = 110\cdot8(\%)$$

$$\text{Average (mean) of price relatives} = \frac{\Sigma p_n/p_o}{N} = \frac{105\cdot4 + 96\cdot5 + 110\cdot8}{3} = 104\cdot2(\%).$$

17.14. Work Problem 17.13 if the median is used in place of the mean.

Solution:

(a) Required index number = median of price relatives 104·6, 97·1 and 111·8 = 104·6.

(b) Required index number = median of price relatives 105·4, 96·5 and 110·8 = 105·4.

17.15. Work Problem 17.13 if the geometric mean is used in place of the mean.

Solution:

(a) Required index number = geometric mean of price relatives 104·6, 97·1 and 111·8
$$= \sqrt[3]{(104\cdot6)(97\cdot1)(111\cdot8)} = 104\cdot3, \text{ using logarithms.}$$

(b) Required index number = geometric mean of price relatives 105·4, 96·5 and 110·8
$$= \sqrt[3]{(105\cdot4)(96\cdot5)(110\cdot8)} = 104\cdot1, \text{ using logarithms.}$$

17.16. Use the simple average (mean) of price relatives to obtain a retail price index number for the commodities in Problem 17.12 using 1949 as base year and 1958 as given year.

Solution:

$$\text{Price relative for coal} \quad = \frac{\text{price of coal in 1958}}{\text{price of coal in 1949}} = \frac{\$28 \cdot 20}{\$20 \cdot 13} = 140 \cdot 1(\%)$$

$$\text{Price relative for petrol} = \frac{\text{price of petrol in 1958}}{\text{price of petrol in 1949}} = \frac{21 \cdot 4 \cancel{c}}{20 \cdot 3 \cancel{c}} = 105 \cdot 4(\%)$$

$$\text{Simple average (mean) of price relatives} = \frac{\Sigma p_n/p_o}{N} = \frac{140 \cdot 1 + 105 \cdot 4}{2} = 122 \cdot 8.$$

Note that the result is *independent* of the units used in price quotations (compare Problem 17.12).

17.17. Work Problem 17.16 if the geometric mean is used.

Solution:

Required index number = geometric mean of price relatives $140 \cdot 1$ and $105 \cdot 4$
$$= \sqrt{(140 \cdot 1)(105 \cdot 4)} = 121 \cdot 5$$

WEIGHTED AGGREGATE METHOD. LASPEYRES' AND PAASCHE'S INDEXES

17.18. Using the data of Problem 17.10 compute a Laspeyres price index number for the year 1958 with (*a*) 1949 and (*b*) 1949–1950 as base.

Solution:

(*a*) Laspeyres' index = weighted aggregate price index with base period quantity weights

$$= \frac{\Sigma p_n q_o}{\Sigma p_o q_o} = \frac{\Sigma(\text{prices in 1958})(\text{quantities in 1949})}{\Sigma(\text{prices in 1949})(\text{quantities in 1949})}$$

$$= \frac{(4 \cdot 13)(9675) + (59 \cdot 7)(117 \cdot 7) + (38 \cdot 9)(77 \cdot 93)}{(3 \cdot 95)(9675) + (61 \cdot 5)(117 \cdot 7) + (34 \cdot 8)(77 \cdot 93)} = 103 \cdot 84, \text{ or } 103 \cdot 8(\%)$$

(*b*) The average quantities of milk, butter and cheese produced in 1949-1950 are $\frac{1}{2}(9675 + 9717) = 9696$, $\frac{1}{2}(117 \cdot 7 + 115 \cdot 5) = 116 \cdot 6$ and $\frac{1}{2}(77 \cdot 93 + 74 \cdot 39) = 76 \cdot 16$ respectively. The average prices in 1949-1950 are shown in Problem 17.10.

$$\text{Laspeyres' index} = \frac{\Sigma p_n q_o}{\Sigma p_o q_o} = \frac{\Sigma(\text{prices in 1958})(\text{average quantities in 1949-50})}{\Sigma(\text{prices in 1949-50})(\text{average quantities in 1949-50})}$$

$$= \frac{(4 \cdot 13)(9696) + (59 \cdot 7)(116 \cdot 6) + (38 \cdot 9)(76 \cdot 16)}{(3 \cdot 92)(9696) + (61 \cdot 85)(116 \cdot 6) + (35 \cdot 1)(76 \cdot 16)} = 104 \cdot 33, \text{ or } 104 \cdot 3(\%)$$

17.19. Using the data of Problem 17.10 compute a Paasche price index number for the year 1958 with (*a*) 1949 and (*b*) 1949–1950 as base.

Solution:

(*a*) Paasche's index = weighted aggregate price index with given year quantity weights

$$= \frac{\Sigma p_n q_n}{\Sigma p_o q_o} = \frac{\Sigma(\text{prices in 1958})(\text{quantities in 1958})}{\Sigma(\text{prices in 1949})(\text{quantities in 1958})}$$

$$= \frac{(4 \cdot 13)(10\,436) + (59 \cdot 7)(115 \cdot 5) + (38 \cdot 9)(82 \cdot 79)}{(3 \cdot 95)(10\,436) + (61 \cdot 5)(115 \cdot 5) + (34 \cdot 8)(82 \cdot 79)} = 103 \cdot 93, \text{ or } 103 \cdot 9(\%).$$

(*b*) Paasche's index $= \dfrac{\Sigma p_n q_n}{\Sigma p_o q_n} = \dfrac{\Sigma(\text{prices in 1958})(\text{quantities in 1958})}{\Sigma(\text{prices in 1949-50})(\text{quantities in 1958})}$

$$= \frac{(4 \cdot 13)(10\,436) + (59 \cdot 7)(115 \cdot 5) + (38 \cdot 9)(82 \cdot 79)}{(3 \cdot 92)(10\,436) + (61 \cdot 85)(115 \cdot 5) + (35 \cdot 1)(82 \cdot 79)} = 104 \cdot 43 \text{ or } 104 \cdot 4(\%).$$

17.20. Find the (*a*) Laspeyres and (*b*) Paasche index numbers for the data of Problem 17.12. (*c*) State an advantage of the Laspeyres index over the Paasche index in case revisions of an index number are to be made from year to year.

Solution:

(*a*) Laspeyres' index $= \dfrac{\Sigma p_n q_o}{\Sigma p_o q_o} = \dfrac{\Sigma(\text{prices in 1958})(\text{quantities in 1949})}{\Sigma(\text{prices in 1949})(\text{quantities in 1949})}$

$= \dfrac{(\$28\cdot20 \text{ per tonne})(3\cdot559 \text{ million tonnes}) + (\$0\cdot214 \text{ per litre})(80\cdot2 \times 159 \text{ million litres})}{(\$20\cdot13 \text{ per tonne})(3\cdot559 \text{ million tonnes}) + (\$0\cdot203 \text{ per litre})(80\cdot2 \times 159 \text{ million litres})}$

$= \dfrac{2829\cdot25 \text{ million dollars}}{2660\cdot26 \text{ million dollars}} = 106\cdot35, \text{ or } 106\cdot4(\%).$

Note that it is *very important* that the units employed are correct and consistent.

(*b*) Paasches' index $= \dfrac{\Sigma p_n q_n}{\Sigma p_o q_n} = \dfrac{\Sigma(\text{prices in 1958})(\text{quantities in 1958})}{\Sigma(\text{prices in 1949})(\text{quantities in 1958})}$

$= \dfrac{(\$28\cdot20 \text{ per tonne})(1\cdot821 \text{ million tonnes}) + (\$0\cdot214 \text{ per litre})(118\cdot6 \times 159 \text{ million litres})}{(\$20\cdot13 \text{ per tonne})(1\cdot821 \text{ million tonnes}) + (\$0\cdot203 \text{ per litre})(118\cdot6 \times 159 \text{ million litres})}$

$= \dfrac{4086\cdot84 \text{ million dollars}}{3864\cdot71 \text{ million dollars}} = 105\cdot747 \text{ or } 105\cdot7(\%).$

In practice, where an index number is to be computed for many commodities it is advisable to arrange the computations in a suitable tabular form (see Problem 17.31, for example).

(*c*) In computing a Laspeyres index, the weight (quantities produced or consumed in the base year, if a price index is computed) do not change from year to year. Thus only information on lastest prices need be obtained.
      In computing a Paasche index, the latest information on weights (quantities) as well as prices must be obtained. Thus computation of a Paasche index involves more labour in collection of data.

17.21. Present an interpretation of the (*a*) Laspeyres and (*b*) Paasche price index numbers in terms of the total value (or total cost) of commodities.

Solution:

(*a*) In computing a Laspeyres price index, $\Sigma p_o q_o$ represents the total value (or total cost) of a set of goods and services or commodities (sometimes called a *market basket*) in the base year or period. The quantity $\Sigma p_n q_o$ represents the total value of *this same market basket* in the given year or period. Thus a Laspeyres price index serves to measure the total costs in any given year of a *fixed market basket* of commodities purchased in the base year.

(*b*) In computing a Paasche price index, $\Sigma p_o q_n$ is the total value (or total cost) of commodities purchased in the given year assuming base year prices, while $\Sigma p_n q_n$ is the total value of commodities purchased in the given year at given year prices. Thus a Paasche price index serves to measure the total cost of a *given year market basket* relative to what the cost would have been if the purchase has been made in the base year.

17.22. It is sometimes stated that the Laspeyres price index tends to *overestimate* price changes while the Paasche price index tends to *underestimate* them. Present a possible reason to substantitate this statement.

Solution:

According to the economic *law of supply and demand*, people tend to purchase *less* when prices are *high* and to purchase *more* when prices are *low*. This is the so-called *elastic demand* which is valid if the need for commodities is not absolutely essential.

In the case of the Laspeyres index, $\Sigma p_n q_o$ will be somewhat higher than it should be, since according to the law of supply and demand people tend to purchase fewer high priced commodities and more lower priced commodities, so that the total cost would be less than predicted by $\Sigma p_n q_o$. Thus the Laspeyres index $\dfrac{\Sigma p_n q_o}{\Sigma p_o q_o}$ tends to be higher than it should be.

In the case of the Paasche index, the roles played by base year and given year quantities are interchanged from what they were in the case of the Laspeyres index. This interchange tends to make the Paasche index lower than it should be.

The above reasoning does not imply that the Laspeyres index is *always* higher than the Paasche index but only that it *tends* to be higher. In practice the Laspeyres index can be greater than, less than or equal to the Paasche index. (See Problems 17.18 and 17.19 where the Laspeyres index is in fact *less* than the Paasche index.)

17.23. Prove that weighted aggregate price index numbers with fixed (quantity) weights satisfy the circular test.

Solution:

Letting q_o denote the fixed weights, we have for any periods a, b and c the index numbers

$$I_{a|b} = \frac{\Sigma\, p_b q_o}{\Sigma\, p_a q_o} \qquad \text{and} \qquad I_{b|c} = \frac{\Sigma\, p_c q_o}{\Sigma\, p_b q_o}$$

Then

$$I_{a|b}\, I_{b|c} = \frac{\Sigma\, p_b q_o}{\Sigma\, p_a q_o} \cdot \frac{\Sigma\, p_c q_o}{\Sigma\, p_b q_o} = \frac{\Sigma\, p_c q_o}{\Sigma\, p_a q_o} = I_{a|c}$$

which shows that the circular test is satisfied.

The Laspeyres and Paasche index numbers do not satisfy the circular test.

FISHER'S IDEAL INDEX

17.24. Show that Fisher's ideal index is the geometric mean of the Laspeyres and Paasche index numbers.

Solution:

Letting F, L and P denote respectively Fisher's, Laspeyres' and Paasche's index numbers, we have

$$F = \sqrt{\left(\frac{\Sigma\, p_n q_o}{\Sigma\, p_o q_o}\right)\left(\frac{\Sigma\, p_n q_n}{\Sigma\, p_o q_n}\right)} = \sqrt{LP}$$

using the definition of L and P. Since $\sqrt{LP}$ is the geometric mean of L and P, the required result follows.

17.25. Prove that Fisher's ideal index lies between the Laspeyres and Paasche index numbers.

Solution:

This follows at once from the fact that $F = \sqrt{LP}$ lies between L and P, since L and P are positive numbers. Note that if $L = P$, then $F = L = P$.

Since by Prob. 17.22, L has a tendency to *overestimate* price changes while P has a tendency to *underestimate* them, it follows that F, which lies between L and P, should provide a better estimate than L or P.

17.26. Determine Fisher's ideal price index for the dairy products in Problem 17.10 for the year 1958 with (a) 1949 and (b) 1949–1950 as base.

Solution:

(a) $F = \sqrt{LP} = \sqrt{(103 \cdot 84)(103 \cdot 93)} = 103 \cdot 9$, by Problems 17.18(a) and 17.19(a).

(b) $F = \sqrt{LP} = \sqrt{(104 \cdot 33)(104 \cdot 43)} = 104 \cdot 4$, by Problems 17.18(b) and 17.19(b).

17.27. Determine Fisher's ideal price index for the data of Problem 17.12.

Solution:

From Problem 17.20, $F = \sqrt{LP} = \sqrt{(106 \cdot 35)(105 \cdot 75)} = 106 \cdot 0$.

Note that a good approximation to $\sqrt{LP}$ when L and P are approximately equal is given by $\frac{1}{2}(L + P)$. This arithmetic mean of L and P can be used as the *definition* of a new index number lying between L and P.

17.28. Prove that Fisher's ideal index satisfies the time reversal test.

Solution:

Let $F_{o|n}$ denote Fisher's ideal index number for a given year with respect to a base year, and $F_{n|o}$ denote Fisher's ideal index when base year and given year are interchanged. Then the time reversal test is satisfied if $F_{o|n} = 1/F_{n|o}$ or $F_{o|n}F_{n|o} = 1$.

By definition, $\quad F_{o|n} = \sqrt{\left(\dfrac{\Sigma\,p_n q_o}{\Sigma\,p_o q_o}\right)\left(\dfrac{\Sigma\,p_n q_n}{\Sigma\,p_o q_n}\right)}. \qquad$ Then $\qquad F_{n|o} = \sqrt{\left(\dfrac{\Sigma\,p_o q_n}{\Sigma\,p_n q_n}\right)\left(\dfrac{\Sigma\,p_o q_o}{\Sigma\,p_n q_o}\right)} \qquad$ and

$$F_{o|n}F_{n|o} = \sqrt{\left(\frac{\Sigma\,p_n q_o}{\Sigma\,p_o q_o}\right)\left(\frac{\Sigma\,p_n q_n}{\Sigma\,p_o q_n}\right)\left(\frac{\Sigma\,p_o q_n}{\Sigma\,p_n q_n}\right)\left(\frac{\Sigma\,p_o q_o}{\Sigma\,p_n q_o}\right)} = 1$$

THE MARSHALL-EDGEWORTH INDEX

17.29. Calculate the Marshall–Edgeworth price index for the data of Problem 17.12.

Solution:

$$\text{Marshall–Edgeworth index} = \frac{\Sigma p_n(q_o + q_n)}{\Sigma p_o(q_o + q_n)}$$

$$= \frac{\Sigma(\text{prices in 1958})(\text{sum of quantities in 1949 and 1958})}{\Sigma(\text{prices in 1949})(\text{sum of quantities in 1949 and 1958})}$$

$$= \frac{(\$28\cdot20)\{(3\cdot559 + 1\cdot821)(10^6)\} + (\$0\cdot214)\{(80\cdot2 + 118\cdot6)(159 + 10^6)\}}{(\$20\cdot13)\{(3\cdot559 + 1\cdot821)(10^6)\} + (\$0\cdot203)\{(80\cdot2 + 118\cdot6)(159 + 10^6)\}} = \frac{6916\cdot0}{6525\cdot0} = 105\cdot9(\%)$$

Note that this lies between the Laspeyres and Paasche index numbers (see Problem 17.20). For a proof that this is always the case, see Problem 17.30.

17.30. (a) Prove that if $\dfrac{X_1}{X_2} < \dfrac{Y_1}{Y_2}$ then $\dfrac{X_1}{X_2} < \dfrac{X_1 + Y_1}{X_2 + Y_2} < \dfrac{Y_1}{Y_2}$, where X_1, X_2, Y_1, Y_2 are any positive numbers.

(b) Use the result in (a) to prove that the Marshall–Edgeworth index number lies between the Laspeyres and Paasche index numbers.

Solution:

(a) If $\dfrac{X_1}{X_2} < \dfrac{Y_1}{Y_2}$, then (1) $X_1 Y_2 < X_2 Y_1$.

Adding $X_1 X_2$ to both sides of (1), we have

$$X_1 X_2 + X_1 Y_2 < X_1 X_2 + X_2 Y_1 \quad \text{or} \quad X_1(X_2 + Y_2) < X_2(X_1 + Y_1) \quad \text{or} \quad (2)\ \frac{X_1}{X_2} < \frac{X_1 + Y_1}{X_2 + Y_2}$$

on dividing both sides by $X_2(X_2 + Y_2)$.

Adding $Y_1 Y_2$ to both sides of (1), we have

$$X_1 Y_2 + Y_1 Y_2 < X_2 Y_1 + Y_1 Y_2 \quad \text{or} \quad Y_2(X_1 + Y_1) < Y_1(X_2 + Y_2) \quad \text{or} \quad (3)\ \frac{X_1 + Y_1}{X_2 + Y_2} < \frac{Y_1}{Y_2}$$

on dividing both sides by $Y_1(X_1 + Y_1)$.

From (2) and (3) the required result follows.

(b) **Case 1:** Laspeyres' index is less than Paasche's index.

Let $X_1 = \Sigma p_n q_o$, $X_2 = \Sigma p_o q_o$, $Y_1 = \Sigma p_n q_n$, $Y_2 = \Sigma p_o q_n$. Then $\dfrac{X_1}{X_2} < \dfrac{Y_1}{Y_2}$ so that from (a),

$$\frac{\Sigma\,p_n q_o}{\Sigma\,p_o q_o} < \frac{\Sigma\,p_n q_o + \Sigma\,p_n q_n}{\Sigma\,p_o q_o + \Sigma\,p_o q_n} < \frac{\Sigma\,p_n q_n}{\Sigma\,p_o q_n}$$

or

$$\frac{\Sigma\,p_n q_o}{\Sigma\,p_o q_o} < \frac{\Sigma\,p_n(q_o + q_n)}{\Sigma\,p_o(q_o + q_n)} < \frac{\Sigma\,p_n q_n}{\Sigma\,p_o q_n}$$

or

$$\text{Laspeyres index} < \text{Marshall–Edgeworth index} < \text{Paasche index}$$

Case 2: Paasche's index is less than Laspeyres' index.

Let $X_1 = \Sigma\, p_n q_n$, $X_2 = \Sigma\, p_o q_n$, $Y_1 = \Sigma\, p_n q_o$, $Y_2 = \Sigma\, p_o q_o$. Then $\dfrac{X_1}{X_2} < \dfrac{Y_1}{Y_2}$ so that from (a),

$$\frac{\Sigma\, p_n q_n}{\Sigma\, p_o q_n} < \frac{\Sigma\, p_n q_n + \Sigma\, p_n q_o}{\Sigma\, p_o q_n + \Sigma\, p_o q_o} < \frac{\Sigma\, p_n q_o}{\Sigma\, p_o q_o}$$

or

$$\frac{\Sigma\, p_n q_n}{\Sigma\, p_o q_n} < \frac{\Sigma\, p_n (q_o + q_n)}{\Sigma\, p_o (q_o + q_n)} < \frac{\Sigma\, p_n q_n}{\Sigma\, p_o q_n}$$

or

Paasche index < Marshall–Edgeworth index < Laspeyres index

It follows from Case 1 and 2 that regardless of whether Laspeyres' index is larger or smaller than Paasche's index, the Marshall–Edgeworth index will lie between them.

WEIGHTED AVERAGE OF RELATIVES

17.31. Compute a weighted arithmetic mean of price relatives for the data of Problem 17.12 using (a) given year value weights and (b) base year value weights, where the base year is 1949 and the given year is 1958.

Solution:

(b) Weighted arithmetic mean of price relatives using given year value weights

$$= \frac{\Sigma(p_n/p_o)(p_n q_n)}{\Sigma\, p_n q_n} = \frac{\Sigma(\text{price relatives})(\text{given year values})}{\Sigma \text{ given year values}}$$

The calculations involved can be organized as in Table 17.12, where the subscript n stands for the given year 1958 and the subscript o stands for the base year 1949, and p and q stand for prices and quantities respectively.

Table 17.12

	p_o	p_n	q_n	p_n/p_o	$p_n q_n$ (millions of dollars)	$(p_n/p_o)(p_n q_n)$ (millions of dollars)
Anthracite Coal	$20·13 (per tonne)	$28·20 (per tonne)	1·821 (million tonnes)	1·4009	51·352	71·939
Petrol	$0·203 (per litre)	$0·214 (per litre)	118·6 × 159 (million litres)	1·0542	4035·484	4254·207
					$\Sigma p_n q_n$ = 4086·836	$\Sigma(p_n/p_o)(p_n q_n)$ = 4326·146

Then the required index number $= \dfrac{\Sigma\,(p_n/p_o)(p_n q_n)}{\Sigma\, p_n q_n} = \dfrac{4326·146}{4035·484} = 107·2(\%)$

(b) Weighted arithmetic mean of price relatives using base year value weights is

$$\frac{\Sigma\,(p_n/p_o)(p_o q_o)}{\Sigma\, p_o q_o} = \frac{\Sigma\, p_n q_o}{\Sigma\, p_o q_o} = \text{Laspeyres' index of Prob. } 17.20(a) = 106·4(\%)$$

We can also proceed by making a table as in part (a).

QUANTITY OR VOLUME INDEX NUMBERS

17.32. Use the data of Problem 17.12 to compute a volume index for 1958 with base year 1949 by means of (a) a simple arithmetic mean of volume relatives, (b) a weighted aggregate volume index with base year price weights, (c) a weighted aggregate volume index with given year price weights.

Solution:

(a) Simple arithmetic mean of relatives volume index

$$= \frac{\Sigma q_n/q_o}{N} = \frac{1·821/3·559 + 118·6/80·2}{2} = \frac{51·17(\%) + 147·88(\%)}{2} = 99·5(\%)$$

(b) Weighted aggregate volume index with base year price weights

$$= \frac{\Sigma p_n p_o}{\Sigma p_o p_o} = \frac{\Sigma(\text{quantities in 1958})(\text{prices in 1949})}{\Sigma(\text{quantities in 1949})(\text{prices in 1949})}$$

$$= \frac{(1 \cdot 821 \text{ million tonnes})(\$20 \cdot 13 \text{ per tonne}) + (118 \cdot 6 \times 159 \text{ million litres})(\$0 \cdot 203 \text{ per litre})}{(3 \cdot 559 \text{ million tonnes})(\$20 \cdot 13 \text{ per tonne}) + (80 \cdot 2 \times 159 \text{ million litres})(\$0 \cdot 203 \text{ per litre})}$$

$$= \frac{3853 \cdot 73 \text{ million dollars}}{2660 \cdot 26 \text{ million dollars}} = 144 \cdot 86, \text{ or } 144 \cdot 9(\%)$$

This is sometimes called *Laspeyres' quantity or volume index.*

(c) Weighted aggregate volume index with given year price weights

$$= \frac{\Sigma q_n p_n}{\Sigma q_o p_n} = \frac{\Sigma(\text{quantities in 1958})(\text{prices in 1958})}{\Sigma(\text{quantities in 1949})(\text{prices in 1958})}$$

$$= \frac{(1 \cdot 821 \text{ million tonnes})(\$28 \cdot 20 \text{ per tonne}) + (118 \cdot 6 \times 159 \text{ million litres})(\$0 \cdot 214 \text{ per litre})}{(3 \cdot 559 \text{ million tonnes})(\$28 \cdot 20 \text{ per tonne}) + (80 \cdot 2 \times 159 \text{ million litres})(\$0 \cdot 214 \text{ per litre})}$$

$$= \frac{4086 \cdot 84 \text{ million dollars}}{2829 \cdot 25 \text{ million dollars}} = 144 \cdot 45, \text{ or } 144 \cdot 4(\%)$$

This is sometimes called *Paasche's quantity or volume index number.*

17.33. From the results of Prob. 17.32 determine Fisher's ideal quantity or volume index number.

Solution:

As for price indexes, Fisher's ideal volume index is given by the geometric mean of the Laspeyres and Paasche volume index numbers. Then from Problem 17.32,

$$\text{Fisher's ideal volume index number} = \sqrt{(144 \cdot 86)(144 \cdot 45)} = 144 \cdot 6$$

VALUE INDEX NUMBERS

17.34. Prove that Fisher's ideal index satisfies the factor reversal test.

Solution:

The factor reversal test is satisfied if the index is such that

$$(\text{price index})(\text{quantity index}) = \text{value index}$$

Let F_P and F_Q be respectively Fisher's ideal price and quantity indexes. Then

$$F_P F_Q = \sqrt{\left(\frac{\Sigma p_n q_o}{\Sigma p_o q_o}\right)\left(\frac{\Sigma p_n q_n}{\Sigma p_o q_n}\right)} \sqrt{\left(\frac{\Sigma q_n p_o}{\Sigma q_o p_o}\right)\left(\frac{\Sigma q_n p_n}{\Sigma q_o p_n}\right)} = \frac{\Sigma p_n q_n}{\Sigma p_o q_o} = \text{value index}$$

so that Fisher's ideal index satisfies the factor reversal test.

17.35. Calculate the value index in Problem 17.34 for the data of Problem 17.12.

Solution:

Since the result

$$\text{value index} = (\text{price index})(\text{quantity index})$$

holds exactly when Fisher's ideal indexes are used, we obtain from Problems 17.27 and 17.33,

$$\text{value index} = (106 \cdot 0\%)(144 \cdot 6\%) = 153 \cdot 3\%$$

This result can also be obtained by direct substitution in $\frac{\Sigma p_n q_n}{\Sigma p_o q_o}$.

CHANGING THE BASE PERIOD OF INDEX NUMBERS

17.36. Establish the validity of the method in Problem 17.3 for obtaining price relatives for a new base period.

Solution:
Assume the periods to be numbered consecutively from 1 to N as in the first row of Table 17.13, and let $p_1, p_2, \ldots, p_N$ denote the prices for these periods as in the second row of the table.

Table 17.13

Period	1	2	3	...	j	...	k	...	N
Prices	p_1	p_2	p_3	...	p_j	...	p_k	...	p_N
Price relatives corresponding to old period j	$p_{j\mid1}$	$p_{j\mid2}$	$p_{j\mid3}$	...	100%	...	$p_{j\mid k}$	...	$p_{j\mid N}$
Price relatives corresponding to new period k	$p_{k\mid1}$	$p_{k\mid2}$	$p_{k\mid3}$	...	$p_{k\mid j}$	...	100%	...	$p_{k\mid N}$

The price relatives corresponding to the periods j and k, which we call the old and new periods respectively, are indicated in the third and fourth rows of the table. Here $p_{j\mid1} = p_1/p_j$, $p_{j\mid2} = p_2/p_j$, etc.

It is clear that the fourth row can be obtained from the third row by dividing each entry in the third row by $p_{j\mid k}$, i.e. the price relative of the kth period with respect to the jth period as base. For example,

$$\frac{p_{j\mid1}}{p_{j\mid k}} = \frac{p_1/p_j}{p_k/p_j} = \frac{p_1}{p_k} = p_{k\mid1}, \quad \text{etc.}$$

The results clearly apply to quantity and value relatives as well as price relatives.

17.37. Prove that the method of Problem 17.36 for changing the base period of index numbers is applicable if and only if the index numbers satisfy the circular test.

Solution:
If we denote the index numbers for the various periods with jth period as base by

(1) $$I_{j\mid1}, I_{j\mid2}, \ldots, I_{j\mid N}$$

and the corresponding index numbers with kth period as base by

(2) $$I_{k\mid1}, I_{k\mid2}, \ldots, I_{k\mid N}$$

we will obtain the sequence *(2)* by dividing each member of the sequence *(1)* by $I_{j\mid k}$ if and only if

$$\frac{I_{j\mid1}}{I_{j\mid k}} = I_{k\mid1}, \quad \frac{I_{j\mid2}}{I_{j\mid k}} = I_{k\mid2}, \quad \ldots$$

or

$$I_{j\mid1} = I_{j\mid k}I_{k\mid1}, \quad I_{j\mid2} = I_{j\mid k}I_{k\mid2}, \quad \ldots$$

which implies that the index numbers satisfy the circular test.

Since the Laspeyres, Paasche, Fisher and Marshall–Edgeworth index numbers do not satisfy the circular test, the method for changing the base does not apply exactly. However, it does apply approximately in practice.

Weighted aggregate index numbers with fixed year weights do satisfy the circular test (see Prob. 17.23). For index numbers computed in this manner the method given for changing the base does apply exactly.

17.38. Table 17.14 shows the index of industrial production of all manufactures for the years 1947–1958 with 1947-1949 as base period. Obtain a new index with (*a*) 1951 and (*b*) 1953–1956 as base.

Table 17.14

Year	1947	1948	1949	1950	1951	1952	1953	1954	1955	1956	1957	1958
Index of industrial production (1947-49 = 100)	100	104	97	112	120	124	134	125	139	143	143	134

Source: Survey of Current Business

Solution:

(a) Divide each index in the table by 120 (the index corresponding to 1951) and express the result as a percentage. The required index numbers with 1951 as base are as in Table 17.15.

Table 17.15

Year	1947	1948	1949	1950	1951	1952	1953	1954	1955	1956	1957	1958
Index of industrial production (1951 = 100)	83	87	81	93	100	103	112	104	116	119	119	112

(b) The average (mean) index for the years 1953–1956 as base is $\frac{1}{4}(134 + 125 + 139 + 143) = 135\cdot25$. Then dividing each index in Table 17.14 by 135·25 and expressing the result as a percentage, we obtain the required index numbers shown in Table 17.16.

Table 17.16

Year	1947	1948	1949	1950	1951	1952	1953	1954	1955	1956	1957	1958
Index of industrial production (1953-56 = 100)	74	77	72	83	89	92	99	92	103	106	106	99

Note that the average index for the new base period 1953–1956 is $\frac{1}{4}(99 + 92 + 103 + 106) = 100$, as it should be.

DEFLATION OF TIME SERIES

17.39. Table 17.17 show the average wages in dollars per hour of railroad workers in the United States during the years 1947–1958. Also shown is a consumer price index for these years with 1947–1949 as base period. Determine the "real" wages of railroad workers during the years 1947–1958 compared with their wages in 1947.

Table 17.17

Year	1947	1948	1949	1950	1951	1952	1953	1954	1955	1956	1957	1958
Average wage of railroad workers (dollars per hour)	1·19	1·33	1·44	1·57	1·75	1·84	1·89	1·94	1·97	2·13	2·28	2·45
Consumer price index (1947-49 = 100)	95·5	102·8	101·8	102·8	111·0	113·5	114·4	114·8	114·5	116·2	120·2	123·5

Source: U.S. Department of Labor

Solution:

(a) We first form a new consumer price index with 1947 as base year by dividing all numbers in the third row of Table 17.17 by 95·5 and expressing the result as a per cent. The result appears in the second row of Table 17.18. We then divide each average wage for the given years (second row of Table 17.17) by the corresponding index number (second row of Table 17.18) to obtain the "real" wage (third row of Table 17.18).

Thus, for example, the real wage corresponding to 1958 is $2·45/129·3% = $1·89. It follows that although the "apparent" wage more than doubled from 1947 to 1958, the "real" wage increased by only 59%, i.e. the *purchasing power* increased by only 59%.

Table 17.18

Year	1947	1948	1949	1950	1951	1952	1953	1954	1955	1956	1957	1958
Consumer price index (1947 = 100)	100	107·6	106·6	107·6	116·2	118·8	119·8	120·2	119·9	121·7	125·9	129·3
"Real" wage of railroad workers (dollars per hour)	1·19	1·24	1·35	1·46	1·51	1·55	1·58	1·61	1·64	1·75	1·81	1·89

17.40. Use the consumer price index in Problem 17.39 to determine the purchasing power of the dollar for the various years assuming that in 1947 the dollar was actually worth a dollar in purchasing power.

Solution:

Dividing $1·00 by each price index in the second row of Table 17.18, we obtain the entries in Table 17.19 showing the purchasing power of the 1947 dollar in each of the given years. In 1958, for example, the entry 0·77 means that a 1958 dollar could buy only 77% of what a 1947 dollar could buy, i.e. the dollar was worth $0·77 in terms of the 1947 dollar.

Data expressed in terms of the value of a dollar at some specified period of time are said to be expressed in *constant dollars* with the given period as reference period or base.

Table 17.19

Year	1947	1948	1949	1950	1951	1952	1953	1954	1955	1956	1957	1958
Purchasing power of the dollar in 1947 dollars	1·00	0·93	0·94	0·93	0·86	0·84	0·83	0·83	0·83	0·82	0·79	0·77

Supplementary Problems

PRICE RELATIVES

17.41. Table 17.20 shows a country's average wholesale prices of wheat for various years. Find the price relative for (a) the year 1958 using 1948 as a base, (b) the years 1949 and 1956 using 1950 as a base, (c) the years 1955-1958 using 1947-1949 = 100.

Table 17.20

Year	1947	1948	1949	1950	1951	1952	1953	1954	1955	1956	1957	1958
Average price of wheat in new pence per kilogramme	2·66	2·50	2·24	2·29	2·41	2·45	2·49	2·56	2·50	2·39	2·35	2·23

Ans. (a) 89·2 (b) 97·8, 104·4 (c) 101·4, 96·9, 95·3, 90·4

17.42. Prove that (a) $p_{a|b}p_{b|c}p_{c|a} = 1$, (b) $p_{a|b}p_{b|c}p_{c|d} = p_{a|d}$.

17.43. Prove that $p_{0|n} = p_{0|1}p_{1|2}p_{2|3} \cdots p_{(n-1)|n}$.

17.44. Prove that the modified circular property follows directly from the circular property and the time reversal property.

17.45. Table 17.21 shows the price relatives of a commodity with 1947-1949 = 100. Determine the price relatives with (a) 1956 = 100, (b) 1955-1956 = 100.

Table 17.21

Year	1955	1956	1957	1958	1959	1960
Price Relative (1947-1949=100)	135	128	120	150	140	162

Ans. (a) 105, 100, 93·8, 117, 109, 127.   (b) 103, 97·3, 91·3, 114, 106, 123

17.46. The price relative for the year 1956 with 1958 as base is $62\frac{1}{2}$, while the price relative for the year 1957 with 1956 as base is $133\frac{1}{3}$. Find the price relative for the year 1958 with (a) 1957, (b) 1956-1957 as base.   *Ans.* (a) 120, (b) 137

17.47. In 1960 the average price of a commodity decreased by 25% of its 1954 value but increased by 50% of its 1946 value. Find the price relative for (a) 1954 and (b) 1960 with 1946 as base.   *Ans.* (a) 200, (b) 150

QUANTITY OR VOLUME RELATIVES

17.48. Table 17.22 shows the electric energy in billions of kilowatt-hours sold to residential or domestic customers in the United States during the years 1947-1958. Reduce the data to quantity relatives with (a) 1953 and (b) 1947-1949 as a base.

Table 17.22

Year	1947	1948	1949	1950	1951	1952	1953	1954	1955	1956	1957	1958
Electric Energy (billions of kW h)	3·68	4·25	4·84	5·59	6·42	7·23	8·09	9·04	10·04	11·15	12·26	13·25

Source: Survey of Current Business

Ans. (a) 45·5, 52·5, 59·8, 69·1, 79·4, 89·4, 100·0, 111·7, 124·1, 137,8, 151·5, 163·8
     (b) 86·5, 99·8, 113·7, 131·3, 150·8, 169·9, 190·1, 212·4, 235·9, 261·9, 288·0, 311·3

17.49. In 1956 the production of a metal ore increased by 40% over that of 1955. In 1957 the production was 20% below that of 1956 but $16\frac{2}{3}$% above that of 1958. Find the price relatives for the years 1955-1958 with base (a) 1955, (b) 1958, (c) 1955-1958.
Ans. (a) 100, 140, 112, 96. (b) 104, 146, 117, 100. (c) 89·3, 125, 100, 85·7

17.50. In the preceding problem if production of metal ore for the year 1957 was 3·20 million tonnes, determine the production for the years (a) 1955, (b) 1956, (c) 1958.
Ans. (a) 2·86, (b) 4·00. (c) 2·74 million tonnes

VALUE RELATIVES

17.51. In 1960 the price of a commodity increased by 50% over that in 1952 while the production of the quantity decreased by 30%. By what percentage did the total value of the commodity in 1960 increase or decrease with respect to the 1952 value?   *Ans.* 5% increase

17.52. Table 17.23 shows the price and value relatives of a commodity for the years 1956-1960 with base periods as indicated. Find the quantity relatives for the commodity with (a) 1956 and (b) 1956-1958 as base. Interpret the results.

Table 17.23

Year	1956	1957	1958	1959	1960
Price Relative (1956 = 100)	100	125	150	175	200
Value Relative (1947-1949=100)	150	180	207	231	252

Ans. (a) 100, 96, 92, 88, 84. (b) 104, 100, 96, 92, 88.

LINK AND CHAIN RELATIVES

17.53. The link relatives for consumption of a commodity during the years 1957-1960 are 90, 120, 125 and 80 respectively. (a) Find the price relative for 1958 with 1960 as base. (b) Chain the link relatives to a 1959 base. (c) Chain the link relatives to a 1957-58 base.
Ans. (a) 100. (b) 74·1, 66·7, 80·0, 100, 80·0 corresponding to 1956–1950 respectively. (c) 101, 99·9, 109, 136, 109 corresponding to 1956–1960 respectively.

17.54. At the end of the first n successive years the production of a commodity was A units. In each successive year the production increased by $r\%$ over that of the previous year. (a) Show that the production during the nth year is $A(1 + r/100)^{n-1}$ units. (b) Show that the total production for all n years is $(100A/r)\ [(1 + r/100)^n - 1]$ units.

INDEX NUMBERS. SIMPLE AGGREGATE METHOD

17.55. Table 17.24 shows a country's prices and quantities of consumption of various non-ferrous metals for the years 1949, 1956 and 1957. Taking 1949 as base year calculate a price index, using the simple aggregate method, for the year (a) 1956, (b) 1957.   *Ans.* 121·7, (b) 110·0

Table 17.24

	Prices (new pence per kg)			Quantities (millions of kg)		
	1949	1956	1957	1949	1956	1957
Aluminum	17·00	26·01	27·52	1357	3707	3698
Copper	19·36	41·88	29·99	2144	2734	2478
Lead	15·18	15·81	14·46	1916	2420	2276
Tin	99·32	101·26	96·17	161	202	186
Zinc	12·15	13·49	11·40	1872	2018	1424

17.56. Prove that the simple aggregate index number satisfies the time reversal and circular tests but does not satisfy the factor reversal test.

SIMPLE AVERAGE OF RELATIVES METHOD

17.57. From the data in Table 17.24 of Prob. 17.55, using a simple average (mean) of price relatives, obtain a price index of non-ferrous metals for (a) 1956 and (b) 1957 with 1949 as base year. Compare with Prob. 17.55.   *Ans.* (a) 137·3, (b) 120·5

17.58. Work Prob. 17.57 using the median.   *Ans.* (a) 111·0, (b) 96·8

17.59. Work Prob. 17.57 using the geometric mean.   *Ans.* (a) 131·3, (b) 116·8

17.60. Work Prob. 17.57 using the harmonic mean.   *Ans.* 126·3, (b) 113·3

WEIGHTED AGGREGATE METHOD.   LASPEYRES' AND PAASCHE'S INDEX

17.61. From the data in Table 17.24 of Prob. 17.55 obtain a Laspeyres price index for (a) 1956 and (b) 1957 with 1949 as base year.   *Ans.* (a) 148·7, (b) 125·5

17.62. From the data in Table 17.24 of Prob. 17.55 obtain a Paasche price index for (a) 1956 and (b) 1957 with 1949 as base year. *Ans.* (a) 150·5, (b) 134·2

17.63. Show that the (a) Laspeyres and (b) Paasche index do not satisfy the time reversal or factor reversal tests.

FISHER'S IDEAL INDEX

17.64. From the data in Table 17.24 of Prob. 17.55 obtain Fisher's ideal price index for (a) 1956 and (b) 1957 with 1949 as base year.   *Ans.* (a) 149·6, (b) 129·8

17.65. Show that Fisher's ideal index does not satisfy the circular test.

MARSHALL-EDGEWORTH INDEX

17.66. From the data in Table 17.24 of Prob. 17.55 obtain the Marshall–Edgeworth price index for (a) 1956 and (b) 1957 with 1949 as base year.   *Ans.* (a) 149·8, (b) 130·5

17.67. Show that the Marshall–Edgeworth index satisfies the time reversal test but not the factor reversal test.

WEIGHTED AVERAGE OF RELATIVES METHOD

17.68. From the data in Table 17.24 of Prob. 17.55 obtain the weighted average of relatives index number for 1956 and 1957 with 1949 as base year using (a) given year values, (b) base year values as weights.
Ans. (a) 163·8, 141·4   (b) 148·7, 125·5

QUANTITY OR VOLUME INDEX NUMBERS

17.69. Use the data in Table 17.24 of Prob. 17.55 to compute volume indexes for 1956 and 1957 with base year 1949 by means of a (a) simple arithmetic mean of volume relatives, (b) simple geometric mean of volume relatives, (c) weighted aggregate volume index with base year price weights (Laspeyres volume index number, (d) weighted aggregate volume index with given year price weights (Paasche volume index number), (e) Fisher's ideal volume index, (f) Marshall-Edgeworth volume index.

VALUE INDEX NUMBERS

17.70. (a) Using 1949 as base year in the data of Prob. 17.55, calculate the value index for each of the years 1956 and 1957. (b) Verify that the value index in (a) is the same as that obtained from the product of Fisher's ideal price and quantity indexes.   *Ans.* (a) 224·4, 183·6

17.71. With 1949 as base year in the data of Prob. 17.55, calculate Price index $\times$ Quantity index for the years 1956 and 1957 using (a) Laspeyres' and (b) Paasche's index numbers. Compare with the true value index.
Ans. (a) 221·6, 171·7; (b) 226·8, 196·3. True values are 224·2 and 183·6 respectively (Prob. 17.70).

17.72. Prove that simple aggregate value index numbers satisfy the time reversal and circular tests.

CHANGING THE BASE PERIOD OF INDEX NUMBERS

17.73. Table 17.25 shows two construction cost indexes for the years 1947-1958. The first, based on an average for 30 cities and compiled by the American Appraisal Company, shows the construction cost index with 1913 = 100. The second, compiled by the Department of Commerce, shows an index with 1947–1949 = 100. (a) Using the data with 1913 = 100, obtain an index with 1947–1949 = 100 by the simple change of base method used for price relatives. (b) Compare the result in (a) with the Department of Commerce compilation, listing possible reasons for any discrepancies observed.

Table 17.25

Year	1947	1948	1949	1950	1951	1952	1953	1954	1955	1956	1957	1958
Amer. Appraisal Co. construction cost index (1913 = 100)	430	490	490	500	532	553	577	591	608	635	663	682
Dept. of Commerce construction cost index (1947-1949 = 100)	93	104	103	107	116	119	122	122	125	132	137	139

Source: Survey of Current Business

DEFLATION OF TIME SERIES

17.74. The United States wholesale price index for the years 1947-1958 with 1947-1949 = 100 is given in Table 17.26. Determine the wholesale purchasing power of a dollar in each of the given years in terms of 1954 dollars.
Ans. 1·14, 1·06, 1·11, 1·07, 0·96, 0·99, 1·00, 1·00, 0·97, 0·94, 0·93

Table 17.26

Year	1947	1948	1949	1950	1951	1952	1953	1954	1955	1956	1957	1958
Wholesale price index (1947-1949 = 100)	96·4	104·4	99·2	103·1	114·8	111·6	110·1	110·3	110·7	114·3	117·6	119·2

Source: Survey of Current Business

17.75. A given time series shows the total annual dollar value of a set of commodities. (*a*) Describe how to adjust the time series so as to eliminate the effect of the changing value of a dollar from year to year. (*b*) Justify theoretically the method used in (*a*). (*c*) Illustrate by an example.

17.76. (*a*) Deflate the time series in the last column of Table 16.45 of Chapter 16 and (*b*) explain the significance of the deflated data.

17.77. Prove that the method of deflating time series, used for example in Problem 17.39, is strictly applicable only if the index numbers satisfy the factor reversal test.

MISCELLANEOUS PROBLEMS

17.78. Prove that if the Laspeyres and Paasche index numbers are equal then they are equal to the Marshall-Edgeworth and Fisher ideal index numbers.

17.79. Construct a table of the various types of index numbers, specifying in each case whether they do or do not satisfy the time reversal, factor reversal and circular tests.

APPENDIX

Appendix I

ORDINATES (Y)
of the
STANDARD
NORMAL CURVE
at z

z	0	1	2	3	4	5	6	7	8	9
0·0	0·3989	0·3989	0·3989	0·3988	0·3986	0·3984	0·3982	0·3980	0·3977	0·3973
0·1	0·3970	0·3965	0·3961	0·3956	0·3951	0·3945	0·3939	0·3932	0·3925	0·3918
0·2	0·3910	0·3902	0·3894	0·3885	0·3876	0·3867	0·3857	0·3847	0·3836	0·3825
0·3	0·3814	0·3802	0·3790	0·3778	0·3765	0·3752	0·3739	0·3725	0·3712	0·3697
0·4	0·3683	0·3668	0·3653	0·3637	0·3621	0·3605	0·3589	0·3572	0·3555	0·3538
0·5	0·3521	0·3503	0·3485	0·3467	0·3448	0·3429	0·3410	0·3391	0·3372	0·3352
0·6	0·3332	0·3312	0·3292	0·3271	0·3251	0·3230	0·3209	0·3187	0·3166	0·3144
0·7	0·3123	0·3101	0·3079	0·3056	0·3034	0·3011	0·2989	0·2966	0·2943	0·2920
0·8	0·2897	0·2874	0·2850	0·2827	0·2803	0·2780	0·2756	0·2732	0·2709	0·2685
0·9	0·2661	0·2637	0·2613	0·2589	0·2565	0·2541	0·2516	0·2492	0·2468	0·2444
1·0	0·2420	0·2396	0·2371	0·2347	0·2323	0·2299	0·2275	0·2251	0·2227	0·2203
1·1	0·2179	0·2155	0·2131	0·2107	0·2083	0·2059	0·2036	0·2012	0·1989	0·1965
1·2	0·1942	0·1919	0·1895	0·1872	0·1849	0·1826	0·1804	0·1781	0·1758	0·1736
1·3	0·1714	0·1691	0·1669	0·1647	0·1626	0·1604	0·1582	0·1561	0·1539	0·1518
1·4	0·1497	0·1476	0·1456	0·1435	0·1415	0·1394	0·1374	0·1354	0·1334	0·1315
1·5	0·1295	0·1276	0·1257	0·1238	0·1219	0·1200	0·1182	0·1163	0·1145	0·1127
1·6	0·1109	0·1092	0·1074	0·1057	0·1040	0·1023	0·1006	0·0989	0·0973	0·0957
1·7	0·0940	0·0925	0·0909	0·0893	0·0878	0·0863	0·0848	0·0833	0·0818	0·0804
1·8	0·0790	0·0775	0·0761	0·0748	0·0734	0·0721	0·0707	0·0694	0·0681	0·0669
1·9	0·0656	0·0644	0·0632	0·0620	0·0608	0·0596	0·0584	0·0573	0·0562	0·0551
2·0	0·0540	0·0529	0·0519	0·0508	0·0498	0·0488	0·0478	0·0468	0·0459	0·0449
2·1	0·0440	0·0431	0·0422	0·0413	0·0404	0·0396	0·0387	0·0379	0·0371	0·0363
2·2	0·0355	0·0347	0·0339	0·0332	0·0325	0·0317	0·0310	0·0303	0·0297	0·0290
2·3	0·0283	0·0277	0·0270	0·0264	0·0258	0·0252	0·0246	0·0241	0·0235	0·0229
2·4	0·0224	0·0219	0·0213	0·0208	0·0203	0·0198	0·0194	0·0189	0·0184	0·0180
2·5	0·0175	0·0171	0·0167	0·0163	0·0158	0·0154	0·0151	0·0147	0·0143	0·0139
2·6	0·0136	0·0132	0·0129	0·0126	0·0122	0·0119	0·0116	0·0113	0·0110	0·0107
2·7	0·0104	0·0101	0·0099	0·0096	0·0093	0·0091	0·0088	0·0086	0·0084	0·0081
2·8	0·0079	0·0077	0·0075	0·0073	0·0071	0·0069	0·0067	0·0065	0·0063	0·0061
2·9	0·0060	0·0058	0·0056	0·0055	0·0053	0·0051	0·0050	0·0048	0·0047	0·0046
3·0	0·0044	0·0043	0·0042	0·0040	0·0039	0·0038	0·0037	0·0036	0·0035	0·0034
3·1	0·0033	0·0032	0·0031	0·0030	0·0029	0·0028	0·0027	0·0026	0·0025	0·0025
3·2	0·0024	0·0023	0·0022	0·0022	0·0021	0·0020	0·0020	0·0019	0·0018	0·0018
3·3	0·0017	0·0017	0·0016	0·0016	0·0015	0·0015	0·0014	0·0014	0·0013	0·0013
3·4	0·0012	0·0012	0·0012	0·0011	0·0011	0·0010	0·0010	0·0010	0·0009	0·0009
3·5	0·0009	0·0008	0·0008	0·0008	0·0008	0·0007	0·0007	0·0007	0·0007	0·0006
3·6	0·0006	0·0006	0·0006	0·0005	0·0005	0·0005	0·0005	0·0005	0·0005	0·0004
3·7	0·0004	0·0004	0·0004	0·0004	0·0004	0·0004	0·0003	0·0003	0·0003	0·0003
3·8	0·0003	0·0003	0·0003	0·0003	0·0003	0·0002	0·0002	0·0002	0·0002	0·0002
3·9	0·0002	0·0002	0·0002	0·0002	0·0002	0·0002	0·0002	0·0002	0·0001	0·0001

Appendix II

AREAS
under the
STANDARD
NORMAL CURVE
from 0 to z

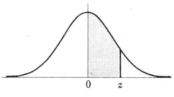

z	0	1	2	3	4	5	6	7	8	9
0·0	0·0000	0·0040	0·0080	0·0120	0·0160	0·0199	0·0239	0·0279	0·0319	0·0359
0·1	0·0398	0·0438	0·0478	0·0517	0·0557	0·0596	0·0636	0·0675	0·0714	0·0754
0·2	0·0793	0·0832	0·0871	0·0910	0·0948	0·0987	0·1026	0·1064	0·1103	0·1141
0·3	0·1179	0·1217	0·1255	0·1293	0·1331	0·1368	0·1406	0·1443	0·1480	0·1517
0·4	0·1554	0·1591	0·1628	0·1664	0·1700	0·1736	0·1772	0·1808	0·1844	0·1879
0·5	0·1915	0·1950	0·1985	0·2019	0·2054	0·2088	0·2123	0·2157	0·2190	0·2224
0·6	0·2258	0·2291	0·2324	0·2357	0·2389	0·2422	0·2454	0·2486	0·2518	0·2549
0·7	0·2580	0·2612	0·2642	0·2673	0·2704	0·2734	0·2764	0·2794	0·2823	0·2852
0·8	0·2881	0·2910	0·2939	0·2967	0·2996	0·3023	0·3051	0·3078	0·3106	0·3133
0·9	0·3159	0·3186	0·3212	0·3238	0·3264	0·3289	0·3315	0·3340	0·3365	0·3389
1·0	0·3413	0·3438	0·3461	0·3485	0·3508	0·3531	0·3554	0·3577	0·3599	0·3621
1·1	0·3643	0·3665	0·3686	0·3708	0·3729	0·3749	0·3770	0·3790	0·3810	0·3830
1·2	0·3849	0·3869	0·3888	0·3907	0·3925	0·3944	0·3962	0·3980	0·3997	0·4015
1·3	0·4032	0·4049	0·4066	0·4082	0·4099	0·4115	0·4131	0·4147	0·4162	0·4177*
1·4	0·4192	0·4207	0·4222	0·4236	0·4251	0·4265	0·4279	0·4292	0·4306	0·4319
1·5	0·4332	0·4345	0·4357	0·4370	0·4382	0·4394	0·4406	0·4418	0·4429	0·4441
1·6	0·4452	0·4463	0·4474	0·4484	0·4495	0·4505	0·4515	0·4525	0·4535	0·4545
1·7	0·4554	0·4564	0·4573	0·4582	0·4591	0·4599	0·4608	0·4616	0·4625	0·4633
1·8	0·4641	0·4649	0·4656	0·4664	0·4671	0·4678	0·4686	0·4693	0·4699	0·4706
1·9	0·4713	0·4719	0·4726	0·4732	0·4738	0·4744	0·4750	0·4756	0·4761	0·4767
2·0	0·4772	0·4778	0·4783	0·4788	0·4793	0·4798	0·4803	0·4808	0·4812	0·4817
2·1	0·4821	0·4826	0·4830	0·4834	0·4838	0·4842	0·4846	0·4850	0·4854	0·4857
2·2	0·4861	0·4864	0·4868	0·4871	0·4875	0·4878	0·4881	0·4884	0·4887	0·4890
2·3	0·4893	0·4896	0·4898	0·4901	0·4904	0·4906	0·4909	0·4911	0·4913	0·4916
2·4	0·4918	0·4920	0·4922	0·4925	0·4927	0·4929	0·4931	0·4932	0·4934	0·4936
2·5	0·4938	0·4940	0·4941	0·4943	0·4945	0·4946	0·4948	0·4949	0·4951	0·4952
2·6	0·4953	0·4955	0·4956	0·4957	0·4959	0·4960	0·4961	0·4962	0·4963	0·4964
2·7	0·4965	0·4966	0·4967	0·4968	0·4969	0·4970	0·4971	0·4972	0·4973	0·4974
2·8	0·4974	0·4975	0·4976	0·4977	0·4977	0·4978	0·4979	0·4979	0·4980	0·4981
2·9	0·4981	0·4982	0·4982	0·4983	0·4984	0·4984	0·4985	0·4985	0·4986	0·4986
3·0	0·4987	0·4987	0·4987	0·4988	0·4988	0·4989	0·4989	0·4989	0·4990	0·4990
3·1	0·4990	0·4991	0·4991	0·4991	0·4992	0·4992	0·4992	0·4992	0·4993	0·4993
3·2	0·4993	0·4993	0·4994	0·4994	0·4994	0·4994	0·4994	0·4995	0·4995	0·4995
3·3	0·4995	0·4995	0·4995	0·4996	0·4996	0·4996	0·4996	0·4996	0·4996	0·4997
3·4	0·4997	0·4997	0·4997	0·4997	0·4997	0·4997	0·4997	0·4997	0·4997	0·4998
3·5	0·4998	0·4998	0·4998	0·4998	0·4998	0·4998	0·4998	0·4998	0·4998	0·4998
3·6	0·4998	0·4998	0·4999	0·4999	0·4999	0·4999	0·4999	0·4999	0·4999	0·4999
3·7	0·4999	0·4999	0·4999	0·4999	0·4999	0·4999	0·4999	0·4999	0·4999	0·4999
3·8	0·4999	0·4999	0·4999	0·4999	0·4999	0·4999	0·4999	0·4999	0·4999	0·4999
3·9	0·5000	0·5000	0·5000	0·5000	0·5000	0·5000	0·5000	0·5000	0·5000	0·5000

Appendix III

PERCENTILE VALUES (t_p)
for
STUDENT'S t DISTRIBUTION
with ν degrees of freedom
(shaded area = p)

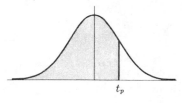

ν	$t_{0.995}$	$t_{0.99}$	$t_{0.975}$	$t_{0.95}$	$t_{0.90}$	$t_{0.80}$	$t_{0.75}$	$t_{0.70}$	$t_{0.60}$	$t_{0.55}$
1	63·66	31·82	12·71	6·31	3·08	1·376	1·000	0·727	0·325	0·158
2	9·92	6·96	4·30	2·92	1·89	1·061	0·816	0·617	0·289	0·142
3	5·84	4·54	3·18	2·35	1·64	0·978	0·765	0·584	0·277	0·137
4	4·60	3·75	2·78	2·13	1·53	0·941	0·741	0·569	0·271	0·134
5	4·03	3·36	2·57	2·02	1·48	0·920	0·727	0·559	0·267	0·132
6	3·71	3·14	2·45	1·94	1·44	0·906	0·718	0·553	0·265	0·131
7	3·50	3·00	2·36	1·90	1·42	0·896	0·711	0·549	0·263	0·130
8	3·36	2·90	2·31	1·86	1·40	0·889	0·706	0·546	0·262	0·130
9	3·25	2·82	2·26	1·83	1·38	0·883	0·703	0·543	0·261	0·129
10	3·17	2·76	2·23	1·81	1·37	0·879	0·700	0·542	0·260	0·129
11	3·11	2·72	2·20	1·80	1·36	0·876	0·697	0·540	0·260	0·129
12	3·06	2·68	2·18	1·78	1·36	0·873	0·695	0·539	0·259	0·128
13	3·01	2·65	2·16	1·77	1·35	0·870	0·694	0·538	0·259	0·128
14	2·98	2·62	2·14	1·76	1·34	0·868	0·692	0·537	0·258	0·128
15	2·95	2·60	2·13	1·75	1·34	0·866	0·691	0·536	0·258	0·128
16	2·92	2·58	2·12	1·75	1·34	0·865	0·690	0·535	0·258	0·128
17	2·90	2·57	2·11	1·74	1·33	0·863	0·689	0·534	0·257	0·128
18	2·88	2·55	2·10	1·73	1·33	0·862	0·688	0·534	0·257	0·127
19	2·86	2·54	2·09	1·73	1·33	0·861	0·688	0·533	0·257	0·127
20	2·84	2·53	2·09	1·72	1·32	0·860	0·687	0·533	0·257	0·127
21	2·83	2·52	2·08	1·72	1·32	0·859	0·686	0·532	0·257	0·127
22	2·82	2·51	2·07	1·72	1·32	0·858	0·686	0·532	0·256	0·127
23	2·81	2·50	2·07	1·71	1·32	0·858	0·865	0·532	0·256	0·127
24	2·80	2·49	2·06	1·71	1·32	0·857	0·685	0·531	0·256	0·127
25	2·79	2·48	2·06	1·71	1·32	0·856	0·684	0·531	0·256	0·127
26	2·78	2·48	2·06	1·71	1·32	0·856	0·684	0·531	0·256	0·127
27	2·77	2·47	2·05	1·70	1·31	0·855	0·684	0·531	0·256	0·127
28	2·76	2·47	2·05	1·70	1·31	0·855	0·683	0·530	0·256	0·127
29	2·76	2·46	2·04	1·70	1·31	0·854	0·683	0·530	0·256	0·127
30	2·75	2·46	2·04	1·70	1·31	0·854	0·683	0·530	0·256	0·127
40	2·70	2·42	2·02	1·68	1·30	0·851	0·681	0·529	0·255	0·126
60	2·66	2·39	2·00	1·67	1·30	0·848	0·679	0·527	0·254	0·126
120	2·62	2·36	1·98	1·66	1·29	0·845	0·677	0·526	0·254	0·126
∞	2·58	2·33	1·96	1·645	1·28	0·842	0·674	0·524	0·253	0·126

Source: R. A. Fisher and F. Yates, *Statistical Tables for Biological, Agricultural and Medical Research* (5th edition), Table III, Oliver and Boyd Ltd., Edinburgh, by permission of the authors and publishers.

Appendix IV

PERCENTILE VALUES (χ_p^2)
for
THE CHI-SQUARE DISTRIBUTION
with ν degrees of freedom
(shaded area $= p$)

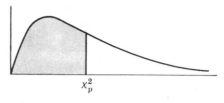

χ_p^2

ν	$\chi^2_{0.995}$	$\chi^2_{0.99}$	$\chi^2_{0.975}$	$\chi^2_{0.95}$	$\chi^2_{0.90}$	$\chi^2_{0.75}$	$\chi^2_{0.50}$	$\chi^2_{0.25}$	$\chi^2_{0.10}$	$\chi^2_{0.05}$	$\chi^2_{0.025}$	$\chi^2_{0.01}$	$\chi^2_{0.005}$
1	7·88	6·63	5·02	3·84	2·71	1·32	0·455	0·102	0·0158	0·0039	0·0010	0·0002	0·0000
2	10·6	9·21	7·38	5·99	4·61	2·77	1·39	0·575	0·211	0·103	0·0506	0·0201	0·0100
3	12·8	11·3	9·35	7·81	6·25	4·11	2·37	1·21	0·584	0·352	0·216	0·115	0·072
4	14·9	13·3	11·1	9·49	7·78	5·39	3·36	1·92	1·06	0·711	0·484	0·297	0·207
5	16·7	15·1	12·8	11·1	9·24	6·63	4·35	2·67	1·61	1·15	0·831	0·554	0·412
6	18·5	16·8	14·4	12·6	10·6	7·84	5·35	3·45	2·20	1·64	1·24	0·872	0·676
7	20·3	18·5	16·0	14·1	12·0	9·04	6·35	4·25	2·83	2·17	1·69	1·24	0·989
8	22·0	20·1	17·5	15·5	13·4	10·2	7·34	5·07	3·49	2·73	2·18	1·65	1·34
9	23·6	21·7	19·0	16·9	14·7	11·4	8·34	5·90	4·17	3·33	2·70	2·09	1·73
10	25·2	23·2	20·5	18·3	16·0	12·5	9·34	6·74	4·87	3·94	3·25	2·56	2·16
11	26·8	24·7	21·9	19·7	17·3	13·7	10·3	7·58	5·58	4·57	3·82	3·05	2·60
12	28·3	26·2	23·3	21·0	18·5	14·8	11·3	8·44	6·30	5·23	4·40	3·57	3·07
13	29·8	27·7	24·7	22·4	19·8	16·0	12·3	9·30	7·04	5·89	5·01	4·11	3·57
14	31·3	29·1	26·1	23·7	21·1	17·1	13·3	10·2	7·79	6·57	5·63	4·66	4·07
15	32·8	30·6	27·5	25·0	22·3	18·2	14·3	11·0	8·55	7·26	6·26	5·23	4·60
16	34·3	32·0	28·8	26·3	23·5	19·4	15·3	11·9	9·31	7·96	6·91	5·81	5·14
17	35·7	33·4	30·2	27·6	24·8	20·5	16·3	12·8	10·1	8·67	7·56	6·41	5·70
18	37·2	34·8	31·5	28·9	26·0	21·6	17·3	13·7	10·9	9·39	8·23	7·01	6·26
19	38·6	36·2	32·9	30·1	27·2	22·7	18·3	14·6	11·7	10·1	8·91	7·63	6·84
20	40·0	37·6	34·2	31·4	28·4	23·8	19·3	15·5	12·4	10·9	9·59	8·26	7·43
21	41·4	38·9	35·5	32·7	29·6	24·9	20·3	16·3	13·2	11·6	10·3	8·90	8·03
22	42·8	40·3	36·8	33·9	30·8	26·0	21·3	17·2	14·0	12·3	11·0	9·54	8·64
23	44·2	41·6	38·1	35·2	32·0	27·1	22·3	18·1	14·8	13·1	11·7	10·2	9·26
24	45·6	43·0	39·4	36·4	33·2	28·2	23·3	19·0	15·7	13·8	12·4	10·9	9·89
25	46·9	44·3	40·6	37·7	34·4	29·3	24·3	19·9	16·5	14·6	13·1	11·5	10·5
26	48·3	45·6	41·9	38·9	35·6	30·4	25·3	20·8	17·3	15·4	13·8	12·2	11·2
27	49·6	47·0	43·2	40·1	36·7	31·5	26·3	21·7	18·1	16·2	14·6	12·9	11·8
28	51·0	48·3	44·5	41·3	37·9	32·6	27·3	22·7	18·9	16·9	15·3	13·6	12·5
29	52·3	49·6	45·7	42·6	39·1	33·7	28·3	23·6	19·8	17·7	16·0	14·3	13·1
30	53·7	50·9	47·0	43·8	40·3	34·8	29·3	24·5	20·6	18·5	16·8	15·0	13·8
40	66·8	63·7	59·3	55·8	51·8	45·6	39·3	33·7	29·1	26·5	24·4	22·2	20·7
50	79·5	76·2	71·4	67·5	63·2	56·3	49·3	42·9	37·7	34·8	32·4	29·7	28·0
60	92·0	88·4	83·3	79·1	74·4	67·0	59·3	52·3	46·5	43·2	40·5	37·5	35·5
70	104·2	100·4	95·0	90·5	85·5	77·6	69·3	61·7	55·3	51·7	48·8	45·4	43·3
80	116·3	112·3	106·6	101·9	96·6	88·1	79·3	71·1	64·3	60·4	57·2	53·5	51·2
90	128·3	124·1	118·1	113·1	107·6	98·6	89·3	80·6	73·3	69·1	65·6	61·8	59·2
100	140·2	135·8	129·6	124·3	118·5	109·1	99·3	90·1	82·4	77·9	74·2	70·1	67·3

Source: Catherine M. Thompson, *Table of percentage points of the χ² distribution,*
Biometrika, Vol. 32 (1941), by permission of the author and publisher.

Appendix V

FOUR-PLACE COMMON LOGARITHMS

N	0	1	2	3	4	5	6	7	8	9	Proportional Parts 1 2 3 4 5 6 7 8 9
10	0000	0043	0086	0128	0170	0212	0253	0294	0334	0374	4 8 12 17 21 25 29 33 37
11	0414	0453	0492	0531	0569	0607	0645	0682	0719	0755	4 8 11 15 19 23 26 30 34
12	0792	0828	0864	0899	0934	0969	1004	1038	1072	1106	3 7 10 14 17 21 24 28 31
13	1139	1173	1206	1239	1271	1303	1335	1367	1399	1430	3 6 10 13 16 19 23 26 29
14	1461	1492	1523	1553	1584	1614	1644	1673	1703	1732	3 6 9 12 15 18 21 24 27
15	1761	1790	1818	1847	1875	1903	1931	1959	1987	2014	3 6 8 11 14 17 20 22 25
16	2041	2068	2095	2122	2148	2175	2201	2227	2253	2279	3 5 8 11 13 16 18 21 24
17	2304	2330	2355	2380	2405	2430	2455	2480	2504	2529	2 5 7 10 12 15 17 20 22
18	2553	2577	2601	2625	2648	2672	2695	2718	2742	2765	2 5 7 9 12 14 16 19 21
19	2788	2810	2833	2856	2878	2900	2923	2945	2967	2989	2 4 7 9 11 13 16 18 20
20	3010	3032	3054	3075	3096	3118	3139	3160	3181	3201	2 4 6 8 11 13 15 17 19
21	3222	3243	3263	3284	3304	3324	3345	3365	3385	3404	2 4 6 8 10 12 14 16 18
22	3424	3444	3464	3483	3502	3522	3541	3560	3579	3598	2 4 6 8 10 12 14 15 17
23	3617	3636	3655	3674	3692	3711	3729	3747	3766	3784	2 4 6 7 9 11 13 15 17
24	3802	3820	3838	3856	3874	3892	3909	3927	3945	3962	2 4 5 7 9 11 12 14 16
25	3979	3997	4014	4031	4048	4065	4082	4099	4116	4133	2 3 5 7 9 10 12 14 15
26	4150	4166	4183	4200	4216	4232	4249	4265	4281	4298	2 3 5 7 8 10 11 13 15
27	4314	4330	4346	4362	4378	4393	4409	4425	4440	4456	2 3 5 6 8 9 11 13 14
28	4472	4487	4502	4518	4533	4548	4564	4579	4594	4609	2 3 5 6 8 9 11 12 14
29	4624	4639	4654	4669	4683	4698	4713	4728	4742	4757	1 3 4 6 7 9 10 12 13
30	4771	4786	4800	4814	4829	4843	4857	4871	4886	4900	1 3 4 6 7 9 10 11 13
31	4914	4928	4942	4955	4969	4983	4997	5011	5024	5038	1 3 4 6 7 8 10 11 12
32	5051	5065	5079	5092	5105	5119	5132	5145	5159	5172	1 3 4 5 7 8 9 11 12
33	5185	5198	5211	5224	5237	5250	5263	5276	5289	5302	1 3 4 5 6 8 9 10 12
34	5315	5328	5340	5353	5366	5378	5391	5403	5416	5428	1 3 4 5 6 8 9 10 11
35	5441	5453	5465	5478	5490	5502	5514	5527	5539	5551	1 2 4 5 6 7 9 10 11
36	5563	5575	5587	5599	5611	5623	5635	5647	5658	5670	1 2 4 5 6 7 8 10 11
37	5682	5694	5705	5717	5729	5740	5752	5763	5775	5786	1 2 3 5 6 7 8 9 10
38	5798	5809	5821	5832	5843	5855	5866	5877	5888	5899	1 2 3 5 6 7 8 9 10
39	5911	5922	5933	5944	5955	5966	5977	5988	5999	6010	1 2 3 4 5 7 8 9 10
40	6021	6031	6042	6053	6064	6075	6085	6096	6107	6117	1 2 3 4 5 6 8 9 10
41	6128	6138	6149	6160	6170	6180	6191	6201	6212	6222	1 2 3 4 5 6 7 8 9
42	6232	6243	6253	6263	6274	6284	6294	6304	6314	6325	1 2 3 4 5 6 7 8 9
43	6335	6345	6355	6365	6375	6385	6395	6405	6415	6425	1 2 3 4 5 6 7 8 9
44	6435	6444	6454	6464	6474	6484	6493	6503	6513	6522	1 2 3 4 5 6 7 8 9
45	6532	6542	6551	6561	6571	6580	6590	6599	6609	6618	1 2 3 4 5 6 7 8 9
46	6628	6637	6646	6656	6665	6675	6684	6693	6702	6712	1 2 3 4 5 6 7 7 8
47	6721	6730	6739	6749	6758	6767	6776	6785	6794	6803	1 2 3 4 5 5 6 7 8
48	6812	6821	6830	6839	6848	6857	6866	6875	6884	6893	1 2 3 4 4 5 6 7 8
49	6902	6911	6920	6928	6937	6946	6955	6964	6972	6981	1 2 3 4 4 5 6 7 8
50	6990	6998	7007	7016	7024	7033	7042	7050	7059	7067	1 2 3 3 4 5 6 7 8
51	7076	7084	7093	7101	7110	7118	7126	7135	7143	7152	1 2 3 3 4 5 6 7 8
52	7160	7168	7177	7185	7193	7202	7210	7218	7226	7235	1 2 2 3 4 5 6 7 7
53	7243	7251	7259	7267	7275	7284	7292	7300	7308	7316	1 2 2 3 4 5 6 6 7
54	7324	7332	7340	7348	7356	7364	7372	7380	7388	7396	1 2 2 3 4 5 6 6 7
N	0	1	2	3	4	5	6	7	8	9	1 2 3 4 5 6 7 8 9

FOUR-PLACE COMMON LOGARITHMS

N	0	1	2	3	4	5	6	7	8	9	Proportional Parts 1 2 3 4 5 6 7 8 9
55	7404	7412	7419	7427	7435	7443	7451	7459	7466	7474	1 2 2 3 4 5 5 6 7
56	7482	7490	7497	7505	7513	7520	7528	7536	7543	7551	1 2 2 3 4 5 5 6 7
57	7559	7566	7574	7582	7589	7597	7604	7612	7619	7627	1 2 2 3 4 5 5 6 7
58	7634	7642	7649	7657	7664	7672	7679	7686	7694	7701	1 1 2 3 4 4 5 6 7
59	7709	7716	7723	7731	7738	7745	7752	7760	7767	7774	1 1 2 3 4 4 5 6 7
60	7782	7789	7796	7803	7810	7818	7825	7832	7839	7846	1 1 2 3 4 4 5 6 6
61	7853	7860	7868	7875	7882	7889	7896	7903	7910	7917	1 1 2 3 4 4 5 6 6
62	7924	7931	7938	7945	7952	7959	7966	7973	7980	7987	1 1 2 3 3 4 5 6 6
63	7993	8000	8007	8014	8021	8028	8035	8041	8048	8055	1 1 2 3 3 4 5 5 6
64	8062	8069	8075	8082	8089	8096	8102	8109	8116	8122	1 1 2 3 3 4 5 5 6
65	8129	8136	8142	8149	8156	8162	8169	8176	8182	8189	1 1 2 3 3 4 5 5 6
66	8195	8202	8209	8215	8222	8228	8235	8241	8248	8254	1 1 2 3 3 4 5 5 6
67	8261	8267	8274	8280	8287	8293	8299	8306	8312	8319	1 1 2 3 3 4 5 5 6
68	8325	8331	8338	8344	8351	8357	8363	8370	8376	8382	1 1 2 3 3 4 4 5 6
69	8388	8395	8401	8407	8414	8420	8426	8432	8439	8445	1 1 2 2 3 4 4 5 6
70	8451	8457	8463	8470	8476	8482	8488	8494	8500	8506	1 1 2 2 3 4 4 5 6
71	8513	8519	8525	8531	8537	8543	8549	8555	8561	8567	1 1 2 2 3 4 4 5 5
72	8573	8579	8585	8591	8597	8603	8609	8615	8621	8627	1 1 2 2 3 4 4 5 5
73	8633	8639	8645	8651	8657	8663	8669	8675	8681	8686	1 1 2 2 3 4 4 5 5
74	8692	8698	8704	8710	8716	8722	8727	8733	8739	8745	1 1 2 2 3 4 4 5 5
75	8751	8756	8762	8768	8774	8779	8785	8791	8797	8802	1 1 2 2 3 3 4 5 5
76	8808	8814	8820	8825	8831	8837	8842	8848	8854	8859	1 1 2 2 3 3 4 5 5
77	8865	8871	8876	8882	8887	8893	8899	8904	8910	8915	1 1 2 2 3 3 4 4 5
78	8921	8927	8932	8938	8943	8949	8954	8960	8965	8971	1 1 2 2 3 3 4 4 5
79	8976	8982	8987	8993	8998	9004	9009	9015	9020	9025	1 1 2 2 3 3 4 4 5
80	9031	9036	9042	9047	9053	9058	9063	9069	9074	9079	1 1 2 2 3 3 4 4 5
81	9085	9090	9096	9101	9106	9112	9117	9122	9128	9133	1 1 2 2 3 3 4 4 5
82	9138	9143	9149	9154	9159	9165	9170	9175	9180	9186	1 1 2 2 3 3 4 4 5
83	9191	9196	9201	9206	9212	9217	9222	9227	9232	9238	1 1 2 2 3 3 4 4 5
84	9243	9248	9253	9258	9263	9269	9274	9279	9284	9289	1 1 2 2 3 3 4 4 5
85	9294	9299	9304	9309	9315	9320	9325	9330	9335	9340	1 1 2 2 3 3 4 4 5
86	9345	9350	9355	9360	9365	9370	9375	9380	9385	9390	1 1 2 2 3 3 4 4 5
87	9395	9400	9405	9410	9415	9420	9425	9430	9435	9440	0 1 1 2 2 3 3 4 4
88	9445	9450	9455	9460	9465	9469	9474	9479	9484	9489	0 1 1 2 2 3 3 4 4
89	9494	9499	9504	9509	9513	9518	9523	9528	9533	9538	0 1 1 2 2 3 3 4 4
90	9542	9547	9552	9557	9562	9566	9571	9576	9581	9586	0 1 1 2 2 3 3 4 4
91	9590	9595	9600	9605	9609	9614	9619	9624	9628	9633	0 1 1 2 2 3 3 4 4
92	9638	9643	9647	9652	9657	9661	9666	9671	9675	9680	0 1 1 2 2 3 3 4 4
93	9685	9689	9694	9699	9703	9708	9713	9717	9722	9727	0 1 1 2 2 3 3 4 4
94	9731	9736	9741	9745	9750	9754	9759	9763	9768	9773	0 1 1 2 2 3 3 4 4
95	9777	9782	9786	9791	9795	9800	9805	9809	9814	9818	0 1 1 2 2 3 3 4 4
96	9823	9827	9832	9836	9841	9845	9850	9854	9859	9863	0 1 1 2 2 3 3 4 4
97	9868	9872	9877	9881	9886	9890	9894	9899	9903	9908	0 1 1 2 2 3 3 4 4
98	9912	9917	9921	9926	9930	9934	9939	9943	9948	9952	0 1 1 2 2 3 3 4 4
99	9956	9961	9965	9969	9974	9978	9983	9987	9991	9996	0 1 1 2 2 3 3 3 4
N	0	1	2	3	4	5	6	7	8	9	1 2 3 4 5 6 7 8 9

Appendix VI

VALUES of $e^{-\lambda}$

$(0 < \lambda < 1)$

λ	0	1	2	3	4	5	6	7	8	9
0·0	1·0000	0·9900	0·9802	0·9704	0·9608	0·9512	0·9418	0·9324	0·9231	0·9139
0·1	0·9048	0·8958	0·8869	0·8781	0·8694	0·8607	0·8521	0·8437	0·8353	0·8270
0·2	0·8187	0·8106	0·8025	0·7945	0·7866	0·7788	0·7711	0·7634	0·7558	0·7483
0·3	0·7408	0·7334	0·7261	0·7189	0·7118	0·7047	0·6977	0·6907	0·6839	0·6771
0·4	0·6703	0·6636	0·6570	0·6505	0·6440	0·6376	0·6313	0·6250	0·6188	0·6126
0·5	0·6065	0·6005	0·5945	0·5886	0·5827	0·5770	0·5712	0·5655	0·5599	0·5543
0·6	0·5488	0·5434	0·5379	0·5326	0·5273	0·5220	0·5169	0·5117	0·5066	0·5016
0·7	0·4966	0·4916	0·4868	0·4819	0·4771	0·4724	0·4677	0·4630	0·4584	0·4538
0·8	0·4493	0·4449	0·4404	0·4360	0·4317	0·4274	0·4232	0·4190	0·4148	0·4107
0·9	0·4066	0·4025	0·3985	0·3946	0·3906	0·3867	0·3829	0·3791	0·3753	0·3716

$(\lambda = 1, 2, 3, \ldots, 10)$

λ	1	2	3	4	5	6	7	8	9	10
$e^{-\lambda}$	0·367 88	0·135 34	0·049 79	0·018 32	0·006 738	0·002 479	0·000 912	0·000 335	0·000 123	0·000 045

Note: To obtain values of $e^{-\lambda}$ for other values of λ, use the laws of exponents.

Example: $e^{-3\cdot48} = (e^{-3\cdot00})(e^{-0\cdot48}) = (0\cdot049\,79)(0\cdot6188) = 0\cdot030\,81$.

Appendix VII

RANDOM NUMBERS

51772	74640	42331	29044	46621	62898	93582	04186	19640	87056
24033	23491	83587	06568	21960	21387	76105	10863	97453	90581
45939	60173	52078	25424	11645	55870	56974	37428	93507	94271
30586	02133	75797	45406	31041	86707	12973	17169	88116	42187
03585	79353	81938	82322	96799	85659	36081	50884	14070	74950
64937	03355	95863	20790	65304	55189	00745	65253	11822	15804
15630	64759	51135	98527	62586	41889	25439	88036	24034	67283
09448	56301	57683	30277	94623	85418	68829	06652	41982	49159
21631	91157	77331	60710	52290	16835	48653	71590	16159	14676
91097	17480	29414	06829	87843	28195	27279	47152	35683	47280
50532	25496	95652	42457	73547	76552	50020	24819	52984	76168
07136	40876	79971	54195	25708	51817	36732	72484	94923	75936
27989	64728	10744	08396	56242	90985	28868	99431	50995	20507
85184	73949	36601	46253	00477	25234	09908	36574	72139	70185
54398	21154	97810	36764	32869	11785	55261	59009	38714	38723
65544	34371	09591	07839	58892	92843	72828	91341	84821	63886
08263	65952	85762	64236	39238	18776	84303	99247	46149	03229
39817	67906	48236	16057	81812	15815	63700	85915	19219	45943
62257	04077	79443	95203	02479	30763	92486	54083	23631	05825
53298	90276	62545	21944	16530	03878	07516	95715	02526	33537

Appendix VIII

DERIVATION OF NORMAL EQUATIONS
FOR LEAST SQUARE LINE

Let the equation of the required least square line be $Y = a_0 + a_1 X$. The values of Y on this line corresponding to $X = X_1, X_2, \ldots, X_N$ are $a_0 + a_1 X_1, a_0 + a_1 X_2, \ldots, a_0 + a_1 X_N$ while the actual values are $Y_1, Y_2, \ldots, Y_N$ respectively. Then the least square line is such that (see Page 219)

$$S = (a_0 + a_1 X_1 - Y_1)^2 + (a_0 + a_1 X_2 - Y_2)^2 + \cdots + (a_0 + a_1 X_N - Y_N)^2 \text{ is a minimum}$$

From the calculus, S is a minimum when the partial derivatives of S with respect to a_0 and a_1 are zero. Then

$$\frac{\partial S}{\partial a_0} = 2\{(a_0 + a_1 X_1 - Y_1) + (a_0 + a_1 X_2 - Y_2) + \cdots + (a_0 + a_1 X_N - Y_N)\} = 0$$

$$\frac{\partial S}{\partial a_1} = 2\{(a_0 + a_1 X_1 - Y_1)X_1 + (a_0 + a_1 X_2 - Y_2)X_2 + \cdots + (a_0 + a_1 X_N - Y_N)X_N\} = 0$$

and these equations give the required normal equations

$$N a_0 + a_1 \Sigma X - \Sigma Y = 0$$

$$a_0 \Sigma X + a_1 \Sigma X^2 - \Sigma XY = 0$$

Index

SCHAUM'S OUTLINE SERIES

COLLEGE PHYSICS
including 625 SOLVED PROBLEMS
Edited by CAREL W. van der MERWE, Ph.D.,
Professor of Physics, New York University

COLLEGE CHEMISTRY
including 385 SOLVED PROBLEMS
Edited by JEROME L. ROSENBERG, Ph.D.,
Professor of Chemistry, University of Pittsburgh

GENETICS
including 500 SOLVED PROBLEMS
By WILLIAM D. STANSFIELD, Ph.D.,
Dept. of Biological Sciences, Calif. State Polytech.

MATHEMATICAL HANDBOOK
including 2400 FORMULAS and 60 TABLES
By MURRAY R. SPIEGEL, Ph.D.,
Professor of Math., Rensselaer Polytech. Inst.

First Yr. COLLEGE MATHEMATICS
including 1850 SOLVED PROBLEMS
By FRANK AYRES, Jr., Ph.D.,
Professor of Mathematics, Dickinson College

COLLEGE ALGEBRA
including 1940 SOLVED PROBLEMS
By MURRAY R. SPIEGEL, Ph.D.,
Professor of Math., Rensselaer Polytech. Inst.

TRIGONOMETRY
including 680 SOLVED PROBLEMS
By FRANK AYRES, Jr., Ph.D.,
Professor of Mathematics, Dickinson College

MATHEMATICS OF FINANCE
including 500 SOLVED PROBLEMS
By FRANK AYRES, Jr., Ph.D.,
Professor of Mathematics, Dickinson College

PROBABILITY
including 500 SOLVED PROBLEMS
By SEYMOUR LIPSCHUTZ, Ph.D.,
Assoc. Prof. of Math., Temple University

STATISTICS
including 875 SOLVED PROBLEMS
By MURRAY R. SPIEGEL, Ph.D.,
Professor of Math., Rensselaer Polytech. Inst.

ANALYTIC GEOMETRY
including 345 SOLVED PROBLEMS
By JOSEPH H. KINDLE, Ph.D.,
Professor of Mathematics, University of Cincinnati

DIFFERENTIAL GEOMETRY
including 500 SOLVED PROBLEMS
By MARTIN LIPSCHUTZ, Ph.D.,
Professor of Mathematics, University of Bridgeport

CALCULUS
including 1175 SOLVED PROBLEMS
By FRANK AYRES, Jr., Ph.D.,
Professor of Mathematics, Dickinson College

DIFFERENTIAL EQUATIONS
including 560 SOLVED PROBLEMS
By FRANK AYRES, Jr., Ph.D.,
Professor of Mathematics, Dickinson College

SET THEORY and Related Topics
including 530 SOLVED PROBLEMS
By SEYMOUR LIPSCHUTZ, Ph.D.,
Assoc. Prof. of Math., Temple University

FINITE MATHEMATICS
including 750 SOLVED PROBLEMS
By SEYMOUR LIPSCHUTZ, Ph.D.,
Assoc. Prof. of Math., Temple University

MODERN ALGEBRA
including 425 SOLVED PROBLEMS
By FRANK AYRES, Jr., Ph.D.,
Professor of Mathematics, Dickinson College

LINEAR ALGEBRA
including 600 SOLVED PROBLEMS
By SEYMOUR LIPSCHUTZ, Ph.D.,
Assoc. Prof. of Math., Temple University

MATRICES
including 340 SOLVED PROBLEMS
By FRANK AYRES, Jr., Ph.D.,
Professor of Mathematics, Dickinson College

PROJECTIVE GEOMETRY
including 200 SOLVED PROBLEMS
By FRANK AYRES, Jr., Ph.D.,
Professor of Mathematics, Dickinson College

GENERAL TOPOLOGY
including 650 SOLVED PROBLEMS
By SEYMOUR LIPSCHUTZ, Ph.D.,
Assoc. Prof. of Math., Temple University

GROUP THEORY
including 600 SOLVED PROBLEMS
By B. BAUMSLAG, B. CHANDLER, Ph.D.,
Mathematics Dept., New York University

VECTOR ANALYSIS
including 480 SOLVED PROBLEMS
By MURRAY R. SPIEGEL, Ph.D.,
Professor of Math., Rensselaer Polytech. Inst.

ADVANCED CALCULUS
including 925 SOLVED PROBLEMS
By MURRAY R. SPIEGEL, Ph.D.,
Professor of Math., Rensselaer Polytech. Inst.

COMPLEX VARIABLES
including 640 SOLVED PROBLEMS
By MURRAY R. SPIEGEL, Ph.D.,
Professor of Math., Rensselaer Polytech. Inst.

LAPLACE TRANSFORMS
including 450 SOLVED PROBLEMS
By MURRAY R. SPIEGEL, Ph.D.,
Professor of Math., Rensselaer Polytech. Inst.

NUMERICAL ANALYSIS
including 775 SOLVED PROBLEMS
By FRANCIS SCHEID, Ph.D.,
Professor of Mathematics, Boston University

DESCRIPTIVE GEOMETRY
including 175 SOLVED PROBLEMS
By MINOR C. HAWK, Head of
Engineering Graphics Dept., Carnegie Inst. of Tech.

ENGINEERING MECHANICS
including 460 SOLVED PROBLEMS
By W. G. McLEAN, B.S. in E.E., M.S.,
Professor of Mechanics, Lafayette College
and E. W. NELSON, B.S. in M.E., M. Adm. E.,
Engineering Supervisor, Western Electric Co.

THEORETICAL MECHANICS
including 720 SOLVED PROBLEMS
By MURRAY R. SPIEGEL, Ph.D.,
Professor of Math., Rensselaer Polytech. Inst.

LAGRANGIAN DYNAMICS
including 275 SOLVED PROBLEMS
By D. A. WELLS, Ph.D.,
Professor of Physics, University of Cincinnati

STRENGTH OF MATERIALS
including 430 SOLVED PROBLEMS
By WILLIAM A. NASH, Ph.D.,
Professor of Eng. Mechanics, University of Florida

FLUID MECHANICS and HYDRAULICS
including 475 SOLVED PROBLEMS
By RANALD V. GILES, B.S., M.S. in C.E.,
Prof. of Civil Engineering, Drexel Inst. of Tech.

FLUID DYNAMICS
including 100 SOLVED PROBLEMS
By WILLIAM F. HUGHES, Ph.D.,
Professor of Mech. Eng., Carnegie Inst. of Tech.
and JOHN A. BRIGHTON, Ph.D.,
Asst. Prof. of Mech. Eng., Pennsylvania State U.

ELECTRIC CIRCUITS
including 350 SOLVED PROBLEMS
By JOSEPH A. EDMINISTER, M.S.E.E.,
Assoc. Prof. of Elec. Eng., University of Akron

ELECTRONIC CIRCUITS
including 160 SOLVED PROBLEMS
By EDWIN C. LOWENBERG, Ph.D.,
Professor of Elec. Eng., University of Nebraska

FEEDBACK & CONTROL SYSTEMS
including 680 SOLVED PROBLEMS
By J. J. DiSTEFANO III, A. R. STUBBERUD,
and I. J. WILLIAMS, Ph.D.,
Engineering Dept., University of Calif., at L.A.

TRANSMISSION LINES
including 165 SOLVED PROBLEMS
By R. A. CHIPMAN, Ph.D.,
Professor of Electrical Eng., University of Toledo

REINFORCED CONCRETE DESIGN
including 200 SOLVED PROBLEMS
By N. J. EVERARD, MSCE, Ph.D.,
Prof. of Eng. Mech. & Struc., Arlington State Col.
and J. L. TANNER III, MSCE;
Technical Consultant, Texas Industries Inc.

MECHANICAL VIBRATIONS
including 225 SOLVED PROBLEMS
By WILLIAM W. SETO, B.S. in M.E., M.S.,
Assoc. Prof. of Mech. Eng., San Jose State College

MACHINE DESIGN
including 320 SOLVED PROBLEMS
By HALL, HOLOWENKO, LAUGHLIN
Professors of Mechanical Eng., Purdue University

BASIC ENGINEERING EQUATIONS
including 1400 BASIC EQUATIONS
By W. F. HUGHES, E. W. GAYLORD, Ph.D.,
Professors of Mech. Eng., Carnegie Inst. of Tech.

ELEMENTARY ALGEBRA
including 2700 SOLVED PROBLEMS
By BARNETT RICH, Ph.D.,
Head of Math. Dept., Brooklyn Tech. H.S.

PLANE GEOMETRY
including 850 SOLVED PROBLEMS
By BARNETT RICH, Ph.D.,
Head of Math. Dept., Brooklyn Tech. H.S.

TEST ITEMS IN EDUCATION
including 3100 TEST ITEMS
By G. J. MOULY, Ph.D., L. E. WALTON, Ph.D.,
Professors of Education, University of Miami